T0145240

Studies in Computational Intelligence

Volume 701

Series editor

Janusz Kacprzyk, Polish Academy of Sciences, Warsaw, Poland
e-mail: kacprzyk@ibspan.waw.pl

About this Series

The series "Studies in Computational Intelligence" (SCI) publishes new developments and advances in the various areas of computational intelligence—quickly and with a high quality. The intent is to cover the theory, applications, and design methods of computational intelligence, as embedded in the fields of engineering, computer science, physics and life sciences, as well as the methodologies behind them. The series contains monographs, lecture notes and edited volumes in computational intelligence spanning the areas of neural networks, connectionist systems, genetic algorithms, evolutionary computation, artificial intelligence, cellular automata, self-organizing systems, soft computing, fuzzy systems, and hybrid intelligent systems. Of particular value to both the contributors and the readership are the short publication timeframe and the worldwide distribution, which enable both wide and rapid dissemination of research output.

More information about this series at http://www.springer.com/series/7092

Sundarapandian Vaidyanathan
Christos Volos
Editors

Advances in Memristors, Memristive Devices and Systems

Editors
Sundarapandian Vaidyanathan
Research and Development Centre
Vel Tech University
Chennai
India

Christos Volos
Department of Physics
Aristotle University of Thessaloniki
Thessaloniki
Greece

ISSN 1860-949X ISSN 1860-9503 (electronic)
Studies in Computational Intelligence
ISBN 978-3-319-84727-6 ISBN 978-3-319-51724-7 (eBook)
DOI 10.1007/978-3-319-51724-7

© Springer International Publishing AG 2017
Softcover reprint of the hardcover 1st edition 2017
This work is subject to copyright. All rights are reserved by the Publisher, whether the whole or part of the material is concerned, specifically the rights of translation, reprinting, reuse of illustrations, recitation, broadcasting, reproduction on microfilms or in any other physical way, and transmission or information storage and retrieval, electronic adaptation, computer software, or by similar or dissimilar methodology now known or hereafter developed.
The use of general descriptive names, registered names, trademarks, service marks, etc. in this publication does not imply, even in the absence of a specific statement, that such names are exempt from the relevant protective laws and regulations and therefore free for general use.
The publisher, the authors and the editors are safe to assume that the advice and information in this book are believed to be true and accurate at the date of publication. Neither the publisher nor the authors or the editors give a warranty, express or implied, with respect to the material contained herein or for any errors or omissions that may have been made. The publisher remains neutral with regard to jurisdictional claims in published maps and institutional affiliations.

Printed on acid-free paper

This Springer imprint is published by Springer Nature
The registered company is Springer International Publishing AG
The registered company address is: Gewerbestrasse 11, 6330 Cham, Switzerland

Preface

About the Subject

Memristor (concatenation of MEMory ResISTOR), is the fourth fundamental circuit element (joining the resistor, the capacitor and the inductor), predicted by Leon Chua in 1971. This element represents one of today's latest technological achievements with a great number of applications. Memristor is a passive two-terminal electronic device which behavior is described by a nonlinear constitutive relation between the voltage drop at its terminal and the current flowing through the device. But the reason why the memristor is substantially different from the other fundamental circuit elements is that, when the applied voltage is turned off, it still remembers how much voltage was applied before and for how long; thus presenting memory of its past. However, this innovative device attracted most of attention worldwide only after 2008 when its practical implementation was announced by Hewlett-Packard, originating intense research activity ever since.

Memristors have brought a revolution in various scientific fields, as many phenomena in systems, such as in thermistors, spintronic devices and molecules could be explained now with the use of the memristor. Also, electronic circuits with memory elements could simulate processes typical of biological systems, such as learning and associative memory and the adaptive behavior of unicellular organisms. Furthermore, neuromorphic computing circuits with memristors can potentially solve problems that are cumbersome or outright intractable by digital computation.

Memristors have been used in cellular neural networks, for performing a number of applications, such as logical operations, image processing operations, complex behavior and higher brain functions, or in designing Boolean logic gates for the AND, OR and NOT operations. In many well-known nonlinear circuits, the nonlinear element has been replaced by memristors and various interesting dynamical phenomena like chaos and hidden attractors have been observed. Therefore, with these wide range of applications, engineering aspects of memristor devices, memristive-based circuits and systems design become significant important.

About the Book

The new Springer book, *Advances in Memristors, Memristive Devices and Systems,* consists of 20 contributed chapters by subject experts who are specialized in the various topics addressed in this book. The special chapters have been brought out in this book after a rigorous review process in the broad areas of modeling and applications of memristors, memristive devices and systems. Special importance was given to chapters offering practical solutions and novel methods for the recent research problems in the modeling and applications of memristors, memristive devices and systems.

This book discusses trends and applications of memristors and memristive devices in engineering.

Objectives of the Book

This volume presents a selected collection of contributions on a focused treatment of recent advances and applications in memristors, memristive devices and systems. The book also discusses multidisciplinary applications in electrical engineering, control engineering, computer science and information technology. These are among those multidisciplinary applications where computational intelligence has excellent potentials for use. Both novice and expert readers should find this book a useful reference in the field of memristors and memristive devices.

Organization of the Book

This well-structured book consists of 20 full chapters.

Book Features

- The book chapters deal with the recent research problems in the areas of memristors and memristive devices.
- The book includes chapters by eminent experts and pioneers of memristors—Leon Chua and R.S. Williams.
- The book chapters contain a good literature survey with a long list of references.
- The book chapters are well-written with a good exposition of the research problem, methodology, block diagrams and circuits.
- The book chapters are lucidly illustrated with numerical examples and simulations.
- The book chapters discuss details of engineering applications and future research areas.

Audience

The book is primarily meant for researchers from academia and industry, who are working on memristors and memristive devices in the research areas—electrical engineering, control engineering, computer science, and information technology. The book can also be used at the graduate or advanced undergraduate level as a textbook or major reference for courses such as power systems, control systems, electrical devices, scientific modeling, computational science and many others.

Chennai, India Sundarapandian Vaidyanathan
Thessaloniki, Greece Christos Volos

Acknowledgements

As the editors, we hope that the chapters in this well-structured book will stimulate further research in memristors, memristive devices and control systems, and utilize them in real-world applications.

We hope sincerely that this book, covering so many different topics, will be very useful for all readers.

We thank eminent Prof. Leon Chua for kindly accepting our invitation and contributing two chapters in this book. We also thank eminent Profs. Alon Ascoli, Ronald Tetzlaff and R.S. Williams for kindly accepting our invitation and contributing a chapter in this book.

We would like to thank all the reviewers for their diligence in reviewing the chapters.

Special thanks go to Springer, especially the book Editorial team.

Chennai, India
Thessaloniki, Greece

Sundarapandian Vaidyanathan
Christos Volos

Contents

Memristor Emulators: A Note on Modeling

A. Ascoli, R. Tetzlaff, L.O. Chua, W. Yi and R.S. Williams

Abstract In a recent publication (Yi et al. 2011) elucidating a possible scheme to write information reliably onto a memory crossbar, Hewlett Packard Labs researchers employed a thyristor-based circuit to emulate the off-to-on switching behaviour of a titanium oxide memristor. The use of a thyristor device allowed them to test inexpensively and reliably the functionalities of the closed-loop crossbar write circuitry by using conventional CMOS components. From a device modeling point of view, however, it is worthy to point out that the aforementioned emulator is not a genuine memristor. The aim of this paper is to demonstrate with an in-depth mathematical analysis that the model of the thyristor does not fall into the class of memristors. The modelling approach adopted in this work may be a source of inspiration for researchers willing to check whether other devices or circuits may be classified as memristors.

Keywords Circuit theory · Memristor · Threshold switching · Thyristor

A. Ascoli (✉) · R. Tetzlaff
Faculty of Electrical Circuit Theory and Information Technology,
Department of Fundamentals of Electrical Circuit Theory and Electronics,
Technische Universität Dresden, Dresden, Germany
e-mail: alon.ascoli@tu-dresden.de

R. Tetzlaff
e-mail: ronald.tetzlaff@tu-dresden.de

L.O. Chua
Department of Electrical Engineering and Computer Sciences,
University of California Berkeley, Berkeley, CA 94720, USA
e-mail: chua@eecs.berkeley.edu

W. Yi
HRL Laboratories, LLC, Malibu, CA 90265, USA
e-mail: wyi@hrl.com

R.S. Williams
Hewlett Packard Labs, Palo Alto, CA 94304, USA
e-mail: stan.williams@hpe.com

© Springer International Publishing AG 2017

S. Vaidyanathan and C. Volos (eds.), *Advances in Memristors, Memristive Devices and Systems*, Studies in Computational Intelligence 701,
DOI 10.1007/978-3-319-51724-7_1

1 Introduction

A substantial amount of work on memristors (Chua 1971) and memristive systems (Chua and Kang 1976) focus on their manufacturing process (Pan et al. 2014), which is engineered so as to shape the electrical characteristics of the devices (Strukov et al. 2008) to enhance their performance as non-volatile memory elements (Waser et al. 2009; Jo et al. 2009; Wylezich et al. 2014) or as biological synapsc emulators (Zamarreño-Ramos et al. 2011). Only a few number of studies is focused on the establishment of solid foundations on the theory of memristor devices, circuits, and systems (Chua 2011, 2014, 2015). However, in our opinion, these theoretical works (Ascoli et al. 2014; Corinto et al. 2011, 2015, 2016; Larentis et al. 2012) are as important as the experimental investigation (Nardi et al. 2012), representing a crucial prerequisite in the ongoing research efforts to explore the full potential of memristors in future electronics (Ascoli et al. 2015c). In fact, gaining a deeper insight into the key mechanisms at the origin of memristive behaviour (Ascoli et al. 2016c) is instrumental to identify advantages and limitations of the adoption of memristors for memory or neuromorphic applications, as well as to understand under which extent may the nonlinear dynamics of these devices (Ascoli et al. 2016a, b) be exploited to develop unconventional forms of sensing (Carrara et al. 2012; Tzouvadaki et al. 2016a) and signal processing (Corino et al. 2012; Yang et al. 2013), as well as novel computing architectures (Talati et al. 2016; Ben-Hur and Kvatinsky 2016). Typically, in our studies, the application of nonlinear circuit theoretic techniques (Chua et al. 1985) to the device models allows the identification of the key factors underlying the emergence of memristive dynamics. On the basis of this knowledge, it is then possible to draw a comprehensive picture of the plethora of nonlinear behaviours a memristor may exhibit under any initial condition/input combination (Ascoli and Corinto 2013). Clearly, the availability of accurate mathematical descriptions is a fundamental preliminary requirement to conduct these theoretical investigations. The derivation of accurate memristor models is in fact one of the most challenging activities in this field of research (Ascoli et al. 2013). The aim of the work presented in this chapter is to clarify an important modeling issue which risks to mislead researchers in the field. In a seminal paper presenting a closed-loop scheme to write data reliably onto a memory crossbar, Hewlett Packard Labs researchers adopted a thyristor-based circuit to emulate the high-to-low resistance switching dynamics of a bipolar titanium dioxide memristor (Yi et al. 2011). This allowed to apply an inexpensive and robust testing and debugging procedure on the proposed data writing scheme. However, from a device modeling point of view, the aforementioned emulator may not be regarded as a memristor. In order to shed light into this important aspect, we derive an accurate model of the thyristor adopted for testing purposes at Hewlett Packard Labs (Yi et al. 2011), proving that it does not fit into the class of memristors (Chua 2015), because it lacks an Ohm-based law (i.e. an algebraic relation expressing the output in terms of the product between the input and a function of the state, and, possibly, the input) (Chua 2015), which necessarily constrains the time evolution of the memristor state, which, otherwise, would be governed solely by the state evolution function. The mathematical analysis presented in this work may be beneficial in those investigations intended to check whether other devices or circuits may be regarded as memristors. The manuscript is structured as follows. Section 2 introduces the thyristor under modeling. Section 3 derives its mathematical model. Section 4 provides the numerical validation for the theoretical results derived in Sect. 3. Finally conclusions and future research developments are drafted in Sect. 5.

2 Emulator

The particular component adopted at Hewlett Packard Labs to emulate the off-to-on switching behaviour of a bipolar titanium oxide memristor (Pickett et al. 2009; Abdalla and Pickett 2011) in the debugging and testing phase of a closed-loop or feedback data writing scheme for memory crossbars is the silicon bilateral switch *BS08D* (Powerex Inc. 2015) manufactured by Powerex Inc., USA. The main building block of the emulator is shown in Fig. 1a, while its circuit symbol is depicted in Fig. 1b. It is a bilateral thyristor consisting of four bipolar junction transistors with substrate node connected to ground—one pair, (Q_1, Q_2), of pnp type, and the other, (Q_3, Q_4), of npn type—as well as of two zener diodes (D_1, D_2) and of a couple of resistors (G_1, G_2). The device has three terminals, i.e. T_1, T_2, and G. Each of the two devices within a box in Fig. 1a represents a unilateral thyristor. It has three terminals, known as anode, cathode, and gate. The base of the pnp transistor is chosen as gate terminal. With such a choice, the unilateral thyristor is also known as *programmable unipolar junction transistor* (PUT) (On Semiconductor 2005). There exists another type of unilateral thyristor, known as *silicon controlled rectifier* (SCR), in which the gate terminal coincides with the base of the npn transistor. With reference to Fig. 1a, coupling the identical left and right PUT devices—let us denote them as cell C and C', enclosed within a rectangle with red and black dashed perimeter and featuring terminal triplets (A, G, K) and (A', G', K'), respectively—through their gates G and G', and connecting the anode A (A') of the first (latter) to the cathode C' (C) of the latter (first), results into the circuit topology of a *BS08D* device. On the other hand, applying the same coupling strategy to two identical SCR components yields the circuit schematics of a TRIAC (On Semiconductor 2005). The *BS08D* is a current-controlled[1] component designed to switch as the flow of the input current results into a device voltage exceeding a threshold value around 8 V. With reference to plot (b) in Fig. 1, in order to match the typical off-to-on switching behaviour of a bipolar titanium oxide memristor, Hewlett Packard Labs' engineers added an additional zener diode with a breakdown voltage around 2.3 V to the bilateral thyristor between its terminals T_2 and G (Yi et al. 2011). As compared to the design in (Yi et al. 2011), here we add yet another zener diode with a breakdown voltage around 2.3 V between

[1] Under quasi-static excitation, the voltage across the device is a single-valued function of the current through it.

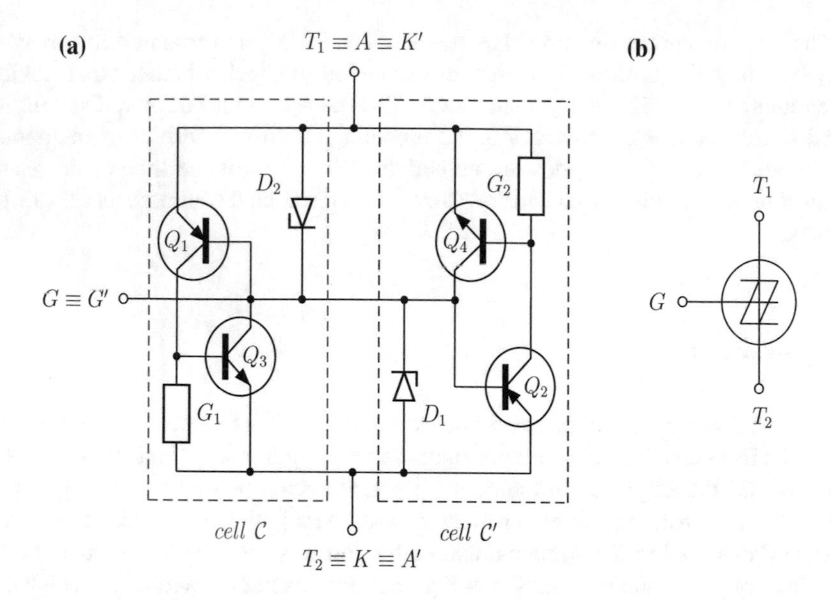

Fig. 1 **a** Circuit schematics of the *BS08D* device, a silicon bilateral thyristor from Powerex, Inc., USA. The anode, cathode, and gate terminals of cell C (C') are identified with symbols A, K, and G (A', K', and G'). **b** Circuit symbol of the three-terminal element

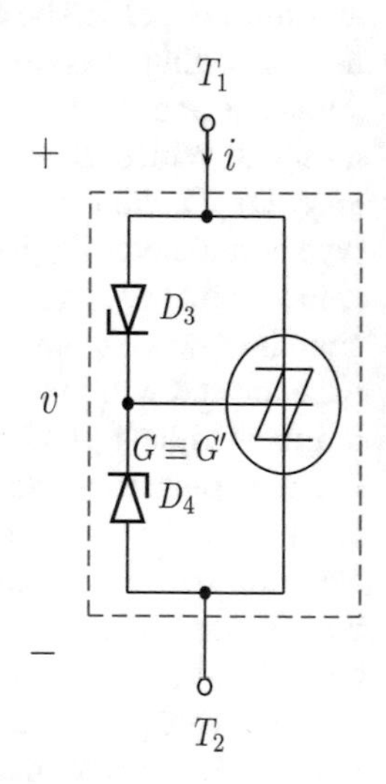

Fig. 2 One-port with off-to-on switching dynamics reminiscent of memristive behaviour under each polarity of the current input. In Yi et al. (2011) the *BS08D* device was coupled to zener diode D_4 only, and the resulting two-terminal element was used to mimic the high-to-low resistance switching of the titanium dioxide memristor under positive stimuli

the terminals T_1 and G of the component in Fig. 1b so as to obtain an odd-symmetric off-to-on switching. The resulting circuit is a one-port with terminals T_1 and T_2, as shown in Fig. 2 within a box with blue dashed perimeter. Let us denote the current through this bipole and the voltage across it as i and v, respectively.

In order to clarify the nature of the two-terminal element in Fig. 2, and avoid an improper device classification, as well as to gain a better understanding of the dynamical phenomena emerging in the current-controlled electronic component, in the next section we shall derive its mathematical model.

3 Model

With reference to the circuit schematics in Fig. 1 and to the emulator topology in Fig. 2, the cell C (C') consists of a complementary transistor pair, namely (Q_1, Q_3)

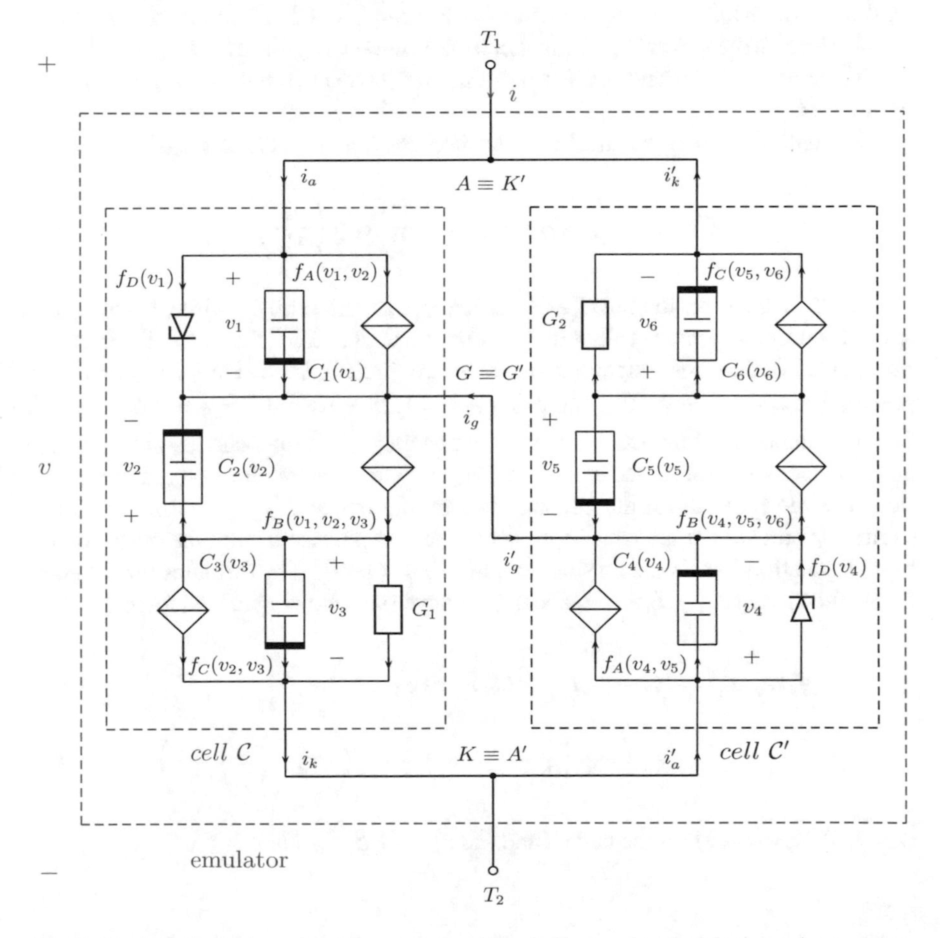

Fig. 3 Equivalent circuit model of the emulator of Fig. 2. The two-terminal device under modeling is encircled within a rectangle with blue dashed perimeter

$((Q_2, Q_4))$, a resistor of conductance G_1 (G_2), as well as a pair of zener diodes,[2] i.e. $(D_2, D_3)(D_1, D_4))$. The circuit model for the two-terminal emulator of Fig. 2, inspired to the theory presented in (Chua 1980), is shown in Fig. 3 within a box with blue dashed perimeter, in line with the colour coding convention adopted in Fig. 2.

3.1 Cell Model

In the characterization of each of the three-terminal cells C and C'—respectively enclosed within a rectangle with red and black dashed perimeter, in analogy to the colour coding scheme adopted in Fig. 1—the bipolar junction transistors are replaced by their Ebers-Moll circuit equivalents. With regards to the cell couplings, as anticipated in Sect. 2 the following terminal pairs are coupled together: (A, K'), (K, A'), and (G, G'). Within each round bracket pair the first (latter) symbol refers to a terminal of cell C (C').

The nonlinear voltage-controlled capacitors are defined as (Chua and Sing 1979)

$$C_j(v_j) = C_{0i} \left(\Psi_{0i} - v_j\right)^{-\frac{1}{m_i}} + \frac{I_{S_{1i}}}{V_T} \exp\left(\frac{v_j}{V_T}\right), \tag{1}$$

where $j \in \{1, 2, 3\}$ for the cell C, and $j \in \{, 4, 5, 6\}$ for the cell C', while, in the corresponding order, i assumes values in the same set, i.e. $\{1, 2, 3\}$, for both cells. In equation (1) C_{0i} defines the junction capacitance coefficient, Ψ_{0i} is the junction contact potential, and $V_T = \frac{kT}{q}$ the thermal voltage, $k = 1.28 \times 10^{-23}$ JK^{-1}, $q = 1.60 \times 10^{-19}$ C, and T denoting Boltzmann constant, elementary electronic charge, and junction absolute temperature, respectively. Further, m_i represents the junction grading coefficient, while τ_i stands for the minority carrier lifetime, and $I_{S_{1i}}$ symbolizes the ideal saturation current component. More details on the physics behind the operation of bipolar junction transistors may be found in Chua (1980). The formulas for the nonlinear functions $f_A(\cdot, \cdot)$, $f_B(\cdot, \cdot, \cdot)$, and $f_C(\cdot, \cdot)$ have the following closed forms:

$$f_A(v_k, v_m) = \left((1 + \gamma_1)I_{S_{11}} + I_{S_{15}}\right)\left(\exp\left(\frac{v_k}{V_T}\right) - 1\right) + I_{S_{21}}\left(\exp\left(\frac{v_k}{2V_T}\right) - 1\right) - I_{S_{12}}\left(\exp\left(\frac{v_m}{V_T}\right) - 1\right), \tag{2}$$

where $(k, m) = (1, 2)$ for the cell C, and $(k, m) = (4, 5)$ for the cell C',

---

[2]The zener diode in each cell is equivalent to the parallel of two zener diodes, one employed within the circuit of the *BS08D* device, refer to Fig. 1, and one adopted to tune the voltage at which the emulator undergoes switching, as it may be evinced by inspection of Fig. 2.

$$
\begin{aligned}
f_B(v_k, v_m, v_n) = {} & I_{S_{11}} \left(\exp\left(\frac{v_k}{V_T}\right) - 1 \right) - I_{S_{22}} \left(\exp\left(\frac{v_m}{2V_T}\right) - 1 \right) \\
& + I_{S_{13}} \left(\exp\left(\frac{v_n}{V_T}\right) - 1 \right) \\
& - \left((1+\gamma_1) I_{S_{12}} + (1+\gamma_2) I_{S_{14}} \right) \left(\exp\left(\frac{v_m}{V_T}\right) - 1 \right),
\end{aligned}
\tag{3}
$$

where $(k, m, n) = (1, 2, 3)$ for the cell C, and $(k, m, n) = (4, 5, 6)$ for the cell C', and

$$
\begin{aligned}
f_C(v_m, v_n) = {} & -I_{S_{14}} \left(\exp\left(\frac{v_m}{V_T}\right) - 1 \right) + I_{S_{23}} \left(\exp\left(\frac{v_n}{2V_T}\right) - 1 \right) \\
& + \left((1+\gamma_2) I_{S_{13}} + I_{S_{16}} \right) \left(\exp\left(\frac{v_n}{V_T}\right) - 1 \right),
\end{aligned}
\tag{4}
$$

where $(m, n) = (2, 3)$ for the cell C, and $(m, n) = (5, 6)$ for the cell C'. In Eqs. (2)–(4) $I_{S_{1j}}$ ($j \in \{1, 2, 3, 4, 5, 6\}$) and $I_{S_{2j}}$ ($j \in \{1, 2, 3\}$) respectively denote ideal and nonlinear saturation current components, while γ_j ($j \in \{1, 2\}$) are recombination factors for the current components. Finally, the nonlinear function $f_D(\cdot)$ is expressed as

$$
\begin{aligned}
f_D(v_j) = {} & I_{S_a} \left(\exp\left(\frac{v_j}{n_a V_T}\right) - 1 \right) - I_{z_a} \exp\left(-\frac{v_j + V_{z_a}}{n_{z_a} V_T} \right) \\
& + I_{S_b} \left(\exp\left(\frac{v_j}{n_b V_T}\right) - 1 \right) - I_{z_b} \exp\left(-\frac{v_j + V_{z_b}}{n_{z_b} V_T} \right),
\end{aligned}
\tag{5}
$$

where $j = 1$ for the cell C, and $j = 4$ for the cell C'. As anticipated earlier, $f_D(\cdot)$ takes into account the currents of zener diode pair $(D_2, D_3)((D_1, D_4))$ for the cell C (C'). Particularly, for each cell, on the right hand side of Eq. 5, the first and last two addends respectively constitute the current of the first and second diode in the aforementioned pair. Applying basic circuit principles, the equations governing the evolution of the voltages across the nonlinear capacitors of the cell C are expressed as

$$
\begin{aligned}
\frac{dv_1}{dt} & = \frac{1}{C_1(v_1)} \left(i_a - f_A(v_1, v_2) - f_D(v_1) \right) \\
& = f_1(v_1, v_2, v_3, i_a, i_g),
\end{aligned}
\tag{6}
$$

$$
\begin{aligned}
\frac{dv_2}{dt} & = \frac{1}{C_2(v_2)} \left(-i_a + f_B(v_1, v_2, v_3) - i_g \right) \\
& = f_2(v_1, v_2, v_3, i_a, i_g),
\end{aligned}
\tag{7}
$$

$$
\begin{aligned}
\frac{dv_3}{dt} & = \frac{1}{C_3(v_3)} \left(i_a + i_g - G_1 v_3 - f_C(v_2, v_3) \right) \\
& = f_3(v_1, v_2, v_3, i_a, i_g),
\end{aligned}
\tag{8}
$$

while the cell C' model is given by

$$\frac{dv_4}{dt} = \frac{1}{C_4(v_4)}\left(i'_a - f_A(v_4, v_5) - f_D(v_4)\right)$$
$$= f_4(v_4, v_5, v_6, i'_a, i'_g), \tag{9}$$
$$\frac{dv_5}{dt} = \frac{1}{C_5(v_5)}\left(-i'_a + f_B(v_4, v_5, v_6) - i'_g\right)$$
$$= f_5(v_4, v_5, v_6, i'_a, i'_g), \tag{10}$$
$$\frac{dv_6}{dt} = \frac{1}{C_6(v_6)}\left(i'_a + i'_g - G_2 v_6 - f_C(v_5, v_6)\right)$$
$$= f_6(v_4, v_5, v_6, i'_a, i'_g). \tag{11}$$

3.2 Interconnection Model

Next, the model of the interconnections between the two cells need to be derived. The application of Kirchhoff's current and voltage laws (Chua et al. 1985) to the coupled cells in Fig. 2 yields:

$$i'_g = -i_g, \tag{12}$$
$$v_1 = v_5 - v_6, \tag{13}$$
$$v_4 = v_2 - v_3. \tag{14}$$

Due to voltage constraints (13)–(14), the order of the dynamical system expressed by Eqs. (6)–(11) is 4. As for the non-redundant state variables, we choose v_2, v_3, v_5, and v_6. Using (13)–(14) into the coupled ordinary differential equations governing the dynamics of the non-redundant state variables, i.e. into Eqs. (7), (8), (10) and (11), the resulting equations become:

$$\frac{dv_2}{dt} = \frac{1}{C_2(v_2)}\left(-i_a + f_B(v_5 - v_6, v_2, v_3) - i_g\right)$$
$$= f_2(v_2, v_3, v_5, v_6, i_a, i'_a, i_g), \tag{15}$$
$$\frac{dv_3}{dt} = \frac{1}{C_3(v_3)}\left(i_a + i_g - G_1 v_3 - f_C(v_2, v_3)\right)$$
$$= f_3(v_2, v_3, v_5, v_6, i_a, i'_a, i_g), \tag{16}$$
$$\frac{dv_5}{dt} = \frac{1}{C_5(v_5)}\left(-i'_a + f_B(v_2 - v_3, v_5, v_6) + i_g\right),$$
$$= f_5(v_2, v_3, v_5, v_6, i_a, i'_a, i_g), \tag{17}$$

$$\frac{dv_6}{dt} = \frac{1}{C_6(v_6)} \left(i'_a - i_g - G_2 v_6 - f_C(v_5, v_6) \right),$$
$$= f_6(v_2, v_3, v_5, v_6, i_a, i'_a, i_g), \tag{18}$$

where we made use of Eq. (12) as well. Next, the variables i_a, i'_a, and i_g need to be expressed in terms of v_2, v_3, v_5, and v_6 as well as of the input current i. Differentiating (13) with respect to the time, inserting the right hand sides of Eqs. (6), (10) and (11) into the resulting expression, and casting the two redundant state variables v_1 and v_4 in terms of the four non-redundant ones, after some algebraic manipulation, the current i_g is found to be given by

$$i_g = \left(\frac{1}{C_5(v_5)} + \frac{1}{C_6(v_6)} \right)^{-1} \left(\frac{i_a}{C_1(v_5 - v_6)} - \frac{f_A(v_5 - v_6, v_2)}{C_1(v_5 - v_6)} - \frac{f_D(v_5 - v_6)}{C_1(v_5 - v_6)} \right.$$
$$\left. - \frac{f_B(v_2 - v_3, v_5, v_6)}{C_5(v_5)} - \frac{G_2 v_6}{C_6(v_6)} - \frac{f_C(v_5, v_6)}{C_6(v_6)} \right) + i'_a$$
$$= i_g(v_2, v_3, v_5, v_6, i_a, i'_a). \tag{19}$$

Let us now compute the time derivative of Eq. (14), and then use the right hand sides of Eqs. (9), (7) and (8) as well as Eq. (19) into the resulting expression. Lengthy calculations provide the following formula for the current i'_a in terms of the current i_a and of the four state variables v_2, v_3, v_5, and v_6:

$$i'_a = \left(\frac{1}{C_4(v_2 - v_3)} + \frac{C_2(v_2) + C_3(v_3)}{C_2(v_2) C_3(v_3)} \right)^{-1} \left(\frac{f_A(v_2 - v_3, v_5)}{C_4(v_2 - v_3)} + \left(\frac{1}{C_2(v_2)} \right. \right.$$
$$\left. + \frac{1}{C_3(v_3)} \right) \left(\frac{1}{C_5(v_5)} + \frac{1}{C_6(v_6)} \right)^{-1} \frac{f_A(v_5 - v_6, v_2)}{C_1(v_5 - v_6)} + \frac{f_B(v_5 - v_6, v_2, v_3)}{C_2(v_2)}$$
$$+ \left(\frac{1}{C_2(v_2)} + \frac{1}{C_3(v_3)} \right) \frac{C_5(v_5) C_6(v_6)}{C_5(v_5) + C_6(v_6)} \frac{f_B(v_2 - v_3, v_5, v_6)}{C_5(v_5)} + \frac{f_C(v_2, v_3)}{C_3(v_3)}$$
$$+ \left(\frac{1}{C_2(v_2)} + \frac{1}{C_3(v_3)} \right) \left(\frac{1}{C_5(v_5)} + \frac{1}{C_6(v_6)} \right)^{-1} \frac{f_C(v_5, v_6)}{C_6(v_6)} + \frac{f_D(v_2 - v_3)}{C_4(v_2 - v_3)}$$
$$+ \left(\frac{1}{C_2(v_2)} + \frac{1}{C_3(v_3)} \right) \frac{C_5(v_5) C_6(v_6)}{C_5(v_5) + C_6(v_6)} \frac{f_D(v_5 - v_6)}{C_1(v_5 - v_6)} + \frac{G_1 v_3}{C_3(v_3)} + \frac{G_2 v_6}{C_6(v_6)}$$
$$\cdot \left(\frac{1}{C_2(v_2)} + \frac{1}{C_3(v_3)} \right) \left(\frac{1}{C_5(v_5)} + \frac{1}{C_6(v_6)} \right)^{-1} - \left(\frac{1}{C_2(v_2)} + \frac{1}{C_3(v_3)} \right) i_a$$
$$\left. \cdot \left(1 + \frac{1}{C_1(v_5 - v_6)} + \left(\frac{1}{C_5(v_5)} + \frac{1}{C_6(v_6)} \right)^{-1} \right) \right)$$
$$= i'_a(v_2, v_3, v_5, v_6, i_a). \tag{20}$$

At this point, inserting this expression for i'_a into Eq. (19), the current i_g is also a function of v_2, v_3, v_5, v_6, and i_a only:

$$\begin{aligned}
i_g = &\left(\frac{1}{C_4(v_2-v_3)} + \frac{C_2(v_2)+C_3(v_3)}{C_2(v_2)C_3(v_3)}\right)^{-1} \frac{f_A(v_2-v_3,v_5)}{C_4(v_2-v_3)} + \Bigg(\Bigg(\frac{1}{C_4(v_2-v_3)}\\
&+ \frac{C_2(v_2)+C_3(v_3)}{C_2(v_2)C_3(v_3)}\Bigg)^{-1} \frac{C_2(v_2)+C_3(v_3)}{C_2(v_2)C_3(v_3)} - 1\Bigg) \frac{f_A(v_5-v_6,v_2)}{C_1(v_5-v_6)} + \Bigg(\frac{1}{C_2(v_2)}\\
&+ \frac{1}{C_3(v_3)} + \frac{1}{C_4(v_2-v_3)}\Bigg)^{-1} \frac{f_B(v_5-v_6,v_2,v_3)}{C_2(v_2)} + \Bigg(\Bigg(\frac{1}{C_2(v_2)} + \frac{1}{C_3(v_3)}\\
&+ \frac{1}{C_4(v_2-v_3)}\Bigg)^{-1} \frac{C_2(v_2)+C_3(v_3)}{C_2(v_2)C_3(v_3)} - 1\Bigg) \frac{f_B(v_2-v_3,v_5,v_6)}{C_5(v_5)} + \Bigg(\frac{1}{C_2(v_2)}\\
&+ \frac{1}{C_3(v_3)} + \frac{1}{C_4(v_2-v_3)}\Bigg)^{-1} \frac{f_C(v_2,v_3)}{C_3(v_3)} + \frac{f_C(v_5,v_6)}{C_6(v_6)} \Bigg(\Bigg(\frac{1}{C_2(v_2)} + \frac{1}{C_3(v_3)}\\
&+ \frac{1}{C_4(v_2-v_3)}\Bigg)^{-1} \left(\frac{1}{C_2(v_2)} + \frac{1}{C_3(v_3)}\right) - 1\Bigg) + \Bigg(\frac{1}{C_4(v_2-v_3)} + \frac{1}{C_2(v_2)}\\
&+ \frac{1}{C_3(v_3)}\Bigg)^{-1} \frac{f_D(v_2-v_3)}{C_4(v_2-v_3)} + \Bigg(-1 + \frac{C_2(v_2)+C_3(v_3)}{C_2(v_2)C_3(v_3)} \Bigg(\frac{1}{C_2(v_2)} + \frac{1}{C_3(v_3)}\\
&+ \frac{1}{C_4(v_2-v_3)}\Bigg)^{-1}\Bigg) \frac{f_D(v_5-v_6)}{C_1(v_5-v_6)} + \left(\frac{1}{C_4(v_2-v_3)} + \frac{C_2(v_2)+C_3(v_3)}{C_2(v_2)C_3(v_3)}\right)^{-1}\\
&\cdot \frac{G_1 v_3}{C_3(v_3)} + \Bigg(\left(\frac{1}{C_4(v_2-v_3)} + \frac{1}{C_2(v_2)} + \frac{1}{C_3(v_3)}\right)^{-1} \left(\frac{1}{C_2(v_2)} + \frac{1}{C_3(v_3)}\right)\\
&-1\Bigg) \frac{G_2 v_6}{C_6(v_6)} - \left(\frac{1}{C_4(v_2-v_3)} + \frac{1}{C_2(v_2)} + \frac{1}{C_3(v_3)}\right)^{-1} i_a - \frac{i_a}{C_1(v_5-v_6)}\\
&\cdot \Bigg(\left(\frac{1}{C_4(v_2-v_3)} + \frac{1}{C_2(v_2)} + \frac{1}{C_3(v_3)}\right)^{-1} \left(\frac{1}{C_2(v_2)} + \frac{1}{C_3(v_3)}\right) - 1\Bigg)\\
= &\, i_g(v_2,v_3,v_5,v_6,i_a). \qquad (21)
\end{aligned}$$

It remains to express the current i_a in terms of the non-redundant state variables as well as of the input current i controlling the emulator operation. Applying the Kirchhoff's Current Law at node T_1 in Fig. 3, the expression for the current i through the bilateral device is found to be given by

$$i = i_a - i'_a(v_2,v_3,v_5,v_6,i_a) + i_g(v_2,v_3,v_5,v_6,i_a). \qquad (22)$$

Inserting the expressions for i'_a and i_g, respectively given in Eqs. (20) and (21), into Eq. (22), the current i_a is found to be described by the following mathematical expression:

$$
\begin{aligned}
i_a = & \left(1 + \frac{1}{C_1(v_5 - v_6)}\left(\frac{C_5(v_5) + C_6(v_6)}{C_5(v_5)C_6(v_6)}\right)^{-1}\right)^{-1}\left(i + \left(\frac{C_5(v_5) + C_6(v_6)}{C_5(v_5)C_6(v_6)}\right)^{-1}\right. \\
& \cdot \left(\frac{f_A(v_5 - v_6, v_2)}{C_1(v_5 - v_6)} + \frac{f_B(v_2 - v_3, v_5, v_6)}{C_5(v_5)} + \frac{f_C(v_5, v_6)}{C_6(v_6)} + \frac{f_D(v_5 - v_6)}{C_1(v_5 - v_6)}\right. \\
& \left.\left. + \frac{G_2 v_6}{C_6(v_6)}\right)\right) \\
= & \; i_a(v_2, v_3, v_5, v_6, i)
\end{aligned} \tag{23}
$$

3.3 Network Model

Equation (23) gives the dependence of the current i_a upon the device current i and the non-redundant state variables v_2, v_3, v_5, and v_6. Inserting this equation into Eq. (20–21) provides the expression for the current i'_a (i_g) in terms of v_2, v_3, v_5, v_6, and i. It follows that the state evolution functions in the state equations (15), (16), (17) and (18) may be expressed only in terms of v_2,v_3,v_5,v_6 and i. The state equations of the emulator fall thus into the following class:

$$\frac{dv_2}{dt} = f_2(v_2, v_3, v_5, v_6, i), \tag{24}$$

$$\frac{dv_3}{dt} = f_3(v_2, v_3, v_5, v_6, i), \tag{25}$$

$$\frac{dv_5}{dt} = f_5(v_2, v_3, v_5, v_6, i), \tag{26}$$

$$\frac{dv_6}{dt} = f_6(v_2, v_3, v_5, v_6, i). \tag{27}$$

The analytical expressions of the functions $f_k(v_2, v_3, v_5, v_6, i)$ for $k \in \{2, 3, 5, 6\}$ are quite long, and are thus omitted to improve readability. This set of coupled ordinary differential equations has the same form as the state equations of current-controlled generic or extended memristors. However, this is not sufficient to classify the emulator as a memristor. In fact, the model of a memristor necessarily includes also an algebraic relation, known as Ohm's law, defining how the output depends on input and states, and, most importantly, imposing the *coincident zero-crossing signature* (Chua 2015), i.e. the constraint for the output to exhibit zeros at the same time instants as the input. With reference to Fig. 3, the input (output) signal of the emulator is the voltage v (current i) across (through) it. Using Eq. (13) the voltage v between terminals T_1 and T_2 may be cast as follows:

$$
\begin{aligned}
v &= -v_2 + v_3 + v_5 - v_6 \\
&= g(v_2, v_3, v_5, v_6).
\end{aligned} \tag{28}
$$

The input current i has impact on the output voltage v, since it is part of the state evolution function defined by the right hand sides of Eqs. (24)–(27), and thus influences the temporal dynamics of the states. However since the right hand side of equation (28) is independent of i, it may not express the Ohm-based law of a current-controlled memristor. In fact the lack of an algebraic relation expressing the device voltage in terms of the product between its current and a function of its states, and, possibly, its current, is at the basis of the violation of the *coincident zero-crossing signature*[3] in the thyristor device under modeling.

4 Numerical Validation

In order to provide some numerical evidence for the similarity between the dynamics of the one-port in Fig. 2 and the typical on-to-off switching behaviour of a memristor, we recur to a numerical simulation carried out in LTSpice. Here a current source is applied directly across the two-terminal device of Fig. 2. Let the source inject a periodic triangular current i of amplitude $i_0 = 10 \cdot 10^{-3}$A and frequency $f = 1$Hz through the bipole (see the red curve in plot (a) of Fig. 4). The device voltage in response to the periodic stimulus is shown in blue on the same plot. Figure 4b illustrates the loci emerging on the i-v plane. This is the quasi-static characteristic of the *BS*08*D* device (Powerex Inc. 2015) (see also the inset in Fig. 8 in (Yi et al. 2011)). Numbered arrows in plots (a) and (b) of Fig. 4 clearly show the evolution of the voltage across the device as the triangular input current ramps up and down during the positive half-cycle. During the phases 1 and 2 (3 and 4), spanning a quarter of the input cycle, the device undergoes an off-to-on (on-to-off) transition. As anticipated earlier, the device dynamics under the positive input half-cycle have been exploited to emulate the off-to-on threshold switching behaviour of a bipolar titanium dioxide memristor in the course of the testing and debugging phase of a closed-loop data writing strategy for memory crossbars (Yi et al. 2011). However, it is instructive to point out that the device of Fig. 2 is not a genuine memristor, as theoretically demonstrated in Sect. 3.1. This is pretty clear from Fig. 4c, showing an enlarged view of the i-v loci of Fig. 4b in the region around the origin: the loci is not pinched in the point $(i, v) = (0, 0)$ (Chua 2014).

Next we shall validate the accuracy of the mathematical model of the silicon bilateral thyristor derived in Sect. 3. The parameters of the model was set as reported in Table 1. Setting the thermal voltage V_T to 26 mV, using a periodic zero-mean triangular input current i of amplitude $i_0 = 10$ mA and frequency $f = 1$Hz, and setting the initial values for the non-redundant state variables to zero, i.e. $v_j(0) = 0$ V for

[3]Let us assume that the two terminals of a one-port made up of arbitrary linear and nonlinear circuit elements, as well as voltage and current sources, are closed onto a current-controlled memristor device. The waveform of the voltage $v(t)$ associated with the current $i(t)$ of any admissible signal pair $(i(t), v(t))$ measured from this circuit set-up must cross the time axis *whenever* $i = 0$. This property of a current-controlled memristor is known as *coincident zero-crossing signature* (Chua 2015).

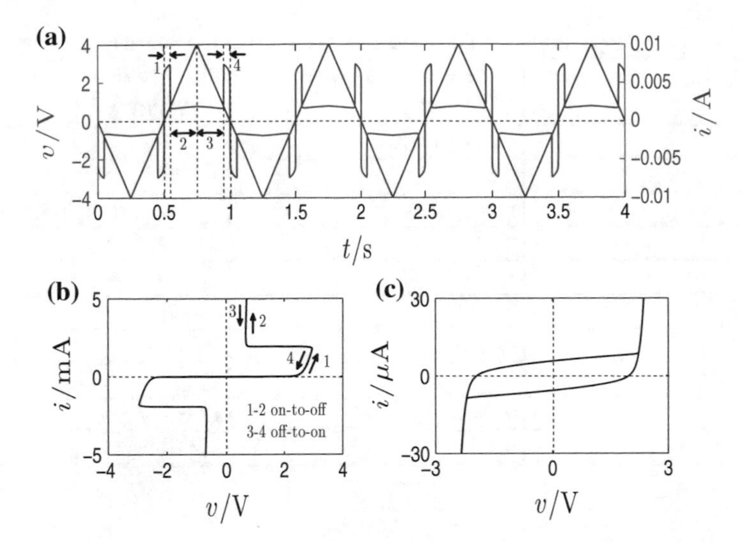

Fig. 4 **a** Time waveforms of the voltage v (in *blue*) falling across the device of Fig. 2 as a result of the current i inserted into it. The frequency of the AC stimulus, whose amplitude is $i_0 = 10 \times 10^{-3}$A, is $f = 1$ Hz, thus the excitation may be referred to as *quasi-static* (Ascoli et al. 2016c). **b** Plot of i versus v. **c** Zoom on the region of the i–v plane within the box defined by $v \in [-3, 3]$V and $i \in [-30, 30]\mu$A

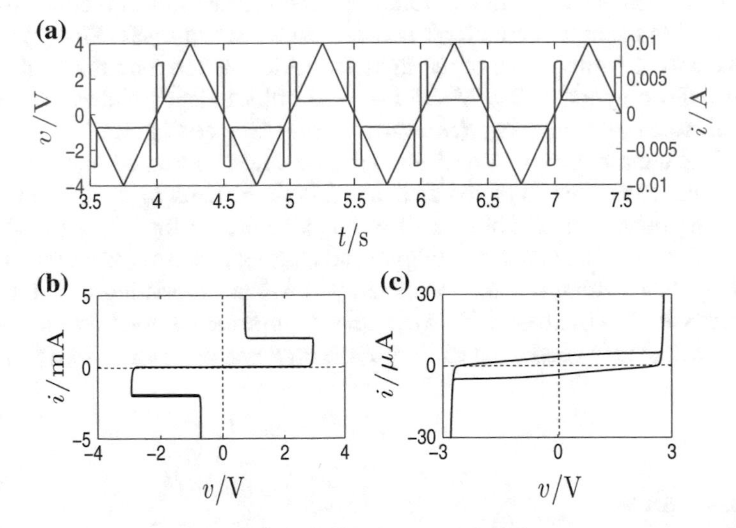

Fig. 5 **a** Time waveforms of the current stimulus (*red curve*) and of the voltage response (*blue curve*). **b** Current-voltage loci of the thyristor. **c** Enlarged view of the i-v loci in the region around the point $(i, v) = (0, 0)$

Table 1 Values of the parameters in the model Eqs. (24)–(27) of the silicon bilateral thyristor

$G_1 = G_2/$ S	$C_{0i}/(\mathrm{FV}^{1/m_i})$ $i \in \{1,3\}$	$C_{02}/(\mathrm{FV}^{1/m_2})$
0.9×10^{-3}	0.02×10^{-8}	0.01×10^{-8}
$\Psi_0/$V	$m_1 = m_2 = m_3$	$\tau_1 = \tau_3/$ s
1	0.5	1×10^{-9}
$\tau_2/$ s	$\gamma_1 = \gamma_2$	$I_{S_{1k}}/$ A, $k \in \{1,2,3,4,6\}$
1×10^{-8}	0.17	1×10^{-14}
$I_{S_{15}}/$ A	$I_{S_{21}}/$ A	$I_{S_{22}} = I_{S_{23}}/$ A
0.25×10^{-14}	5×10^{-12}	5×10^{-11}
$I_{S_a} = I_{S_b}/$ A	$n_a = n_b$	$V_{z_a}/$ V
467.04×10^{-18}	2	8
$V_{z_b}/$ V	$n_{z_a} = n_{z_b}$	$I_{z_a} = I_{z_b}/$ A
2.3	0.3	5×10^{-3}

$j \in \{2,3,5,6\}$, the numerical integration of the state equations (24)–(27) resulted in simulation results capturing quantitatively the dynamics observed in the LTSpice-based investigation of Fig. 4. Figure 5a depicts in red the time waveform of the current inserted into the two-terminal device of Fig. 3, and in blue the resulting voltage drop between terminals T_1 and T_2. Referring to the latter signal, note the off-to-on-to-off transition the thyristor undergoes over each input half-cycle. Plot (b) in Fig. 5 shows the current-voltage loci of the thyristor device. A zoom on the region of the i-v plane defined by $v \in [-3,3]$V and $i \in [-30,30]\mu$A clearly shows the violation of the *coincident zero-crossing signature* (Chua 2015) (see Fig. 5c).

As a final remark, note that a similar analysis could be carried out to show that the particular component adopted at Hewlett Packard Labs to emulate the on-to-off switching behaviour of a bipolar titanium oxide memristor (Pickett et al. 2009; Abdalla and Pickett 2011) in the debugging and testing phase of the aforementioned closed-loop data writing scheme for memory crossbars, consisting of two back-to-back lambda diodes (Kano et al. 1975), is not a genuine memristor. Since the analysis is similar to the study conducted earlier for the thyristor, we omit it from this manuscript.

5 Conclusions

The close synergy between experimental (Tzouvadaki et al. 2016a, b) and theoretical (Ascoli et al. 2015a, b) studies is a crucial requirement to foster progress in memristor research (Vallero et al. 2016; Ascoli et al. 2015a, b; Slesazeck et al. 2015; Levy et al. 2014), particularly since the theory may explain (Ascoli et al. 2015c) and/or predict (Ascoli et al. 2016c) the mechanisms behind the experimental observation of complex dynamical phenomena (Vaidyanathan and Volos 2016a) in nonlinear systems

(Vaidyanathan and Voles 2016a, b). This chapter clarifies an important modelling issue regarding a silicon bilateral thyristor-based circuit used to emulate the off-to-on switching dynamics of a bipolar titanium dioxide memristor in the testing and debugging phase of a feedback data writing scheme for memory crossbars. Despite the thyristor exhibits off-to-on dynamics reminiscent of the switching process a memristor undergoes under positive stimuli, thus representing a valid tool for testing and debugging memristive circuits, from a device modeling point of view it may not be regarded as a genuine memory resistor (Chua 2015). The present manuscript first derives the mathematical model of the thyristor-based emulator, adopted in the seminal paper from Hewlett Packard Labs (Yi et al. 2011) to verify the mechanisms underlying a closed-loop crossbar data writing process, and then proves that it does not strictly fall into the most general class of memristors (Chua 2015). Numerical simulation results are then provided to support the conclusions from the theoretical analysis. All in all, this work is meant to warn the uninitiated against an improper circuit theoretic classification of the thyristor. The mathematical analysis presented in the chapter may inspire other studies intended to check whether or not a device or circuit may be classified as memristor, and fits well in the framework of our activities aimed at establishing robust foundations on memristor theory.

Acknowledgements The support from EU COST Action IC1401 and Czech Science Foundation (grant no. $14 - 19865S$) are acknowledged. We also thank the Deutsche Forschung Gesellschaft (DFG) for their financial contribution to the research project “Locally active memristive data processing (LAMP)” under grant number $TE257/22 - 1$. L. Chua's research is supported by AFOSR Grant FA 9550-13-1-0136.

References

Abdalla, H., & Pickett, M. (2011). Spice modeling of memristors. In *IEEE International Symposium on Circuits and Systems (ISCAS)* (pp. 1832–1835).

Ascoli, A., & Corinto, F. (2013). Memristor models in a chaotic neural circuit. *International Journal of Bifurcation and Chaos (IJBC)*, *23*(1350052), 28.

Ascoli, A., Corinto, F., Senger, V., & Tetzlaff, R. (2013). Memristor model comparison. *IEEE Circuits and Systems Magazine*, *13*, 89–105.

Ascoli, A., Corinto, F., & Tetzlaff, R. (2015a). A class of versatile circuits, made up of standard electrical components, are memristors. *International Journal of Circuit Theory and Applications (IJCTA)*, *44*, 127–146.

Ascoli, A., Corinto, F., & Tetzlaff, R. (2015b). Generalized boundary condition memristor model. *International Journal of Circuit Theory and Applications (IJCTA)*, *44*, 60–84.

Ascoli, A., Schmidt, T., Corinto, F., & Tetzlaff, R. (2014). *Application of the Volterra Series paradigm to memristive systems*, Chap. 5 (pp. 163–191). New York: Springer.

Ascoli, A., Slesazeck, S., Mähne, H., Tetzlaff, R., & Mikolaijck, T. (2015c). Nonlinear dynamics of a locally-active memristor. *IEEE Transactions on Circuits and Systems–I (TCAS–I): Regular Papers*, *62*, 1165–1174.

Ascoli, A., Tetzlaff, R., & Chua, L. (2016a). The first ever real bistable memristors-part i: Theoretical insights on local fading memory. *IEEE Transactions on Circuits and Systems-II (TCAS-II): Express Briefs*.

Ascoli, A., Tetzlaff, R., & Chua, L. (2016b). The first ever real bistable memristors-part ii: Design and analysis of a local fading memory system. *IEEE Transactions on Circuits and Systems-II (TCAS-II): Express Briefs*.

Ascoli, A., Tetzlaff, R., Chua, L., Strachan, J., & Williams, R. (2016c). History erase effect in a non-volatile memristor. *IEEE Transactions on Circuits and Systems–I (TCAS–I): Regular Papers, 63*, 389–400.

Ben-Hur, R., & Kvatinsky, S. (2016). Memristive memory processing unit (mpu) controller for in-memory processing. In *Proceedings of International Conference on the Science of Electrical Engineering (ICSEE)*.

Carrara, S., Sacchetto, D., Doucey, M.-A., Baj-Rossi, C., Micheli, G. D., & Leblebici, Y. (2012). Memristive-biosensors: A new detection method by using nanofabricated memristors. *Sensors and Actuators B: Chemical, 171–172*, 449–457.

Chua, L. O. (1971). Memristor-The missing circuit element. *IEEE Transactions on Circuit Theory, 18*, 507–519.

Chua, L. (1980). Device modelling via basic nonlinear circuit elements. *IEEE Transactions on Circuits and Systems–I: Regular Papers, CAS, 27*, 1014–1044.

Chua, L. (2011). Resistance switching memories are memristors. *Applied Physics A, 102*(4), 765–783.

Chua, L. (2014). If it's pinched, it's a memristor. *Semiconductor Science and Technology, 29*(104001), 42.

Chua, L. (2015). Everything you wish to know about memristors but are afraid to ask. *Radioengineering, 24*, 319–368.

Chua, L., Desoer, C., & Kuh, E. (1985). *Linear and nonlinear circuits*. New York, USA: McGraw-Hill.

Chua, L., & Sing, Y. W. (1979). Nonlinear lumped-circuit model for s.c.r. *Electronic Circuits and Systems, 3*, 5–14.

Chua, L. O., & Kang, S. M. (1976). Memristive devices and system. *Proceedings of IEEE, 64*, 209–223.

Corinto, F., Ascoli, A., & Gilli, M. (2011). Nonlinear dynamics of memristive oscillators. *IEEE Transactions on Circuits Systems I Regular Papers, 58*, 1323–1336.

Corinto, F., Ascoli, A., & Gilli, M. (2012). Analysis of current-voltage characteristics for memristive elements in pattern recognition systems. *International Journal of Circuit Theory and Applications (IJCTA), 40*, 1277–1320.

Corinto, F., Chua, L., & Civalleri, P. (2015). A theoretical approach to memristor devices. *IEEE Journal on Emerging and Selected Topics in Circuits and Systems (JETCAS), 5*, 123–132.

Corinto, F., & Forti, M. (2016). Memristor circuits: Flux-charge analysis method. *IEEE Transactions on Circuits and Systems–I (TCAS–I): Regular Papers*, 1–13.

Jo, S., Kim, K.-H., & Lu, W. (2009). High-density crossbar arrays based on a si memristive system. *Nanoletters, 9*, 870–874.

Kano, G., Iwasa, H., Takagi, H., & Teramoto, I. (1975). The lambda diode: Versatile negative-resistance device. *Electronics, 13*, 105–109.

Larentis, S., Nardi, F., Balatti, S., Gilmer, D., & Ielmini, D. (2012). Resistive switching by voltage-driven ion migration in bipolar rram-part ii: Modeling. *IEEE Transactions on Electron Devices, 59*, 2468–2475.

Levy, Y., Bruck, J., Cassuto, Y., Friedman, E., Kolodny, A., Yaacobi, E., et al. (2014). Logic operation in memory using a memristive akers array. *Microelectronics Journal, 45*, 1429–1437.

Nardi, F., Larentis, S., Balatti, S., Gilmer, D., & Ielmini, D. (2012). Resistive switching by voltage-driven ion migration in bipolar rram-part i: Experimental study. *IEEE Transactions on Electron Devices, 59*, 2461–2467.

On Semiconductor, U. (2005). *Thyristor theory and design considerations*. http://onsemi.com.

Pan, F., Gao, S., Chen, C., Song, C., & Zeng, F. (2014). Recent progress in resistive random access memories: Materials, switching mechanisms and performance. *Materials Science and Engineering R, Elsevier, 83*, 1–59.

Pickett, M., Strukov, D., Borghetti, J., Yang, J., Snider, G., Stewart, D., et al. (2009). Switching dynamics in titanium dioxide memristive devices. *Journal of Applied Physics, 106*, 074508-1–074508-6.

Powerex Inc, U. (2015). *BS08D-T112 silicon bilateral switch*. http://www.pwrx.com/Product/BS08D-T112.

Slesazeck, S., Mähne, H., Wylezich, H., Wachowiak, A., Radhakrishnan, J., & Ascoli, A., et al. (2015). Physical model of threshold switching in nbo_2 based memristors. *RSC Advances, Royal Society of Chemistry*.

Strukov, D. B., Snider, G. S., Stewart, D. R., & Williams, R. S. (2008). The missing memristor found. *Nature, 453*, 80–83.

Talati, N., Gupta, S., Mane, P., & Kvatinsky, S. (2016). Logic design within memristive memories using memristor aided logic (magic). *IEEE Transactions on Nanotechnology, 15*, 635–650.

Tzouvadaki, I., Jolly, P., Lu, X., Ingebrandt, S., de Micheli, G., Estrela, P., & Carrara, S. (2016a). Label-free ultrasensitive memristive aptasensor. *Nanoletters, 16*, 4472–4476.

Tzouvadaki, I., Madaboosi, N., Taurino, I., Chu, V., Conde, J., de Micheli, G., & Carrara, S. (2016b). Study on the bio-functionalization of memristive nanowires for optimum memristive biosensors. *RSC Journal of Materials Chemistry B, 4*, 2153–2162.

Vaidyanathan, S., & Volos, C. (2016a). *Advances and applications in chaotic systems*. Berlin, Germany: Springer.

Vaidyanathan, S., & Volos, C. (2016b). *Advances and applications in nonlinear control systems*. Berlin, Germany: Springer.

Vallero, A., Tzouvadaki, I., Puppo, F., Doucey, M. -A., Delaloye, J. -F., Micheli, G. D., & Carrara, S. (2016). Memristive biosensors integration with microfluidic platform. *IEEE Transactions on Circuits and Systems-I: Regular Papers*.

Waser, R., Dittmann, R., Staikov, G., & Szot, K. (2009). Redox-based resistive switching memories—Nanoionic mechanisms, prospects, and challenges. *Advanced Materials, 21*, 2632–2663.

Wylezich, H., Mähne, H., Rensberg, J., Ronning, C., Zahn, P., & Slesazeck, S., et al. (2014). Local ion irradiation induced resistive threshold and memory switching in nb_2o_5/nbo_x films. *ACS Applied Materials & Interfaces: American Chemical Society*, 1–23.

Yang, J. J., Strukov, D. B., & Stewart, D. R. (2013). Memristive devices for computing. *Nature Nanotechnology, 8*, 13–24.

Yi, W., Perner, F., Qureshi, M., Abdalla, H., Pickett, M., Yang, J., et al. (2011). Feedback write scheme for memristive switching devices. *Applied Physics A, 102*, 973–982.

Zamarreño-Ramos, C., Camuñas-Mesa, L. A., Pérez-Carrasco, J. A., Masquelier, T., Serrano-Gotarredona, T., & Linares-Barranco, B. (2011). On spike-timing-dependent-plasticity, memristive devices, and building a self-learning visual cortex. *Frontiers in Neuroscience, 5*, 1–22.

A Simple Oscillator Using Memristor

Maheshwar Pd. Sah, Vetriveeran Rajamani, Zubaer Ibna Mannan, Abdullah Eroglu, Hyongsuk Kim and Leon Chua

Abstract This paper presents a simple oscillator using a battery and a second order memristor without the energy storage elements inductor and capacitor. The oscillating mechanism of the proposed circuit has been explained via Hopf bifurcation theorem, small signal model, local activity principle and edge of chaos theorem. This paper can be also used as a reference for explaining the intimate relationship between the super-critical *Hopf bifurcation phenomenon* and the *edge of chaos*.

Keywords Memristor · Pinched hysteresis loop · Oscillator · Local activity · Edge of chaos · Super-critical Hopf-bifurcation

---

M.Pd. Sah · A. Eroglu
School of Applied Sciences and Engineering Technology, Ivy Tech Community College, Evansville, IN 47710, USA
e-mail: sahmaheshwar@gmail.com

A. Eroglu
e-mail: eroglua@ipfw.edu

M.Pd. Sah
Department of Electrical and Computer Engineering, Purdue University Fort Wayne, Fortwayne, IN 46805, USA

M.Pd.Sah · V. Rajamani · Z.I. Mannan · H. Kim (✉)
Division of Electronics and Information Engineering, Chonbuk National University, Jeonju, Jeonbuk 54896, South Korea
e-mail: hskim@jbnu.ac.kr

V. Rajamani
e-mail: vetriece86@gmail.com

Z.I. Mannan
e-mail: zimannan@gmail.com

L. Chua
Department of Electrical Engineering and Computer Sciences, University of California, Berkeley, CA 94720, USA
e-mail: chua@berkeley.edu

L. Chua
TUM Fakultat fur Elektrotechnik und Informationstechnik, Technische Universität München, Arcisstrasse 21, Munich 80333, Germany

© Springer International Publishing AG 2017

S. Vaidyanathan and C. Volos (eds.), *Advances in Memristors, Memristive Devices and Systems*, Studies in Computational Intelligence 701,
DOI 10.1007/978-3-319-51724-7_2

1 Introduction

An electronic oscillator circuit is generally designed by using one linear capacitor and one linear inductor, or two linear capacitors, or two linear inductors, along with a locally-active nonlinear 2-terminal resistor having a negative slope region in the DC *V-I* curve (e.g., a tunnel diode), or a locally-passive nonlinear 2-terminal resistor (e.g., p-n junction diode, zener diode, varistor, etc.), and a locally-active 3-terminal resistor, such as a transistor, in addition to the ubiquitous battery, required to satisfy the *first law of thermodynamics*. Examples of tunnel diode oscillator (Mehta and Mehta 2005) and the well-known Wien-bridge oscillator[1] are shown in Fig. 1a and b respectively.

Figure 1c represents the world's simplest electronic oscillator containing *only one memristor* connected in parallel with a battery.

The *memristor* in this circuit is a generic (Chua 2014, 2015; Mannan et al. 2016; Rajamani et al. 2016) *2nd-order locally-active memristor* described by the following state-dependent Ohm's law and state equations:

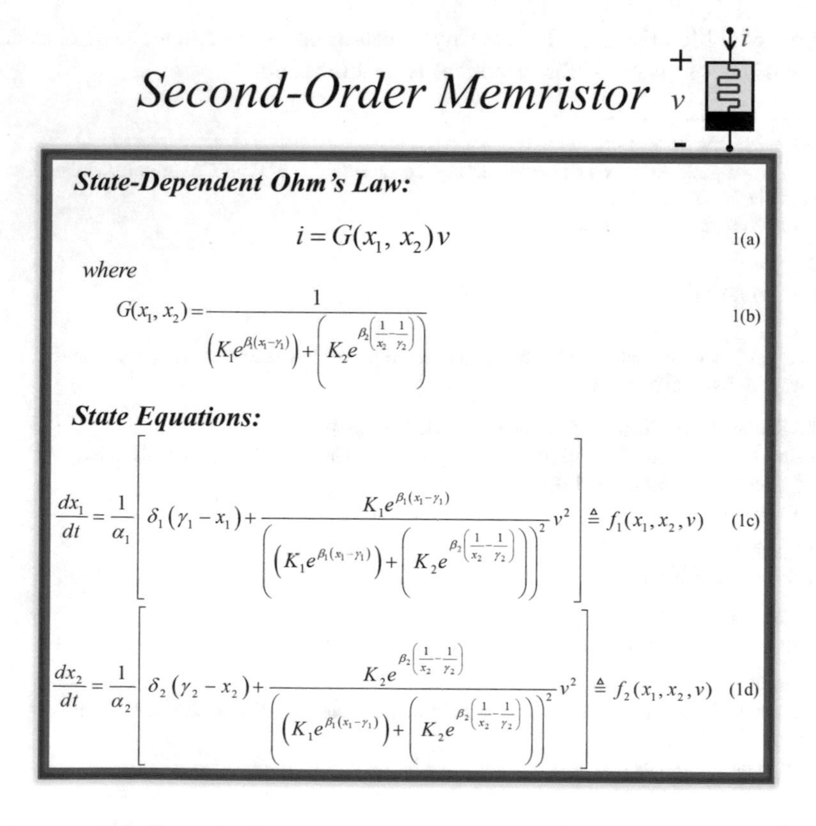

[1]The circuit diagram of the Wien-bridge oscillator can be found from the following link http://www.circuitstoday.com/wien-bridge-oscillator.

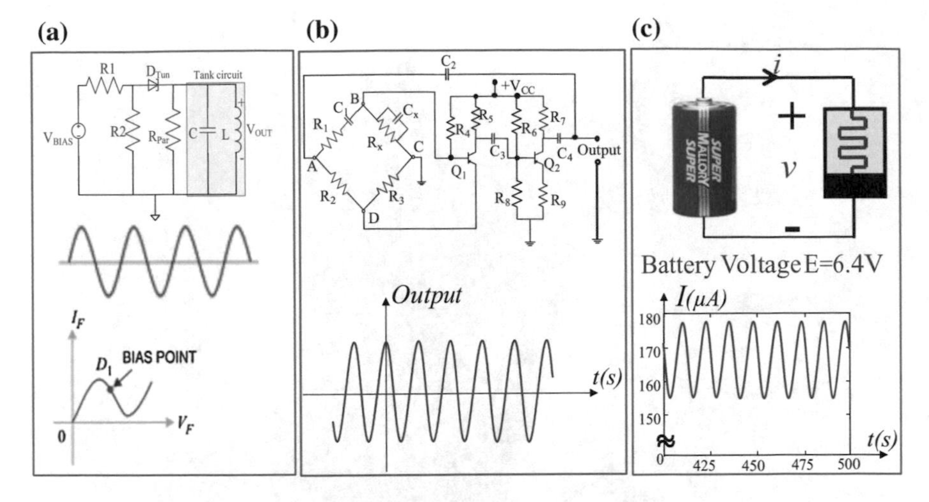

Fig. 1 **a** Simplest oscillator using a tunnel diode and an LC tank circuit (Mehta 2005). **b** Wien-bridge oscillator using resistors, capacitors and transistors (see footnote 1). **c** World's simplest oscillator using only one memristor. The *blue* near-sinusoidal waveform is obtained by computer simulation of (1) with the parameters listed in Table 1, and initial states $x_1(0) = 300.002$ *and* $x_2(0) = 300.004$

The parameters chosen in this paper are summarized in Table 1. The 3D cross section of the surface $f_1(x_1, x_2, v)$ and $f_2(x_1, x_2, v)$ are shown in Fig. 2a and b respectively at $V = 6.4$ V. Although (1) can be implemented in hardware by several methods, all results in this paper are obtained by computer simulations to avoid ambiguities in modeling the physical devices.

2 Pinched Hysteresis Loop and DC *V-I* Curve of the Second-Order-Generic Memristor from Fig. 1c

2.1 Pinched Hysteresis Loops Under Bipolar Periodic Signal

The memristor exhibits a unique fingerprint called a pinched hysteresis loop under excitation of any bipolar periodic signal with zero average. To illustrate the

Table 1 Parameter values of the second-order generic memristor

$K1 = 10^3$	$K2 = 10^5$
$\beta_1 = 10^4$	$\beta_2 = 10^7$
$\gamma_1 = 300$	$\gamma_2 = 300$
$\alpha_1 = 0.8$	$\alpha_2 = 0.2$
$\delta_1 = 0.8$	$\delta_2 = 0.1$

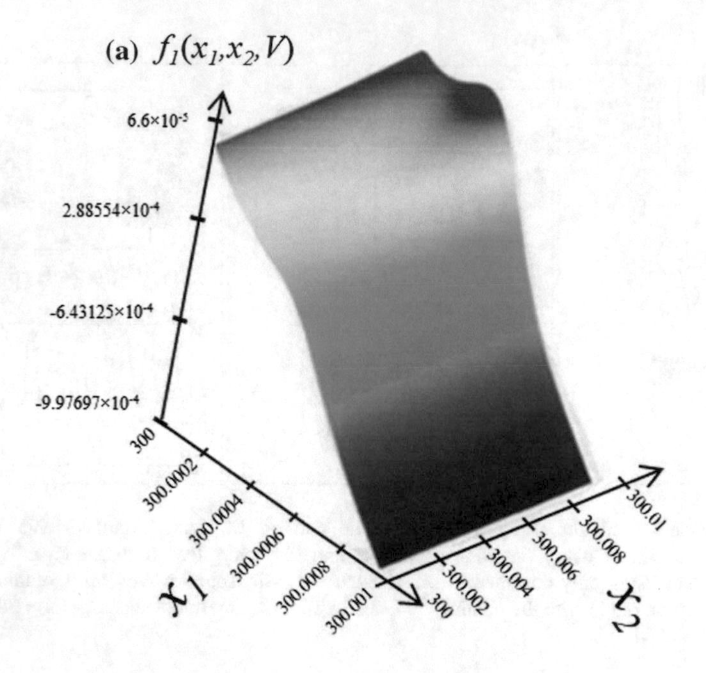

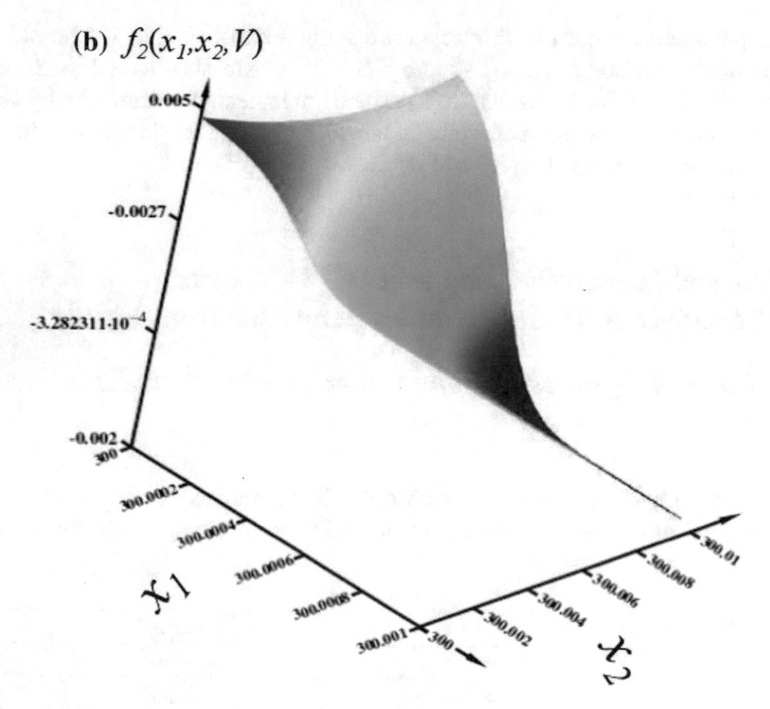

Fig. 2 The cross section of the surfaces **a** $f_1(x_1, x_2, V)$ and **b** $f_2(x_1, x_2, V)$ at $V = 6.4$ V

memristor in (1) exhibits this fingerprints, we apply a sinusoidal voltage signal *v (t)* = *A sin(2πft)* with amplitude *A* = 12 V, and frequency *f* = 0.1 Hz across this memristor. Figure 3a shows the output current *i(t),* the state variables $x_1(t)$, $x_2(t)$ and the memductance *G(t)* with respect to time *t,* respectively. Observe from Fig. 3a that *i(t)* always passes through the origin whenever *v(t)* is zero at point 1, and 3. Observe also the memductance $G(t) \geq 0$. The upper figure in Fig. 3b is a double-valued Lissajous figure plotted on the *i* versus *v* plane. Such a multi-valued Lissajous figure of *v(t), i(t),* which passes through the origin is called a pinched hysteresis loop (Chua 2003). This unique feature is the characteristic property of a memristor that distinguishes it from non-memristive devices. The lower figure in Fig. 3b shows the variation of memductance with respect to applied voltage *v(t).*

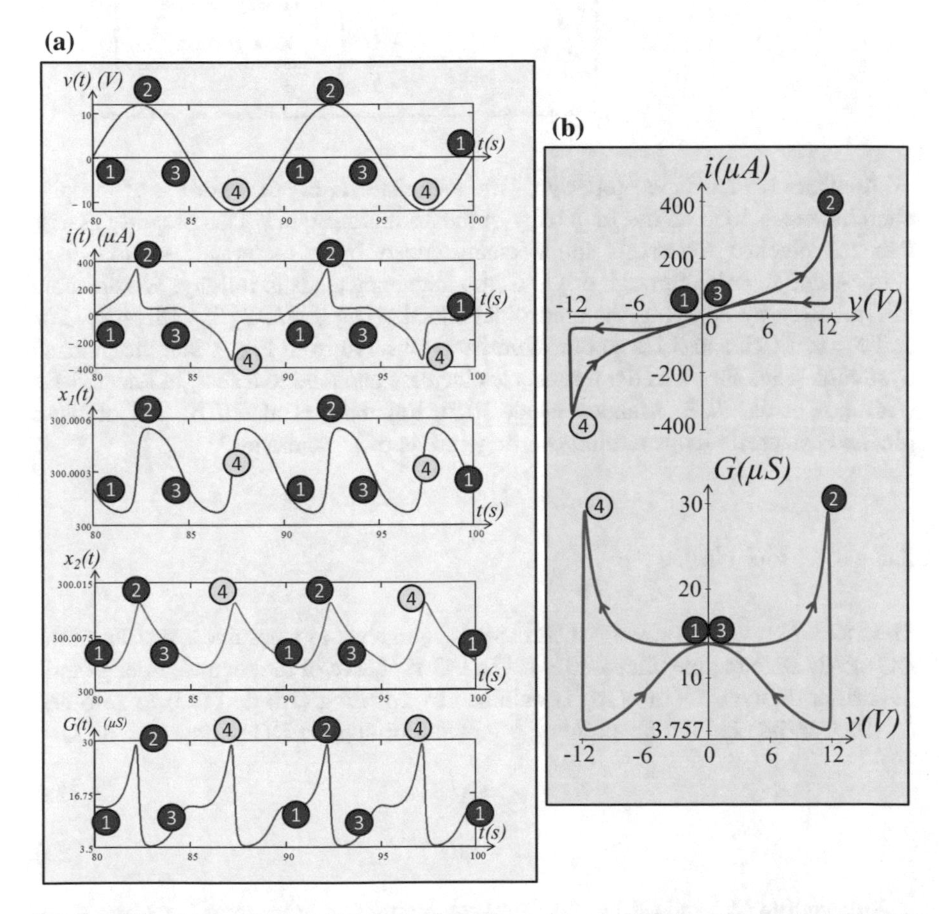

Fig. 3 **a** Waveforms of the applied sinusoidal voltage *v(t)* = *A sin(2πft)*, output current *i(t),* state variables $x_1(t)$, $x_2(t)$, and memductance *G*(*t*) of the second-order generic memristor. **b** Pinched hysteresis loop plotted on the *i* versus *v* plane and memductance hysteresis loop plotted on the *G* versus *v* plane, respectively. The simulations were performed at *A* = 12 V, *f* = 0.1 Hz, $x_1(0) = 300.002$ *and* $x_2(0) = 300.004$

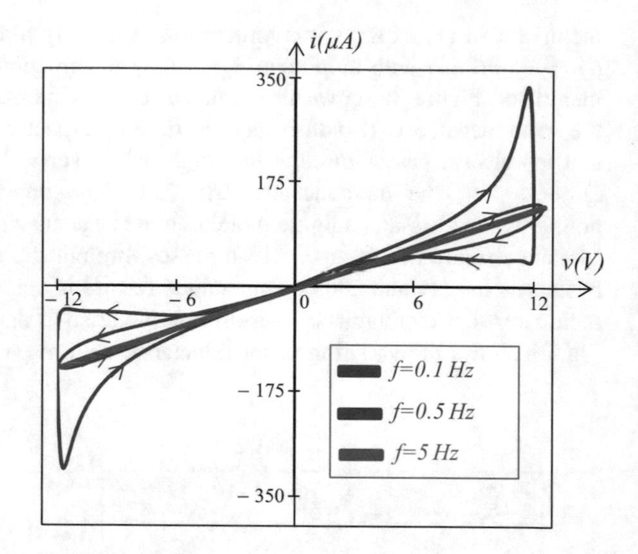

Fig. 4 Pinched hysteresis loops of the second-order generic memristor at frequencies $f = 0.1$, 0.5 and 5 Hz. The input is a sinusoidal signal $v(t) = A\ sin(2\pi ft)$, with $A = 12$ V, and the initial states are $x_1(0) = 300.002$ and $x_2(0) = 300.004$

Another characteristic property of the memristor is the dependence of the pinched hysteresis loop on the frequency of the excitation signal. This property asserts that the pinched hysteresis loops characterized by a memristor shrinks to a single-value function through origin as the frequency tends to infinity. We illustrate this property by applying the sinusoidal signal $v(t) = A\ sin(2\pi ft)$ with $A = 12$ V and $f = 0.1$, 0.5, and 5 Hz to our memristor. Observe from Fig. 4 that the pinched hysteresis loops shrink as the frequencies increase and tend to a straight line at 5 Hz (Adhikari et al. 2013; Mannan et al. 2016; Rajamani et al. 2016). All of these pinched hysteresis loops exhibit the fingerprints of a memristor.

2.2 *DC V-I Curve*

The DC *V-I* curve of a generic memristor is equivalent to a nonlinear resistor at the DC steady state regime (Chua 2014). The DC *V-I* curve of the second-order generic memristor defined in (1a)–(1d) is obtained by equating (1c) and (1d) to zero and solving for the equilibrium point as a function of applied DC voltage $v = V$, i.e.

$$x_1 = \hat{x}_1(V) \tag{2a}$$

$$x_2 = \hat{x}_2(V) \tag{2b}$$

Substituting (2a) and (2b) in (1b), and solving for the DC current $i = I$ from (1a), we obtain

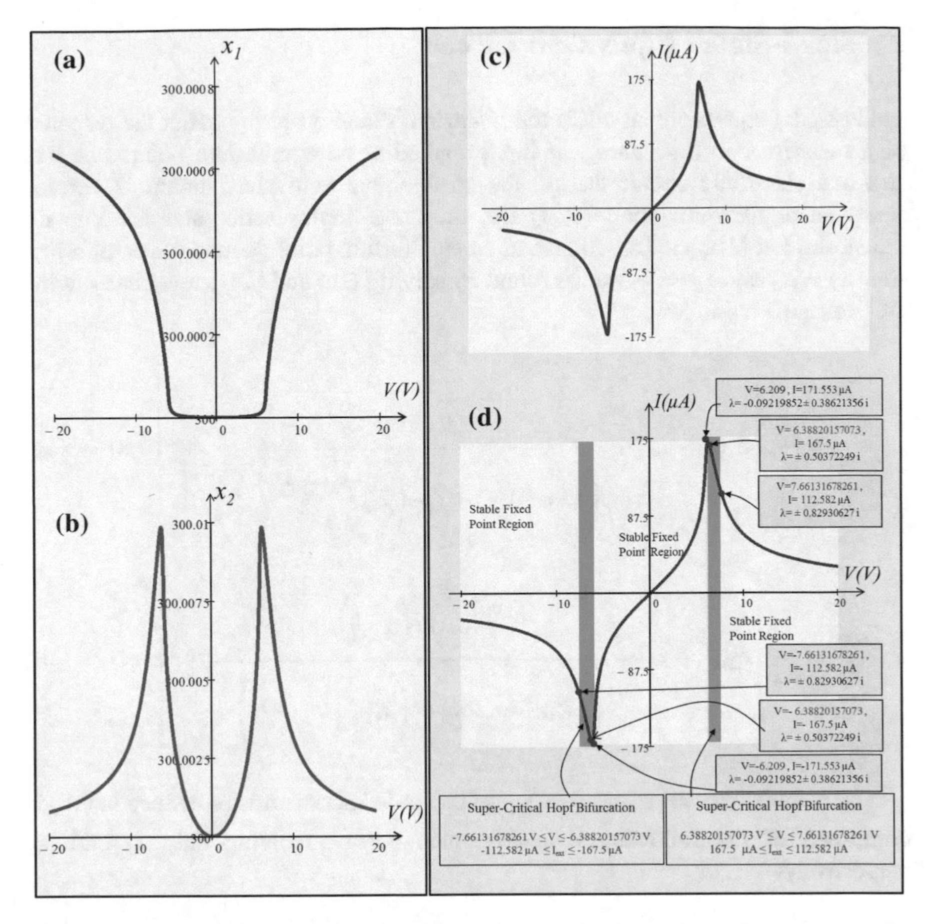

Fig. 5 The DC equilibrium of **a** state x_1, **b** state x_2, and **c** DC *V-I* curve in steady state regime, for $-20\ \text{V} \leq V \leq 20\ \text{V}$, **d** portions of the DC *V-I* curve which give rise to two distinct super-critical Hopf bifurcations

$$I = G(x_1, x_2)\ V \tag{3}$$

Applying (2a), (2b) and (3), for $-20\ \text{V} \leq V \leq 20\ \text{V}$, we obtain the red DC *V-I* curve of our second-order generic memristor shown in Fig. 5c whereas the state variables x_1 and x_2 are shown in Fig. 5a and b, respectively. Note that at *steady state* the *V-I* curve in Fig. 5c is equivalent to the *V-I* curve of a *nonlinear resistor* (Chua 1969). Figure 5d shows the portions of the DC *V-I* curve which give rise to two distinct super-critical Hopf bifurcations.

3 Small-Signal Equivalent Circuit

Small-signal equivalent circuit is the *linearized* circuit used to predict the response of a memristor to a *small-signal* input applied at an equilibrium point. Just like standard electronic circuit theory, the small-signal equivalent circuit is derived about an equilibrium point (V, I) by using the Taylor series and the Laplace transform. Let V be the DC voltage at an equilibrium point Q, then the equilibrium state $x_1 = X_1$ and $x_2 = X_2$ can be found by solving (1c) and (1d) numerically at the DC voltage V as follow:

$$\frac{dx_1}{dt} = \frac{1}{\alpha_1}\left[\delta_1(\gamma_1 - x_1) + \frac{K_1 e^{\beta_1(x_1-\gamma_1)}}{\left((K_1 e^{\beta_1(x_1-\gamma_1)}) + \left(K_2 e^{\beta_2\left(\frac{1}{x_2}-\frac{1}{\gamma_2}\right)}\right)\right)^2} V^2\right] = 0 \tag{4a}$$

$$\frac{dx_2}{dt} = \frac{1}{\alpha_2}\left[\delta_2(\gamma_2 - x_2) + \frac{K_2 e^{\beta_2\left(\frac{1}{x_2}-\frac{1}{\gamma_2}\right)}}{\left((K_1 e^{\beta_1(x_1-\gamma_1)}) + \left(K_2 e^{\beta_2\left(\frac{1}{x_2}-\frac{1}{\gamma_2}\right)}\right)\right)^2} V^2\right] = 0 \tag{4b}$$

The memristance $M(x_1, x_2) \triangleq \frac{1}{G(x_1,x_2)}$ of the 2nd-order memristor defined in (1b) is composed of the following two decoupled terms involving only x_1 and x_2, respectively:

$$M(x_1, x_2) = \left(K_1 e^{\beta_1(x_1-\gamma_1)}\right) + \left(K_2 e^{\beta_2\left(\frac{1}{x_2}-\frac{1}{\gamma_2}\right)}\right) \tag{5}$$

We can synthesize the $M(x_1, x_2)$ by two first-order memristors connected in series, as shown in Fig. 6. The memristance of the upper and lower memristors in Fig. 6 are defined by the first and second terms of Eq. (5) respectively, where the memductance $G_1(x_1) = 1/R_1(x_1)$ and $G_2(x_2) = 1/R_2(x_2)$ are defined in (6a) and (6b), respectively:

$$G_1(x_1) = \frac{1}{(K_1 e^{\beta_1(x_1-\gamma_1)})} \tag{6a}$$

$$G_2(x_2) = \frac{1}{K_2 e^{\beta_2\left(\frac{1}{x_2}-\frac{1}{\gamma_2}\right)}} \tag{6b}$$

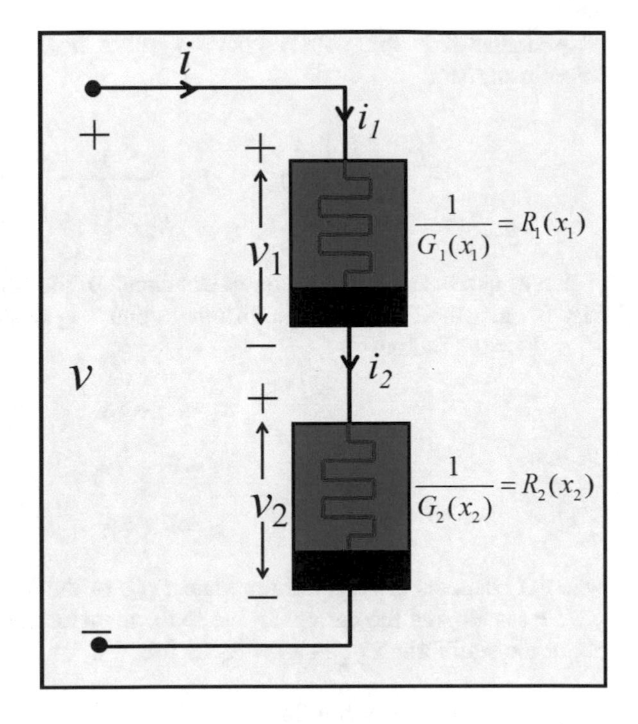

Fig. 6 The second-order memristor defined in (1) can be realized by connecting two "uncoupled" first-order voltage-controlled memristors in series. The memductance $G_1(x_1)$ of the first memristor is defined by (6a), and the memductance $G_2(x_2)$ of the second memristor is defined by (6b). The corresponding state equation is given by (9) and (10), respectively

Note that $i = i_1 = i_2$ and $v = v_1 + v_2$ in Fig. 6. Using (6a) and (6b), we have following relationships:

$$G_1(x_1)v_1 = G_2(x_2)v_2 \tag{7a}$$

$$v_1 + v_2 = v \tag{7b}$$

It follows from (7a) and (7b) that

$$v = \frac{G_1(x_1) + G_2(x_2)}{G_2(x_2)} v_1 \tag{7c}$$

$$= \left(\frac{K_1 e^{\beta_1(x_1 - \gamma_1)} + K_2 e^{\beta_2\left(\frac{1}{x_2} - \frac{1}{\gamma_2}\right)}}{K_1 e^{\beta_1(x_1 - \gamma_1)}} \right) v_1 \tag{8}$$

From (8) and (1c), we obtain the following state equation of the upper memristor:

$$\frac{dx_1}{dt} = \frac{1}{\alpha_1}\left[\delta_1(\gamma_1 - x_1) + \frac{1}{K_1 e^{\beta_1(x_1 - \gamma_1)}} v_1^2\right] \triangleq f(x_1, v_1) \tag{9}$$

A similar derivation with respect to v_2 gives the following state equation for the lower memristor:

$$\frac{dx_2}{dt} = \frac{1}{\alpha_2}\left[\delta_2(\gamma_2 - x_2) + \frac{1}{K_2 e^{\beta_2\left(\frac{1}{x_2}-\frac{1}{\gamma_2}\right)}} v_2^2\right] \triangleq f(x_2, v_2) \tag{10}$$

Let us derive small-signal equivalent circuit of the upper and lower memristor in Fig. 6 at their DC equilibrium point $v_1 = V_1$ and $v_2 = V_2$ where $V_1 + V_2 = V$. Define,

$$x_1 = X_1 + \delta x_1 \tag{11a}$$

$$v_1 = V_1 + \delta v_1 \tag{11b}$$

$$i_1 = I_1 + \delta i_1 \tag{11c}$$

where X_1 denotes the equilibrium state $x_1(Q)$ of the upper memristor at $v_1 = V_1$.

We can expand the current i_1 due to the memductance $G_1(x_1)$ in a Taylor series about the equilibrium point $x_1 = X_1$ as follows:

$$\begin{aligned} i_1 &= I_1 + \delta i_1 \\ &= a'_{00}(Q) + a'_{11}(Q)\delta x_1 + a'_{12}(Q)\delta v_1 + h.o.t \end{aligned} \tag{12a}$$

where,

$$I_1 = a'_{00}(Q) = G_1(X_1)V_1 \tag{12b}$$

$$a'_{11}(Q) = \dot{G}_1(x_1)v_1\big|_Q = -\beta_1\left(K_1 e^{\beta_1(X_1-\gamma_1)}\right)^{-1} V_1 \tag{12c}$$

$$a'_{12}(Q) = G_1(x_1)\big|_Q = \left(K_1 e^{\beta_1(X_1-\gamma_1)}\right)^{-1} \tag{12d}$$

and $h.o.t$ denotes the higher-order terms in δx_1 and δv_1. Assuming $|\delta x_1| \ll 1$ and $|\delta v_1| \ll 1$, we can neglect the $h.o.t$ term in (12a) to obtain the following *linear* equation,

$$\delta i_1 = a'_{11}(Q)\delta x_1 + a'_{12}(Q)\delta v_1 \tag{13}$$

Let us expand state equation $f(x_1, v_1)$ of (9) in Taylor series about the equilibrium point $(x_1(Q), V_1(Q))$:

$$f(X_1 + \delta x_1, V_1 + \delta v_1) = f(X_1, V_1) + b_{11}(Q)\delta x_1 + b_{12}(Q)\delta v_1 + h.o.t \tag{14a}$$

where,

$$b'_{11}(Q) = \left.\frac{\partial f(x_1, v_1)}{\partial x_1}\right|_Q = -\left[\frac{\delta_1}{\alpha_1} + \frac{\beta_1 V_1^2}{\alpha_1}\left(K_1 e^{\beta_1(X_1 - \gamma_1)}\right)^{-1}\right] \tag{14b}$$

$$b'_{12}(Q) = \left.\frac{\partial f(x_1, v_1)}{\partial v_1}\right|_Q = \frac{2\left(K_1 e^{\beta_1(X_1 - \gamma_1)}\right)^{-1}}{\alpha_1} V_1 \tag{14c}$$

Note that $f(X_1, V_1) = 0$ since (X_1, V_1) is a point on the DC $V_1 - I_1$ curve. Let us linearize the non-linear state equation $\dot{x}_1 = f(x_1, v_1)$ by neglecting the *h.o.t* from (14a) as follows:

$$\frac{d(\delta x_1)}{dt} = b'_{11}(Q)\delta x_1 + b'_{12}(Q)\delta v_1 \tag{15}$$

Taking the Laplace transform of (13) and (15) (Chua and Kang 1976) we obtain,

$$\hat{\imath}_1(s) = a'_{11}(Q)\,\hat{x}_1(s) + a'_{12}(Q)\hat{v}_1(s) \tag{16}$$

$$s\hat{x}_1(s) = b'_{11}(Q)\hat{x}_1(s) + b'_{12}(Q)\hat{v}_1(s) \tag{17}$$

where the Laplace transform of $\delta x_1(t), \delta i_1(t)$ and $\delta v_1(t)$ are denoted by $\hat{x}_1(s), \hat{\imath}_1(s)$ and $\hat{v}_1(s)$ respectively. From (17), we obtain

$$\hat{x}_1(s) = \frac{b'_{12}(Q)\hat{v}_1(s)}{s - b'_{11}(Q)} \tag{18}$$

From (16) and (18), the admittance function $Y_1(s, Q)$ of the upper memristor is

$$Y_1(s, Q) \triangleq \frac{\hat{\imath}_1(s)}{\hat{v}_1(s)} = \frac{a'_{11}(Q)b'_{12}(Q)}{s - b'_{11}(Q)} + a'_{12}(Q) \tag{19}$$

Rearranging (19), we obtain

$$Y_1(s, Q) = \frac{1}{s\frac{1}{a'_{11}(Q)b'_{12}(Q)} + \frac{(-b'_{11}(Q))}{a'_{11}(Q)b'_{12}(Q)}} + a'_{12}(Q) \tag{20}$$

Let us recast (20) into the form

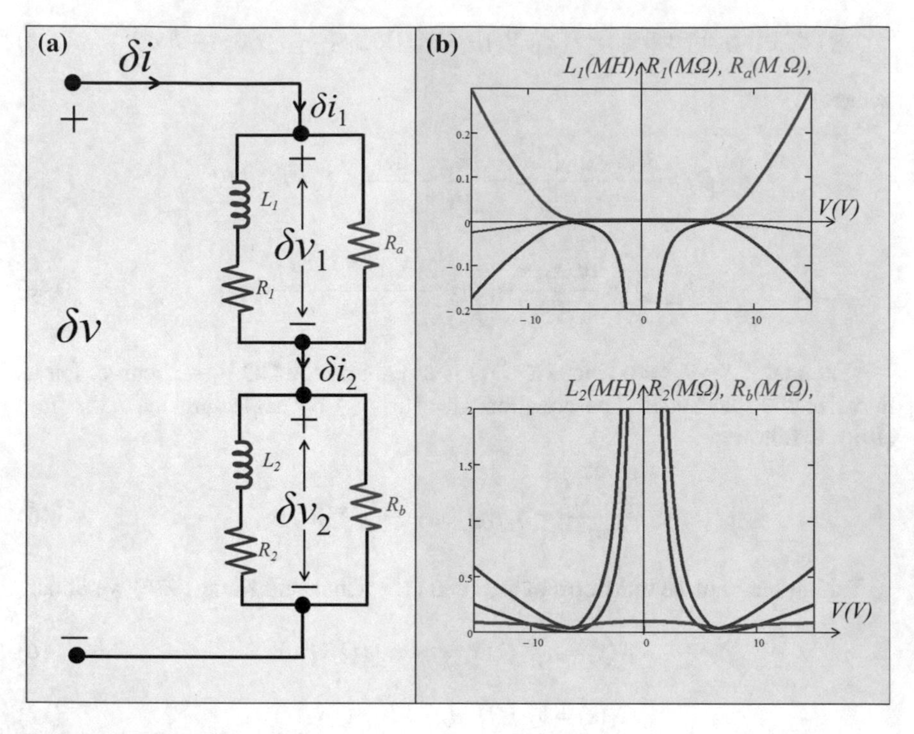

Fig. 7 **a** Small-signal equivalent circuit of the second-order memristor. **b** Inductances and resistances in the small-signal equivalent circuit of the second-order memristor calculated at the DC voltage V

$$Y_1(s,Q) = \frac{1}{sL_1 + R_1} + \frac{1}{R_a} \tag{21}$$

where $Y_1(s, Q)$ denotes the small-signal admittance of the upper memristor at Q, whose circuit as shown in Fig. 7a, where the parameters L_1, R_1 and R_a are defined by:

$$L_1 = \frac{1}{a'_{11}(Q)b'_{12}(Q)} \tag{22a}$$

$$R_1 = \frac{-b'_{11}(Q)}{a'_{11}(Q)b'_{12}(Q)} \tag{22b}$$

$$R_a = \frac{1}{a'_{12}(Q)} \tag{22c}$$

and state variable x_1 at Q can be computed numerically by solving the following equation:

$$\frac{dx_1}{dt} = \frac{1}{\alpha_1}\left[\delta_1(\gamma_1 - x_1) + G_1(x_1)V_1^2\right] = 0 \tag{22d}$$

Similarly, the small-signal admittance of the lower memristor is given by

$$Y_2(s,Q) \triangleq \frac{\hat{i}_2(s)}{\hat{v}_2(s)} = \frac{c'_{11}(Q)d'_{12}(Q)}{s - d'_{11}(Q)} + c'_{12}(Q) \tag{23}$$

Rearranging (23), we have

$$Y_2(s,Q) = \frac{1}{s\frac{1}{c'_{11}(Q)d'_{12}(Q)} + \frac{(-d'_{11}(Q))}{c'_{11}(Q)d'_{12}(Q)}} + c'_{12}(Q) = \frac{1}{sL_2 + R_2} + \frac{1}{R_b} \tag{24}$$

where $Y_2(s,Q)$ denotes the small-signal admittance of the lower memristor of the circuit of Fig. 7a where the parameters L_2, R_2 and R_b are given by:

$$L_2 = \frac{1}{c'_{11}(Q)d'_{12}(Q)} \tag{25a}$$

$$R_2 = \frac{-d'_{11}(Q)}{c'_{11}(Q)\,d'_{12}(Q)} \tag{25b}$$

$$R_b = \frac{1}{c'_{12}(Q)} \tag{25c}$$

The state variable x_2 at the equilibrium point Q can be computed numerically by solving the following equation:

$$\frac{dx_2}{dt} = \frac{1}{\alpha_2}\left[\delta_2(\gamma_2 - x_2) + G_2(x_2)V_2^2\right] = 0 \tag{25d}$$

For the convenience of readers, the explicit formulas for computing L_1, R_1, R_a and L_2, R_2, R_b as a function of V_1 and V_2 are given in Table 2 along with the state equations $f(x_1, v_1)$ and $f(x_2, v_2)$, respectively. The corresponding small-signal equivalent circuit due to L_1, R_1, R_a and L_2, R_2, R_b and plots of inductances and resistances are shown in Fig. 7a and b, respectively, for the memristor. Observe that the inductance L_1 and resistance R_1 are always negative for any DC equilibrium voltage V. The small-signal equivalent circuit of the second-order generic memristor with its inductances and resistances calculated at $V = 6.4$ V is shown in Fig. 8.

Table 2 Formulas for calculating L_1, R_1, R_a and L_2, R_2, R_b of the second-order memristor

Computation of L_1	Computation of R_1	Computation of R_a and x_1
$a'_{11}(Q)=\dot{G}_1(x_1)v_1\Big\|_Q$ $G_1(x_1)=\left(K_1e^{\beta_1(x_1-\gamma_1)}\right)^{-1}$ $\dot{G}_1(x_1)\Big\|_Q=-\frac{\beta_1}{K_1e^{\beta_1(X_1-\gamma_1)}}$ $a'_{11}(Q)=-\frac{\beta_1}{K_1e^{\beta_1(X_1-\gamma_1)}}V_1$ $b'_{12}(Q)=\frac{\partial f(x_1,v_1)}{\partial v_1}\Big\|_Q$ $=\frac{2G_1(X_1)}{\alpha_1}V_1$ $\boxed{L_1=\frac{1}{a'_{11}(Q)b'_{12}(Q)}}$	$b'_{11}(Q)=\frac{\partial f(x_1,v_1)}{\partial x_1}\Big\|_Q$ $=-\left[\frac{\delta_1}{\alpha_1}+\frac{\beta_1V_1^2}{\alpha_1}\left(K_1e^{\beta_1(X_1-\gamma_1)}\right)^{-1}\right]$ $\boxed{R_1=-\frac{b'_{11}(Q)}{a'_{11}(Q)b'_{12}(Q)}}$	$a'_{12}(Q)=G_1(x_1)\Big\|_Q$ $=\left(K_1e^{\beta_1(X_1-\gamma_1)}\right)^{-1}$ $R_a=\frac{1}{a'_{12}(Q)}$ Numerically, x_1 can be computed as $\frac{dx_1}{dt}=\frac{1}{\alpha_1}\left[\delta_1(\gamma_1-x_1)+G_1(x_1)V_1^2\right]=0$
Computation of L_2	**Computation of R_2**	**Computation of R_b and x_2**
$c'_{11}(Q)=\dot{G}_2(x_2)v_2\Big\|_Q$ $G_2(x_2)=\left(K_2e^{\beta_2\left(\frac{1}{x_2}-\frac{1}{\gamma_2}\right)}\right)^{-1}$ $\dot{G}_2(x_2)\Big\|_Q=\frac{\beta_2}{K_2X_2^2}\left(e^{\beta_2\left(\frac{1}{X_2}-\frac{1}{\gamma_2}\right)}\right)^{-1}$ $c'_{11}(Q)=\frac{\beta_2}{K_2X_2^2}\left(e^{\beta_2\left(\frac{1}{X_2}-\frac{1}{\gamma_2}\right)}\right)^{-1}V_2$ $d'_{12}(Q)=\frac{\partial f(x_2,v_2)}{\partial v_2}\Big\|_Q$ $=\frac{2G_2(X_2)}{\alpha_2}V_2$ $\boxed{L_2=\frac{1}{c'_{11}(Q)d'_{12}(Q)}}$	$d'_{11}(Q)=\frac{\partial f(x_2,v_2)}{\partial x_2}\Big\|_Q$ $=\left(\frac{-\delta_2}{\alpha_2}+\left(\frac{\beta_2V_2^2}{K_2\alpha_2X_2^2}\left(e^{\beta_2\left(\frac{1}{X_2}-\frac{1}{\gamma_2}\right)}\right)^{-1}\right)\right)$ $\boxed{R_2=-\frac{d'_{11}(Q)}{c'_{11}(Q)d'_{12}(Q)}}$	$c'_{12}(Q)=G_2(x_2)\Big\|_Q$ $=\left(K_2e^{\beta_2\left(\frac{1}{X_2}-\frac{1}{\gamma_2}\right)}\right)^{-1}$ $R_b=\frac{1}{c'_{12}(Q)}$ Numerically, x_2 can be computed as $\frac{dx_2}{dt}=\frac{1}{\alpha_2}\left[\delta_2(\gamma_2-x_2)+G_2(x_2)V_2^2\right]=0$

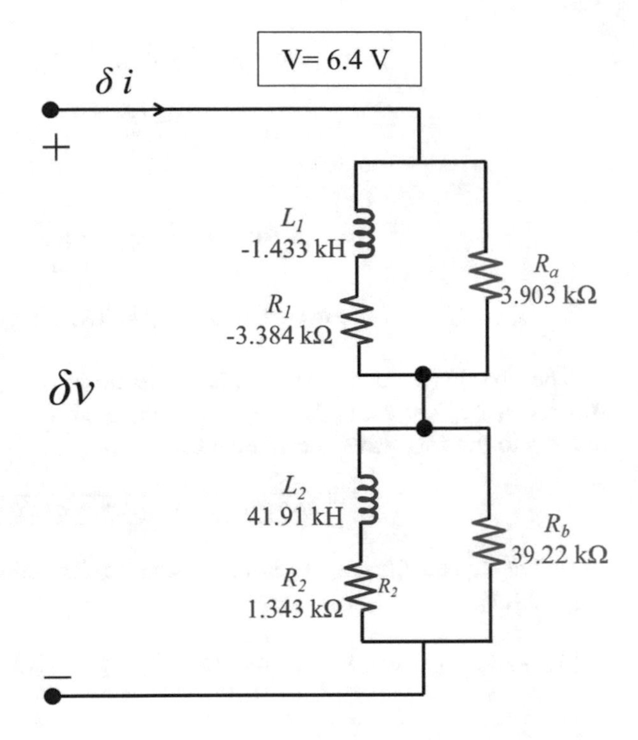

Fig. 8 Small-signal equivalent circuit of the second-order generic memristor calculated at $V = 6.4$ V

3.1 *Admittance and Pole-Zero Diagram of Second-Order Memristor*

Let $Q(V_Q, I_Q)$ be any point on the DC *V-I* curve of a second-order generic memristor and let (X_{1_Q}, X_{2_Q}) be the corresponding equilibrium state.

Define,

$$x_1 = X_{1_Q} + \Delta x_1 \tag{26a}$$

$$x_2 = X_{2_Q} + \Delta x_2 \tag{26b}$$

$$v = V_Q + \Delta v \tag{26c}$$

$$i = I_Q + \Delta i \tag{26d}$$

Let us expand the current $i = G(x_1, x_2)\, v$ in (1a) using Taylor series at the equilibrium point (X_{1_Q}, X_{2_Q}, V_Q)

$$i = a_{00} + a_{11}\Delta x_1 + a_{12}\Delta x_2 + a_{13}\Delta v + h.o.t. \tag{27}$$

where,

$$a_{00} = G(X_{1_Q}, X_{2_Q})\, V_Q = I_Q \tag{28a}$$

$$a_{11} = V_Q \left.\frac{\partial G(x_1, x_2)}{\partial x_1}\right|_Q \tag{28b}$$

$$a_{12} = V_Q \left.\frac{\partial G(x_1, x_2)}{\partial x_2}\right|_Q \tag{28c}$$

$$a_{13} = G(x_1, x_2)|_Q = G(X_{1_Q}, X_{2_Q}) \tag{28d}$$

The h.o.t. in (27) denotes the higher-order terms of Δx_1, Δx_2, and Δv_1. Assuming $\Delta x_1 << 1$, $\Delta x_2 << 1$ and $\Delta v_1 << 1$ then the h.o.t. term can be neglected and (27) reduces to the following linear equation

$$\boxed{\Delta i = a_{11}\Delta x_1 + a_{12}\Delta x_2 + a_{13}\Delta v} \tag{29}$$

Let us expand $f_1(x_1, x_2, v)$ in (1c) using Taylor series at the equilibrium point (X_{1_Q}, X_{2_Q}, V_Q)

$$f_1(X_{1_Q} + \Delta x_1, X_{2_Q} + \Delta x_2, V_Q + \Delta v) = f_1(X_{1_Q}, X_{2_Q}, V_Q) + b_{11}\Delta x_1 + b_{12}\Delta x_2 + b_{13}\Delta v + h.o.t. \tag{30}$$

where,

$$b_{11} = \left.\frac{\partial f_1(x_1, x_2, v)}{\partial x_1}\right|_Q \tag{31a}$$

$$b_{12} = \left.\frac{\partial f_1(x_1, x_2, v)}{\partial x_2}\right|_Q \tag{31b}$$

$$b_{13} = \left.\frac{\partial f_1(x_1, x_2, v)}{\partial v}\right|_Q \tag{31c}$$

Since (X_{1_Q}, X_{2_Q}, V_Q) specifies a point on the DC *V-I* curve, the term $f_1(X_{1_Q}, X_{2_Q}, V_Q) = 0$. Neglecting the h.o.t. term in (30), we obtain the following linear equation:

$$\boxed{\frac{d(\Delta x_1)}{dt} = b_{11}\Delta x_1 + b_{12}\Delta x_2 + b_{13}\Delta v} \tag{32}$$

Similarly, let us expand $f_2(x_1, x_2, v)$ in (1d) using Taylor series at the equilibrium point (X_{1_Q}, X_{2_Q}, V_Q)

$$f_2(X_{1_Q} + \Delta x_1, X_{2_Q} + \Delta x_2, V_Q + \Delta v) = f_2(X_{1_Q}, X_{2_Q}, V_Q) + c_{11}\Delta x_1 + c_{12}\Delta x_2 + c_{13}\Delta v + h.o.t. \tag{33}$$

where,

$$c_{11} = \left. \frac{\partial f_2(x_1, x_2, v)}{\partial x_1} \right|_Q \tag{34a}$$

$$c_{12} = \left. \frac{\partial f_2(x_1, x_2, v)}{\partial x_2} \right|_Q \tag{34b}$$

$$c_{13} = \left. \frac{\partial f_2(x_1, x_2, v)}{\partial v} \right|_Q \tag{34c}$$

Since (X_{1_Q}, X_{2_Q}, V_Q) specifies a point on the DC *V-I* curve, the term $f_2(X_{1_Q}, X_{2_Q}, V_Q) = 0$. Neglecting the h.o.t. term in (33), we obtain the following linear equation:

$$\boxed{\frac{d(\Delta x_2)}{dt} = c_{11}\Delta x_1 + c_{12}\Delta x_2 + c_{13}\Delta v} \tag{35}$$

Taking Laplace transform of (29), (32) and (35), we obtain

$$\hat{\imath}(s) = a_{11}\,\hat{x}_1(s) + a_{12}\hat{x}_2(s) + a_{13}\hat{v}(s) \tag{36}$$

$$s\hat{x}_1(s) = b_{11}\hat{x}_1(s) + b_{12}\hat{x}_2(s) + b_{13}\hat{v}(s) \tag{37}$$

$$s\hat{x}_2(s) = c_{11}\hat{x}_1(s) + c_{12}\hat{x}_2(s) + c_{13}\hat{v}(s) \tag{38}$$

where $\hat{\imath}(s)$, $\hat{x}_1(s), \hat{x}_2(s)$ and $\hat{v}(s)$ denote the Laplace transform of Δi, Δx_1, Δx_2, and Δv, respectively. Solving (37) and (38) for $\hat{x}_1(s)$ and $\hat{x}_2(s)$, we obtain

$$\hat{x}_1(s) = \frac{[b_{12}c_{13} + b_{13}(s - c_{12})]}{[(s - b_{11})(s - c_{12}) - b_{12}c_{11}]}\hat{v}(s) \tag{39a}$$

$$\hat{x}_2(s) = \frac{[b_{13}c_{11} + c_{13}(s - b_{11})]}{[(s - b_{11})(s - c_{12}) - b_{12}c_{11}]}\hat{v}(s) \tag{39b}$$

The explicit formula for computing the parameters in (39a) and (39b) are given in Table 3. By substituting $\hat{x}_1(s)$ and $\hat{x}_2(s)$ from (39a, 39b) into (36), we obtain

$$Y(s; Q) \triangleq \frac{\hat{\imath}(s)}{\hat{v}(s)} \tag{40}$$

Table 3 Explicit formula for computing the parameters in (40)

$$a_{11}=\frac{-V K_1\beta_1 e^{\beta_1(X_1-\gamma_1)}}{\left[\left(K_1 e^{\beta_1(X_1-\gamma_1)}\right)+\left(K_2 e^{\beta_2\left(\frac{1}{X_2}-\frac{1}{\gamma_2}\right)}\right)\right]^2}$$

$$a_{12}=\frac{V K_2\beta_2 e^{\beta_2\left(\frac{1}{X_2}-\frac{1}{\gamma_2}\right)}}{X_2^2\left[\left(K_1 e^{\beta_1(X_1-\gamma_1)}\right)+\left(K_2 e^{\beta_2\left(\frac{1}{X_2}-\frac{1}{\gamma_2}\right)}\right)\right]^2}$$

$$a_{13}=\frac{1}{\left(K_1 e^{\beta_1(X_1-\gamma_1)}\right)+\left(K_2 e^{\beta_2\left(\frac{1}{X_2}-\frac{1}{\gamma_2}\right)}\right)}$$

$$b_{11}=\frac{-1}{\alpha_1}\left[\delta_1-\frac{K_1\beta_1 e^{\beta_1(X_1-\gamma_1)}}{\left(\left(K_1 e^{\beta_1(X_1-\gamma_1)}\right)+\left(K_2 e^{\beta_2\left(\frac{1}{X_2}-\frac{1}{\gamma_2}\right)}\right)\right)^2}V^2+\frac{2K_1^2\beta_1 e^{2\beta_1(X_1-\gamma_1)}}{\left(\left(K_1 e^{\beta_1(X_1-\gamma_1)}\right)+\left(K_2 e^{\beta_2\left(\frac{1}{X_2}-\frac{1}{\gamma_2}\right)}\right)\right)^3}V^2\right]$$

$$b_{12}=\frac{2K_1K_2\beta_2 e^{\beta_1(X_1-\gamma_1)}e^{\beta_2\left(\frac{1}{X_2}-\frac{1}{\gamma_2}\right)}}{\alpha_1 X_2^2\left[\left(K_1 e^{\beta_1(X_1-\gamma_1)}\right)+\left(K_2 e^{\beta_2\left(\frac{1}{X_2}-\frac{1}{\gamma_2}\right)}\right)\right]^3}V^2$$

$$b_{13}=\frac{2K_1 e^{\beta_1(X_1-\gamma_1)}}{\alpha_1\left[\left(K_1 e^{\beta_1(X_1-\gamma_1)}\right)+\left(K_2 e^{\beta_2\left(\frac{1}{X_2}-\frac{1}{\gamma_2}\right)}\right)\right]^2}V$$

$$c_{11}=\frac{-2K_1K_2\beta_1 e^{\beta_1(X_1-\gamma_1)}e^{\beta_2\left(\frac{1}{X_2}-\frac{1}{\gamma_2}\right)}}{\alpha_2\left[\left(K_1 e^{\beta_1(X_1-\gamma_1)}\right)+\left(K_2 e^{\beta_2\left(\frac{1}{X_2}-\frac{1}{\gamma_2}\right)}\right)\right]^3}V^2$$

$$c_{13}=\frac{2K_2 e^{\beta_2\left(\frac{1}{X_2}-\frac{1}{\gamma_2}\right)}}{\alpha_2\left[\left(K_1 e^{\beta_1(X_1-\gamma_1)}\right)+\left(K_2 e^{\beta_2\left(\frac{1}{X_2}-\frac{1}{\gamma_2}\right)}\right)\right]^2}V$$

$$c_{12}=\frac{-1}{\alpha_2}\left[\delta_2+\frac{K_2\beta_2 e^{\beta_2\left(\frac{1}{X_2}-\frac{1}{\gamma_2}\right)}}{X_2^2\left[\left(K_1 e^{\beta_1(X_1-\gamma_1)}\right)+\left(K_2 e^{\beta_2\left(\frac{1}{X_2}-\frac{1}{\gamma_2}\right)}\right)\right]^2}V^2-\frac{2K_2^2\beta_2 e^{2\beta_2\left(\frac{1}{X_2}-\frac{1}{\gamma_2}\right)}}{X_2^2\left[\left(K_1 e^{\beta_1(X_1-\gamma_1)}\right)+\left(K_2 e^{\beta_2\left(\frac{1}{X_2}-\frac{1}{\gamma_2}\right)}\right)\right]^3}V^2\right]$$

$x_1 = X_1$ and $x_2 = X_2$ can be computed by solving the following equilibrium equations at V

$$\delta_1(\gamma_1-x_1)+\frac{K_1 e^{\beta_1(x_1-\gamma_1)}}{\left(\left(K_1 e^{\beta_1(x_1-\gamma_1)}\right)+\left(K_2 e^{\beta_2\left(\frac{1}{x_2}-\frac{1}{\gamma_2}\right)}\right)\right)^2}V^2=0$$

$$\delta_2(\gamma_2-x_2)+\frac{K_2 e^{\beta_2\left(\frac{1}{x_2}-\frac{1}{\gamma_2}\right)}}{\left(\left(K_1 e^{\beta_1(x_1-\gamma_1)}\right)+\left(K_2 e^{\beta_2\left(\frac{1}{x_2}-\frac{1}{\gamma_2}\right)}\right)\right)^2}V^2=0$$

$$Y(s;Q)=\frac{b_2s^2+b_1s+b_0}{a_2s^2+a_1s+a_0} \tag{41}$$

The expression $Y(s; Q)$ in (41) is called the small-signal admittance of the second-order memristor about the equilibrium point Q, where a_2, a_1, a_0, b_2, b_1 and b_0 are given by

$$\left.\begin{aligned} a_2 &= 1 \\ a_1 &= -(b_{11}+c_{12}) \\ a_0 &= b_{11}c_{12}-b_{12}c_{11} \end{aligned}\right\} \tag{42a}$$

$$\left.\begin{aligned} b_2 &= a_{13} \\ b_1 &= a_{11}b_{13}+a_{12}c_{13}-a_{13}(b_{11}+c_{12}) \\ b_0 &= a_{11}(b_{12}c_{13}-b_{13}c_{12})+a_{12}(b_{13}c_{11}-b_{11}c_{13})+a_{13}(b_{11}c_{12}-b_{12}c_{11}) \end{aligned}\right\} \tag{42b}$$

The *pole-zero diagram* of the small-signal admittance $Y(s; V)$ is computed by factorizing the denominator and numerator of (41):

$$Y(s;V)=\frac{k(s-z_1)(s-z_2)}{(s-p_1)(s-p_2)} \tag{43}$$

where p_i and z_i denote the poles and zeros of admittance function $Y(s; V)$ respectively.

The loci of the zeros and poles are shown in the Fig. 9a and b respectively, over the applied DC voltage $-20\ \text{V} \leq V \leq 20\ \text{V}$. In Fig. 9b, the arrowheads indicate the direction of pole movements in the interval of $-20\ \text{V} \leq V \leq 0\ \text{V}$ whereas the direction of pole movements for $0\ \text{V} \leq V \leq 20\ \text{V}$ is in reverse direction. Note that the reverse arrowheads for $0\ \text{V} \leq V \leq 20\ \text{V}$ are omitted to avoid the clutter.

Observe from Fig. 9a that $Im\ z_1$, $Im\ z_2$ are always zero, and $Re\ z_2$ is always negative. The poles diagram in Fig. 9b shows the real part of the poles p_1 and p_2 are zero at $V = 6.38820157073$ V and $V = 7.66131678261$ V, respectively which are also called as *Hopf-bifurcation points* in bifurcation theory. Observe that the real parts $\text{Re}p_1$ and $\text{Re}p_2$ of the poles are always positive between the bifurcation points $6.38820157073\ \text{V} < V < 7.66131678261\ \text{V}$. In the horizontal segment where the $Im\ p_1$ and $Im\ p_2$ are zero, at that point several poles of $\text{Re}p_1$ and $\text{Re}p_2$ exist due to different input voltages as shown in Fig. 9b and at $V = 0$ V, the value of $\text{Re}p_1$ and $\text{Re}p_2$ is -0.5 and -1, respectively. As the value of the voltage V increases the value of the poles of p_1 and p_2 increases. For $V \leq 6.03481$ V and $V \geq 11.1305$ V the $Im\ p_1$ and $Im\ p_2$ become zero.

3.2 *Frequency Response of Second-Order Generic Memristor*

The frequency response of the second-order generic memristor at an equilibrium point Q is computed by substituting $s = i\omega$, for the complex frequency s in (41) at the equilibrium point Q, where the angular frequency $\omega = 2\pi f$. The corresponding

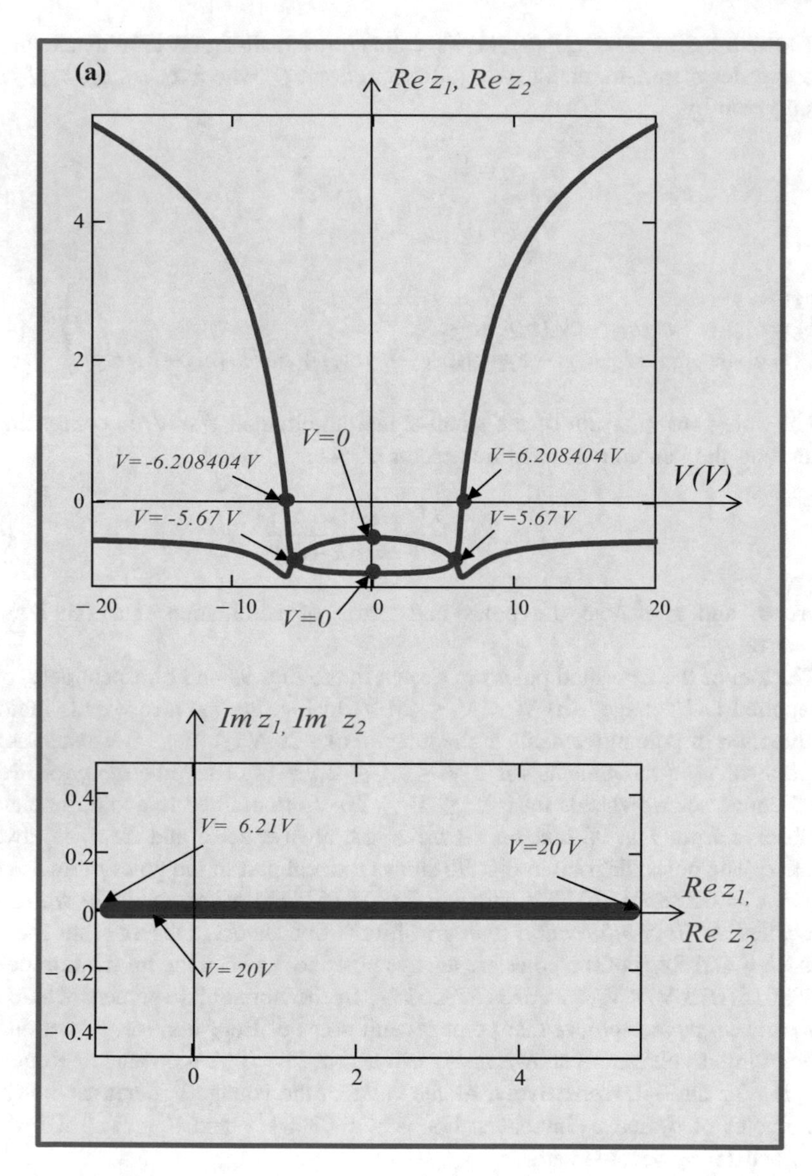

Fig. 9 Poles and zeros diagram of the admittance function $Y(s;V) = \frac{k(s-z_1)(s-z_2)}{(s-p_1)(s-p_2)}$ for $-20\text{ V} \leq V \leq 20\text{ V}$. **a** Zeros Diagram. **b** Poles Diagram. *Arrowheads* indicate the direction of pole movements in the interval of $-20\text{ V} \leq V \leq 0\text{ V}$. The movements of poles in the interval of $0\text{ V} \leq V \leq 20\text{ V}$, which are the reverse direction of $-20\text{ V} \leq V \leq 0\text{ V}$ interval, are omitted to avoid the clutter. When voltage V is infinitive, the locations of poles p_1 and p_2 are -0.5 and -77.4759895, respectively

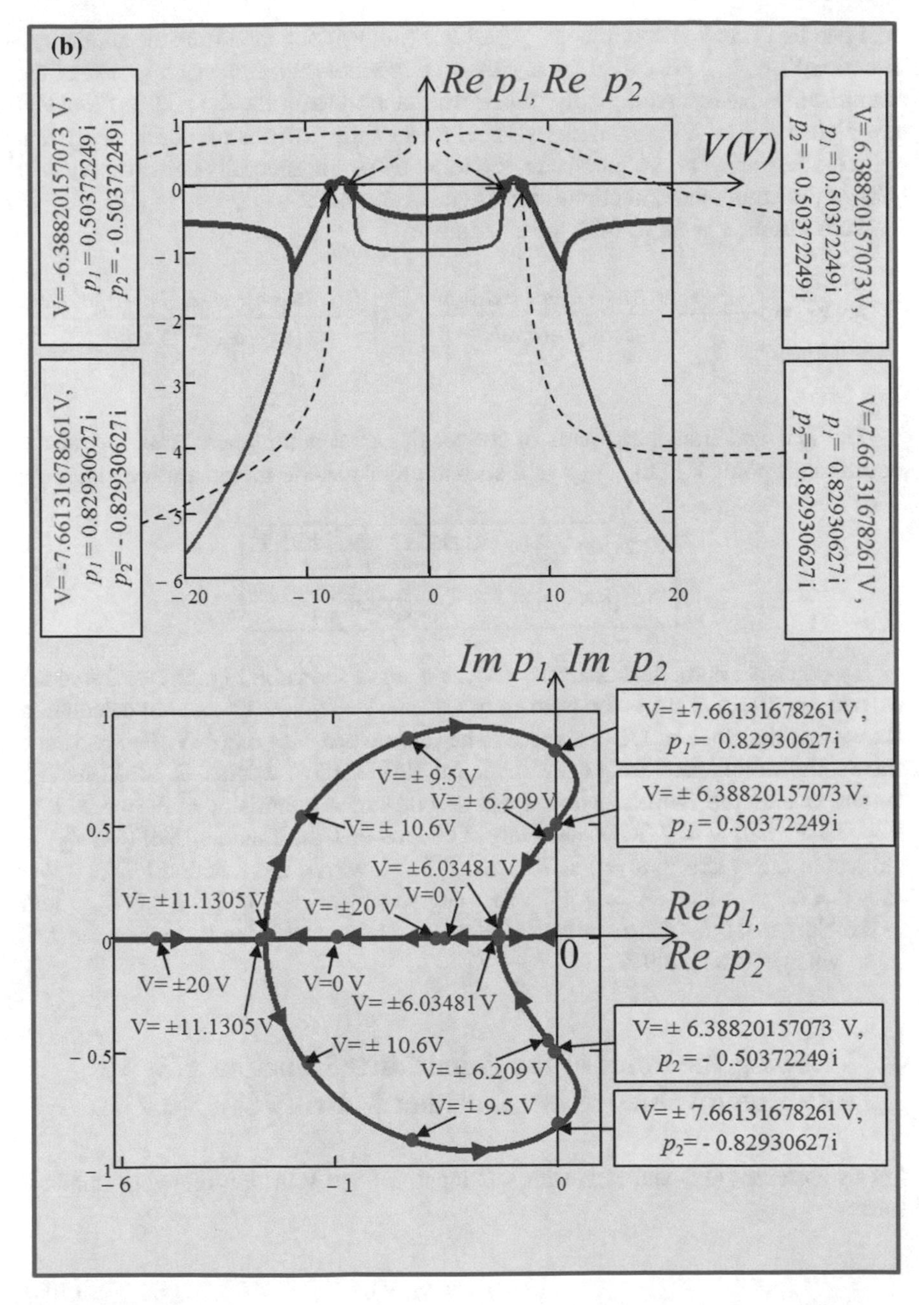

Fig. 9 (continued)

real part $\mathrm{Re}Y(i\omega; V_Q)$ and imaginary part $\mathrm{Im}Y(i\omega; V_Q)$ obtained from the admittance function $Y(i\omega; V_Q)$ are called the *small-signal admittance frequency response* of the memristor in basic circuit theory. When the function $\mathrm{Re}Y(i\omega; V_Q)$ and $\mathrm{Im}Y(i\omega; V_Q)$ are plotted on the horizontal and vertical axes of the Cartesian co-ordinate system with the frequency ω as a parameter, the resulting plot is generally known as *Nyquist plot* of the admittance functions at the equilibrium point Q.

Substituting $s = i\omega$ in (41), we obtain

$$Y(i\omega; V_Q) = \left[\frac{(a_0 - a_2\omega^2)(b_0 - b_2\omega^2) + a_1 b_1 \omega^2}{(a_0 - a_2\omega^2)^2 + a_1^2\omega^2}\right] + i\left[\frac{[(a_0 - a_2\omega^2)b_1 - a_1(b_0 - b_2\omega^2)]\omega}{(a_0 - a_2\omega^2)^2 + a_1^2\omega^2}\right] \tag{44}$$

The real and imaginary parts of the small-signal admittance $Y(i\omega; V_Q)$ at the equilibrium point V_Q (X_{1_Q}, X_{2_Q}) of a second-order generic memristor are given by:

$$\boxed{\begin{aligned} \mathrm{Re}Y(i\omega; V_Q) &= \frac{(a_0 - a_2\omega^2)(b_0 - b_2\omega^2) + a_1 b_1 \omega^2}{(a_0 - a_2\omega^2)^2 + a_1^2\omega^2} \\ \mathrm{Im}Y(i\omega; V_Q) &= \frac{[(a_0 - a_2\omega^2)b_1 - a_1(b_0 - b_2\omega^2)]\omega}{(a_0 - a_2\omega^2)^2 + a_1^2\omega^2} \end{aligned}} \tag{45}$$

By extensive numerical analysis of DC *V-I* curve shown in Fig. 5d, we found the current $I = 171.553$ μA is the maximum value at $V = 6.209$ V[2] and our calculation shows that the slope of DC *V-I* curve is negative when $V > 6.209$ V. Figure 10a–c shows the admittance frequency response $\mathrm{Re}Y(i\omega; V_Q)$ versus ω, $\mathrm{Im}Y(i\omega; V_Q)$ versus ω and the *Nyquist plot* of the second-order memristor at $V = 6.209$ V, $V = 6.3$ V, and $V = 7$ V, respectively. Observe that the function $\mathrm{Re}Y(i\omega; V_Q)$ is tangent to the ω axis at $\omega = 0$ at $V = 6.209$ V. However, the function $\mathrm{Re}Y(i\omega; V_Q)$ at $V = 6.3$ V and $V = 7$ V are negative for $-0.42 \le \omega \le 0.42$, and $-0.6908 \le \omega \le 0.6908$, memristor defined in (1) is *locally active* when the DC input voltage $V > 6.209$ V.

4 Mapping the Poles of the Admittance Function *Y(S; V)* with Eigen values of the Jacobian Matrix $J(X_1, X_2; V)$

Let us represent (1c) and (1d) with DC input voltage V in the following standard form:

$$\frac{dx_1}{dt} = f_1(x_1, x_2, V) \tag{46a}$$

[2]The DC *V-I* curve in Fig. 5d for negative voltage $(V \le 0)$ is just the reflected (odd-symmetric) mirror image about the origin $V = 0$ over the positive input voltage $(V \ge 0)$ region.

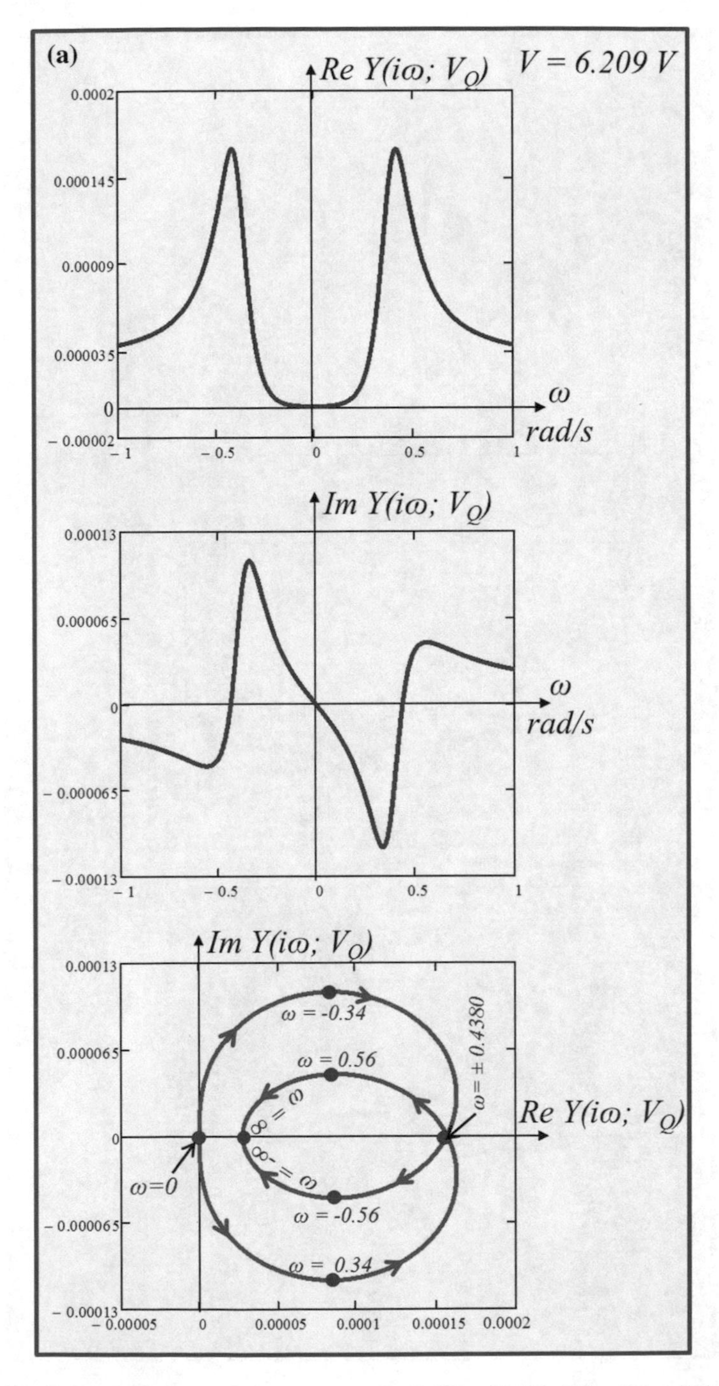

Fig. 10 *Small-signal admittance frequency response Re Y(iω; V_Q), Im Y(iω; V_Q)* and Nyquist plot of our second-order memristor at **a** $V = 6.209$ V, **b** $V = 6.3$ V, and **c** $V = 7$ V

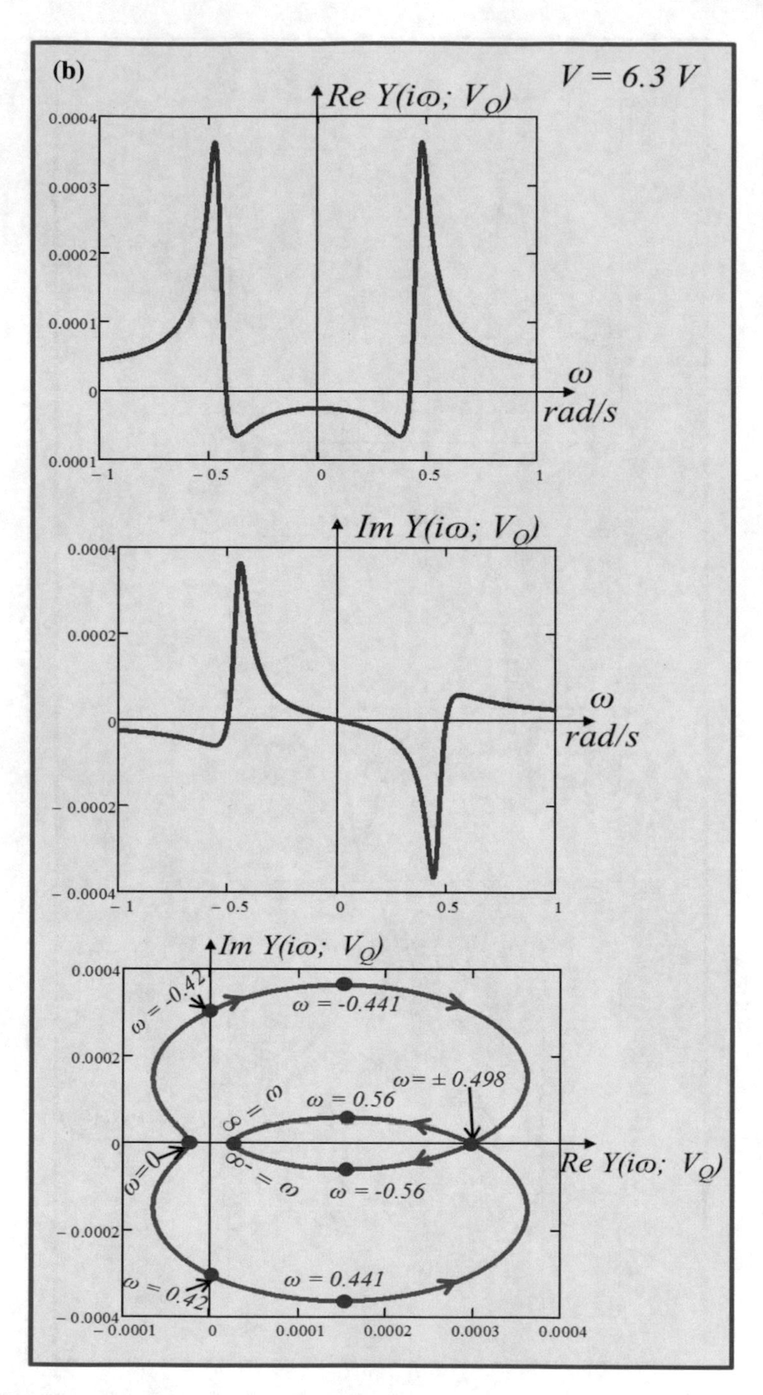

Fig. 10 (continued)

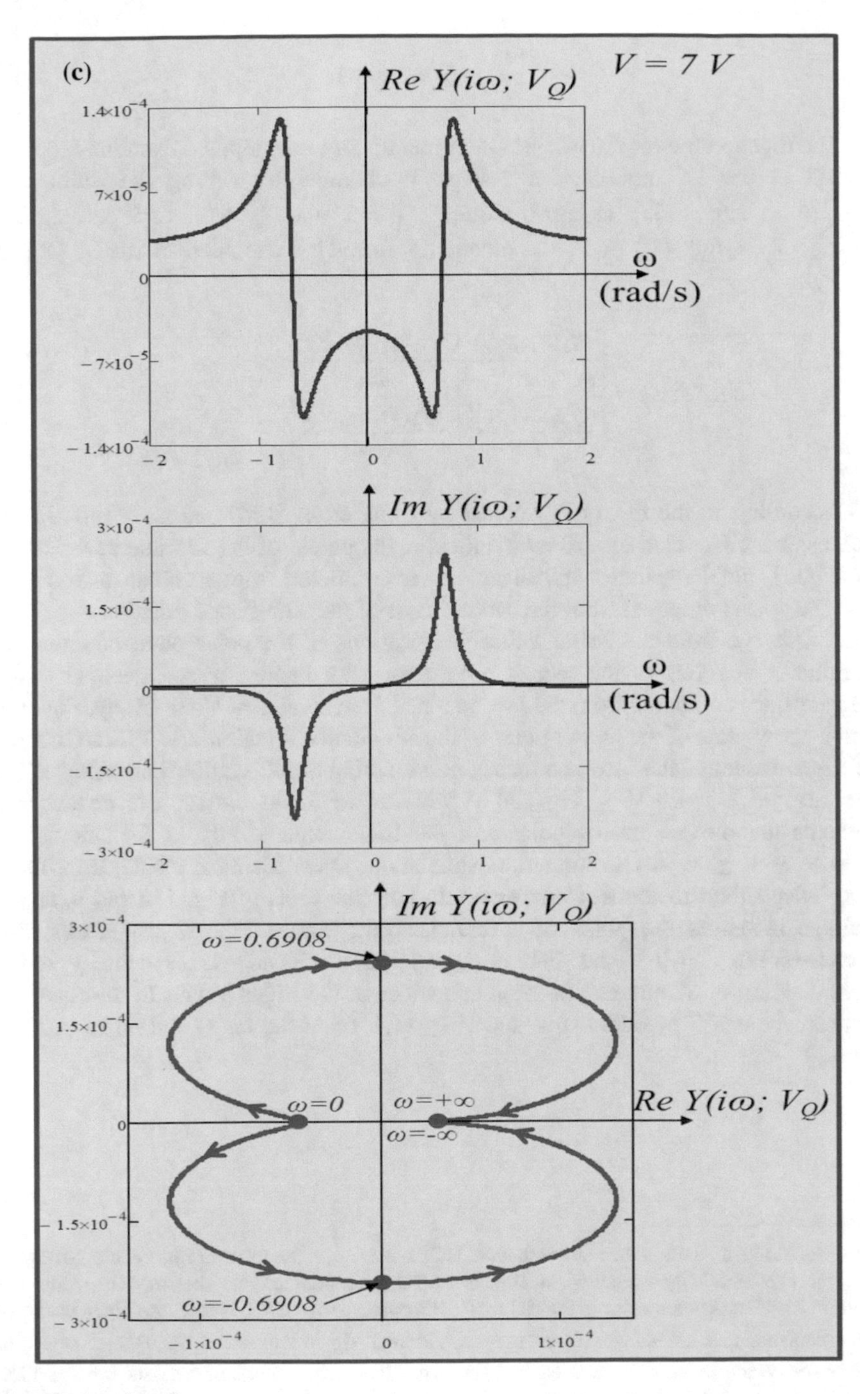

Fig. 10 (continued)

$$\frac{dx_2}{dt} = f_2(x_1, x_2, V) \tag{46b}$$

The Eigen values of the second-order memristor are computed from the Jacobian matrix at the DC equilibrium voltage V obtained by setting the differential Eqs. (46a) and (46b) to zero. Setting $\frac{dx_1}{dt} = 0$ and $\frac{dx_2}{dt} = 0$, and solving for $x_1 = X_1(V), x_2 = X_2(V)$ at V, we obtain the following Jacobian matrix at $(X_1(V), X_2(V))$:

$$J(X_1, X_2; V) = \begin{bmatrix} \frac{\partial f_1(x_1, x_2; V)}{\partial x_1} & \frac{\partial f_1(x_1, x_2; V)}{\partial x_2} \\ \frac{\partial f_2(x_1, x_2; V)}{\partial x_1} & \frac{\partial f_2(x_1, x_2; V)}{\partial x_2} \end{bmatrix}_{(x_1 = X_1(V), x_2 = X_2(V))} \tag{47}$$

According to the theory developed by Chua et al. (1987, 2012a, b) the Eigen values of the Jacobian matrix are identical to the poles[3] of the admittance functions $Y(s; V)$. Table 4 illustrates the Eigen values computed from Jacobian matrix (47) and the poles computed from the denominator of the admittance function $Y(s; V)$ in (43). Observe from the Table 4 that the locations of the poles obtained from the admittance function of the second order memristor in Fig. 1c are identical to the Eigen values computed from the Jacobian matrix evaluated at V. Similarly, Fig. 11a and b show plots of the loci of poles of the admittance function $Y(s; V)$, and the loci of Eigen values of the Jacobian matrix as a function of DC equilibrium voltage V in the interval of -20 V $\leq V \leq 20$ V whereas to avoid clutter, the arrowheads indicate the movements of poles and the Eigen values only in the interval of -20 V $\leq V \leq 0$ V. Our numerical simulations show identical results from these two independent methods, as expected. In both the plots of Fig. 11a and b, as the voltage increases the poles of admittance function and the Eigen values also increases. For $V = 0$ V, the value of p_1 and p_2 is -0.5 and -1, respectively, in the pole diagram of admittance function as well as in the Eigen values of the Jacobian matrix whereas $V \leq 6.03481$ V and $V \geq 11.1305$ V the $Im\ p_1$ and $Im\ p_2$ become zero.

[3]We would like to caution the readers that the DC *current* I_{ext} is the *input* in Chua et al. (2012a, b), and the two small-signal equivalent circuits of the potassium ion-channel memristor and the sodium ion-channel memristor in the HH model are connected parallel. Hence, the Eigen values of the Jacobian matrix are identical to the *poles* of the small-signal impedance $Z(s, I) \triangleq \frac{V(s)}{I(s)} = \frac{1}{Y(s)}$, or equivalently, the *zeros* of the admittance $Y(s)$. In the 2nd-order memristor case, the *input* is a DC *voltage* V and the two small-signal circuit components shown in Fig. 8 are connected in series. It follows that the *poles* of the *admittance function* $Y(s, V) \triangleq \frac{I(s)}{V(s)}$ of the second-order memristor in Fig. 1c are equivalent to the *Eigen values* of the Jacobian matrix (47).

Table 4 Comparison of the *poles* of the admittance function *Y(s; V)* and the *Eigen values* of the Jacobian matrix $J(X_1, X_2; V)$

V	Poles of the admittance function $Y(s; V)$	Eigen values of the Jacobian Matrix $J(X_1,X_2;V)$
0	-0.5, -1	-0.5, -1
1	-0.4946059, -0.9987707	-0.4946059, -0.9987707
2	-0.4777081, -0.9947101	-0.4777081, -0.9947101
3	-0.4467612, -0.9863768	-0.4467612, -0.9863768
4	-0.3955726, -0.9695863	-0.3955726, -0.9695863
5	-0.3066828, -0.9280549	-0.3066828, -0.9280549
6	-0.1341699, -0.4813821	-0.1341699, -0.4813821
6.38820157073	± 0.50372249 i	± 0.50372249 i
7	0.0958965 ± 0.6948223 i	0.0958965 ± 0.6948223 i
7.66131678261	± 0.82930627 i	± 0.82930627 i
8	-0.0995709 ± 0.8814635 i	-0.0995709 ± 0.8814635 i
9	-0.4856009 ± 0.9272146 i	-0.4856009 ± 0.9272146 i
10	-0.8945248 ± 0.7769816 i	-0.8945248 ± 0.7769816 i

5 Local Activity and Edge of Chaos

The *local activity theorem* provides the fundamental concept for predicting whether the nonlinear system can exhibit complexity or not, whereas a small neighborhood of the *edge of chaos* in the parameter space of a dynamical system is where *complex phenomena and information processing* will most likely emerge (Chua et al. 1987; Chua 1998; Dogaru and Chua 1998; Vaidyanathan and Volos 2016a, b). Applying the above theorem in this paper for the second-order memristor, we found from Figs. 10a and 5d that the memristor is locally active only when $V > 6.209$ V, i.e.

$$\mathrm{Re}Y(i\omega; V_Q) < 0, \ for \ V > 6.209\mathrm{V} \tag{48}$$

Observe from Figs. 9b and 11a that the real part of the poles of $Y(s; V_Q)$ vanishes at $V = 6.38820157073$ V, i.e. the poles of the admittance functions $Y(s; V)$ has a pair of complex poles $p_1 = i0.50372249$ and $p_2 = -i0.50372249$ located on the imaginary axis ($Re\ P_i = 0$) at the above applied DC voltage. It follows that the corresponding equilibrium $(X_1(V), X_2(V))$ point is no longer *asymptotically* stable

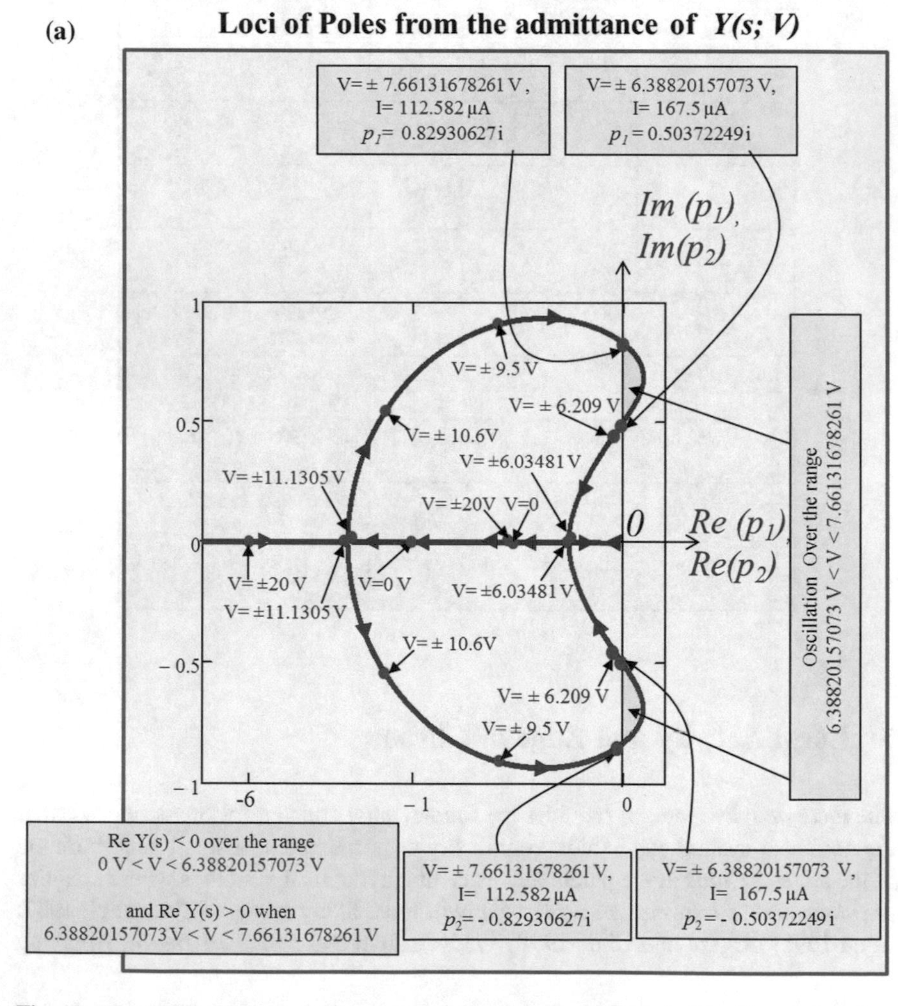

Fig. 11 **a** Loci of the Poles from the admittance function *Y(s; V)*. **b** Loci of the Eigen values from the Jacobian matrix $J(X_1, X_2; V)$. *Arrowheads* indicate the movements of poles and the Eigen values in the interval of -20 V $\leq V \leq 0$ V

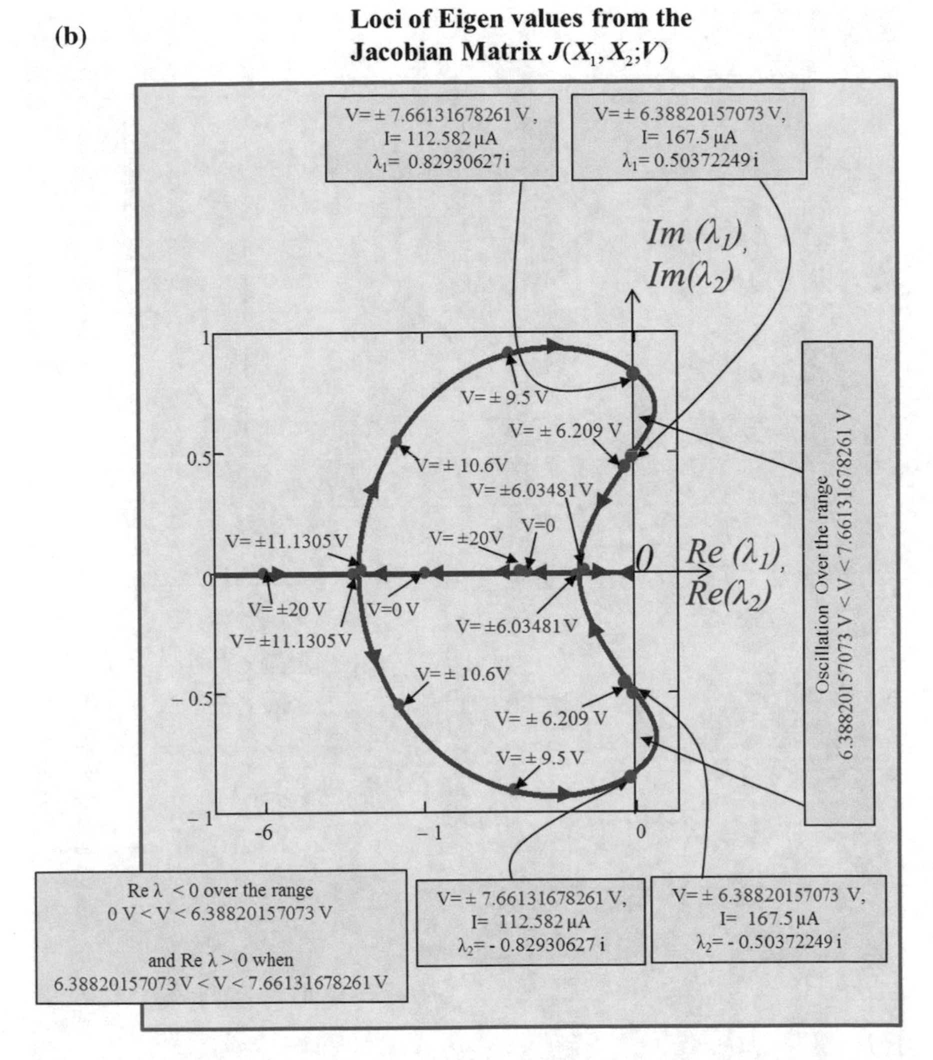

Fig. 11 (continued)

at the above parameter value of V, and becomes unstable thereafter. In other words, the *edge of chaos* regime which started at $V = 6.209$ V (*resp.* $I = 171.553$ μA) exists only over the following the tiny interval (see Fig. 5d):

$$\boxed{\begin{array}{l}\textit{Edge of chaos domain } 1:\\ 6.209\ \mathrm{V} < V < 6.38820157073\ \mathrm{V}\\ 171.553\ \mu\mathrm{A} > I > 167.5\ \mu\mathrm{A}\end{array}} \tag{49}$$

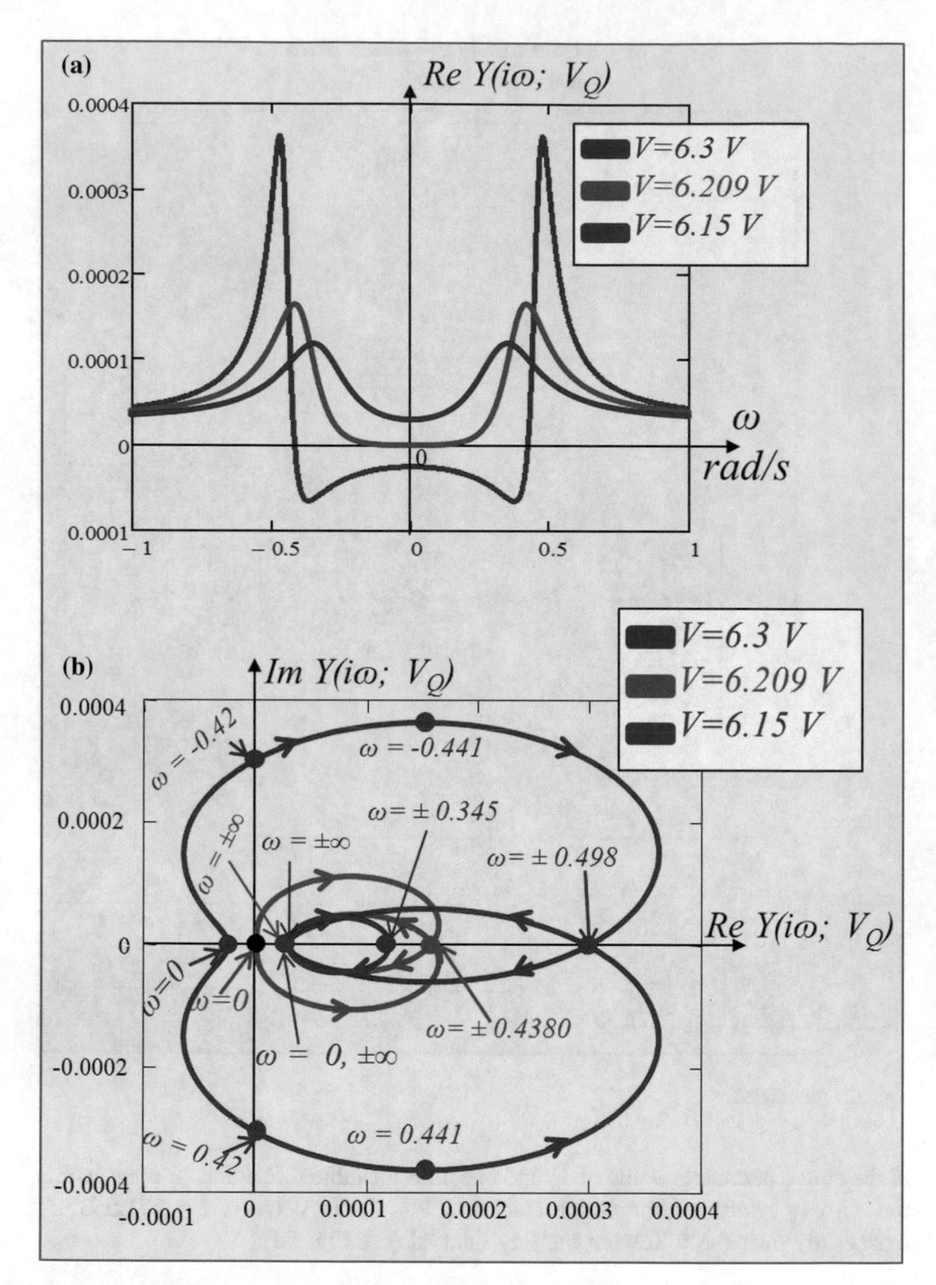

Fig. 12 **a** Illustration of the *principle of local activity* in the second-order memristor at DC input voltage $V = 6.15$ V, $V = 6.209$ V *and* $V = 6.3$ V. **b** The corresponding Nyquist plot for the input DC voltage $V = 6.15$ V, $V = 6.209$ V and $V = 6.3$ V. **c** *Edge of chaos domain 1* and *edge of chaos domain 2* on the zoom DC *V-I* curve of our second-order memristor

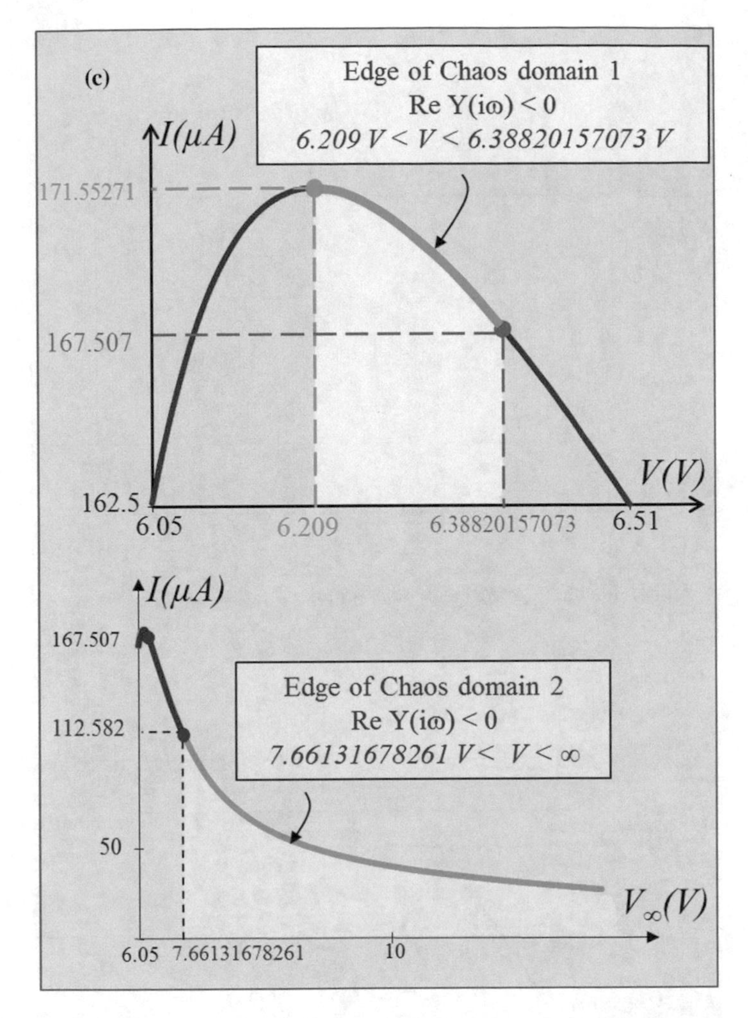

Fig. 12 (continued)

The loci of the two complex poles p_1 and p_2 of $Y(s;\ V)$ in Fig. 11a reveals the pole p_1 migrated along the red curve in the *right-half plane* to the *left-half plane* as V increases beyond $V = 6.38820157073$ V and crosses the imaginary axis at $V = 7.66131678261$ V (*resp.* $I = 112.582$ μA) where the real part of the pole of $Y(s;\ V_Q)$ vanishes at $p_1 = -i0.82930627$. Any further increase in the voltage V moves the pole p_1 back into the *left-half plane*. This confirms the existence of a second *edge of chaos* regime starting from $V = 7.66131678261$ V (*resp.* 112.582 μA), and which extend, over all $V > 7.6613$ V (see Fig. 5d); namely,

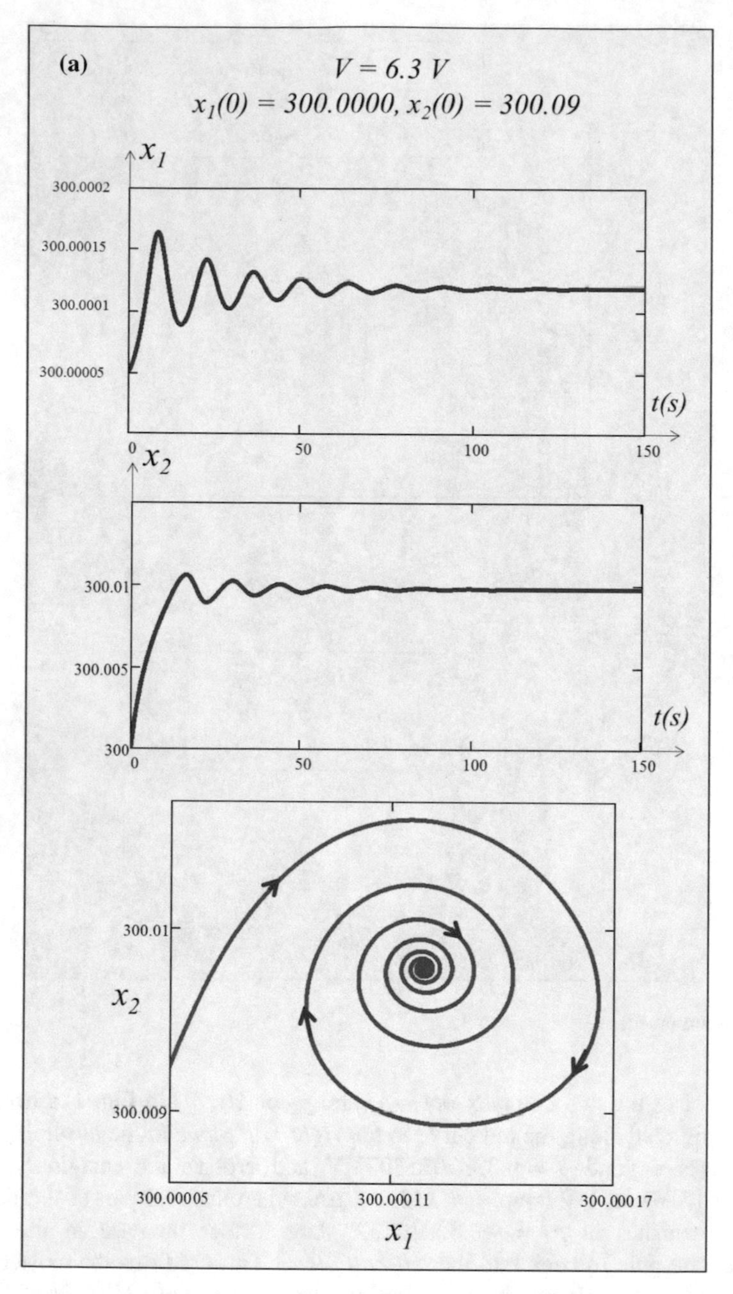

Fig. 13 Numerical Simulations to confirm *super-critical Hopf-bifurcation* theorem at the *first Hopf-bifurcation* $V = 6.38820157073$ V and at the second Hopf-bifurcation $V = 7.66131678261$ V. **a** Transient waveform converging to DC equilibrium point when the DC voltage $V = 6.3$ V was chosen near but just to the *left* of the *first Hopf-bifurcation* (**b**). Transient waveform converging to stable oscillation, when $V = 6.4$ V was chosen within the *Hopf super-critical region.* **c** Transient waveform converging to DC equilibrium point when the DC voltage $V = 7.7$ V was chosen near but just to the left of the *second Hopf-bifurcation* (see Fig. 11b)

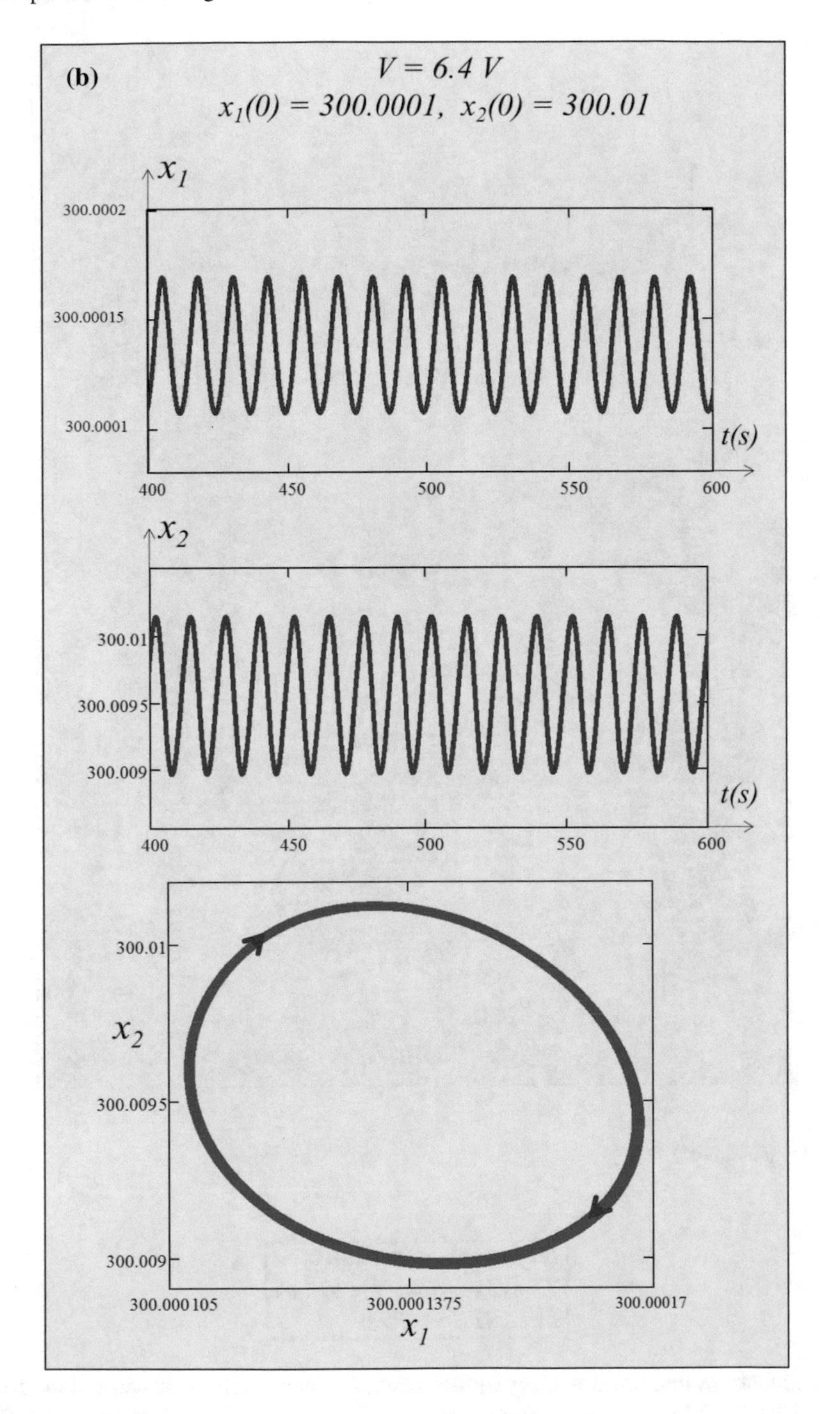

Fig. 13 (continued)

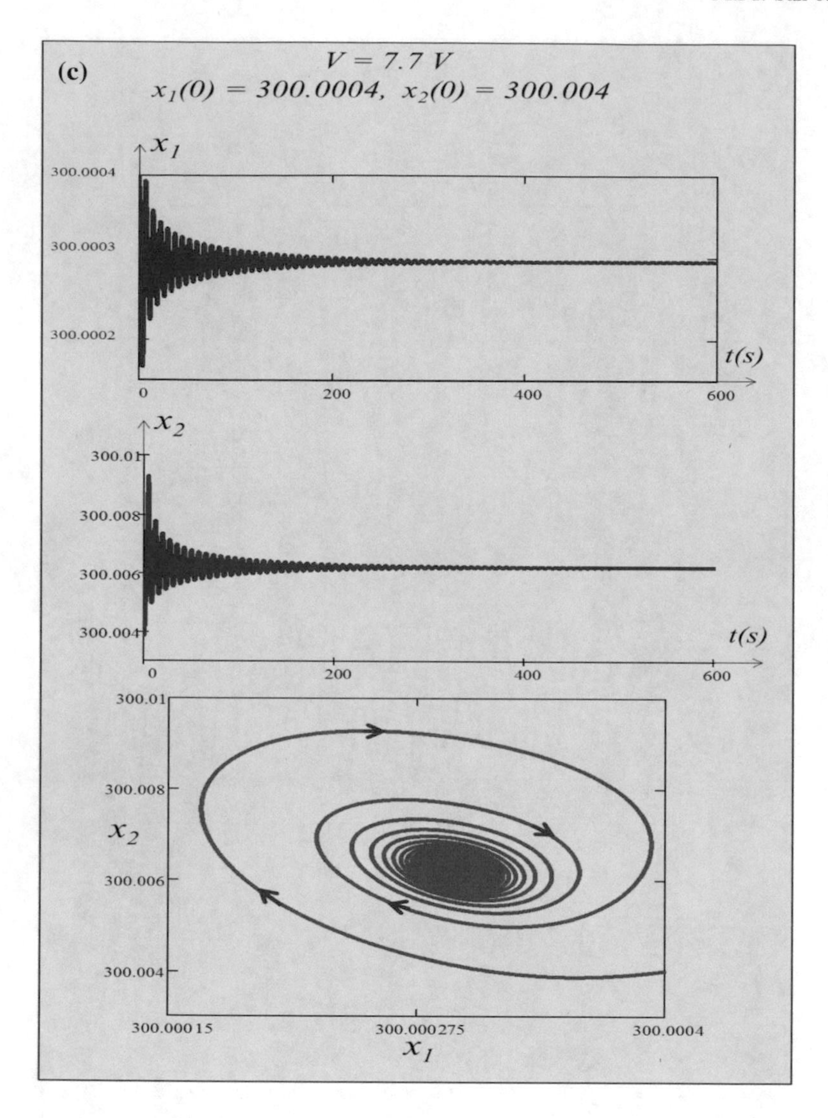

Fig. 13 (continued)

$$\boxed{\begin{array}{l}\textit{Edge of chaos domain } 2\text{:}\\ 7.66131678261\text{ V} < V < \infty\\ 112.582\,\mu\text{A} > I > 0\end{array}} \tag{50}$$

To illustrate that *local activity* in the second-order generic memristor starts from $V = 6.209$ V, Fig. 12a and b show the plot of $\text{Re}Y(i\omega; V_Q)$ and the corresponding Nyquist plot for the DC input voltage $V = 6.15$ V, $V = 6.209$ V and $V = 6.3$ V, respectively. Observe, the Nyquist plot of the admittance function $Y(i\omega; V_Q)$ is tangent to the ω axis at $\omega = 0$ for $V = 6.209$ V. Also, observe that for $V = 6.15$ V, the real part of the Nyquist plot of the admittance function is positive, i.e., $\mathrm{Re}Y(i\omega; V_Q) > 0$, confirming the memristor is *not locally active*. However, for $V = 6.3$ V, the real part of the Nyquist plot of the admittance function $Y(i\omega; V_Q)$ is negative confirming that the memristor is *locally active*. The corresponding Nyquist plots in $ImY(i\omega; V_Q)$ versus $\mathrm{Re}Y(i\omega; V_Q)$ plane is shown in Fig. 12b. Figure 12c shows the *edge of chaos domain 1* and *edge of chaos domain 2* on the zoom DC *V-I* curve of our second-order memristor. Observe that in both *edge of chaos domain 1* (49), and *domain 2* (50), we have $\mathrm{Re}Y(i\omega; V_Q) < 0, \mathrm{Re}p_1 < 0,$ and $\mathrm{Re}p_2 < 0$ (see Figs. 5d and 11a).

6 Hopf Bifurcation

Hopf bifurcation is a local bifurcation generated by non-linear dynamical systems in which an equilibrium point changes stability at some critical parameter value μ , under certain conditions. The bifurcation can be *super-critical* or *sub-critical* resulting in a stable or unstable limit cycle respectively, and is confirmed by the computation of a Hopf coefficient "*a*" at the equilibrium point when a pair of *eigen values* of the associated Jacobian matrix are purely imaginary. The standard *Hopf coefficient* "*a*" for a *second-order ODE* is given by:

$$\begin{aligned} a = & \frac{1}{16}\left[f_{xxx} + f_{xyy} + g_{xxy} + g_{yyy}\right] \\ & + \frac{1}{16\omega_0}\left[f_{xy}\left(f_{xx} + f_{yy}\right) - g_{xy}\left(g_{xx} + g_{yy}\right) - f_{xx}g_{xx} + f_{yy}g_{yy}\right] \end{aligned} \tag{51}$$

The plot of *Im(λ)* versus *Re(λ)* shown in Fig. 11b shows, the two Hopf-bifurcation points occur at $V = 6.38820157073$ V and $V = 7.66131678261$ V, respectively where the real parts of the Eigen values of Jacobian matrix at these two points are zero (*pure imaginary Eigen values)*. The eigen values within these two bifurcation points lie on the right-half plane (*Re(λ)* > 0), confirming the second-order memristor could generate oscillation. To confirm that these two Hopf-bifurcation points are *super-critical*, let us compute the Hopf coefficient "*a*" at $V = 6.38820157073$ V, where the functions f_1 and f_2 in (46a) and (46b) are denoted by f and g, respectively, in (51).

$$\boxed{\begin{array}{l} a = 1.826 \times 10^3 > 0 \, for \, V = 6.38820157073 \text{ V and } \omega_0 = 0.50372249 \\ a = 1.59 \times 10^3 > 0 \, for \, V = 6.38820157073 \text{ V and } \omega_0 = -0.50372249 \end{array}} \tag{52}$$

The coefficient $a > 0$ at the first Hopf-bifurcation point ($V = 6.38820157073$ V) implies the bifurcation is *super-critical* because the parameter $\mu = V$ enters the *unstable region* ($Re\ (\lambda) > 0$) by crossing the imaginary axis from *left to right* in Fig. 11a. Similarly, the Hopf coefficient "a" at the second Hopf-bifurcation point at $V = 7.66131678261$ V is found to be

$$\boxed{\begin{array}{l} a = -1.155 \times 10^3 < 0 \, for \, V = 7.66131678261 \text{ and } \omega_0 = 0.82930627 \\ a = -1.239 \times 10^3 < 0 \, for \, V = 7.66131678261 \text{ and } \omega_0 = -0.82930627 \end{array}} \tag{53}$$

The coefficient $a < 0$ at the second Hopf-Bifurcation point ($V = 7.66131678261$ V) actually implies the bifurcation is *super-critical* because the parameter $\mu = V$ returns to the *stable region* ($Re\ (\lambda) < 0$) by crossing the imaginary axis from *right to left*, as the parameter $\mu = V$ increases beyond the bifurcation value $V = 7.66131678261$ V. The formula given in all standard

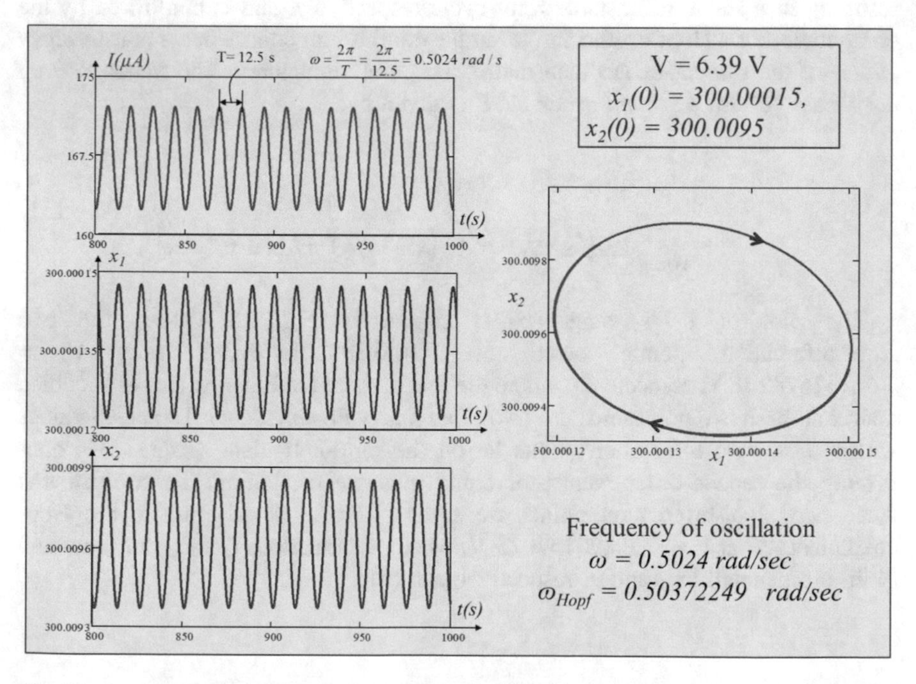

Fig. 14 An example illustrating the frequency of oscillation ω is very close to the predicted Hopf frequency ω_{Hopf} at $V = 6.39$ V

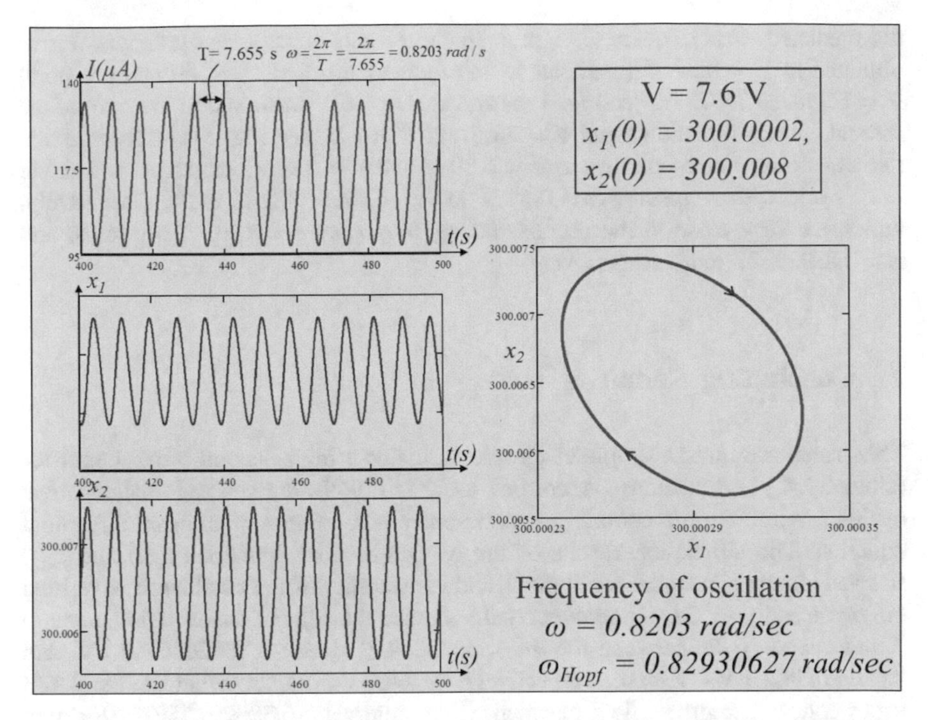

Fig. 15 An example illustrating the frequency of oscillation ω is very close to the predicted Hopf frequency ω_{Hopf} at $V = 7.6$ V

textbooks (Meiss 2007) for the Hopf coefficient "*a*" was derived by *assuming* the system becomes unstable (Re (λ) > 0) as the parameter μ crosses the imaginary axis from *left to right*, as μ *increases* beyond the Hopf-bifurcation point. It follows from the *super-critical Hopf-bifurcation theorem* that there exists a small sinusoidal oscillation for any value of the DC voltage V chosen sufficiently near but greater than the right boundary at $V = 6.38820157073$ V, and the right boundary at $V = 7.66131678261$ V where the equilibrium point is unstable. However, any initial state beyond *super-critical region* converges to another *stable equilibrium* point. We verified this phenomenon in our second-order memristor by choosing voltage $V = 6.3$ V, which is near but slightly to the left of the *first Hopf-bifurcation* point at $V = 6.38820157073$ V. Observe from Fig. 13a that the transient waveform converges to an asymptotically *stable* equilibrium point. An *identical* phenomenon was observed as shown in Fig. 13c, where the DC input voltage $V = 7.7$ V was chosen near but to the left of the *second Hopf bifurcation* point at $V = 7.66131678261$ V. However, the transient waveform converges to a *stable limit cycle* as shown in Fig. 13b, when $V = 6.4$ V was chosen within the *Hopf super-critical region.*

Numerical simulations are performed within the super-critical Hopf region near the two bifurcation points to confirm that the frequency of the oscillation is close to the predicted Hopf frequency. Figures 14 and 15 show examples of the waveforms obtained at $V = 6.39$ V (near but to the right of the first Hopf-bifurcation point $V = 6.38820157073$ V, see Fig. 11b) and at $V = 7.6$ V (near and to the right of the second Hopf bifurcation point $V = 7.66131678261$ V, see Fig. 11b). Observe that the waveforms corresponding to $V = 6.39$ V and $V = 7.6$ V converged to a stable limit cycle with *frequency* $\omega = 0.5024$ rad/s, and $\omega = 0.8203$ rad/s, respectively, which are very close to the predicted *Hopf frequency* $\omega = 0.50372249$ rad/s, and $\omega = 0.82930627$ rad/s, respectively.

7 Concluding Remarks

This paper presented a simple electronic oscillator using a second-order memristor (Chua 2014), and a battery. According to Chua (1969), the simplest mathematical oscillator circuit must contain a second-order autonomous nonlinear differential equation. The simulation results of the two differential equations $x_1(t)$ and $x_2(t)$ showed almost sinusoidal oscillations and the stability of the oscillation is verified via phase analysis. Our simulation results showed the edge of chaos regime *domain 1* and *domain 2* lie between the intervals $6.209 \text{ V} < V < 6.38820157073$ V, and $7.66131678261 \text{ V} < V < \infty$, respectively in *DC V-I curve,* whereas the Hopf super-critical regime lie between the interval $6.38820157073 \text{ V} < V < 7.66131678261$ V. Beyond both ends of the super-critical interval, the circuit tends to a DC equilibrium point on the DC *V-I* curve. A small-signal equivalent circuit was derived by choosing a DC equilibrium point Q and are found to consist of two identical linear resistor-inductor (RL) sub-circuits (with different resistance and inductance values) connected in series. The poles of the admittance function $Y(s;\ V)$ are shown to be identical to the *Eigen values* of the Jacobian matrix of this small-signal equivalent circuit describing the memristor-battery circuit. At $V = 6.38820157073$ V and at $V = 7.66131678261$ V, the loci of the *Eigen value* as a function of the DC voltage V crosses the imaginary axis. We found, all the initial conditions decay to a DC operating point, if we increase the battery voltage V from $V = 0$ V to the left boundary of the edge of chaos at $V = 6.209$ V. A further increase in V causes the pair of complex-conjugate *Eigen values* to cross the imaginary axis, while spawning a small sinusoidal oscillation whose amplitude increases rapidly (like the square root of 2) with increasing battery voltage. This phenomenon of a super-critical Hopf bifurcation (Meiss 2007) serves as a textbook example, which we had confirmed analytically by showing the Hopf bifurcation coefficient "a" is positive. On the other hand, the amplitude of the oscillation begins to decrease, with further increase in battery voltage V, illustrating the prediction of the Hopf bifurcation theorem no longer exists.

Similarly, when the battery voltage is decreased from far beyond the right boundary of the edge of chaos *domain 2* at $V = 7.66131678261$ V, we found once again all initial conditions converge to a DC operating point at the corresponding battery voltage. Furthermore, when the battery voltage reaches the left boundary of the edge of chaos *domain 2* at $V = 7.66131678261$ V then the pair of complex-conjugate *Eigen values* crosses the imaginary axis from *right to left* while spawning another small sinusoidal oscillation, whose amplitude increases rapidly like before, as we continue to *decrease* the battery voltage. Soon the sinusoidal waveform merges seamlessly with the earlier sinusoidal waveform spawned from the right boundary of the edge of chaos domain1!

Indeed, we have also proved that the second-order memristor oscillator could generate sinusoidal oscillation via a *super-critical* Hopf bifurcation. In this paper, the computation of Hopf bifurcation coefficient "*a*" gives $a < 0$ in contrast to the standard Hopf bifurcation condition which satisfies that $a > 0$. The reason for the above difference in the sign of "*a*" is due to the fact that the calculation of Hopf bifurcation coefficient "*a*" described in nonlinear dynamics textbooks (Meiss 2007) is based on the assumption that the pair of complex-conjugate *Eigen values* crosses the imaginary axis from *left to right* as the bifurcation parameter *increases*.

Finally, we conclude that the memristor-battery oscillator gives rise to two sinusoidal oscillations originating from either boundary of the edge of chaos regime of the memristor via *super-critical* Hopf bifurcation and it also provides the textbook example for detail understanding of *super-critical* Hopf bifurcation phenomenon.[4] In this paper, the second order memristor represents the model of a physical device called *Positive Temperature Coefficient* (*PTC)* and *Negative Temperature Coefficient (NTC) thermistor* connected in series. So, as a future work, it is possible to generate oscillations in a real circuit by connecting *PTC and NTC thermistors* in series across the battery via *super-critical* Hopf bifurcation phenomenon.

Acknowledgements The authors would like to acknowledge financial support from the USA Air force office of Scientific Research under Grant number FA9550-13-1-0136 and from the European Commission Marie Curie Fellowship and two National Research Foundation of Korea (NRF) grants funded by the Korea government (2013R1A2A2A01068683 and 2012R1A1A2044078).

References

Adhikari, S. P., Sah, M. P., Kim, H., & Chua, L. O. (2013). Three fingerprints of memristors. *IEEE Transactions on Circuits and Systems I: Regular Papers, 60,* 3008–3021.

Chua, L. O. (1969). *Introduction to nonlinear network theory*. New York: McGraw-Hill.

Chua, L. O., & Kang, S. M. (1976). Memristive devices and systems. *Proceedings of the IEEE, 64,* 209–223.

Chua, L. O., Desoer, C. A., & Kuh, E. S. (1987). *Linear and nonlinear circuits*. New York: McGraw-Hill.

---

[4]In contrast, the second-order sodium ion-channel memristor oscillator circuit presented in Fig. 40 of Chua (2014) spawns an *unstable* sinusoidal oscillation near a boundary of the edge of chaos of the sodium memristor via a *sub-critical* Hopf bifurcation, a much more subtle bifurcation phenomenon.

Chua, L. O. (1998). *CNN: A paradigm for complexity*. Singapore: World Scientific.
Chua, L. O. (2003). Nonlinear circuit foundations for nanodevices—Part I: The four element torus. *Proceedings of the IEEE, 91,* 1830–1859.
Chua, L., Sbitnev, V., & Kim, H. (2012a). Hodgkin-Huxley axon is made of memristors. *International Journal of Bifurcation and Chaos, 22*, 1230011/1-48.
Chua, L., Sbitnev, V., & Kim, H. (2012b). Neurons are poised near the edge of chaos. *International Journal of Bifurcation and Chaos, 22*, 1250098/1-49.
Chua, L. (2014). If it's pinched it's a memristor. *Semiconductor Science and Technology, 29,* 104001/1-42.
Chua, L. (2015). Everything you wish to know about memristors but are afraid to ask. *Radio Engineering, 24,* 2/319-368.
Dogaru, R., & Chua, L. O. (1998). Edge of chaos and local activity domain of FitzHugh-Nagumo equation. *International Journal of Bifurcation and Chaos, 8,* 211–257.
Mannan, Z. I., Choi, H., & Kim, H. (2016). Chua corsage memristor oscillator via Hopf bifurcation. *International Journal of Bifurcation and Chaos, 26,* 1630009/1–28.
Mehta, V. K., & Mehta, R. (2005). *Principle of electronics*. India: S. Chand and Co., Ltd.
Meiss, J. D. (2007). *Differential dynamical systems*. USA: Society of Industrial and Applied Mathematics.
Rajamani, V., Yang, C., Kim, H., & Chua, L. (2016). Design of a low-frequency oscillator with ptc memristor and an inductor. *International Journal of Bifurcation and Chaos, 26*, 1630021/1-27.
Vaidyanathan, S., & Volos, C. (2016a). *Advances and applications in nonlinear control systems*. Berlin: Springer.
Vaidyanathan, S., & Volos, C. (2016b). *Advances and applications in chaotic systems*. Berlin: Springer.

A Hyperjerk Memristive System with Hidden Attractors

Viet-Thanh Pham, Sundarapandian Vaidyanathan, Christos Volos, Xiong Wang and Duy Vo Hoang

Abstract After the introduction by Leonov and Kuznetsov of a new classification of nonlinear dynamics with kinds of attractors (self-excited attractors and hidden attractors), this subject has received a significant interest. From an engineering point of view, hidden attractors are important and can lead to unexpected behavior. Various chaotic systems with the presence of hidden attractors have been discovered recently. Especially, memristor, the fourth basic circuit element, can be used to construct such chaotic systems. This chapter presents a new memristive system which can display hidden chaotic attractor. Interestingly, this memristive system is a hyperjerk system because it involves time derivatives of a jerk function. The fundamental dynamics properties of such memristive system are discovered by calculating the number of equilibrium points, using phase portraits, Poincaré map, bifurcation diagram, maximum Lyapunov exponents, and Kaplan–Yorke fractional dimension. Also, we have investigated the multi–stability in the memristive system by varying the value of its initial condition. In addition, adaptive synchronization for the hyperjerk memristive system is also studied. The proposed memristive system can be applied into chaos–based engineering applications because of its chaotic behavior.

V.-T. Pham (✉)
School of Electronics and Telecommunications, Hanoi University of Science and Technology, Hanoi, Vietnam
e-mail: pvt3010@gmail.com

S. Vaidyanathan
Research and Development Centre, Vel Tech University, Chennai, Tamil Nadu, India
e-mail: sundarcontrol@gmail.com

C. Volos
Physics Department, Aristotle University of Thessaloniki, Thessaloniki, Greece
e-mail: volos@physics.auth.gr

X. Wang
Institute for Advanced Study, Shenzhen University, Shenzhen 518060, Guangdong, People's Republic of China
e-mail: wangxiong8686@szu.edu.cn

D.V. Hoang
Ton Duc Thang University, Ho Chi Minh, Vietnam
e-mail: vohoangduy@tdt.edu.vn

© Springer International Publishing AG 2017
S. Vaidyanathan and C. Volos (eds.), *Advances in Memristors, Memristive Devices and Systems*, Studies in Computational Intelligence 701, DOI 10.1007/978-3-319-51724-7_3

Keywords Chaos · Hidden attractor · Hyperjerk · Equilibrium · Memristive system · Bifurcation · Multi–stability

1 Introduction

In the last decades, there was a growing interest in studying chaotic systems and their applications (Lorenz 1963; Strogatz 1994; Chen and Ueta 1999; Chen and Yu 2003; Sprott 2003; Yalcin et al. 2005; Azar and Vaidyanathan 2015a, b). Beside conventional chaotic systems such as Lorenz's system (Lorenz 1963), Rössler's system (Rössler 1976), Arneodo's system (Arneodo et al. 1981), Chen's system (Chen and Ueta 1999), and Lü's system (Lü and Chen 2002), numerous studies concerning new chaotic systems have been carried out in last years (Wang and Chen 2012; Vaidyanathan 2013; Jafari et al. 2013; Molaei et al. 2013; Kingni et al. 2014; Pehlivan et al. 2014; Pham et al. 2014b).

Recently, a new classification of attractors has been proposed by Leonov and Kuznetsov, who consider two kinds of attractors: self–excited attractor and hidden attractor (Leonov et al. 2011a, b, 2012, 2014). We can localize self–excited attractors by the standard computational procedure. For example, we select a point from the unstable manifold in a neighborhood of an unstable equilibrium and follow up the state of the attractor. Therefore most of classical chaotic systems display self–excited attractors. Hidden attractor, in contrast to self–excited attractor, cannot be computed by the standard procedure because its basin of attraction does not contain neighborhoods of any equilibria. There has been increasing interest in chaotic systems with the presence of hidden attractors (Jafari and Sprott 2013). It is worth noting that memristors have been applied in the investigation of chaotic systems with hidden attractors (Driscoll et al. 2011; Muthuswamy and Kokate 2009; Muthuswamy 2010). Itoh and Chua derived memristor oscillators by replacing Chua's diodes in Chua's oscillator by memristors (Itoh and Chua 2008). Buscarino et al. introduced memristive chaotic circuits based on cellular nonlinear networks (Buscarino et al. 2012). Simplest chaotic memristor–based circuits were discovered in Muthuswamy and Chua (2010). Moreover, Iu et al. controlled chaos in a memristor based circuit using a twin–T notch filter (Iu et al. 2011).

In this chapter, we introduce a hyperjerk system based on a memristive device. Owing the presence of a memristive device, this particular hyperjerk system has an infinite number of equilibrium points. The rest of chapter is organized as follows. Related works are summarized in Sect. 2. Section 3 provides the mathematical model of the memristive hyperjerk system, while dynamics and properties of the memristive system are presented in Sect. 4. We propose the adaptive synchronization for achieving global chaos synchronization of the identical novel hyperjerk systems with two unknown parameters in Sect. 5. Finally, conclusions are drawn in Sect. 6.

2 Related Works

It is shown that tremendous research efforts have been devoted to jerk chaotic systems (Sprott 2010). On one hand, from the mathematical point of view a jerk system is based on a jerk equation that involves a third–time derivative of a single variable, for example x. Therefore, a jerk system is described by the following form:

$$\frac{d^3x}{dt^3} = f\left(\frac{d^2x}{dt^2}, \frac{dx}{dt}, x\right) \tag{1}$$

On the other hand, from the mechanical point of view, the nonlinear $f(.)$ in system (1) is called the "jerk" due to the fact that it describes the third–time derivative of the scalar x. Thus, it represents the first–time derivative of acceleration in a mechanical system (Schot 1978). Especially, well–known chaotic systems, i.e. Lorenz and Rössler systems can be described by using a mathematical model of a jerk system (Linz 1997; Lainscsek et al. 2003).

Various chaotic jerk systems were introduced in the literature. Eichhorn et al. proposed simple polynomial classes of chaotic jerky dynamics (Eichhorn et al. 2002). Sun and Sprott investigated a piecewise exponential jerk system (Sun and Sprott 2009). Malasoma indicated the simplest dissipative chaotic jerk equation which was parity invariant (Malasoma 2000). Another simple chaotic jerk system with exponential nonlinearity was presented in Munmuangsaen et al. (2011), while its elegant electronic circuital implementation, including six resistors, three capacitors, four operational amplifiers and a silicon diode only, was introduced in Sprott (2011). Louodop and collaborators presented a linear transformation of Model MO5 and its practical finite–time synchronization (Louodop et al. 2014). A six–term 3–D novel jerk chaotic system with two hyperbolic sinusoidal nonlinearities was proposed by Vaidyanathan et al. (2014). The finding of a window in the parameter space in which the jerk system displayed the unusual and striking feature of multiple attractors (e.g. coexistence of four disconnected periodic and chaotic attractors) was reported in Kengne et al. (2016). Multi–scroll chaotic attractors could be generated in the jerk mode (Liu et al. 2012) or jerk circuits (Yu et al. 2005; Ma et al. 2014), while multi–scroll and hypercube attractors were also achieved from a general jerk circuit using Josephson junctions (Yalcin and Ozoguz 2007). Moreover, dynamics and delayed feedback control for a 3–D jerk system with only one stable equilibria was discussed by Wang et al. (2015).

By generalizing the definition of a jerk system Sprott (1997), a hyperjerk system is given as

$$\frac{d^{(n)}x}{dt^n} = f\left(\frac{d^{(n-1)}x}{dt^{n-1}}, \ldots, \frac{dx}{dt}, x\right), \tag{2}$$

with $n \geq 4$ (Sprott 2010). It is noting that Elhadj and Sprott indicated that hyperjerk form can describe all periodically forced oscillators and many of the coupled oscillators (Elhadj and Sprott 2013). Chaotic hyperjerk system including fourth and fifth derivatives was introduced by Chlouverakis and Sprott Chlouverakis and Sprott (2006). In addition, a 4–D novel hyperchaotic hyperjerk system was proposed by Sundarapandian (Vaidyanathan et al. 2015).

Interestingly, there are chaotic jerk/hyperjerk systems without equilibria or with an infinite number of equilibrium points. Wang and Chen proposed a special autonomous jerk system with no equilibrium point when constructing a chaotic system with any number of equilibria (Wang and Chen 2013). Some simple jerk systems without equilibrium were found by using a systematic searching procedure (Jafari et al. 2013). A chaotic memory system with infinitely many equilibria was designed by using the concept of memory element (Bao et al. 2013). Studying such jerk/hyperjerk systems with special features is still an open research direction.

3 Model of the Memristive System

Memristor was invented by L.O. Chua and was considered as the fourth basic circuit element beside the three conventional circuit elements (the resistor, the inductor and the capacitor) (Chua 1971). Recently, numerous applications of memristor and memristive systems have been proposed (Tetzlaff 2014). Memristor–based systems have been used in different areas such as memristive adaptive filters (Driscoll et al. 2010), memristive model of amoeba learning (Pershin et al. 2009), resistance switching memories (Chua 2011), memristor cellular automata (Itoh and Chua 2009), or programmable analog integrated circuits (Shin et al. 2011) etc.

It has been known that memristor presents the relationship between two fundamental circuit variables, the charge (q) and the flux (φ) (Tetzlaff 2014). Therefore there are two kinds of memristor: charge–controlled memristor and flux–controlled memristor (Chua 1971; Tetzlaff 2014). A charge–controlled memristor is described by

$$v_M = M(q)\, i_M, \tag{3}$$

where v_M is the voltage across the memristor and i_M is the current through the memristor. Here the memristance (M) is defined by

$$M(q) = \frac{d\varphi(q)}{dq}, \tag{4}$$

while the flux–controlled memristor is given by

$$i_M = W(\varphi)\, v_M, \tag{5}$$

where $W(\varphi)$ is the memductance, which is defined by

$$W(\varphi) = \frac{dq(\varphi)}{d\varphi}. \tag{6}$$

Moreover, by generalizing the original definition of a memristor (Chua and Kang 1976; Tetzlaff 2014), a memristive system is given as:

$$\begin{cases} \dot{x} = F(x, u, t) \\ y = G(x, u, t)\, u, \end{cases} \tag{7}$$

where u, y, and x denote the input, output and state of the memristive system, respectively. The function F is a continuously differentiable, n–dimensional vector field and G is a continuous scalar function.

Based on the definition of memristive system (Chua and Kang 1976; Pershin et al. 2009; Tetzlaff 2014; Bao et al. 2013), a memristive device is introduced in this section and used in other sections of our chapter. The memristive device is described in the following form:

$$\begin{cases} \dot{x}_1 = x_2 \\ y = \left(1 - x_1\right) x_2. \end{cases} \tag{8}$$

Here x_2, y, and x_1 are the input, output and state of the memristive device, respectively.

In this work, a novel 4–D memristive system is constructed by applying the memristive device (8) and using the approach in Bao et al. (2013). The novel memristive system is described in the following form:

$$\begin{cases} \dot{x}_1 = x_2 \\ \dot{x}_2 = x_3 \\ \dot{x}_3 = x_4 \\ \dot{x}_4 = -x_3 - a x_4 - b \sin\left(x_3\right) x_4 - y, \end{cases} \tag{9}$$

where a, b are two positive parameters and $y = \left(1 - x_1\right) x_2$ is the output of memristive device (8).

The novel memristive system (9) can be rewritten by

$$\frac{d^4 x_1}{dt^4} = f\left(\frac{d^3 x_1}{dt^3}, \frac{d^2 x_1}{dt^2}, \frac{d x_1}{dt}, x_1\right), \tag{10}$$

where

$$f = -\frac{d^2 x_1}{dt^2} - a\frac{d^3 x_1}{dt^3} - b \sin\left(\frac{d^2 x_1}{dt^2}\right)\frac{d^3 x_1}{dt^3} - \left(1 - x_1\right)\frac{d x_1}{dt}. \tag{11}$$

Therefore, the memristive system (9) is called a hyperjerk system because it involves time derivatives of a jerk function (Sprott 1997, 2010). In this chapter, the memristive system (9) is chaotic when the parameters a, and b take the values

$$a = 0.55, \quad b = 0.2. \tag{12}$$

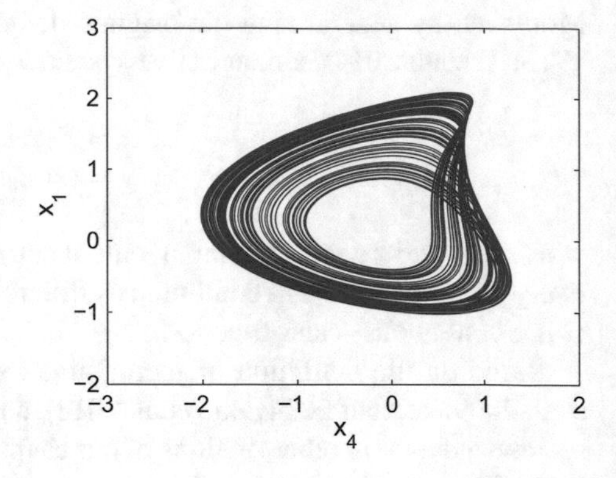

Fig. 1 2–D projection of the chaotic hyperjerk memristive system (9) in (x_4, x_1)–plane

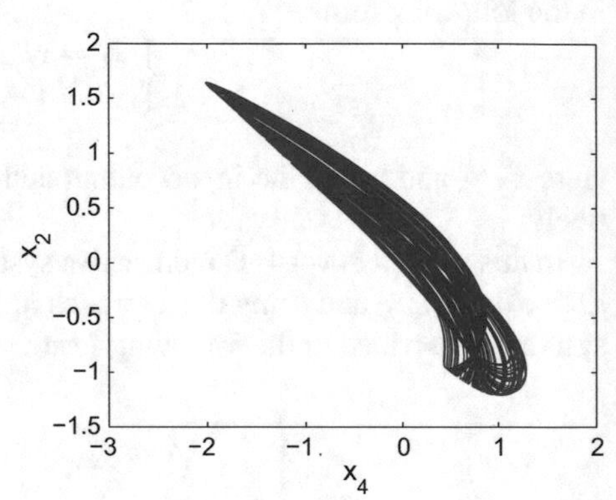

Fig. 2 2–D projection of the chaotic hyperjerk memristive system (9) in (x_4, x_2)–plane

For the selected parameter values in (12), the Lyapunov exponents of the novel memristive system (9) are obtained as

$$L_1 = 0.0578, \ \ L_2 = 0.0010, \ \ L_3 = 0, \ \ L_4 = -0.6069. \tag{13}$$

For numerical simulations, we take the initial conditions of the novel memristive system (9) as

$$x_1(0) = 0, \ \ x_2(0) = 0.01, \ \ x_3(0) = 0, \ \ x_4(0) = 0. \tag{14}$$

Figures 1, 2 and 3 illustrate the 2–D projections of the new hyperjerk memristive system (9). In addition, the Poincaré map in Fig. 4 shows the folding properties of chaos.

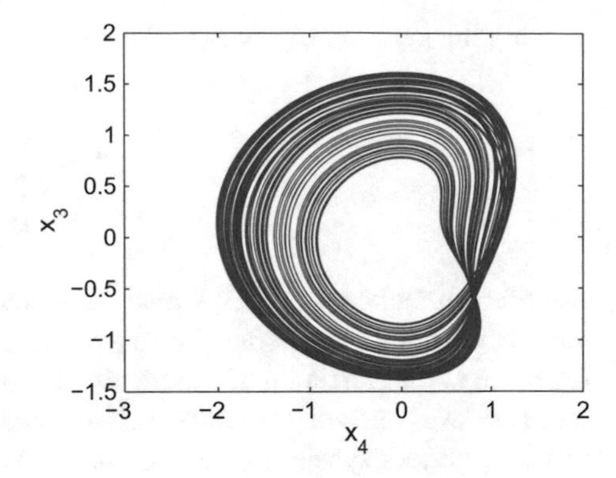

Fig. 3 2–D projection of the chaotic hyperjerk memristive system (9) in (x_4, x_3)–plane

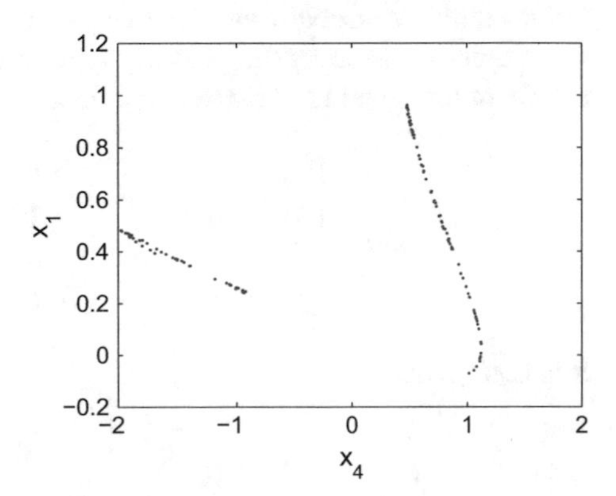

Fig. 4 Poincaré map the chaotic hyperjerk memristive system (9) in (x_4, x_1)–plane for $x_3 = 0$

4 Dynamics and Properties of the Memristive System

4.1 Equilibrium Points

The equilibrium points of the 4–D novel memristive hyperjerk system (9) are obtained by solving the equations

$$\begin{cases} f_1(x_1, x_2, x_3, x_4) = x_2 & = 0 \\ f_2(x_1, x_2, x_3, x_4) = x_3 & = 0 \\ f_3(x_1, x_2, x_3, x_4) = x_4 & = 0 \\ f_4(x_1, x_2, x_3, x_4) = -x_3 - ax_4 - bsin\left(x_3\right)\, x_4 - y = 0 \end{cases} . \tag{15}$$

Thus, the equilibrium points of the system (9) are characterized by the equations

$$y = (1 - x_1)x_2 = 0, \;\; x_2 = 0, \;\; x_3 = 0, \;\; x_4 = 0 \tag{16}$$

Solving the system (16), we get the equilibrium points of the hyperjerk system (9) as

$$E_c = \begin{bmatrix} c \\ 0 \\ 0 \\ 0 \end{bmatrix}, \tag{17}$$

where c is a real constant. Obviously, the novel hyperjerk system (9) has an infinite number of equilibrium points due to the presence of a memristive device (8). According to a new classification of chaotic dynamics (Leonov et al. 2011a, b, 2012, 2014), there are two kinds of attractors: self–excited attractors and hidden attractors. As a result, hyperjerk system (9) can be considered as a chaotic memristive system with hidden attractor (Leonov and Kuznetsov 2011, 2013; Jafari and Sprott 2013).

In order to discover the stability type of the equilibrium points E_c the Jacobian matrix of the novel memristive hyperjerk system (9) is calculated at any point $\mathbf{x}$ as

$$J(\mathbf{x}) = \begin{bmatrix} 0 & 1 & 0 & 0 \\ 0 & 0 & 1 & 0 \\ 0 & 0 & 0 & 1 \\ x_2 & x_1 - 1 & -1 - b\cos(x_3)\, x_4 & -a - b\sin(x_3) \end{bmatrix}. \tag{18}$$

It is noting that

$$J_0 \triangleq J(E_c) = \begin{bmatrix} 0 & 1 & 0 & 0 \\ 0 & 0 & 1 & 0 \\ 0 & 0 & 0 & 1 \\ 0 & c-1 & -1 & -0.55 \end{bmatrix}, \tag{19}$$

which has the characteristic equation

$$\lambda\left(\lambda^3 + 0.55\lambda^2 + \lambda + 1 - c\right) = 0. \tag{20}$$

When $c = 0$ the characteristic equation (20) has a zero eigenvalue and three nonzero eigenvalues

$$\lambda_1 = 0, \;\; \lambda_2 = -0.8193, \;\; \lambda_{3,4} = 0.1346 \pm 1.0966i \tag{21}$$

This shows that the equilibrium point E_c is an unstable saddle–focus point.

4.2 *Lyapunov Exponents and Kaplan–Yorke Dimension*

For the parameter values $a = 0.55, b = 0.2$ and $c = 0$, the calculated Lyapunov exponents of the novel memristive hyperjerk system (9) are

$$L_1 = 0.0578, \;\; L_2 = 0.0010, \;\; L_3 = 0 \;\text{ and }\; L_4 = -0.6069 \tag{22}$$

There is one positive Lyapunov exponents in the LE spectrum (22), thus the novel memristive hyperjerk system (9) exhibits chaotic behavior.

In addition, since $L_1 + L_2 + L_3 + L_4 = -0.5481 < 0$, it indicates that the novel memristive system (9) is dissipative.

The Kaplan–Yorke fractional dimension, that presents the complexity of attractor (Strogatz 1994; Sprott 2003), is defined by

$$D_{KY} = j + \frac{1}{\left|L_{j+1}\right|} \sum_{i=1}^{j} L_i \tag{23}$$

where j is the largest integer satisfying $\sum_{i=1}^{j} L_i \geq 0$ and $\sum_{i=1}^{j+1} L_i < 0$. Therefore, the Kaplan–Yorke dimension of the novel memristive hyperjerk system (9) is calculated as

$$D_{KY} = 3 + \frac{L_1 + L_2 + L_3}{|L_4|} = 3.0969, \tag{24}$$

which is fractional.

4.3 Bifurcation Diagram and Maximum Lyapunov Exponents

We use bifurcation diagram and Maximum Lyapunov exponents (MLE) to discover further the dynamics of the memristive hyperjerk system (9). Bifurcation diagram and Maximum Lyapunov exponents of memristive hyperjerk system (9) are reported in Figs. 5 and 6. As shown in Figs. 5 and 6, memristive hyperjerk system (9) exhibits chaotic behavior and periodical states. Also, the system exhibits the well-known route to chaos through the mechanism of period doubling when the parameter a is decreased.

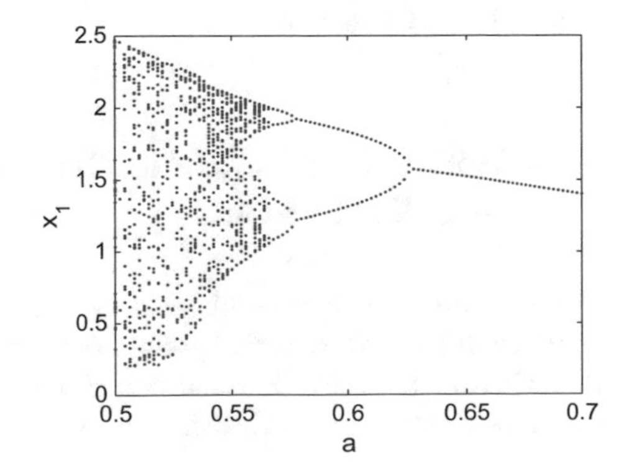

Fig. 5 Bifurcation diagram of the hyperjerk memristive system (9) when changing the value of parameter a for $b = 0.2$, and the initial conditions $(x_1(0), x_2(0), x_3(0), x_4(0)) = (0, 0.01, 0, 0)$

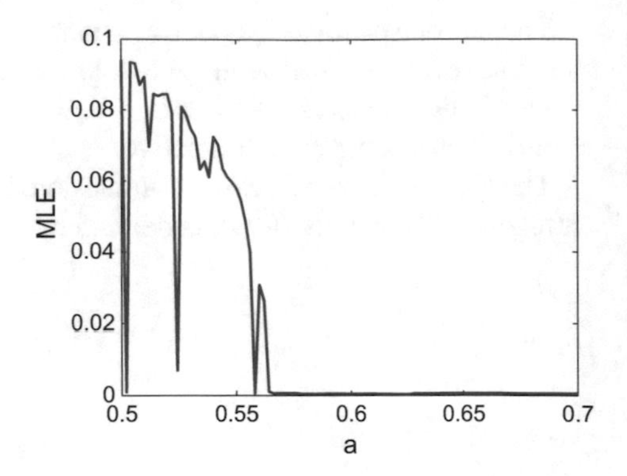

Fig. 6 Maximum Lyapunov exponents of the hyperjerk memristive system (9) when varying the value of parameter a for $b = 0.2$, and the initial conditions $(x_1(0), x_2(0), x_3(0), x_4(0)) = (0, 0.01, 0, 0)$

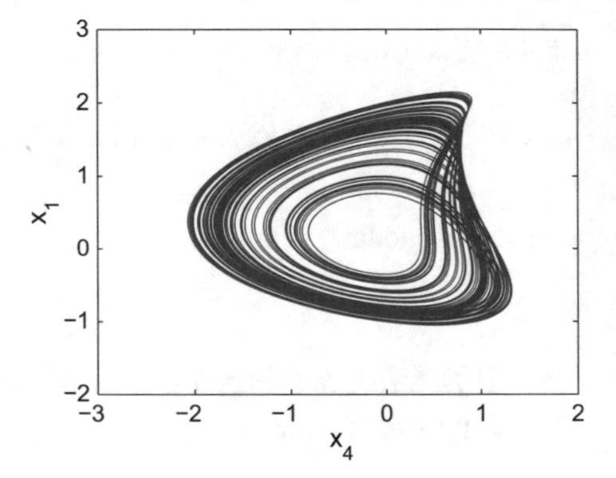

Fig. 7 Chaotic behavior of the hyperjerk memristive system (9) in (x_4, x_1)–plane for $a = 0.55$, $b = 0.2$, and the initial conditions $(x_1(0), x_2(0), x_3(0), x_4(0)) = (-0.01, 0.01, 0, 0)$

4.4 Multistability

As seen in Eq. (17), there is an infinite number of equilibrium points in hyperjerk memristive system (9). Moreover, these equilibrium points depend on the internal state of the memristive device (x_1). By changing the initial condition of the internal state x_1, we observed multi–stability of system (9). Chaos coexists with various periodical states, for example period–4 state, period–2 state, or period–1 state (see Figs. 7, 8, 9, 10 and 11).

5 Adaptive Synchronization for the Hyperjerk Memristive System

The possibility of synchronization is one of the most important characteristics when studying new chaotic systems (Pecora and Carroll 1990; Kapitaniak 1994; Boccaletti et al. 2002; Fortuna and Frasca 2007). There are numerous research activities relating to synchronization of nonlinear systems in the literature (Bus-

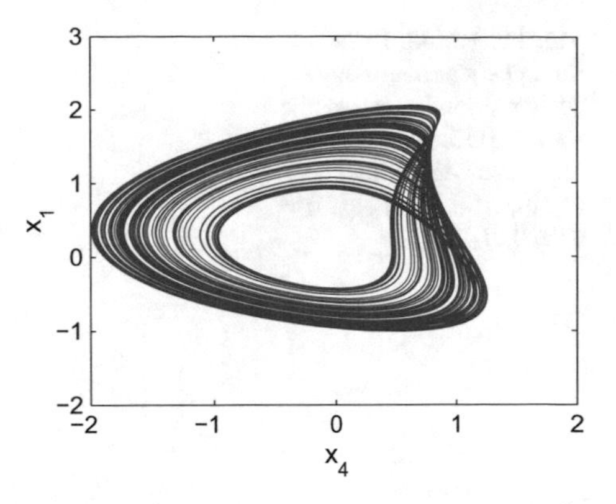

Fig. 8 Chaotic behavior of the hyperjerk memristive system (9) in (x_4, x_1)–plane for $a = 0.55$, $b = 0.2$, and the initial conditions $(x_1(0), x_2(0), x_3(0), x_4(0)) = (0.005, 0.01, 0, 0)$

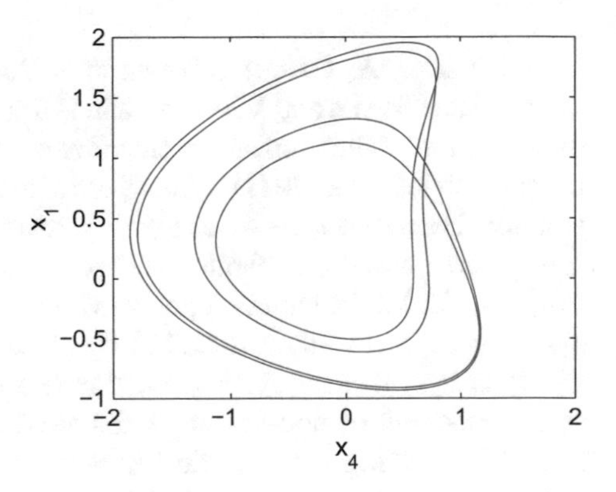

Fig. 9 Period–4 state of the hyperjerk memristive system (9) in (x_4, x_1)–plane for $a = 0.55$, $b = 0.2$, and the initial conditions $(x_1(0), x_2(0), x_3(0), x_4(0)) = (0.03, 0.01, 0, 0)$

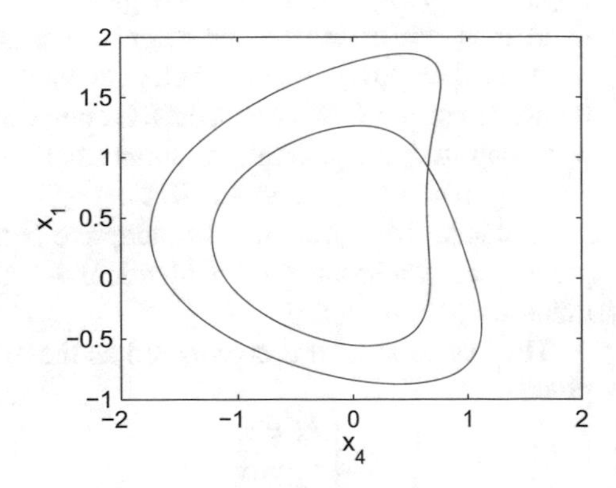

Fig. 10 Period–2 state of the hyperjerk memristive system (9) in (x_4, x_1)–plane for $a = 0.55$, $b = 0.2$, and the initial conditions $(x_1(0), x_2(0), x_3(0), x_4(0)) = (0.05, 0.01, 0, 0)$

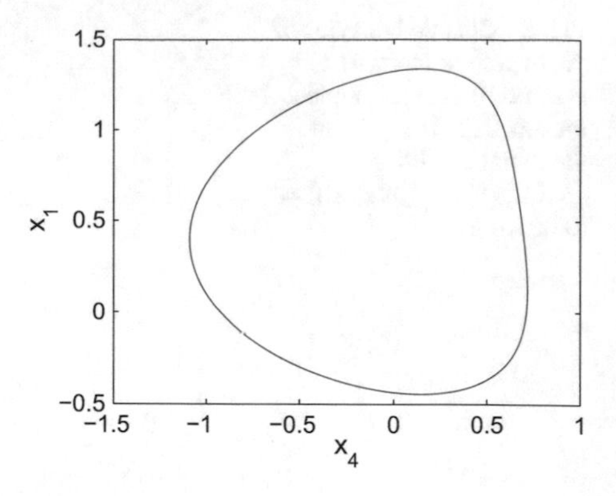

Fig. 11 Period–1 state of the hyperjerk memristive system (9) in (x_4, x_1)–plane for $a = 0.55$, $b = 0.2$, and the initial conditions $(x_1(0), x_2(0), x_3(0), x_4(0)) = (0.2, 0.01, 0, 0)$

carino et al. 2009; Gamez-Guzman et al. 2009; Srinivasan et al. 2011; Pham et al. 2014a; Karthikeyan and Vaidyanathan 2014; Vaidyanathan 2014). Volos et al. introduced various synchronization phenomena in bidirectionally coupled double scroll circuits (Volos et al. 2011). Huang et al. implemented shape synchronization control for three–dimensional chaotic systems (Huang et al. 2016). Image encryption process based on chaotic synchronization phenomena was proposed in Volos et al. (2013). Vaidyanathan reported various synchronization schemes such as anti–synchronization (Vaidyanathan 2012), adaptive synchronization (Vaidyanathan et al. 2014), or hybrid chaos synchronization (Karthikeyan and Vaidyanathan 2014). Fast synchronization of non–identical chaotic modulation–based secure systems using a modified sliding mode controller was investigated in Kajbaf et al. (2016). Rasappan and Vaidyanathan studied global chaos synchronization of WINDMI and Coullet chaotic systems using adaptive backstepping control design (Rasappan and Vaidyanathan 2014). As have been known that the backstepping control method is a recursive approach that connects the choice of a Lyapunov function with the design of a controller and guarantees global asymptotic stability of strict feedback systems (Rasappan and Vaidyanathan 2014).

In this section, we study an adaptive backstepping controller to achieve complete chaos synchronization of identical 4–D memristive hyperjerk systems with two unknown parameters.

The master system is considered as the 4–D novel memristive hyperjerk system given by

$$\begin{cases} \dot{x}_1 = x_2 \\ \dot{x}_2 = x_3 \\ \dot{x}_3 = x_4 \\ \dot{x}_4 = -x_3 - ax_4 - b\sin\left(x_3\right)\, x_4 - x_2 + x_1 x_2 \end{cases} \tag{25}$$

where x_1, x_2, x_3, x_4 are the states of the system, and a, b are unknown constant parameters.

The slave system is considered as the 4–D novel memristive hyperjerk system given by

$$\begin{cases} \dot{y}_1 = y_2 \\ \dot{y}_2 = y_3 \\ \dot{y}_3 = y_4 \\ \dot{y}_4 = -y_3 - ay_4 - b\sin\left(y_3\right) y_4 - y_2 + y_1 y_2 + u \end{cases} \tag{26}$$

where y_1, y_2, y_3, y_4 are the states of the system, and u is a backstepping control to be determined using estimates $\hat{a}(t)$ and $\hat{b}(t)$ for a and b, respectively.

The synchronization errors between the states of the master system (25) and the slave system (26) are defined as

$$\begin{cases} e_1 = y_1 - x_1 \\ e_2 = y_2 - x_2 \\ e_3 = y_3 - x_3 \\ e_4 = y_4 - x_4 \end{cases} \tag{27}$$

Thus, the error dynamics is easily obtained as follows

$$\begin{cases} \dot{e}_1 = e_2 \\ \dot{e}_2 = e_3 \\ \dot{e}_3 = e_4 \\ \dot{e}_4 = -e_3 - ae_4 - e_2 - b(\sin\left(y_3\right) y_4 - \sin\left(x_3\right) x_4) + y_1 y_2 - x_1 x_2 + u \end{cases} \tag{28}$$

The parameter estimation errors are defined as:

$$\begin{cases} e_a(t) = a - \hat{a}(t) \\ e_b(t) = b - \hat{b}(t) \end{cases} \tag{29}$$

Differentiating (29) with respect to t, we obtain the following equations:

$$\begin{cases} \dot{e}_a(t) = -\dot{\hat{a}}(t) \\ \dot{e}_b(t) = -\dot{\hat{b}}(t) \end{cases} \tag{30}$$

Next, the main result of this section will be presented and proved.

Theorem 1 *The identical 4-D novel memristive hyperjerk systems (25) and (26) with unknown parameters a and b are globally and exponentially synchronized by the adaptive control law*

$$\begin{cases} u(t) = -5e_1 - 9e_2 - 8e_3 - (4 - \hat{a}(t))e_4 + \hat{b}(t)\left(\sin\left(y_3\right) y_4 - \sin\left(x_3\right) x_4\right) \\ \qquad - \left(y_1 y_2 - x_1 x_2\right) - kz_4 \end{cases} \tag{31}$$

where $k > 0$ is a gain constant,

$$z_4 = 3e_1 + 5e_2 + 3e_3 + e_4, \tag{32}$$

and the update law for the parameter estimates $\hat{a}(t), \hat{b}(t), \hat{c}(t)$ *is given by*

$$\begin{cases} \dot{\hat{a}}(t) = -e_4 z_4 \\ \dot{\hat{b}}(t) = -\left(\sin\left(y_3\right) y_4 - \sin\left(x_3\right) x_4\right) z_4 \end{cases} \tag{33}$$

Proof We prove this result via backstepping control method and Lyapunov stability theory.

First, we define a quadratic Lyapunov function

$$V_1(z_1) = \frac{1}{2} z_1^2 \tag{34}$$

where

$$z_1 = e_1 \tag{35}$$

Differentiating V_1 along the error dynamics (28), we get

$$\dot{V}_1 = z_1 \dot{z}_1 = e_1 e_2 = -z_1^2 + z_1(e_1 + e_2) \tag{36}$$

Here, we define

$$z_2 = e_1 + e_2 \tag{37}$$

Using (37), we can simplify the Eq. (36) as

$$\dot{V}_1 = -z_1^2 + z_1 z_2 \tag{38}$$

Secondly, we define a quadratic Lyapunov function

$$V_2(z_1, z_2) = V_1(z_1) + \frac{1}{2} z_2^2 = \frac{1}{2}\left(z_1^2 + z_2^2\right) \tag{39}$$

Differentiating V_2 along the error dynamics (28), we get

$$\dot{V}_2 = -z_1^2 - z_2^2 + z_2(2e_1 + 2e_2 + e_3) \tag{40}$$

Now, we define

$$z_3 = 2e_1 + 2e_2 + e_3 \tag{41}$$

Using (41), we can simplify the Eq. (40) as

$$\dot{V}_2 = -z_1^2 - z_2^2 + z_2 z_3 \tag{42}$$

Thirdly, we define a quadratic Lyapunov function

$$V_3(z_1, z_2, x_3) = V_2(z_1, z_2) + \frac{1}{2} z_3^2 = \frac{1}{2}\left(z_1^2 + z_2^2 + z_3^2\right) \tag{43}$$

Differentiating V_3 along the error dynamics (28), we get

$$\dot{V}_3 = -z_1^2 - z_2^2 - z_3^2 + z_3(3e_1 + 5e_2 + 3e_3 + e_4) \tag{44}$$

Now, we define

$$z_4 = 3e_1 + 5e_2 + 3e_3 + e_4 \tag{45}$$

Using (45), we can simplify the equation (44) as

$$\dot{V}_3 = -z_1^2 - z_2^2 - z_3^2 + z_3 z_4 \tag{46}$$

Finally, we define a quadratic Lyapunov function

$$V(z_1, z_2, z_3, z_4, e_a, e_b) = V_3(z_1, z_2, z_3) + \frac{1}{2} z_4^2 + \frac{1}{2} e_a^2 + \frac{1}{2} e_b^2 \tag{47}$$

which is a positive definite function on R^6.

Differentiating V along the error dynamics (28), we get

$$\dot{V} = -z_1^2 - z_2^2 - z_3^2 - z_4^2 + z_4(z_4 + z_3 + \dot{z}_4) - e_a\dot{\hat{a}} - e_b\dot{\hat{b}} \tag{48}$$

Equation (48) can be written compactly as

$$\dot{V} = -z_1^2 - z_2^2 - z_3^2 - z_4^2 + z_4 S - e_a\dot{\hat{a}} - e_b\dot{\hat{b}} \tag{49}$$

where

$$S = z_4 + z_3 + \dot{z}_4 = z_4 + z_3 + 3\dot{e}_1 + 5\dot{e}_2 + 3\dot{e}_3 + \dot{e}_4 \tag{50}$$

A simple calculation gives

$$S = 5e_1 + 9e_2 + 8e_3 + (4 - a)e_4 - b\left(\sin\left(y_3\right) y_4 - \sin\left(x_3\right) x_4\right) + \left(y_1 y_2 - x_1 x_2\right) + u \tag{51}$$

Substituting the adaptive control law (31) into (51), we obtain

$$S = -\left[a - \hat{a}(t)\right] e_4 - \left[b - \hat{b}(t)\right] \left(\sin\left(y_3\right) y_4 - \sin\left(x_3\right) x_4\right) - k z_4 \tag{52}$$

Using the definitions (30), we can simplify (52) as

$$S = -e_a e_4 - e_b\left(\sin\left(y_3\right) y_4 - \sin\left(x_3\right) x_4\right) - k z_4 \tag{53}$$

Substituting the value of S from (53) into (49), we obtain

$$\begin{cases} \dot{V} = -z_1^2 - z_2^2 - z_3^2 - (1+k)z_4^2 + e_a(-e_4 z_4 - \dot{\hat{a}}) \\ \quad +e_b \left[-\left(\sin\left(y_3\right) y_4 - \sin\left(x_3\right) x_4\right) z_4 - \dot{\hat{b}} \right] \end{cases} \tag{54}$$

Substituting the update law (33) into (54), we get

$$\dot{V} = -z_1^2 - z_2^2 - z_3^2 - (1+k)z_4^2, \tag{55}$$

which is a negative semi–definite function on R^6. Therefore, according to the Lyapunov stability theory (Sastry 1999; Khalil 2002) we obtain $e_1(t) \to 0$, $e_2(t) \to 0$, $e_3(t) \to 0$, $e_4(t) \to 0$ exponentially when $t \to 0$ that is, synchronization between master and slave system.

In order to confirm and demonstrate the effectiveness of the proposed synchronization scheme, we consider a numerical example. In the numerical simulations, the fourth–order Runge–Kutta method is used to solve the systems. The parameters of the memristive hyperjerk systems are selected as $a = 0.55$, $b = 0.2$ and the positive gain constant as $k = 4$. The initial conditions of the master system (25) and the slave system (26) have been chosen as $x_1(0) = 0$, $x_2(0) = 0.01$, $x_3(0) = 0$, $x_4(0) = 0$ and $y_1(0) = 0.005$, $y_2(0) = 0.001$, $y_3(0) = 0.02$, $y_4(0) = 0.01$, respectively. We assumed that the initial values of the parameter estimates are $\hat{a}(0) = 0.02$ and $\hat{b}(0) = 0.01$.

When adaptive control law (31) and the update law for the parameter estimates (33) are applied, the master (25) and slave system (26) are synchronized completely as shown in Figs. 12, 13, 14 and 15. In such figures, time series of master states are denoted as blue solid lines while corresponding slave states are plotted as red dash–dot lines. In addition, the time–history of the complete synchronization errors e_1, e_2, e_3, and e_4 are reported in Fig. 16. The obtained results illustrate the correctness of the used approach.

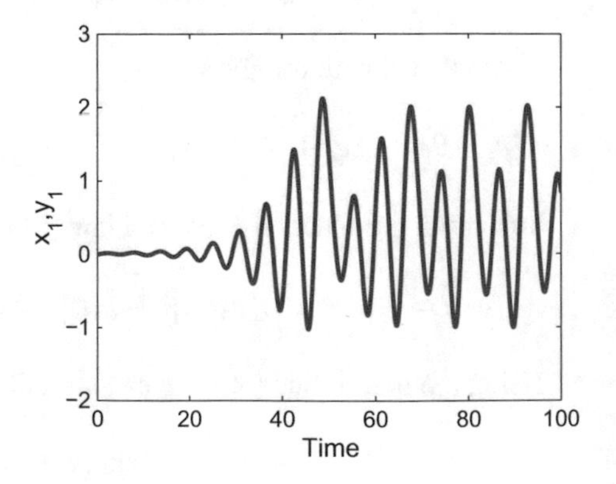

Fig. 12 Synchronization of the states $x_1(t)$ and $y_1(t)$

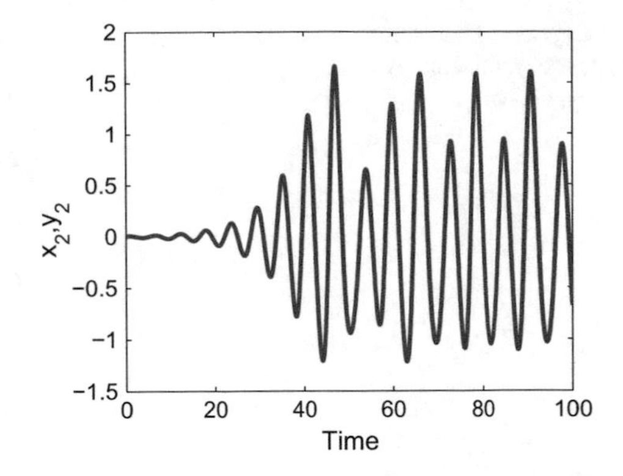

Fig. 13 Synchronization of the states $x_2(t)$ and $y_2(t)$

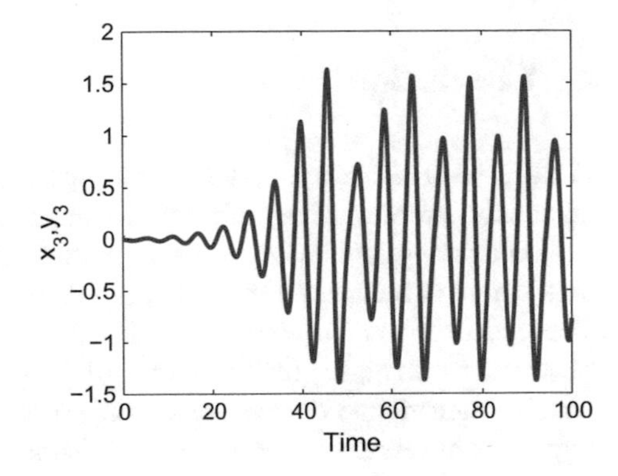

Fig. 14 Synchronization of the states $x_3(t)$ and $y_3(t)$

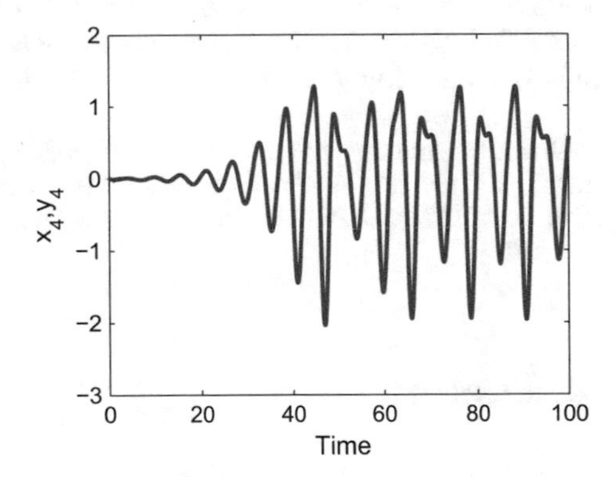

Fig. 15 Synchronization of the states $x_4(t)$ and $y_4(t)$

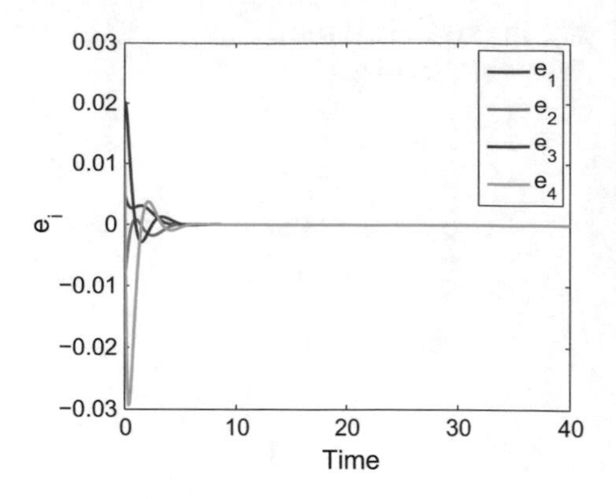

Fig. 16 Time series of the synchronization errors e_1, e_2, e_3, and e_4

6 Conclusion

A novel 4–D hyperjerk system is proposed in this work. The hyperjerk system has an infinite number of equilibrium, therefore it belongs to a newly introduced class of nonlinear systems with hidden attractors. The presence of memristive device creates complex behavior in such hyperjerk system such as chaos and multi–stability. The coexistence of chaotic attractor with periodical attractors of different states is observed by changing the value of initial conditions, which is related to the internal state of memristive device. Synchronization scheme for identical 4–D memristive hyperjerk systems with two unknown parameters has been presented via backstepping control method and Lyapunov stability theory. The simulation results confirm the effectiveness of the synchronization scheme. The hyperjerk system can used in chaos–based engineering applications due to its complex dynamics.

Acknowledgements The author Xiong Wang was supported by the National Natural Science Foundation of China (No. 61601306) and Shenzhen Overseas High Level Talent Peacock Project Fund (No. 20150215145C). V.-T. Pham is grateful to Le Thi Van Thu, Philips Electronics—Vietnam, for her help.

References

Arneodo, A., Coullet, P., & Tresser, C. (1981). Possible new strange attractors with spiral structure. *Communications in Mathematical Physics*, *79*, 573–579.

Azar, A. T., & Vaidyanathan, S. (2015a). *Chaos modeling and control systems design*. New York: Springer.

Azar, A. T., & Vaidyanathan, S. (2015b). *Computational intelligence applications in modeling and control*. New York: Springer.

Bao, B., Zou, X., Liu, Z., & Hu, F. (2013). Generalized memory element and chaotic memory system. *International Journal of Bifurcation and Chaos*, *23*, 1350135–12.

Boccaletti, S., Kurths, J., Osipov, G., Valladares, D. L., & Zhou, C. S. (2002). The synchronization of chaotic systems. *Physics Reports*, *366*, 1–101.

Buscarino, A., Fortuna, L., & Frasca, M. (2009). Experimental robust synchronization of hyperchaotic circuits. *Physica D*, *238*, 1917–1922.

Buscarino, A., Fortuna, L., Frasca, M., Gambuzza, L. V., & Sciuto, G. (2012). Memristive chaotic circuits based on cellular nonlinear networks. *International Journal of Bifurcation and Chaos*, *22*, 1250070-1–13.

Chen, G., & Ueta, T. (1999). Yet another chaotic attractor. *International Journal of Bifurcation and Chaos*, *9*, 1465–1466.

Chen, G., & Yu, X. (2003). *Chaos control: Theory and applications*. Berlin: Springer.

Chlouverakis, K. E., & Sprott, J. C. (2006). Chaotic hyperjerk systems. *Chaos, Solitons Fractals*, *28*, 739–746.

Chua, L. O. (1971). Memristor-the missing circuit element. *IEEE Transactions on Circuit Theory*, *18*, 507–519.

Chua, L. O. (2011). Resistance switching memories are memristors. *Applied Physics A*, *102*(4), 765–783.

Chua, L. O., & Kang, S. M. (1976). Memristive devices and system. *Proceedings of the IEEE*, *64*, 209–223.

Driscoll, T., Pershin, Y. V., Basov, D. N., & Ventra, M. D. (2011). Chaotic memristor. *Applied Physics A*, *102*, 885–889.

Driscoll, T., Quinn, J., Klien, S., Kim, H. T., Kim, B. J., Pershin, Y. V., et al. (2010). Memristive adaptive filters. *Applied Physics Letters*, *97*, 093502.

Eichhorn, R., Linz, S. J., & Hanggi, P. (2002). Simple polynomial classes of chaotic jerky dynamics. *Chaos Solitons Fractals*, *13*, 1–15.

Elhadj, Z., & Sprott, J. C. (2013). Transformation of 4-D dynamical systems to hyperjerk form. *Palestine Journal of Mathematics*, *2*, 38–45.

Fortuna, L., & Frasca, M. (2007). Experimental synchronization of single-transistor-based chaotic circuits. *Chaos*, *17*, 043118-1–5.

Gamez-Guzman, L., Cruz-Hernandez, C., Lopez-Gutierrez, R., & Garcia-Guerrero, E. E. (2009). Synchronization of chua's circuits with multi-scroll attractors: Application to communication. *Communications in Nonlinear Science and Numerical Simulation*, *14*, 2765–2775.

Huang, Y., Wang, Y., Chen, H., & Zhang, S. (2016). Shape synchronization control for three-dimensional chaotic systems. *Chaos, Solitons & Fractals*, *87*, 136–145.

Itoh, M., & Chua, L. O. (2008). Memristor oscillators. *International Journal of Bifurcation and Chaos*, *18*, 3183–3206.

Itoh, M., & Chua, L. O. (2009). Memristor cellular automata and memristor discrete time cellular neural networks. *International Journal of Bifurcation and Chaos*, *19*, 3605–3656.

Iu, H. H. C., Yu, D. S., Fitch, A. L., Sreeram, V., & Chen, H. (2011). Controlling chaos in a memristor based circuit using a twin-T notch filter. *IEEE Transactions on Circuits and Systems I, Regular Paper*, *58*, 1337–1344.

Jafari, S., & Sprott, J. C. (2013). Simple chaotic flows with a line equilibrium. *Chaos, Solitons Fractals*, *57*, 79–84.

Jafari, S., Sprott, J. C., & Golpayegani, S. M. R. H. (2013). Elementary quadratic flows with no equilibria. *Physics Letters A*, 377:699–702.

Kajbaf, A., Akhaee, M. A., & Sheikhan, M. (2016). Fast synchronization of non-identical chaotic modulation-based secure systems using a modified sliding mode controller. *Chaos, Solitons & Fractals*, *84*, 49–57.

Kapitaniak, T. (1994). Synchronization of chaos using continuous control. *Physical Review E*, *50*, 1642–1644.

Karthikeyan, R., & Vaidyanathan, S. (2014). Hybrid chaos synchronization of four-scroll systems via active control. *Journal of Electrical Engineering*, *65*, 97–103.

Kengne, J., Njitacke, Z. T., & Fotsin, H. (2016). Dynamical analysis of a simple autonomous jerk system with multiple attractors. *Nonlinear Dynamics*, *83*, 751–765.

Khalil, H. (2002). *Nonlinear systems*. New Jersey, USA: Prentice Hall.
Kingni, S. T., Jafari, S., Simo, H., & Woafo, P. (2014). Three-dimensional chaotic autonomous system with only one stable equilibrium: Analysis, circuit design, parameter estimation, control, synchronization and its fractional-order form. *The European Physical Journal Plus*, *129*, 76.
Lainscsek, C., Lettellier, C., & Gorodnitsky, I. (2003). Global modeling of the rössler system from the z-variable. *Physics Letters A*, *314*, 409–427.
Leonov, G. A., & Kuznetsov, N. V. (2011). Algorithms for searching for hidden oscillations in the Aizerman and Kalman problems. *Doklady Mathematics*, *84*, 475–481.
Leonov, G. A., & Kuznetsov, N. V. (2013). Hidden attractors in dynamical systems: From hidden oscillation in Hilbert-Kolmogorov, Aizerman and Kalman problems to hidden chaotic attractor in Chua circuits. *International Journal of Bifurcation and Chaos*, *23*, 1330002.
Leonov, G. A., Kuznetsov, N. V., Kiseleva, M. A., Solovyeva, E. P., & Zaretskiy, A. M. (2014). Hidden oscillations in mathematical model of drilling system actuated by induction motor with a wound rotor. *Nonlinear Dynamics*, *77*, 277–288.
Leonov, G. A., Kuznetsov, N. V., Kuznetsova, O. A., Seldedzhi, S. M., & Vagaitsev, V. I. (2011a). Hidden oscillations in dynamical systems. *Transactions on Systems and Control*, *6*, 54–67.
Leonov, G. A., Kuznetsov, N. V., & Vagaitsev, V. I. (2011b). Localization of hidden Chua's attractors. *Physics Letters A*, *375*, 2230–2233.
Leonov, G. A., Kuznetsov, N. V., & Vagaitsev, V. I. (2012). Hidden attractor in smooth Chua system. *Physica D*, *241*, 1482–1486.
Linz, S. J. (1997). Nonlinear dynamical models and jerky motion. *American Journal of Physics*, *65*, 523–526.
Liu, C., Yi, J., Xi, X., An, L., & Fu, Y. (2012). Research on the multi-scroll chaos generation based on Jerk mode. *Procedia Engineering*, *29*, 957–961.
Lorenz, E. N. (1963). Deterministic non-periodic flow. *Journal of the Atmospheric Sciences*, *20*, 130–141.
Louodop, P., Kountchou, M., Fotsin, H., & Bowong, S. (2014). Practical finite-time synchronization of jerk systems: Theory and experiment. *Nonlinear Dynamics*, *78*, 597–607.
Lü, J., & Chen, G. (2002). A new chaotic attractor coined. *International Journal of Bifurcation and Chaos*, *12*, 659–661.
Ma, J., Wu, X., Chu, R., & Zhang, L. (2014). Selection of multi-scroll attractors in Jerk circuits and their verification using Pspice. *Nonlinear Dynamics*, *76*, 1951–1962.
Malasoma, J. M. (2000). What is the simplest dissipative chaotic jerk equation which is parity invariant. *Physics Letters A*, *264*, 383–389.
Molaei, M., Jafari, S., Sprott, J. C., & Golpayegani, S. (2013). Simple chaotic flows with one stable equilibrium. *International Journal of Bifurcation and Chaos*, *23*, 1350188.
Munmuangsaen, B., Srisuchinwong, B., & Sprott, J. C. (2011). Generalization of the simplest autonomous chaotic system. *Physics Letters A*, *375*, 1445–1450.
Muthuswamy, B. (2010). Implementing memristor based chaotic circuits. *International Journal of Bifurcation and Chaos*, *20*, 1335–1350.
Muthuswamy, B., & Chua, L. O. (2010). Simplest chaotic circuits. *Journal of Bifurcation and Chaos*, *20*, 1567–1580.
Muthuswamy, B., & Kokate, P. P. (2009). Memristor-based chaotic circuits. *IETE Technical Review*, *26*, 415–426.
Pecora, L. M., & Carroll, T. L. (1990). Synchronization in chaotic signals. *Physical Review A*, *64*, 821–824.
Pehlivan, I., Moroz, I., & Vaidyanathan, S. (2014). Analysis, synchronization and circuit design of a novel butterfly attractor. *Journal of Sound and Vibration*, *333*, 5077–5096.
Pershin, Y. V., Fontaine, S. L., & Ventra, M. D. (2009). Memristive model of amoeba learning. *Physical Review E*, *80*, 021926.

Pham, V. T., Rahma, F., Frasca, M., & Fortuna, L. (2014a). Dynamics and synchronization of a novel hyperchaotic system without equilibrium. *International Journal of Bifurcation and Chaos*, *24*, 1450087–11.

Pham, V.-T., Volos, C., Jafari, S., Wang, X., & Vaidyanathan, S. (2014b). Hidden hyperchaotic attractor in a novel simple memristive neural network. *Optoelectronics and Advanced Materials, Rapid Communications*, *8*, 1157–1163.

Rasappan, S., & Vaidyanathan, S. (2014). Global chaos synchronization of WINDMI and Coullet chaotic systems using adaptive backstepping control design. *Kyungpook Mathematical Journal*, *54*, 293–320.

Rössler, O. E. (1976). An equation for continuous chaos. *Physics Letters A*, *57*, 397–398.

Sastry, S. (1999). *Nonlinear systems: Analysis, stability, and control*. USA: Springer.

Schot, S. (1978). Jerk: The time rate of change of acceleration. *American Journal of Physics*, *46*, 1090–1094.

Shin, S., Kim, K., & Kang, S. M. (2011). Memristor applications for programmable analog ICs. *IEEE Transactions on Nanotechnology*, *410*, 266–274.

Sprott, J. C. (1997). Some simple chaotic jerk functions. *American Journal of Physics*, *65*, 537–543.

Sprott, J. C. (2003). *Chaos and times-series analysis*. Oxford: Oxford University Press.

Sprott, J. C. (2010). *Elegant chaos: Algebraically simple chaotic flows*. Singapore: World Scientific.

Sprott, J. C. (2011). A new chaotic jerk circuit. *IEEE Transactions on Circuits and Systems II: Express Briefs*, *58*, 240–243.

Srinivasan, K., Senthilkumar, D. V., Murali, K., Lakshmanan, M., & Kurths, J. (2011). Synchronization transitions in coupled time-delay electronic circuits with a threshold nonlinearity. *Chaos*, *21*, 023119.

Strogatz, S. H. (1994). *Nonlinear dynamics and chaos: With applications to Physics, Biology, Chemistry, and Engineering*. Massachusetts: Perseus Books.

Sun, K. H., & Sprott, J. C. (2009). A simple jerk system with piecewise exponential nonlinearity. *International Journal of Nonlinear Sciences and Numerical Simulation*, *10*, 1443–1450.

Tetzlaff, R. (2014). *Memristor and memristive systems*. New York: Springer.

Vaidyanathan, S. (2012). Anti-synchronization of four-wing chaotic systems via sliding mode control. *International Journal of Automation and Computing*, *9*, 274–279.

Vaidyanathan, S. (2013). A new six-term 3-D chaotic system with an exponential nonleariry. *Far East Journal of Mathematical Sciences*, *79*, 135–143.

Vaidyanathan, S. (2014). Analysis and adaptive synchronization of eight-term novel 3-D chaotic system with three quadratic nonlinearities. *European Physical Journal Special Topics*, *223*, 1519–1529.

Vaidyanathan, S., Volos, C., Pham, V. T., & Madhavan, K. (2015). Analysis, adaptive control and synchronization of a novel 4-D hyperchaotic hyperjerk system and its SPICE implementation. *Archives of Control Sciences*, *25*, 135–158.

Vaidyanathan, S., Volos, C., Pham, V. T., Madhavan, K., & Idowo, B. A. (2014). Adaptive backstepping control, synchronization and circuit simualtion of a 3-D novel jerk chaotic system with two hyperbolic sinusoidal nonlinearities. *Archives of Control Sciences*, *33*, 257–285.

Volos, C. K., Kyprianidis, I. M., & Stouboulos, I. N. (2011). Various synchronization phenomena in bidirectionally coupled double scroll circuits. *Communications in Nonlinear Science and Numerical Simulation*, *71*, 3356–3366.

Volos, C. K., Kyprianidis, I. M., & Stouboulos, I. N. (2013). Image encryption process based on chaotic synchronization phenomena. *Signal Processing*, *93*, 1328–1340.

Wang, X., & Chen, G. (2012). A chaotic system with only one stable equilibrium. *Communications in Nonlinear Science and Numerical Simulation*, *17*, 1264–1272.

Wang, X., & Chen, G. (2013). Constructing a chaotic system with any number of equilibria. *Nonlinear Dynamics*, *71*, 429–436.

Wang, Z., Sun, W., Wei, Z., & Zhang, S. (2015). Dynamical and delayed feedbacl control for a 3D jerk system with hidden attractor. *Nonlinear Dynamics*, *82*, 577–588.

Yalcin, M. E., & Ozoguz, S. (2007). n-scroll chaotic attractors from a first-order time-delay differential equation. *Chaos*, *17*, 033112-1–8.
Yalcin, M. E., Suykens, J. A. K., & Vandewalle, J. (2005). *Cellular neural networks, multi-scroll chaos and synchronization*. Singapore: World Scientific.
Yu, S., Lü, J., Leung, H., & Chen, G. (2005). Design and implementation of n-scroll chaotic attractors from a general Jerk circuit. *IEEE Transactions on Circuits and Systems I*, *52*, 1459–1476.

A Memristive System with Hidden Attractors and Its Engineering Application

Viet-Thanh Pham, Sundarapandian Vaidyanathan, Christos Volos, Esteban Tlelo-Cuautle and Fadhil Rahma Tahir

Abstract After the successful fabrication of memristor at Hewlett–Packard Laboratories, memristor—based systems and their potential applications have been getting a great deal of attention in different areas from associative memory, neural networks, programmable analog ICs to low–power computing and so on. It is well known that the presence of memristor in a dynamical system may yield novel features because it is both a nonlinear element and a memory element. In this chapter, we present a memristive system with an infinite number of equilibrium points. From the computing view of point, such system belongs to a class of systems with hidden attractors according to a new classification of nonlinear dynamics. This classification has proposed by Leonov and Kuznetsov and played a significant role in engineering applications. In this work, we study the complex dynamics of the introduced memristive system. It is worth noting that the proposed system can generate hyperchaotic behavior which will be used for image encryption to illustrate its engineering application. The chaos–based image encryption has many applications in digital image storing, medical image databases, video conferencing or military transmit systems.

V.-T. Pham (✉)
School of Electronics and Telecommunications, Hanoi University of Science and Technology, Hanoi, Vietnam
e-mail: pvt3010@gmail.com

S. Vaidyanathan
Research and Development Centre, Vel Tech University, Chennai, Tamil Nadu, India
e-mail: sundarcontrol@gmail.com

C. Volos
Physics Department, Aristotle University of Thessaloniki, Thessaloniki, Greece
e-mail: volos@physics.auth.gr

E. Tlelo-Cuautle
Department of Electronics, INAOE, Tonantzintla Puebla 72840, Mexico
e-mail: etlelo@inaoep.mx

F.R. Tahir
Department of Electrical Engineering, College of Engineering University of Basrah, Basrah, Iraq
e-mail: fadhilrahma.creative@gmail.com

© Springer International Publishing AG 2017

S. Vaidyanathan and C. Volos (eds.), *Advances in Memristors, Memristive Devices and Systems*, Studies in Computational Intelligence 701,
DOI 10.1007/978-3-319-51724-7_4

Keywords Chaos · Hyperchaos · Lyapunov exponents · Bifurcation · Hidden attractor · Equilibrium · Memristor · Encryption · Security

1 Introduction

In the past few decades, various chaotic systems have been investigated, for example: Lorenz system (Lorenz 1963), Rössler system (Rössler 1976), Arneodo system (Arneodo et al. 1981), Chen system (Chen and Ueta 1999), Lü system (Lü and Chen 2002), Vaidyanathan system (Vaidyanathan 2013), time–delay systems (Barnerjee et al. 2012), Tacha system (Tacha et al. 2016), jerk systems (Vaidyanathan et al. 2014). In addition, hyperchaotic systems have been discovered (Rössler 1979). Hyperchaotic system is characterized by more than one positive Lyapunov exponent. Therefore hyperchaotic system can exhibit a higher level of complexity with respect to chaotic system (Vaidyanathan and Azar 2015). Hyperchaos is better than conventional chaos in a variety of areas, for example, hyperchaos increases the security of chaotic–based communication systems significantly (Udaltsov et al. 2003; Sadoudi et al. 2013) or encryption algorithm based on hyperchaos is safer than one based on chaos (Gao and Chen 2008). Moreover hyperchaos has been applied in different areas such as cryptosystems (Grassi and Mascolo 1999), neural networks (Huang and Yang 2006), secure communications (Udaltsov et al. 2003; Sadoudi et al. 2013), or laser design (Vicente et al. 2005).

The realization of memristor at Hewlett–Packard Labs promotes potential memristor—based applications (Strukov et al. 2008). Some attractive memristor—based applications are high–speed low–power processors (Yang et al. 2013), adaptive filter (Driscoll et al. 2010), pattern recognition systems (Corinto et al 2012), associative memory (Pershin and Ventra 2010), neural networks (Adhikari et al. 2012; Ascoli and Corinto 2013), and programmable analog integrated circuits (Shin et al. 2011). Especially, the intrinsic nonlinear characteristic of memristor has been exploited in designing hyperchaotic oscillators (Buscarino et al. 2012a; Fitch et al. 2012). Hyperchaos was generated by combining a memristor with cubic nonlinear characteristics and a modified canonical Chua's circuit (Fitch et al. 2012). This memristor—based modified canonical Chua's circuit is a five–dimensional hyperchaotic oscillator. By extending the HP memristor—based canonical Chua's oscillator, a six–dimensional hyperchaotic oscillator was designed (Buscarino et al. 2012b). Authors used a configuration based on two HP memristors in antiparallel (Buscarino et al. 2012a). Four–dimensional hyperchaotic memristive systems were discovered by Li et al. (2014, 2015). A 4D memristive system with a line of equilibrium points was presented in Li et al. (2014) while another memristive system with an uncountable infinite number of stable and unstable equilibria was reported in Li et al. (2015). Interestingly, memristor—based hyperchaotic systems without equilibrium were introduced in Pham et al. (2014b; 2015). These memristive systems belong to a new category of chaotic systems with hidden attractors (Leonov et al. 2011; Leonov and Kuznetsov 2013). Hidden attractor cannot be found by using a numerical method in which a trajectory started from a point on the unstable manifold in the neighbourhood of an unstable equilibrium (Jafari and Sprott 2013). Thus hidden attractors play an important role in many fields such as in mechanics, secure communication and electronics (Kuznetsov et al. 2011; Leonov et al. 2011; Pham et al. 2014a, c; Sharma et al. 2015).

In this chapter, we study a system based on a memristive device and its application. It is interesting that the memristive system has an infinite number of equilibrium points and can generate hyperchaos. This chapter is organized as follows. Section 2 presents the description of the memristive system. Dynamics and properties of such memristive system are investigated in Sect. 3. We implement an image encryption scheme based on the memeristive system in Sect. 4 and discuss its security in Sect. 5. Finally, conclusions are drawn in Sect. 6.

2 Description of the Proposed System

According to studies of (Chua 1971; Chua and Kang 1976), a memristive system is defined by

$$\begin{cases} \dot{w} = f(w, y, t) \\ h(w, y) = g(w, y, t)\, y, \end{cases} \tag{1}$$

where y, $h(w, y)$, w are the input, output, and internal state of the memristive device. The functions f and g are a continuous n–dimensional vector function and a continuous scalar function.

Based on the definition of memristive system, recently authors have introduced a novel memrisitve system (Pham et al. 2014b) in the following form:

$$\begin{cases} \dot{x} = -10x - ay - yz \\ \dot{y} = -6x + 1.2xz + 0.1h(w, y) + b \\ \dot{z} = -z - 1.2xy \\ \dot{w} = y, \end{cases} \tag{2}$$

where a and b are two positive real parameters. Here $h(w, y)$ is the output of the memristive device described by

$$\begin{cases} \dot{w} = y \\ h(w, y) = \left(1 + 0.24w^2 - 0.0016w^4\right) y. \end{cases} \tag{3}$$

As have been known system (2) can display hyperchaotic attractors without equilibrium for $b \neq 0$ (Pham et al. 2014b). In addition, dynamics of the memristive system without equilibrium have been investigated by using numerical simulations and circuital implementation (Pham et al. 2014b).

When $b = 0$, the system (2) can be rewritten by

$$\begin{cases} \dot{x} = -10x - ay - yz \\ \dot{y} = -6x + 1.2xz + 0.1\left(1 + 0.24w^2 - 0.0016w^4\right)y \\ \dot{z} = -z - 1.2xy \\ \dot{w} = y. \end{cases} \tag{4}$$

It is easy to get the equilibrium points of system (4) by solving $\dot{x} = 0$, $\dot{y} = 0$, $\dot{z} = 0$, and $\dot{w} = 0$, that is

$$-10x - ay - yz = 0, \tag{5}$$

$$-6x + 1.2xz + 0.1\left(1 + 0.24w^2 - 0.0016w^4\right)y = 0, \tag{6}$$

$$-z - 1.2xy = 0, \tag{7}$$

$$y = 0, \tag{8}$$

From Eq. (8), we have $y = 0$. By substituting $y = 0$ in Eq. (5), it leads to $x = 0$. As result, we get $z = 0$ from Eq. (7). In addition, Eq. (6) insists and does not depend on w. In other words, system (4) has an infinite number of equilibrium points

$$E\left(0, 0, 0, w\right). \tag{9}$$

Moreover, the equilibrium points are located on a line.

System (4) belongs to a new class of systems with hidden attractor from a computational point of view (Jafari and Sprott 2013). The basin of the system may intersect the line equilibrium in some sections. But there are uncountable points on the line that are outside the basin of attraction. Thus the knowledge about equilibrium points does not help in their localization (Jafari and Sprott 2013).

In the next section, we present complex dynamics of the memristive system with infinite equilibria (4).

3 Dynamics of the Memristive System

When choosing the parameter $a = 5$ and the initial condition

$$\left(x\left(0\right), y\left(0\right), z\left(0\right), w\left(0\right)\right) = \left(0, 0.01, 0.01, 0\right), \tag{10}$$

the calculated Lyapunov exponents of system (4) are

$$L_1 = 0.1364,\ L_2 = 0.0071,\ L_3 = 0,\ L_4 = -10.8584. \tag{11}$$

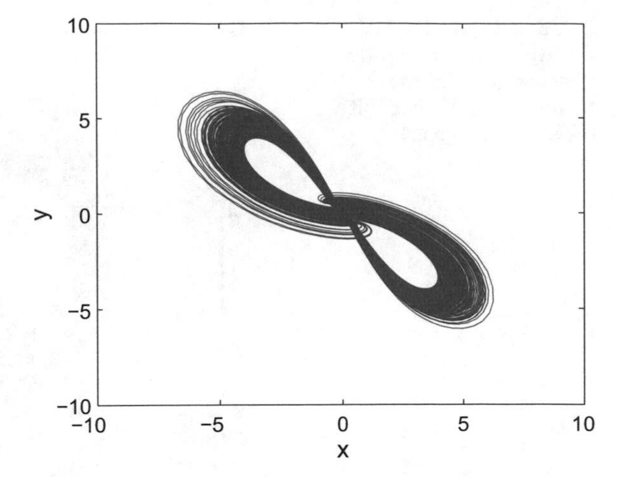

Fig. 1 2–D projection of the hyperchaotic memristive system with an infinite number of equilibrium points (4) in the (x, y)–plane

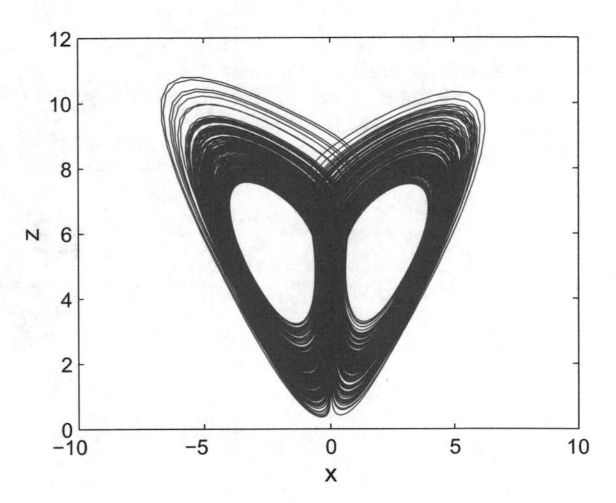

Fig. 2 2–D projection of the hyperchaotic memristive system with an infinite number of equilibrium points (4) in the (x, z)–plane

Therefore, system (4) is a four–dimensional hyperchaotic system because there are two positive Lyapunov exponents, one zero and one negative Lyapunov exponent. Figures 1, 2, 3 and 4 display 2–D projections of the hyperchaotic attractors with an infinite number of equilibrium points.

It has been known that the Kaplan–Yorke fractional dimension presenting the complexity of attractor is given by

$$D_{KY} = j + \frac{1}{\left|L_{j+1}\right|} \sum_{i=1}^{j} L_i, \tag{12}$$

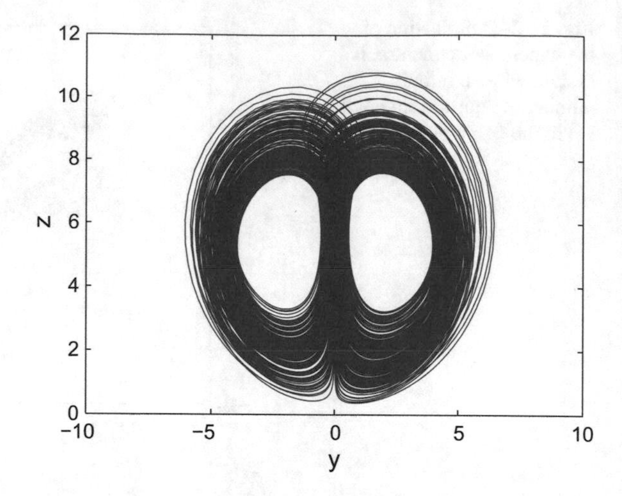

Fig. 3 2–D projection of the hyperchaotic memristive system with an infinite number of equilibrium points (4) in the (y, z)–plane

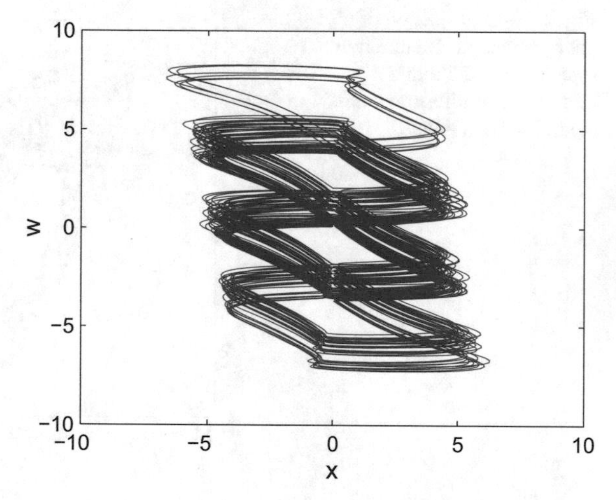

Fig. 4 2–D projection of the hyperchaotic memristive system with an infinite number of equilibrium points (4) in the (x, w)–plane

where j is the largest integer satisfying $\sum_{i=1}^{j} L_i \geq 0$ and $\sum_{i=1}^{j+1} L_i < 0$. The calculated Kaplan–Yorke fractional dimension of system (4) for $a = 5$ is

$$D_{KY} = 3 + \frac{L_1 + L_2 + L_3}{|L_4|} = 3.0132, \tag{13}$$

which indicated a strange attractor. Poincaré maps of system (4) are also illustrated in Figs. 5, 6 and 7. As can be seen from the Poincaré maps, the memristive system (4) has complex dynamics.

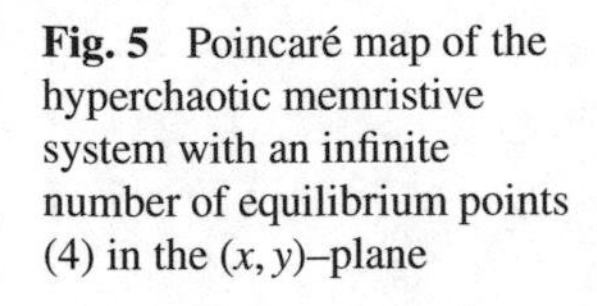

Fig. 5 Poincaré map of the hyperchaotic memristive system with an infinite number of equilibrium points (4) in the (x, y)–plane

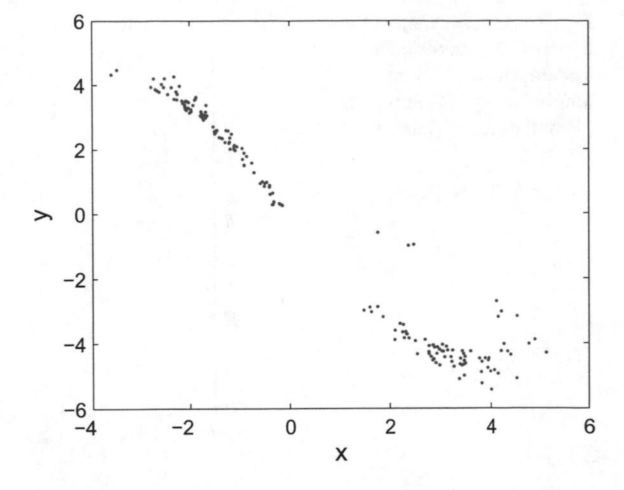

Fig. 6 Poincaré map of the hyperchaotic memristive system with an infinite number of equilibrium points (4) in the (x, z)–plane

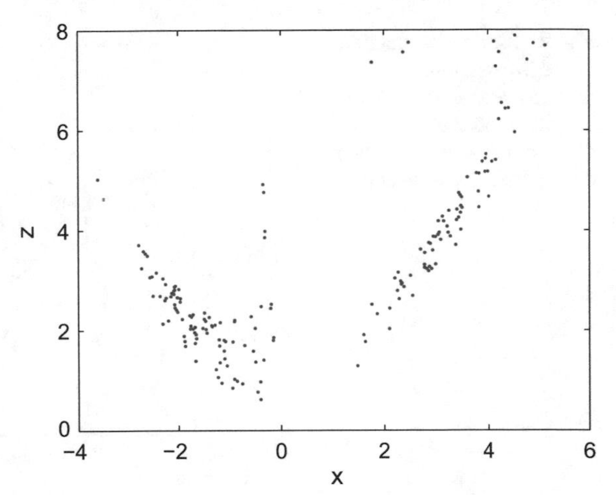

We discover detail dynamics of system (4) by using bifurcation diagram and Lyapunov exponents. The bifurcation diagram of the variable z and the corresponding Lyapunov exponents are reported in Figs. 8, 9 and 10. The system (4) performs periodic state, chaos, and hyperchaos when varying the parameter a from 1 to 6. The system displays limit cycle for $a \in [1, 1.46]$, $[1.5, 1.96]$, $[2.48, 2.88]$. For example, periodical states of system (4) for $a = 2.6$ are presented in Figs. 11, 12, 13 and 14. Chaotic behavior can be observed for $a \in [3.06, 3.34]$, $(4.66, 4.78)$. The system can exhibits hyperchaotic behavior for $a \in [4.08, 4.66]$, $[4.78, 6]$.

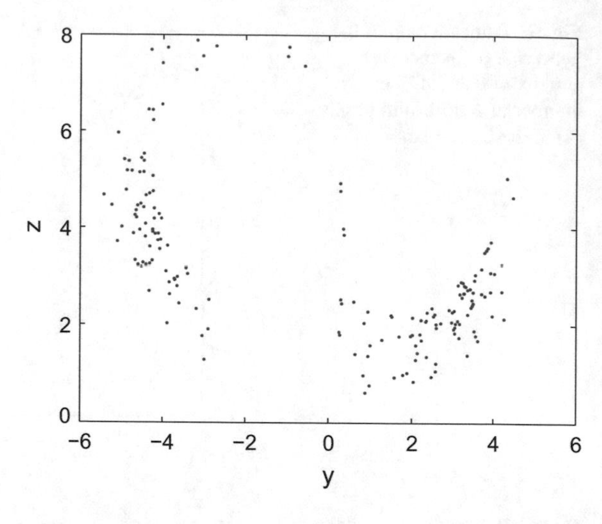

Fig. 7 Poincaré map of the hyperchaotic memristive system with an infinite number of equilibrium points (4) in the (y, z)–plane

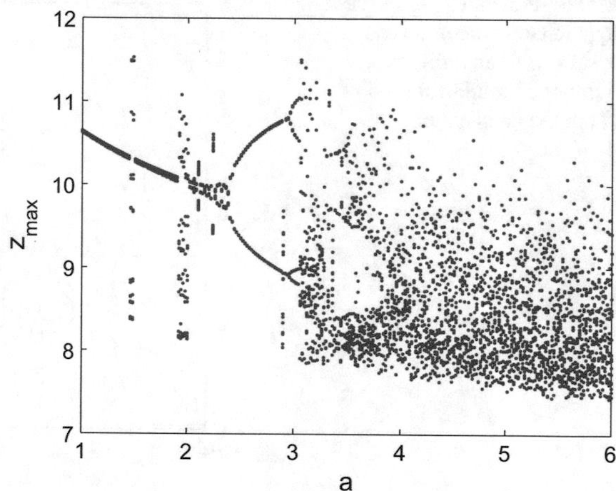

Fig. 8 Bifurcation diagram of the hyperchaotic memristive system with an infinite number of equilibrium points (4) when changing the parameter a

4 Application of the Proposed System

Nowadays, digital image information has become popular in the world because of the rapid development of Internet. In many applications such as military images, online personal photographs, or fingerprint images of authentication systems, it has to meet the requirements of safety and security (Volos et al. 2013). Therefore, numerous encryption techniques, especially chaos–based encryption, have been proposed and implemented (Liao et al. 2010; Matthews 1989; Seyedzadeh et al. 2012; Tong and Cui 2009; Wang et al. 2012; Yeung and Pankanti 2000; Zhang et al. 2005).

In this section, we use the encryption scheme suggested by Gao and Chen (Gao and Chen 2008) to illustrate a possible application of the proposed memristive sys-

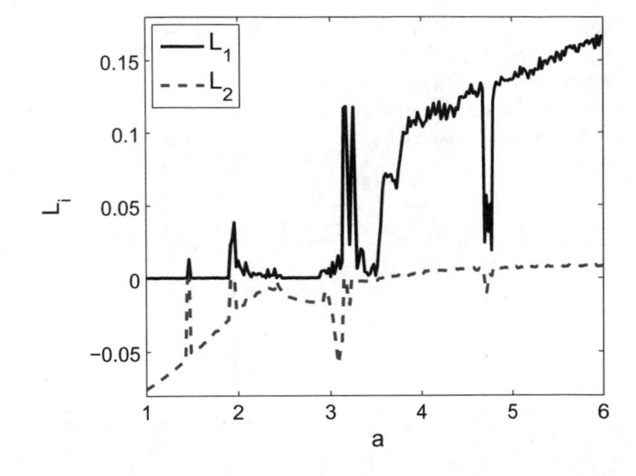

Fig. 9 Two largest Lyanpunov exponents of the hyperchaotic memristive system with an infinite number of equilibrium points (4) when varying the parameter a

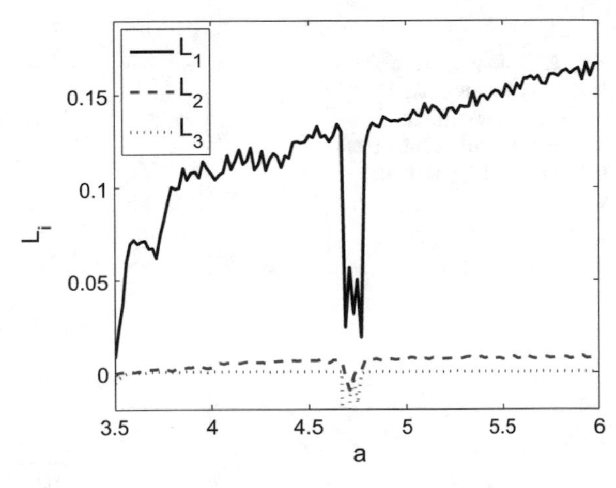

Fig. 10 Three largest Lyanpunov exponents of the hyperchaotic memristive system with an infinite number of equilibrium points (4) when varying the parameter a

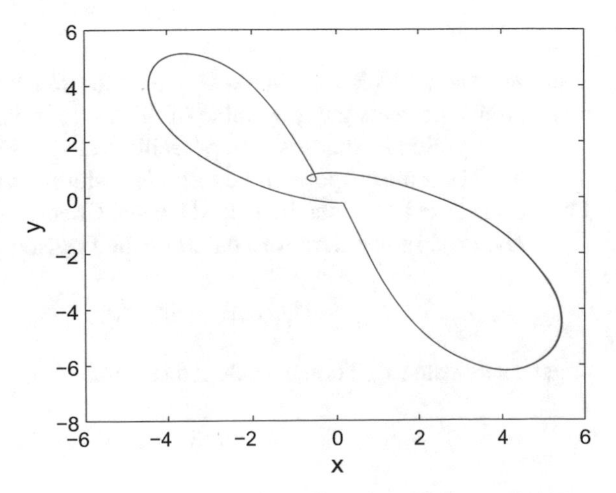

Fig. 11 Limit cycle of the hyperchaotic memristive system with an infinite number of equilibrium points (4) in the (x, y)–plane for $a = 2.6$

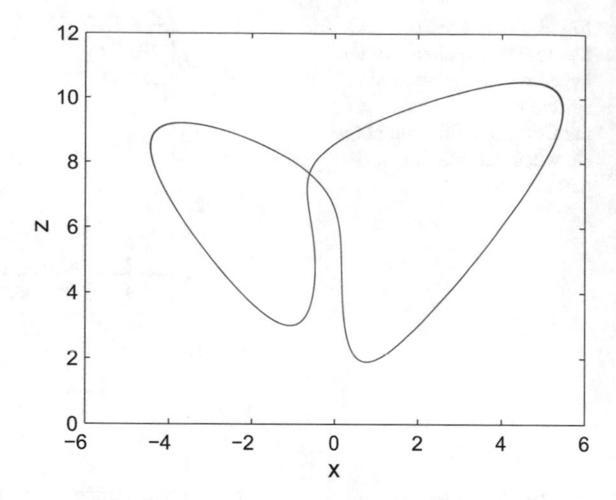

Fig. 12 Limit cycle of the hyperchaotic memristive system with an infinite number of equilibrium points (4) in the (x, z)–plane for $a = 2.6$

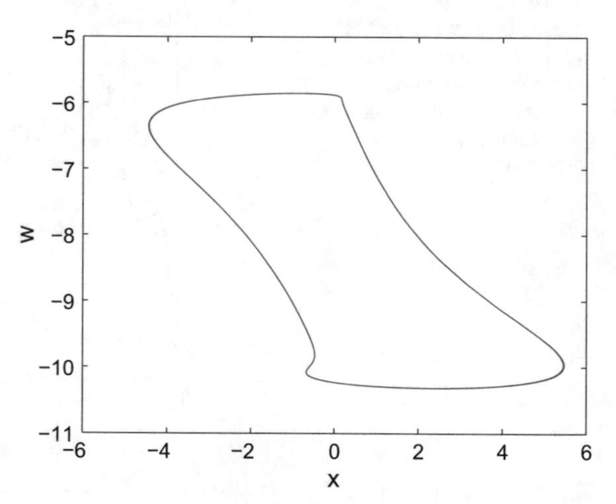

Fig. 13 Limit cycle of the hyperchaotic memristive system with an infinite number of equilibrium points (4) in the (x, w)–plane for $a = 2.6$

tem. We consider a plain–image with the dimension $N \times M$. The position matrix of pixels, which presents the grey value of the image is denoted as $P_{i,j}(I)$.

The encryption includes two steps as illustrated in Fig. 15.

Step 1: The main purpose of the step 1 is to shuffle the position of the plain image. This step is based on a chaotic map (Gao and Chen 2008).

Firstly, we do some iterations based on the Logistic map

$$x_{n+1} = 4x_n \left(1 - x_n\right), \tag{14}$$

to get a new value x_0. Then we calculate a value of h:

$$h = \text{mod}\left(x_0 \times 10^{14}, M\right), \tag{15}$$

in which the function mod(.) returns the remainder after division.

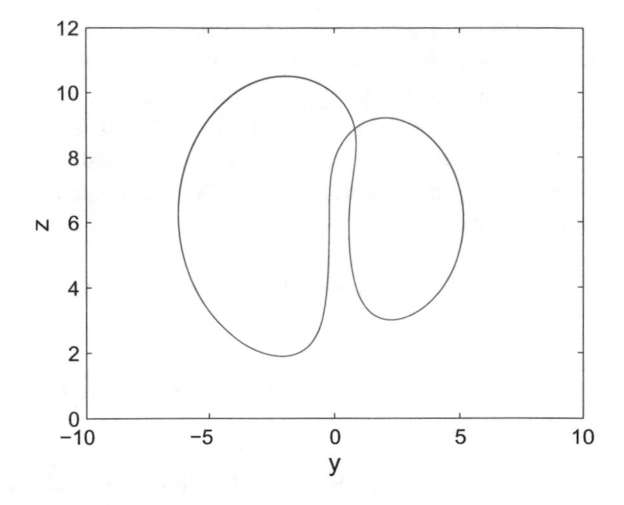

Fig. 14 Limit cycle of the hyperchaotic memristive system with an infinite number of equilibrium points (4) in the (y, z)–plane for $a = 2.6$

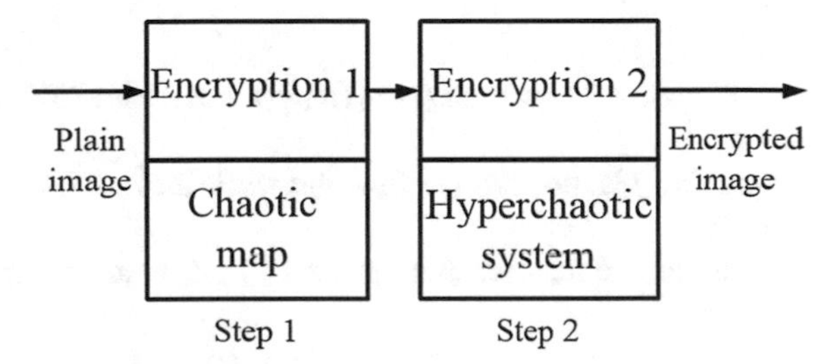

Fig. 15 Block diagram of the encryption scheme including two steps. The first step is based on a chaotic map while the second step is based on a hyperchaotic system

Secondly, we obtain M different data by repeating (15). These obtained data are reordered in $\{h_i,\ i = 1, 2, \dots, M\}$ where $h_i \neq h_j$ if $i \neq j$. Then the rows of position matrix $P_{i,j}$ are rearranged by using $\{h_i,\ i = 1, 2, \dots, M\}$. In other words, we create a new image position matrix $P^h_{i,j}$ based on row transformation. After that we shuffle the column position of the image for every row of the new position matrix $P^h_{i,j}$ with the same approach.

Thirdly, we do the iteration of Logistic map to calculate the value of l by using:

$$l = \mathrm{mod}\left(x_0 \times 10^{14}, N\right). \tag{16}$$

Fourthly, we repeat the iteration of Logistic map and (16) to get N different data. These data are reordered in $\{l_i,\ i = 1, 2, \dots, N\}$ where $l_i \neq l_j$ if $i \neq j$. Next the data of every column of position matrix $P^h_{i,j}$ are rearranged by using $\{l_i,\ i = 1, 2, \dots, N\}$. In other words, we create a new column transformation of the first row of image position matrix $P^{hl}_{i,j}$.

Finally, by completing the column transformation for all rows, an image total shuffling matrix $P^{hl}_{i,j}$ is derived.

Step 2: The main purpose of the step 2 is to encrypt the shuffled image. This step is based on a hyperchaotic system (Gao and Chen 2008).

Firstly, we iterate the hyperchaotic memristive system (4) for N_0 times by applying the Runge–Kutta algorithm to eliminate the effect of transient procedure.

Secondly, we iterate the hyperchaotic memristive system (4) to get four state variables x, y, z, and w at the N_0 time. Four corresponding decimal fractions x, y, z, w are generated as

$$x = \mathrm{mod}\left((\mathrm{abs}\,(x) - \mathrm{floor}\,(\mathrm{abs}\,(x))) \times 10^{14}, 256\right), \tag{17}$$

$$y = \mathrm{mod}\left((\mathrm{abs}\,(y) - \mathrm{floor}\,(\mathrm{abs}\,(y))) \times 10^{14}, 256\right), \tag{18}$$

$$z = \mathrm{mod}\left((\mathrm{abs}\,(z) - \mathrm{floor}\,(\mathrm{abs}\,(z))) \times 10^{14}, 256\right), \tag{19}$$

$$w = \mathrm{mod}\left((\mathrm{abs}\,(w) - \mathrm{floor}\,(\mathrm{abs}\,(w))) \times 10^{14}, 256\right), \tag{20}$$

where abs(.) is the absolute function while the function floor(.) calculates the nearest integer.

Thirdly, the new serial numbers X, Y, Z, W are given by

$$X = \mathrm{mod}\,(x, 4), \tag{21}$$

$$Y = \mathrm{mod}\,(y, 4), \tag{22}$$

$$Z = \mathrm{mod}\,(z, 4), \tag{23}$$

$$W = \mathrm{mod}\,(w, 4). \tag{24}$$

Depending on the values of these new serial numbers, there are corresponding groups of states (B_1, B_2, B_3) to perform encryption. For instance, the combination of states (B_1, B_2, B_3) are (x, y, z), (x, y, w), (x, z, w), and (y, z, w) for the serial numbers 0, 1, 2, and 3, respectively. Then we apply the XOR operation between three bytes of the image total shuffling matrix $P_{i,j}^{hl}$ and three bytes of the selected group of three states as follows

$$\begin{cases} C_{3\times(i-1)+1} = P_{3\times(i-1)+1} \oplus B_1 \\ C_{3\times(i-1)+2} = P_{3\times(i-1)+2} \oplus B_2 \\ C_{3\times(i-1)+3} = P_{3\times(i-1)+3} \oplus B_3, \end{cases} \tag{25}$$

Fig. 16 Presentation of the plain image

Fig. 17 Presentation of the encrypted image

in which P_j and $C_j, j = 1, 2, \ldots, N \times M$ indicate the pixels of the plain shuffled image and the ciphered image.

Finally, we continue doing the encryption until the whole image is encrypted.

We have applied the image encryption scheme to the plain–image with the size 256×256 (Fig. 16). We assume that the secret key is

$$\left(x(0), y(0), z(0), w(0), N_0\right) = (0, 0.01, 0.01, 0, 3000). \quad (26)$$

The encrypted image is obtained as illustrated in Fig. 17 while the decrypted image is shown in Fig. 18.

Fig. 18 Presentation of the decrypted image

5 Security Analysis

In this section, we consider the security of the image encryption scheme.

5.1 Key Space Analysis

Secret keys in the encryption scheme are the initial values of chaotic map and the hyperchaotic memristive system, as well as the systems' parameter values. In addition, secret keys can include the iteration number N_0. Thus, the key space is enough large to resist brute–force attacks.

5.2 Key Sensitivity

An intruder, who does not know the secret key, cannot recover the original plain image. In order to show the sensitivity of the encryption scheme to the secret key, we take an example where the intruder decrypts the encrypted image in Fig. 17 with the following secret key:

$$\left(x(0), y(0), z(0), w(0), N_0\right) = \left(0, 0.01, 0.01, 10^{-6}, 3000\right). \tag{27}$$

The failure of recovering the plain image by the intruder is illustrated in Fig. 19.

Fig. 19 The recovered image by an intruder

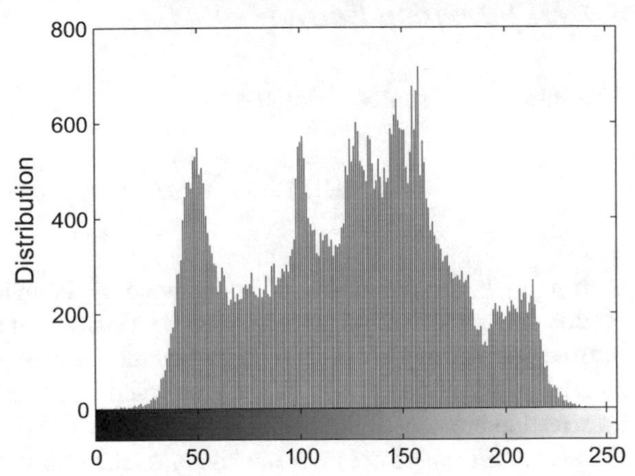

Fig. 20 Histogram of the plain image

5.3 *Histogram Analysis*

As have been known, two methods preventing the statistical attacks are the diffusion and confusion Shannon (1949). The histograms of the plain–image and the encrypted image are presented in Figs. 20, 21. It is easy to see that the histogram of the ciphered image is different from one of the plain–image. The histogram of the ciphered image has a uniform distribution which indicates the security of the encryption scheme from a statistical attack.

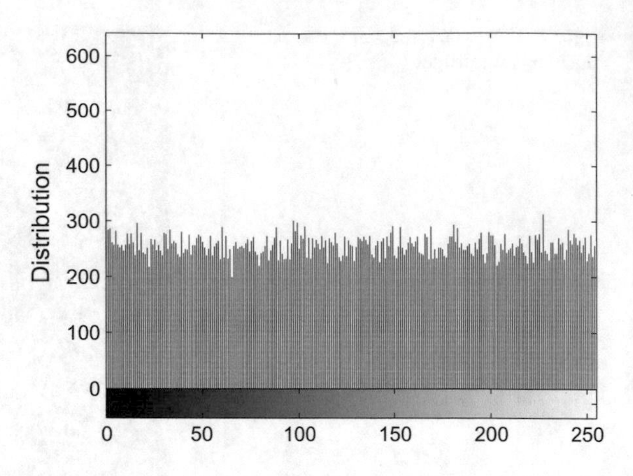

Fig. 21 Histogram of the encrypted image

5.4 *Information Entropy*

The entropy of a source is defined by

$$E(S) = -\sum_{i=0}^{N-1} p\left(s_i\right) \log_2\left(p\left(s_i\right)\right), \tag{28}$$

with $p\left(s_i\right)$ is the possibility of appearance of the symbol s_i. Therefore, the information entropy of an image indicates the distribution of the gray scale values. The information entropy is much bigger when the distribution is much uniform. The calculated information entropy of the encrypted image is 7.9965. It is higher than the information entropy of the plain image (7.4888). The higher information entropy of the encrypted image presents the safety of the encryption scheme from an entropy attack (Volos et al. 2013).

6 Conclusion

In this chapter, a dynamical system with a memristive device has been studied. The system has many special features such as hyperchaos, a infinite number of equilibrium points, and hidden attractors due to the presence of the memristive device. Fundamental dynamical behaviors of the memristive system are discovered through calculating equilibrium points, phase portraits of hyperchaotic attractors, Poincaré maps, bifurcation diagram, Lyapunov exponents and Kaplan–Yorke dimension. The memristive system can be used in potential applications in secure communications and cryptography because of its complex behavior. In particular, we have implemented an image encryption scheme based on the hyperchaotic memristive system. In addition, security analysis of image encryption scheme are discussed. Further studies related to applications of this system will be presented in our future works.

Acknowledgements Viet–Thanh Pham would like to thank Le Thi Van Thu, Philips Electronics – Vietnam, for her help.

References

Adhikari, S. P., Yang, C., Kim, H., & Chua, L. O. (2012). Memristor bridge synapse-based neural network and its learning. *IEEE Transactions on Neural Networks and Learning Systems*, *23*, 1426–1435.

Arneodo, A., Coullet, P., & Tresser, C. (1981). Possible new strange attractors with spiral structure. *Communications in Mathematical Physics*, *79*, 573–579.

Ascoli, A., & Corinto, F. (2013). Memristor models in a chaotic neural circuit. *International Journal of Bifurcation and Chaos*, *23*, 1350052.

Barnerjee, T., Biswas, D., & Sarkar, B. C. (2012). Design and analysis of a first order time-delayed chaotic system. *Nonlinear Dynamics*, *70*, 721–734.

Buscarino, A., Fortuna, L., Frasca, M., & Gambuzza, L. V. (2012a). A chaotic circuit based on Hewlett–Packard memristor. *Chaos*, *22*, 023136.

Buscarino, A., Fortuna, L., Frasca, M., & Gambuzza, L. V. (2012b). A gallery of chaotic oscillators based on hp memristor. *International Journal of Bifurcation and Chaos*, *22*, 1330015–14.

Chen, G., & Ueta, T. (1999). Yet another chaotic attractor. *International Journal of Bifurcation and Chaos*, *9*, 1465–1466.

Chua, L. O. (1971). Memristor-The missing circuit element. *IEEE Transactions Circuit Theory*, *18*, 507–519.

Chua, L. O., & Kang, S. M. (1976). Memristive devices and system. *Proceedings of the IEEE*, *64*, 209–223.

Corinto, F., Ascoli, A., & Gilli, M. (2012). Analysis of current-voltage characteristics for memristive elements in pattern recognition systems. *International Journal of Circuit Theory and Applications*, *40*, 1277–1320.

Driscoll, T., Quinn, J., Klien, S., Kim, H. T., Kim, B. J., Pershin, Y. V., et al. (2010). Memristive adaptive filters. *Applied Physics Letters*, *97*, 093502.

Shannon, C. E. (1949). Communication theory of secrecy system. *Journal of Bell Systems Technology*, *28*, 656–715.

Fitch, A. L., Yu, D., Iu, H. H. C., & Sreeram, V. (2012). Hyperchaos in an memristor-based modified canonical Chua's circuit. *International Journal of Bifurcation and Chaos*, *22*, 1250133–8.

Gao, T., & Chen, Z. (2008). A new image encryption algorithm based on hyper-chaos. *Physics Letters A*, *372*, 394–400.

Grassi, G., & Mascolo, S. (1999). A system theory approach for designing cryptosystems based on hyperchaos. *IEEE Transactions on Circuits and Systems I Fundamental Theory and Applications*, *46*, 1135–1138.

Huang, Y., & Yang, X. (2006). Hyperchaos and bifurcation in a new class of four-dimensional hopfield neural networks. *Neurocomputing*, *69*, 1787–1795.

Jafari, S., & Sprott, J. C. (2013). Simple chaotic flows with a line equilibrium. *Chaos, Solitons and Fractals*, *57*, 79–84.

Kuznetsov, N. V., Leonov, G. A., & Seledzhi, S. M. (2011). Hidden oscillations in nonlinear control systems. *IFAC Proceedings*, *18*, 2506–2510.

Leonov, G. A., & Kuznetsov, N. V. (2013). Hidden attractors in dynamical systems: From hidden oscillation in Hilbert-Kolmogorov, Aizerman and Kalman problems to hidden chaotic attractor in chua circuits. *International Journal of Bifurcation and Chaos*, *23*, 1330002.

Leonov, G. A., Kuznetsov, N. V., Kuznetsova, O. A., Seldedzhi, S. M., & Vagaitsev, V. I. (2011). Hidden oscillations in dynamical systems. *Transactions on Systems Control*, *6*, 54–67.

Li, Q., Hu, S., Tang, S., & Zeng, G. (2014). Hyperchaos and horseshoe in a 4D memristive system with a line of equilibria and its implementation. *International Journal of Circuit Theory and Applications, 42*, 1172–1188.

Li, Q., Zeng, H., & Li, J. (2015). Hyperchaos in a 4D memristive circuit with infinitely many stable equilibria. *Nonlinear Dynamics, 79*, 2295–2308.

Liao, X., Lai, S., & Zhou, Q. (2010). A novel image encryption algorithm based on self-adaptive wave transmission. *Signal Processing, 90*, 2714–2722.

Lorenz, E. N. (1963). Deterministic non-periodic flow. *Journal of the Atmospheric Sciences, 20*, 130–141.

Lü, J., & Chen, G. (2002). A new chaotic attractor coined. *International Journal of Bifurcation and Chaos, 12*, 659–661.

Matthews, R. (1989). One the derivation of a chaotic encryption algorithm. *Cryptologia, 8*, 29–42.

Pershin, Y. V., & Ventra, M. D. (2010). Experimental demonstration of associative memory with memristive neural networks. *Neural Networks, 23*, 881–886.

Pham, V.-T., Jafari, S., Volos, C., Wang, X., & Golpayegani, S. M. R. H. (2014a). Is that really hidden? The presence of complex fixed-points in chaotic flows with no equilibria. *International Journal of Bifurcation and Chaos, 24*, 1450146.

Pham, V.-T., Volos, C., & Gambuzza, L. V. (2014b). A memristive hyperchaotic system withou equilibrium. *The Scientific World Journal, 2014*, 368986–9.

Pham, V.-T., Volos, C. K., Jafari, S., Wei, Z., & Wang, X. (2014c). Constructing a novel no-equilibrium chaotic system. *International Journal of Bifurcation and Chaos, 24*, 1450073.

Pham, V. T., Volos, C. K., Vaidyanathan, S., Le, T. P., & Vu, V. Y. (2015). A memristor-based hyperchaotic system with hidden attractors: Dyamics, sychronization and circuital emulating. *Journal of Engineering Science and Technology Review, 8*, 205–214.

Rössler, O. E. (1976). An equation for continuous chaos. *Physics Letters A, 57*, 397–398.

Rössler, O. E. (1979). An equation for hyperchaos. *Physics Letters A, 71*, 155–157.

Sadoudi, S., Tanougast, C., Azzaz, M. S., & Dandache, A. (2013). Design and FPGA implementation of a wireless hyperchaotic communication system for secure realtime image transmission. *EURASIP Journal on Image and Video Processing, 943*, 1–18.

Seyedzadeh, S. M., Mirzakuchaki, S., & Fast, A. (2012). Color image encryption algorithm based on coupled two-dimensional piecewise chaotic map. *Signal Processing, 92*, 1201–1215.

Sharma, P. R., Shrimali, M. D., Prasad, A., Kuznetsov, N. V., & Leonov, G. A. (2015). Control of multistability in hidden attractors. *The European Physical Journal Special Topics, 224*, 1485–1491.

Shin, S., Kim, K., & Kang, S. M. (2011). Memristor applications for programmable analog ICs. *IEEE Transactions on Nanotechnology, 410*, 266–274.

Strukov, D. B., Snider, G. S., Stewart, D. R., & Williams, R. S. (2008). The missing memristor found. *Nature, 453*, 80–83.

Tacha, O., Volos, C. K., Kyprianidis, I. M., Stouboulos, I. N., Vaidyanathan, S., & Pham, V.-T. (2016). Analysis, adaptive control and circuit simulation of a novel nonlinear finance system. *Applied Mathematics and Computation, 276*, 200–217.

Tong, X., & Cui, M. (2009). Image encryption scheme based on 3d baker with dynamical compound chaotic sequence cipher generator. *Signal Processing, 89*, 480–491.

Udaltsov, V. S., Goedgebuer, J. P., Larger, L., Cuenot, J. B., Levy, P., & Rhodes, W. T. (2003). Communicating with hyperchaos: the dynamics of a DNLF emitter and recovery of transmitted information. *Optics and Spectroscopy, 95*, 114–118.

Vaidyanathan, S. (2013). A new six-term 3-D chaotic system with an exponential nonlineariry. *Far East Journal of Mathematical Sciences, 79*, 135–143.

Vaidyanathan, S., & Azar, A. T. (2015). Analysis and control of a 4-D novel hyperchaotic system. In A. T. Azar & S. Vaidyanathan (Eds.), *Chaos Modeling and Control Systems Design* (Vol. 581, pp. 19–38). Studies in Computational Intelligence Germany: Springer

Vaidyanathan, S., Volos, C., Pham, V. T., Madhavan, K., & Idowo, B. A. (2014). Adaptive back-stepping control, synchronization and circuit simualtion of a 3-D novel jerk chaotic system with two hyperbolic sinusoidal nonlinearities. *Archives of Control Sciences*, *33*, 257–285.
Vicente, R., Dauden, J., Colet, P., & Toral, R. (2005). Analysis and characterization of the hyper-chaos generated by a semiconductor laser subject to a delayed feedback loop. *IEEE Journal of Quantum Electronics*, *41*, 541–548.
Volos, C. K., Kyprianidis, I. M., & Stouboulos, I. N. (2013). Image encryption process based on chaotic synchronization phenomena. *Signal Processing*, *93*, 1328–1340.
Wang, X., Teng, L., Qin, X., & Novel, A. (2012). Colour image encryption algorithm based on chaos. *Signal Processing*, *92*, 1101–1108.
Yang, J. J., Strukov, D. B., & Stewart, D. R. (2013). Memristive devices for computing. *Nature Nanotechnology*, *8*, 13–24.
Yeung, M. M., & Pankanti, S. (2000). Verification cryptosystems: issues and challenges. *Journal of Electronic Imaging*, *9*, 468–476.
Zhang, L., Liao, X., & Wang, X. (2005). An image encryption approach based on chaotic map. *Chaos, Solitons and Fractals*, *24*, 759–756.

Adaptive Control, Synchronization and Circuit Simulation of a Memristor-Based Hyperchaotic System With Hidden Attractors

Sundarapandian Vaidyanathan, Viet-Thanh Pham and Christos Volos

Abstract Memristor-based systems and their potential applications, in which memristor is both a nonlinear element and a memory element, have been received significant attention in the control literature. In this work, we study a memristor-based hyperchaotic system with hidden attractors. First, we study the dynamic properties of the memristor-based hyperchaotic system such as equilibria, Lyapunov exponents, Poincaré map, etc. We obtain the Lyapunov exponents of the memristor-based system as $L_1 = 0.1244$, $L_2 = 0.0136$, $L_3 = 0$ and $L_4 = -10.8161$. Since there are two positive Lyapunov exponents, the memristor-based system is hyperchaotic. Also, the Kaplan-Yorke fractional dimension of the memristor-based hyperchaotic system is obtained as $D_{KY} = 3.0128$. Next, we design adaptive control and synchronization schemes for the memristor-based hyperchaotic system. The main adaptive control and synchronization results are established using Lyapunov stability theory. MATLAB simulations are shown to illustrate all the main results of this work. Finally, an electronic circuit emulating the memristor-based hyperchaotic system has been designed using off-the-shelf components.

Keywords Memristor · Hidden attractor · Chaos · Hyperchaos · Adaptive control · Synchronization · Circuit

---

S. Vaidyanathan (✉)
Research and Development Centre, Vel Tech University,
Avadi, Chennai 600062, Tamil Nadu, India
e-mail: sundarcontrol@gmail.com

V.-T. Pham
School of Electronics and Telecommunications, Hanoi University
of Science and Technology, Hanoi, Vietnam
e-mail: pvt3010@gmail.com

C. Volos
Physics Department, Aristotle University of Thessaloniki, Thessaloniki, Greece
e-mail: volos@physics.auth.gr

© Springer International Publishing AG 2017

S. Vaidyanathan and C. Volos (eds.), *Advances in Memristors, Memristive Devices and Systems*, Studies in Computational Intelligence 701,
DOI 10.1007/978-3-319-51724-7_5

1 Introduction

Chua's circuit (Matsumoto 1984), the Cellular Neural Networks (CNNs) (Chua and Yang 1988a, b) and the memristor (Chua 1971) are three attractive inventions of Prof. Leon O. Chua and these inventions are widely regarded as the major breakthroughs in the literature of the nonlinear control systems. Chua's circuit has been applied in various areas in engineering (Liu et al. 2004; Fortuna et al. 2009; Chua 1994; Albuquerque et al. 2008; Tang and Wang 2005). Cellular Neural Networks have been applied in various areas such as chaos (Vaidyanathan 2016), secure communications (Wang et al. 2012b), cryptosystem (Cheng and Cheng 2013), etc. The studies on memristor (Joglekar and Wolf 2009; Shin et al. 2011; Wang et al. 2012a; Shang et al. 2012; Adhikari et al. 2012, 2013; Yang et al. 2013) have received significant attention only recently after the realization of a solid-state thin film two-terminal memristor at Hewlett-Packard Laboratories (Strukov et al. 2008).

Memristor was proposed by L.O. Chua as the fourth basic circuit element besides the three conventional ones (resistor, inductor and capacitor) (Tetzlaff 2014).

Memristor depicts the relationship between two fundamental circuit variables, *viz.* the charge (q) and the flux (φ). Hence, there are two kinds of memristors: (1) *charge-controlled memristor*, and (2) *flux-controlled memristor*.

A charge-controlled memristor is described by

$$v_M = M(q) i_M \tag{1}$$

where v_M is the *voltage* across the memristor and i_M is the *current* through the memristor. Here, the *memristance* (M) is defined by

$$M(q) = \frac{d\varphi(q)}{dq} \tag{2}$$

A flux-controlled memristor is given by

$$i_M = W(\varphi) v_M \tag{3}$$

where $W(\varphi)$ is the *memductance*, which is defined by

$$W(q) = \frac{dq(\varphi)}{d\varphi} \tag{4}$$

By generalizing the original definition of a memristor (Chua 1971; Tetzlaff 2014), a *memristive system* is defined as

$$\begin{cases} \dot{x} = f(x, u, t) \\ y = g(x, u, t)u \end{cases} \tag{5}$$

where x is the *state*, u is the *input* and y is the *output* of the system (5). We assume that the function f is a continuously differentiable, n-dimensional vector field and g is a continuous scalar function.

The intrinsic nonlinear characteristic of memristor has applications in implementing chaotic systems with complex dynamics as well as special features (Itoh and Chua 2008; Muthuswamy and Kokate 2009). For example, a simple memristor-based chaotic system including only three elements (an inductor, a capacitor and a memristor) was introduced in Muthuswamy and Chua (2010). Also, a system containing an HP memristor model and triangular wave sequence can generate multi-scroll chaotic attractors (Li et al. 2014). Moreover, a four-dimensional hyperchaotic memristive system with a line equilibrium was presented by Li et al. (2013).

Chaos theory deals with the qualitative study of chaotic dynamical systems and their applications in science and engineering. A dynamical system is called *chaotic* if it satisfies the three properties: boundedness, infinite recurrence and sensitive dependence on initial conditions (Azar and Vaidyanathan 2015).

Some classical paradigms of 3-D chaotic systems in the literature are Lorenz system (Lorenz 1963), Rössler system (Rössler 1976), ACT system (Arneodo et al. 1981), Sprott systems (Sprott 1994), Chen system (Chen and Ueta 1999), Lü system (Lü and Chen 2002), Cai system (Cai and Tan 2007), Tigan system (Tigan and Opris 2008), etc.

Many new chaotic systems have been discovered in the recent years such as Zhou system (Zhou et al. 2008), Zhu system (Zhu et al. 2010), Li system (Li 2008), Wei-Yang system (Wei and Yang 2010), Sundarapandian systems (Sundarapandian 2013; Sundarapandian and Pehlivan 2012), Vaidyanathan systems (Vaidyanathan 2013a, b, 2014a, b, c, d, 2015b, n; Vaidyanathan and Azar 2015b; Vaidyanathan and Madhavan 2013; Vaidyanathan and Pakiriswamy 2015; Vaidyanathan et al. 2014c, 2015b, d, f, g; Vaidyanathan and Volos 2015; Vaidyanathan 2015m), Pehlivan system (Pehlivan et al. 2014), Sampath system (Sampath et al. 2015), etc.

Chaos theory has many applications in science and engineering such as chemical systems (Vaidyanathan 2015i, g, s, o, t, c, k, u), biological systems (Vaidyanathan 2015d, e, a, j, p, f, x, q, y, r, z, h, v, l, w), memristors (Pham et al. 2015; Volos et al. 2015; Abdurrahman et al. 2015), etc.

The study of control of a chaotic system investigates feedback control methods that globally or locally asymptotically stabilize or regulate the outputs of a chaotic system. Many methods have been designed for control and regulation of chaotic systems such as active control (Sundarapandian 2010, 2011; Vaidyanathan 2011b), adaptive control (Vaidyanathan et al. 2014a, 2015e, h), backstepping control (Li et al. 2007; Wang and Ge 2008), sliding mode control (Vaidyanathan 2012c, e), etc.

Synchronization of chaotic systems is a phenomenon that occurs when two or more chaotic systems are coupled or when a chaotic system drives another chaotic system. Because of the butterfly effect which causes exponential divergence of the trajectories of two identical chaotic systems started with nearly the same initial conditions, the synchronization of chaotic systems is a challenging research problem in the chaos literature (Azar and Vaidyanathan 2015, 2016; Azar et al. 2017; Vaidyanathan and Volos 2016a, b).

Pecora and Carroll pioneered the research on synchronization of chaotic systems with their seminal papers (Carroll and Pecora 1991; Pecora and Carroll 1990). The active control method (Karthikeyan and Sundarapandian 2014; Sarasu and Sundarapandian 2011a, b; Sundarapandian and Karthikeyan 2012b; Vaidyanathan 2011a, 2012d; Vaidyanathan and Rajagopal 2011a, b; Vaidyanathan and Rasappan 2011) is typically used when the system parameters are available for measurement. Adaptive control method (Sarasu and Sundarapandian 2012a, b, c; Sundarapandian and Karthikeyan 2011a, b, 2012a; Vaidyanathan 2013c, 2012b; Vaidyanathan and Azar 2015a; Vaidyanathan and Pakiriswamy 2013; Vaidyanathan and Rajagopal 2011c; Vaidyanathan et al. 2014b, 2015c) is typically used when some or all the system parameters are not available for measurement and estimates for the uncertain parameters of the systems.

Sampled-data feedback control method (Gan and Liang 2012; Xiao et al. 2014) and time-delay feedback control method (Chen et al. 2014; Jiang et al. 2004) are also used for synchronization of chaotic systems. Backstepping control method (Rasappan and Vaidyanathan 2012a, b, c, 2013, 2014; Suresh and Sundarapandian 2013; Vaidyanathan and Rasappan 2014; Vaidyanathan et al. 2015a, i) is also applied for the synchronization of chaotic systems. Backstepping control is a recursive method for stabilizing the origin of a control system in strict-feedback form (Khalil 2001). In this research work, we apply backstepping control method for the adaptive control and synchronization of the novel hyperjerk system.

Sliding mode control method (Sundarapandian and Sivaperumal 2011; Vaidyanathan 2012a, 2014e; Vaidyanathan and Azar 2015c, d; Vaidyanathan and Sampath 2011, 2012) is also a popular method for the synchronization of chaotic systems.

According to a new classification of chaotic dynamics (Kuznetsov and Leonov 2014; Leonov et al. 2011; Dudkowski et al. 2016), there are two kinds of chaotic attractors: (1) *self-cited attractors*, and (2) *hidden attractors*. The classical attractors of Lorenz, Rössler, Chua, Chen, and other widely-known attractors are those excited from unstable equilibria (Azar and Vaidyanathan 2015, 2016; Azar et al. 2017; Vaidyanathan and Volos 2016a, b). From the computational point of view, this allows one to use numerical method, in which after transient process a trajectory, started from a point of unstable manifold in the neighborhood of equilibrium, reaches an attractor and identifies it. However there are attractors of another type: hidden attractors, a basin of attraction of which does not contain neighborhoods of equilibria. Hidden attractors cannot be reached by trajectory from neighborhoods of equilibria.

In this work, we discuss the dynamics and properties of our recent memristor-based hyperchaotic system (Pham et al. 2015). We also derive new results for the adaptive control and synchronization of this memristor-based hyperchaotic system with unknown system parameters. The main adaptive control results are derived using Lyapunov stability theory (Khalil 2001).

This work is organized as follows. Section 2 describes the model of memristor-based system. Section 3 describes the qualitative properties of the memristor-based hyperchaotic system. Section 4 derives new results for the adaptive control of the memristor-based hyperchaotic system. Section 5 discusses new results for the global hyperchaos synchronization of the memristor-based hyperchaotic systems via adaptive control. In Sect. 6, a circuit implementation of the memristor-based hyperchaotic system is studied in detail. Finally, Sect. 7 concludes this work with a summary of the main results.

2 Model of Memristor-Based System

This work describes a flux-controlled memristor. For this construction, we use the following memductance function (Bao et al. 2010; Fitch et al. 2012; Muthuswamy 2010)

$$W(\varphi) = 1 + 6\varphi^2 \tag{6}$$

Based on this memristor, a four-dimensional system is introduced as follows:

$$\begin{cases} \dot{x} = -10x - 5y - 5yz \\ \dot{y} = -6x + 6xz + ayW(\varphi) + b \\ \dot{z} = -z - 6xy \\ \dot{\varphi} = y \end{cases} \tag{7}$$

where a, b are real parameters and $W(\varphi)$ is the memductance as defined in (6).

When $b = 0$, the memristor-based system (7) has the line equilibrium $E(0, 0, 0, \varphi)$.

Also, when $b = 0$, the system (7) is hyperchaotic for different values of the parameter a (Li et al. 2013). For example, when $a = 0.1, b = 0$, and initial conditions are selected as $(x(0), y(0), z(0), \varphi(0)) = (0, 0.01, 0.01, 0)$, the system (7) exhibits hyperchaotic behavior. In this case, the memristor-based system (7) is similar to the system reported in (Li et al. 2013). Hence, we will not discuss the case $b = 0$ for the rest of this work.

3 Dynamics of the Memristor-Based System

We consider the memristor-based system (7) when $b \neq 0$.

We find the equilibrium points of the system (7) by solving the following system of equations

$$-10x - 5y - 5yz = 0 \tag{8a}$$

$$-6x + 6xz + ay(1 + 6\varphi^2) + b = 0 \tag{8b}$$

$$-z - 6xy = 0 \tag{8c}$$

$$y = 0 \tag{8d}$$

Solving (8a), (8c) and (8d), we get

$$x = 0, \; y = 0, \; z = 0 \tag{9}$$

Thus, Eq. (8b) reduces to $b = 0$, which is a contradiction.

Hence, there is no equilibrium for the memristor-based system (7).

Next, we take the parameters of the memristor-based system (7) as

$$a = 0.1, \; b = -0.001 \tag{10}$$

We choose the initial conditions of the system (7) as

$$x(0) = 0, \; y(0) = 0.01, \; z(0) = 0.01, \; \varphi(0) = 0 \tag{11}$$

For the parameter values (10) and the initial values (11), the Lyapunov exponents of the memristor-based system (7) are obtained as

$$L_1 = 0.1244, \; L_2 = 0.0136, \; L_3 = 0, \; L_4 = -10.8161 \tag{12}$$

Thus, the memristor-based system (7) is a hyperchaotic system because it has more than one positive Lyapunov exponent (Azar and Vaidyanathan 2016; Azar et al. 2017; Vaidyanathan and Volos 2016a, b).

Since the system (7) has no equilibrium point, it can be classified as a hyperchaotic system with hidden strange attractor.

We also note that the system (7) has been proposed briefly in Pham et al. (2014), but the behavior of such a system has not been fully investigated. In this work, we detail the properties of (7) in detail.

The Kaplan-Yorke fractional dimension describes the complexity of a chaotic attractor (Frederickson et al. 1983). Suppose that a chaotic system of order n has n Lyapunov exponents $L_1, L_2, \dots, L_n$, which are arranged in decreasing order, i.e.

$$L_1 \geq L_2 \geq \cdots \geq L_n \tag{13}$$

Then the Kaplan-Yorke dimension of the chaotic system of order n is defined by

$$D_{KY} = j + \frac{1}{|L_{j+1}|} \sum_{i=1}^{j} L_j \tag{14}$$

where j is the largest integer satisfying $\sum_{i=1}^{j} L_i \geq 0$ and $\sum_{i=1}^{j+1} L_i < 0$.

The Kaplan-Yorke dimension of the memristor-based hyperchaotic system (7) is calculated as

$$D_{KY} = 3 + \frac{L_1 + L_2 + L_3}{|L_4|} = 3.0128 \tag{15}$$

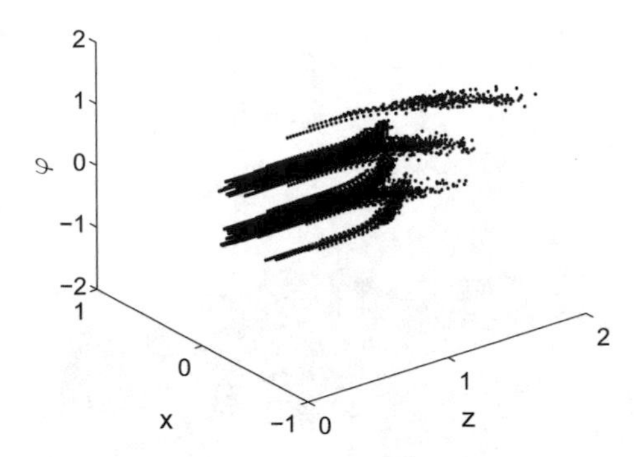

Fig. 1 Poincaré map in the (x, z, φ) space when $y = 0$ for $a = 0.1$ and $b = -0.001$

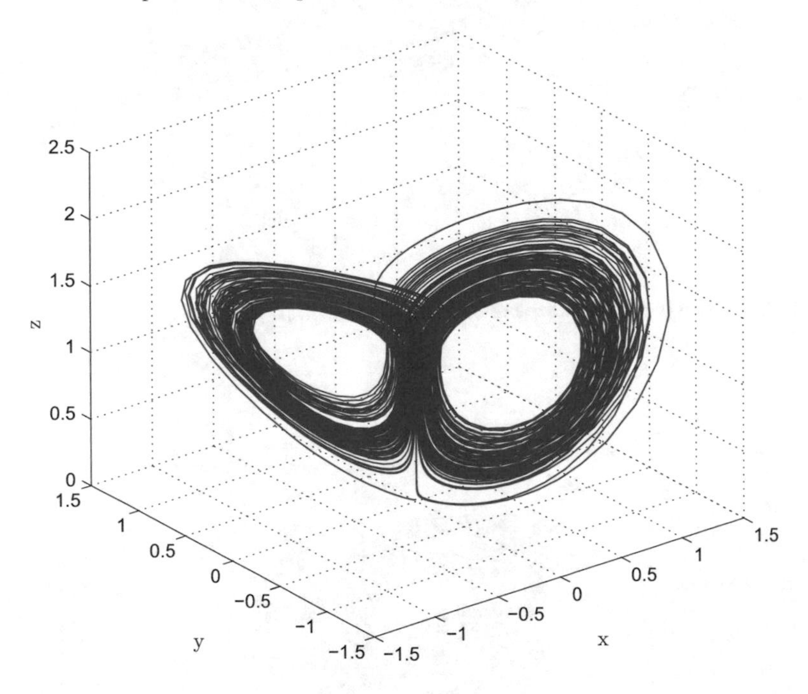

Fig. 2 3-D phase portrait of the memristor-based system in (x, y, z) space

which is fractional. Moreover, it can be seen from the Poincaré map Fig. 1 that the memristor-based hyperchaotic system (7) exhibits a rich dynamical behavior.

Figures 2, 3, 4 and 5 show the 3-D projections of the memristor-based hyperchaotic system in (x, y, z), (x, y, φ), (x, z, φ) and (y, z, φ) spaces, respectively.

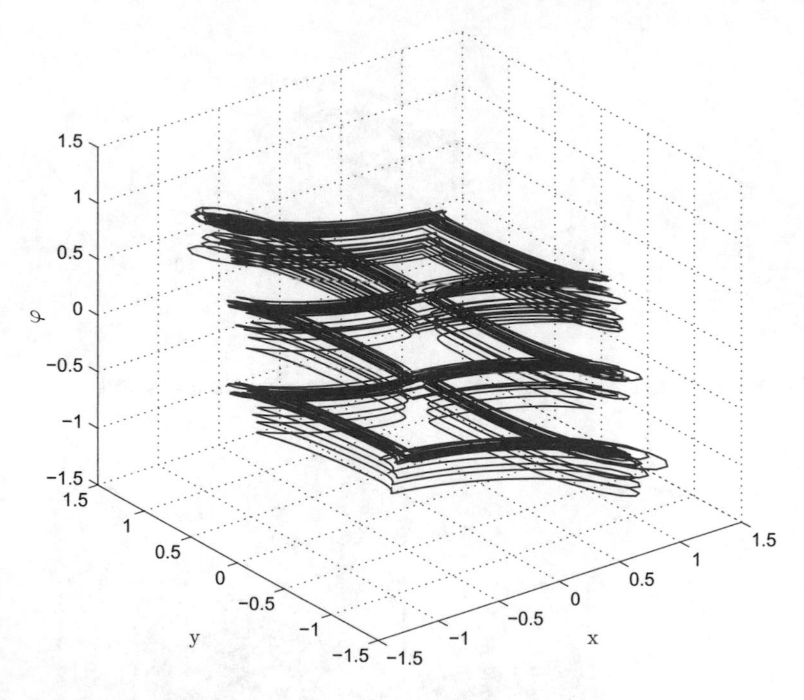

Fig. 3 3-D phase portrait of the memristor-based system in (x, y, φ) space

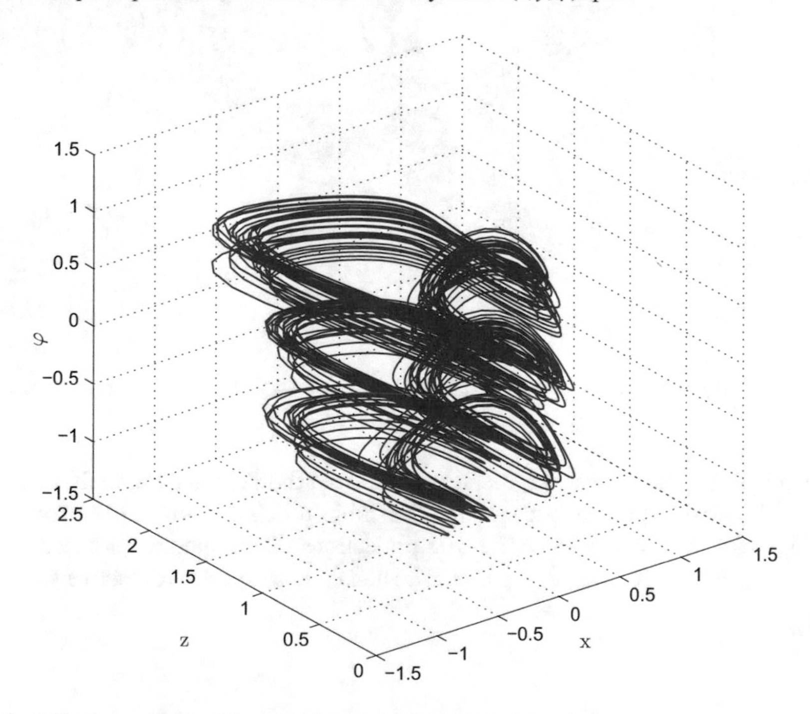

Fig. 4 3-D phase portrait of the memristor-based system in (x, z, φ) space

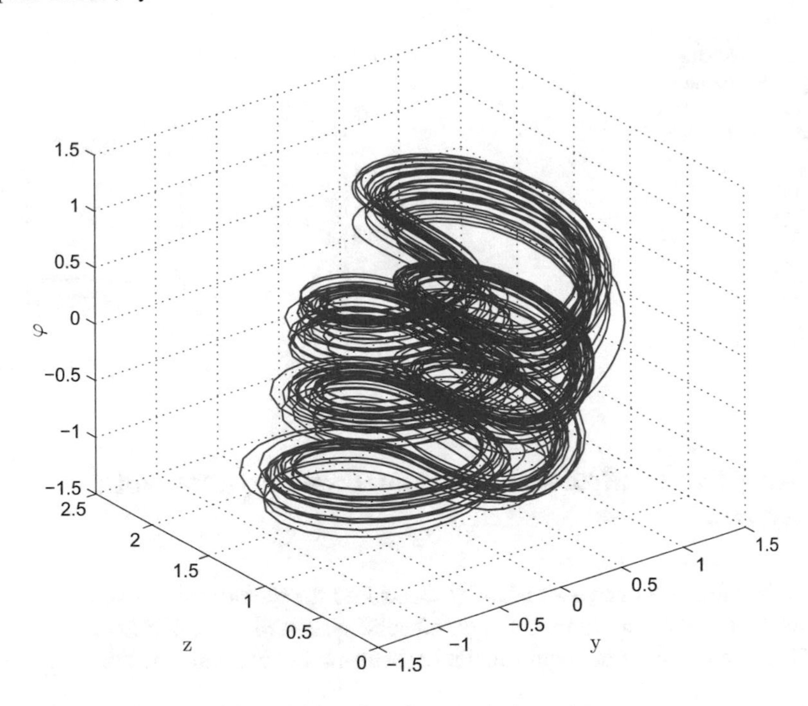

Fig. 5 3-D phase portrait of the memristor-based system in (y, z, φ) space

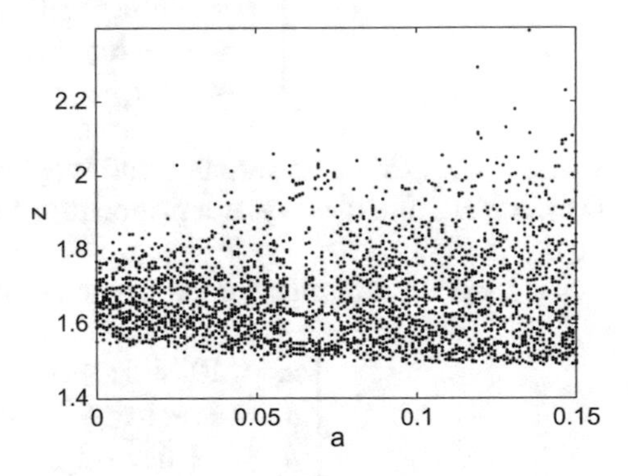

Fig. 6 Bifurcation diagram of $z_{\max}$ with $b = -0.001$ and a as varying parameter

In our work, the parameter b is fixed as $b = -0.001$, while the parameter a indicating the strength of the memristor is varied. The bifurcation diagram is presented in Fig. 6 by plotting the local maxima of the state variable $z(t)$ when changing the value of the parameter a. The spectrum of the corresponding Lyapunov exponents is depicted in Fig. 7.

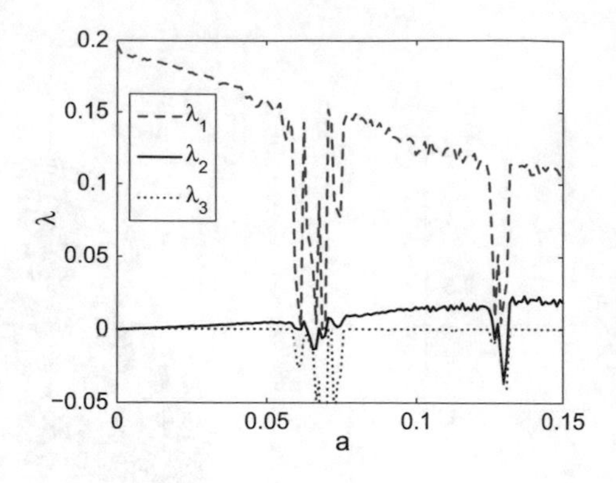

Fig. 7 Three largest Lyapunov exponents of memristor-based system versus a for $b = -0.001$

4 Adaptive Control of Memristor-Based Hyperchaotic System

In this section, we derive an adaptive controller for globally stabilizing all the trajectories of the memristor-based hyperchaotic system discussed in Sect. 2.

Thus, we consider the controlled memristor-based hyperchaotic system given by

$$\begin{cases} \dot{x} = -10x - 5y - 5yz + u_x \\ \dot{y} = -6x + 6xz + ayW(\varphi) + b + u_y \\ \dot{z} = -z - 6xy + u_z \\ \dot{\varphi} = y + u_\varphi \end{cases} \tag{16}$$

where x, y, z, φ are state variables and $W(\varphi)$ is the memductance as defined in (6). In (16), a, b are unknown system parameters, and u_x, u_y, u_z, u_φ are adaptive controls to be determined.

As adaptive controller for the memristor-based system (16), we take

$$\begin{cases} u_x = 10x + 5y + 5yz - k_x x \\ u_y = 6x - 6xz - \hat{a}(t)yW(\varphi) - \hat{b}(t) - k_y y \\ u_z = z + 6xy - k_z z \\ u_\varphi = -y - k_\varphi \varphi \end{cases} \tag{17}$$

where k_x, k_y, k_z, k_φ are positive constants and $\hat{a}(t), \hat{b}(t)$ are estimates of unknown parameters a, b, respectively.

Substituting (17) into (16), we get the closed-loop system

$$\begin{cases} \dot{x} = -k_x x \\ \dot{y} = [a - \hat{a}(t)]yW(\varphi) + b - \hat{b}(t) - k_y y \\ \dot{z} = -k_z z \\ \dot{\varphi} = -k_\varphi \varphi \end{cases} \tag{18}$$

We define parameter estimation errors as

$$\begin{cases} e_a(t) = a - \hat{a}(t) \\ e_b(t) = b - \hat{b}(t) \end{cases} \tag{19}$$

Differentiating the parameter estimation errors, we get

$$\begin{cases} \dot{e}_a(t) = -\dot{\hat{a}}(t) \\ \dot{e}_b(t) = -\dot{\hat{b}}(t) \end{cases} \tag{20}$$

Using (19), we can simplify the dynamics (18) as

$$\begin{cases} \dot{x} = -k_x x \\ \dot{y} = e_a y W(\varphi) + e_b - k_y y \\ \dot{z} = -k_z z \\ \dot{\varphi} = -k_\varphi \varphi \end{cases} \tag{21}$$

Next, we define the quadratic Lyapunov function

$$V(x, y, z, \varphi, e_a, e_b) = \frac{1}{2}\left(x^2 + y^2 + z^2 + \varphi^2 + e_a^2 + e_b^2\right), \tag{22}$$

which is positive definite on $\mathbf{R}^6$.

Differentiating V along the trajectories of (21) and (20), we get

$$\dot{V} = -k_x x^2 - k_y y^2 - k_z z^2 - k_\varphi \varphi^2 + e_a\left[y^2 W(\varphi) - \dot{\hat{a}}\right] + e_b\left[y - \dot{\hat{b}}\right] \tag{23}$$

In view of (23), we take the parameter update law as follows:

$$\begin{cases} \dot{\hat{a}} = y^2 W(\varphi) \\ \dot{\hat{b}} = y \end{cases} \tag{24}$$

Next, we state the main result of this section.

Theorem 1 *The memristor-based hyperchaotic system (16) with unknown system parameters is globally and exponentially stabilized for all initial values by the adaptive control law (17) and the parameter update law (24), where k_x, k_y, k_z, k_φ are positive gain constants.*

Proof This result is proved via Lyapunov stability theory (Khalil 2001).

For this purpose, we consider the quadratic Lyapunov function V defined by (22), which is positive definite on $\mathbf{R}^6$.

Substituting the parameter update law (24) into (23), we obtain

$$\dot{V} = -k_x x^2 - k_y y^2 - k_z z^2 - k_\varphi \varphi^2, \tag{25}$$

Clearly, $\dot{V}$ is a negative semi-definite function on $\mathbf{R}^6$.

Thus, we conclude that the state $X(t) = [x(t), y(t), z(t), \varphi(t)]^T$ and the parameter estimation error $[e_a(t), e_b(t)]^T$ are globally bounded.

We define $k = \min\{k_x, k_y, k_z, k_\varphi\}$.

Then it is clear from (25) that

$$\dot{V} \leq -k \parallel X(t) \parallel^2 \tag{26}$$

or

$$k \parallel X(t) \parallel^2 \leq -\dot{V} \tag{27}$$

Integrating the inequality (27) from 0 to t, we get

$$k \int_0^t \parallel X(\tau) \parallel^2 d\tau \leq - \int_0^t V(\tau) d\tau = V(0) - V(t) \tag{28}$$

From (28), we conclude that $X(t) \in L_2$.

Using (21), we conclude that $\dot{X}(t) \in L_2$.

Using Barbalat's lemma (Khalil 2001), it follows that $X(t) \to 0$ exponentially as $t \to \infty$ for all initial conditions $X(0) \in \mathbf{R}^4$. This completes the proof. ■

For numerical simulations, we take the parameter values of the memristor-based system (16) as in the hyperchaotic case, i.e. $a = 0.1$ and $b = -0.001$.

We take the gain constants as $k_x = 5, k_y = 5, k_z = 5$ and $k_\varphi = 5$. Also, we take the initial conditions of the estimates of the parameters as

$$\hat{a}(0) = 7.4 \;\; \hat{b}(0) = -4.3 \tag{29}$$

We take the initial conditions of the memristor-based system (16) as

$$x(0) = 2.4, \;\; y(0) = 5.2, \;\; z(0) = 3.9, \;\; \varphi(0) = 1.6 \tag{30}$$

Figure 8 shows the exponential stabilization of the states of the system (16).

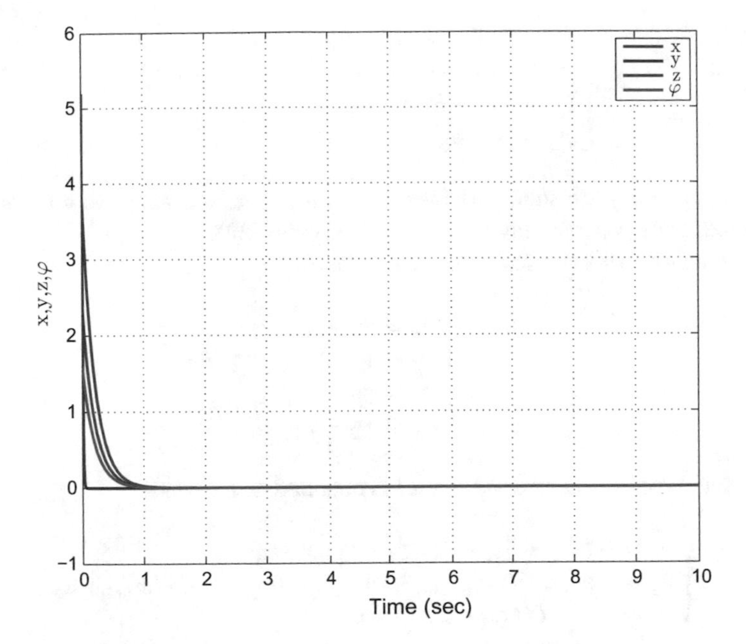

Fig. 8 Exponential stabilization of the states of the memristor-based hyperchaotic system

5 Adaptive Synchronization of Memristor-Based Hyperchaotic Systems

In this section, we derive an adaptive controller for globally synchronizing the respective trajectories of the identical memristor-based hyperchaotic systems discussed in Sect. 2.

As the master system, we consider the memristor-based hyperchaotic system given by

$$\begin{cases} \dot{x}_1 = -10x_1 - 5y_1 - 5y_1z_1 \\ \dot{y}_1 = -6x_1 + 6x_1z_1 + ay_1W(\varphi_1) + b \\ \dot{z}_1 = -z_1 - 6x_1y_1 \\ \dot{\varphi}_1 = y_1 \end{cases} \tag{31}$$

where x_1, y_1, z_1, φ_1 are state variables and $W(\varphi)$ is the memductance as defined in (6). In (31), a and b are unknown system parameters.

As the slave system, we consider the controlled memristor-based hyperchaotic system given by

$$\begin{cases} \dot{x}_2 = -10x_2 - 5y_2 - 5y_2 z_2 + u_x \\ \dot{y}_2 = -6x_2 + 6x_2 z_2 + a y_2 W(\varphi_2) + b + u_y \\ \dot{z}_2 = -z_2 - 6x_2 y_2 + u_z \\ \dot{\varphi}_2 = y_2 + u_\varphi \end{cases} \tag{32}$$

where x_2, y_2, z_2, φ_2 are state variables and u_x, u_y, u_z, u_φ are adaptive controls to be determined for synchronizing the states of (31) and (32).

The synchronization error is defined as follows:

$$\begin{cases} e_x = x_2 - x_1 \\ e_y = y_2 - y_1 \\ e_z = z_2 - z_1 \\ e_\varphi = \varphi_2 - \varphi_1 \end{cases} \tag{33}$$

The synchronization error dynamics is obtained as follows:

$$\begin{cases} \dot{e}_x = -10e_x - 5e_y - 5(y_2 z_2 - y_1 z_1) + u_x \\ \dot{e}_y = -6e_x + 6(x_2 z_2 - x_1 z_1) + a[y_2 W(\varphi_2) - y_1 W(\varphi_1)] + u_y \\ \dot{e}_z = -e_z - 6(x_2 y_2 - x_1 y_1) + u_z \\ \dot{e}_\varphi = e_y + u_\varphi \end{cases} \tag{34}$$

Next, we define the adaptive controller

$$\begin{cases} u_x = 10e_x + 5e_y + 5(y_2 z_2 - y_1 z_1) - k_x e_x \\ u_y = 6e_x - 6(x_2 z_2 - x_1 z_1) - \hat{a}(t)[y_2 W(\varphi_2) - y_1 W(\varphi_1)] - k_y e_y \\ u_z = e_z + 6(x_2 y_2 - x_1 y_1) - k_z e_z \\ u_\varphi = -e_y - k_\varphi e_\varphi \end{cases} \tag{35}$$

where k_x, k_y, k_z, k_φ are positive constants and $\hat{a}(t)$ is an estimate of the unknown parameter a.

Substituting (35) into (34), we get the closed-loop error dynamics

$$\begin{cases} \dot{e}_x = -k_x e_x \\ \dot{e}_y = [a - \hat{a}(t)][y_2 W(\varphi_2) - y_1 W(\varphi_1)] - k_y e_y \\ \dot{e}_z = -k_z e_z \\ \dot{e}_\varphi = -k_\varphi e_\varphi \end{cases} \tag{36}$$

We define the parameter estimation error as

$$e_a(t) = a - \hat{a}(t) \tag{37}$$

Differentiating the parameter estimation error, we get

$$\dot{e}_a(t) = -\dot{\hat{a}}(t) \tag{38}$$

Using (37), we can simplify the dynamics (36) as

$$\begin{cases} \dot{e}_x = -k_x e_x \\ \dot{e}_y = e_a[y_2 W(\varphi_2) - y_1 W(\varphi_1)] - k_y e_y \\ \dot{e}_z = -k_z e_z \\ \dot{e}_\varphi = -k_\varphi e_\varphi \end{cases} \tag{39}$$

Next, we define the quadratic Lyapunov function

$$V(e_x, e_y, e_z, e_\varphi, e_a) = \frac{1}{2}\left(e_x^2 + e_y^2 + e_z^2 + e_\varphi^2 + e_a^2\right), \tag{40}$$

which is positive definite on $\mathbf{R}^5$.

Differentiating V along the trajectories of (39) and (38), we get

$$\dot{V} = -k_x e_x^2 - k_y e_y^2 - k_z e_z^2 - k_\varphi e_\varphi^2 + e_a\left[e_y[y_2 W(\varphi_2) - y_1 W(\varphi_1)] - \dot{\hat{a}}\right] \tag{41}$$

In view of (41), we take the parameter update law as follows:

$$\dot{\hat{a}} = e_y[y_2 W(\varphi_2) - y_1 W(\varphi_1)] \tag{42}$$

Next, we state the main result of this section.

Theorem 2 *The memristor-based hyperchaotic systems (31) and (32) with unknown system parameters are globally and exponentially synchronized for all initial values by the adaptive control law (35) and the parameter update law (42), where* k_x, k_y, k_z, k_φ *are positive gain constants.*

Proof This result is proved via Lyapunov stability theory (Khalil 2001).

For this purpose, we consider the quadratic Lyapunov function V defined by (40), which is positive definite on $\mathbf{R}^5$.

Substituting the parameter update law (42) into (41), we obtain

$$\dot{V} = -k_x e_x^2 - k_y e_y^2 - k_z e_z^2 - k_\varphi e_\varphi^2, \tag{43}$$

Clearly, $\dot{V}$ is a negative semi-definite function on $\mathbf{R}^5$.

Thus, we conclude that the error $\mathbf{e}(t) = [e_x(t), e_y(t), e_z(t), e_\varphi(t)]^T$ and the parameter estimation error $e_a(t)$ are globally bounded. We define $k = \min\{k_x, k_y, k_z, k_\varphi\}$.

Then it is clear from (43) that

$$\dot{V} \leq -k \,\|\, \mathbf{e}(t) \,\|^2 \tag{44}$$

or

$$k \parallel \mathbf{e}(t) \parallel^2 \leq -\dot{V} \tag{45}$$

Integrating the inequality (45) from 0 to t, we get

$$k \int_0^t \parallel \mathbf{e}(\tau) \parallel^2 d\tau \leq -\int_0^t V(\tau) d\tau = V(0) - V(t) \tag{46}$$

From (46), we conclude that $\mathbf{e}(t) \in L_2$. Using (39), we conclude that $\dot{\mathbf{e}}(t) \in L_2$.

Using Barbalat's lemma (Khalil 2001), it follows that $\mathbf{e}(t) \to 0$ exponentially as $t \to \infty$ for all initial conditions $\mathbf{e}(0) \in \mathbf{R}^4$. This completes the proof. ■

For numerical simulations, we take the parameter values of the memristor-based systems (31) and (32) as in the hyperchaotic case, *i.e.* $a = 0.1$ and $b = -0.001$.

We take the gain constants as $k_x = 5, k_y = 5, k_z = 5$ and $k_\varphi = 5$. Also, we take the initial conditions of the parameter estimate as $\hat{a}(0) = 5.4$.

We take the initial conditions of the master system (31) as

$$x_1(0) = 9.1, \quad y_1(0) = -8.5, \quad z_1(0) = 4.7, \quad \varphi_1(0) = -2.9 \tag{47}$$

We take the initial conditions of the slave system (32) as

$$x_2(0) = 7.8, \quad y_2(0) = 6.3, \quad z_2(0) = 2.8, \quad \varphi_2(0) = 1.4 \tag{48}$$

Figures 9, 10, 11 and 12 show the complete synchronization of the states of the memristor-based hyperchaotic systems (31) and (32). Figure 13 shows the exponential convergence of the synchronization errors e_x, e_y, e_z, e_φ.

6 Circuit Design of the Memristor-Based Hyperchaotic System

Using electronic circuits emulating chaotic and hyperchaotic systems is an effective tool for investigating the dynamics of such systems (Fortuna et al. 2009; Sprott 2011). From the point of view of practical applications, the realization of chaotic electronic circuits based on theoretical models is an important topic. Such circuits are main parts in diverse chaos-based applications such as image encryption scheme, path planning generator for autonomous mobile robots, random bit generator, etc. (Chen and Ueda 2002; Volos et al. 2012, 2013; Yalcin et al. 2004).

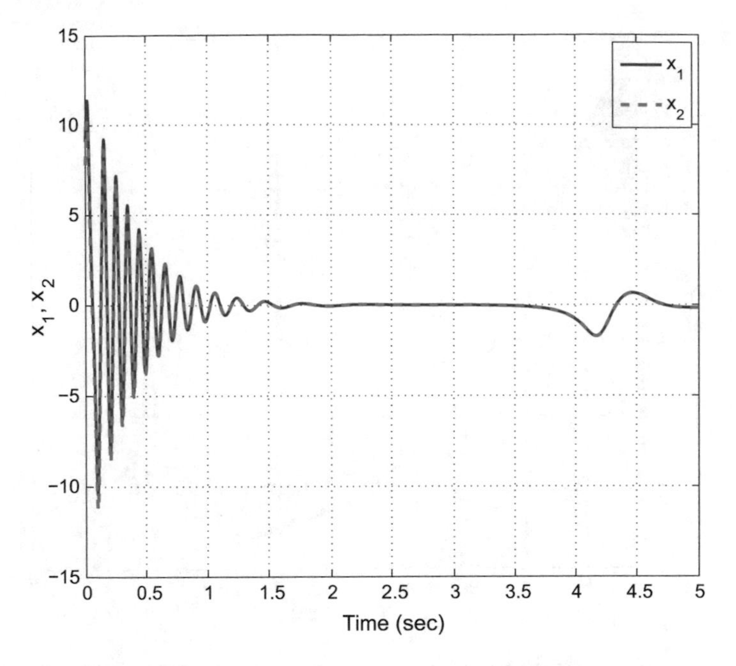

Fig. 9 Synchronization of the states x_1 and x_2

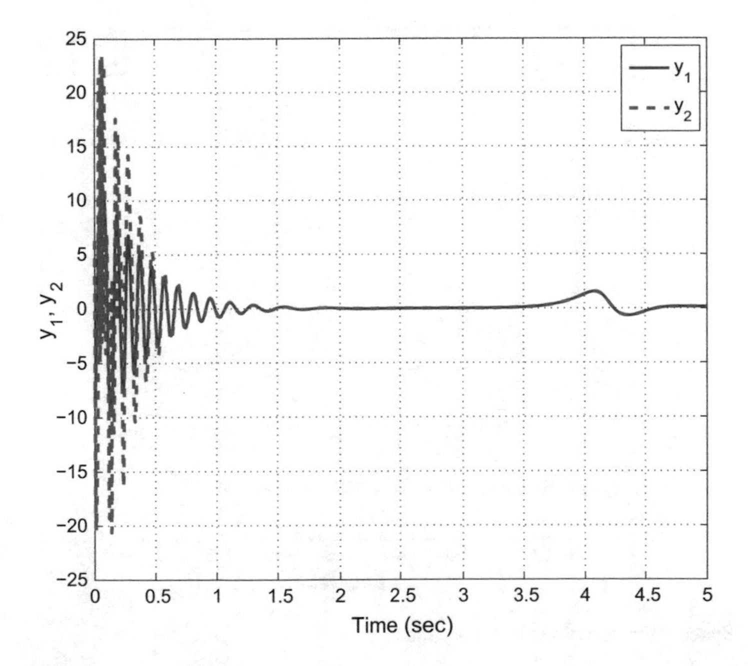

Fig. 10 Synchronization of the states y_1 and y_2

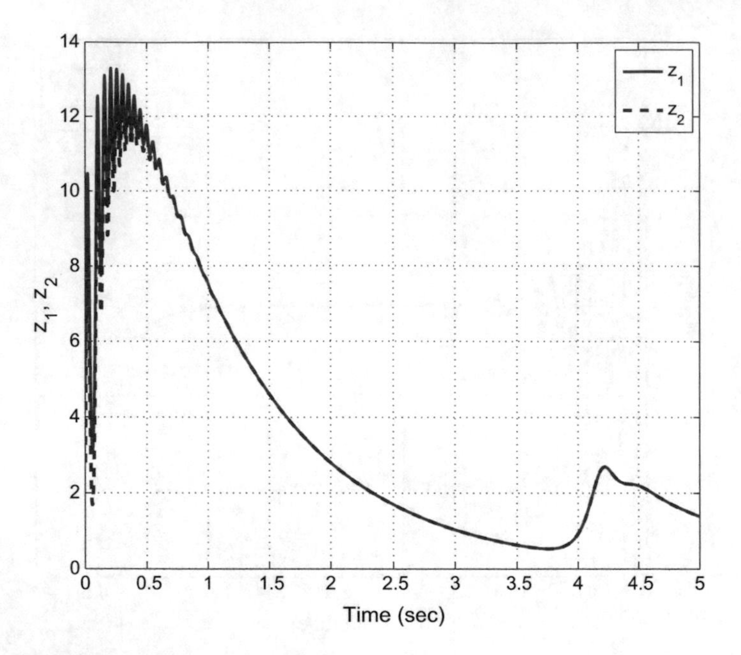

Fig. 11 Synchronization of the states z_1 and z_2

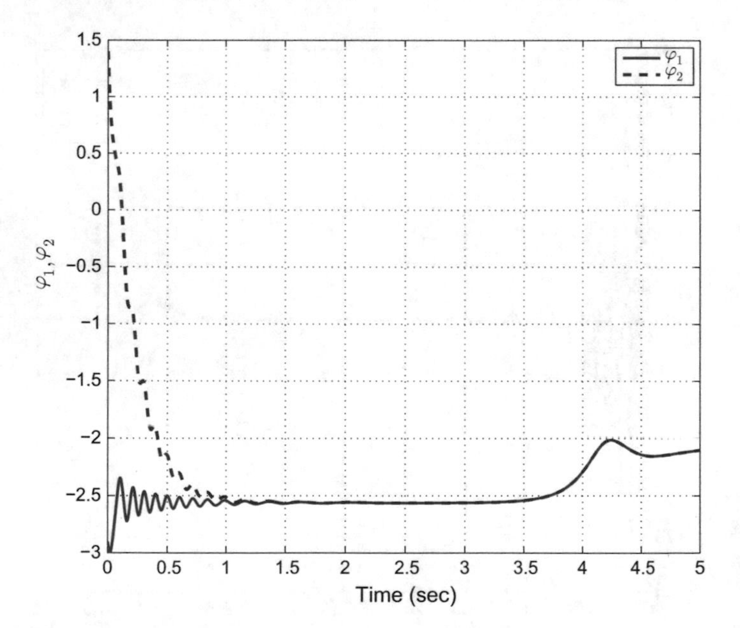

Fig. 12 Synchronization of the states φ_1 and φ_2

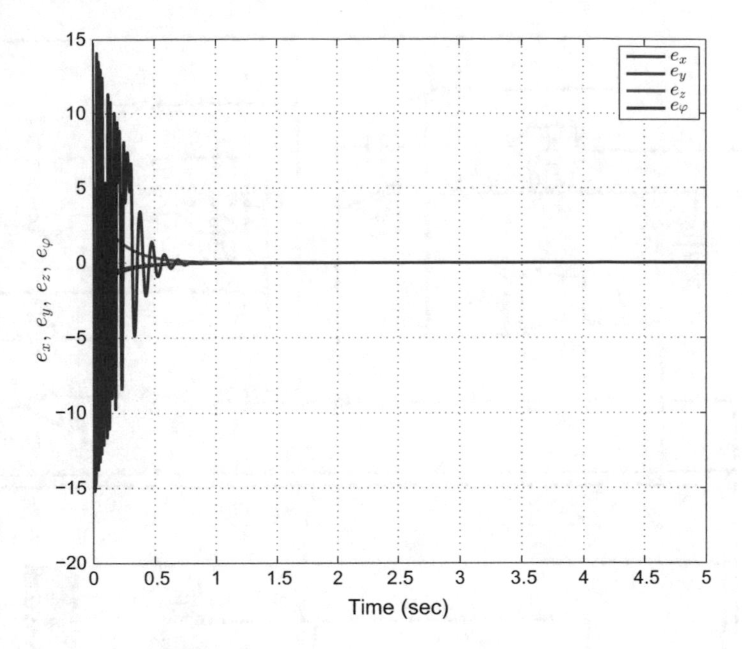

Fig. 13 Time-history of the synchronization errors e_x, e_y, e_z, e_φ

In this section, an electronic circuit is designed to implement the memristor-based hyperchaotic system (7). The circuit in Fig. 14 has been designed following a general approach based on operational amplifiers (Fortuna et al. 2009).

The variables x, y, z, φ of the system (7) are the voltages across the capacitor C_1, C_2, C_3 and C_4, respectively. As shown in Fig. 14, the memristor is realized by common electronic components. Indeed, the sub-circuit of memristor in Fig. 14 only emulates the memristor because there are not any commercial off-the-shelf memristors in the market yet.

By applying Kirchhoff's circuit laws, the corresponding circuital equations of the circuit can be written as follows:

$$\begin{cases} \dot{v}_{C_1} = -\dfrac{1}{R_1C_1}v_{C_1} - \dfrac{1}{R_2C_1}v_{C_2} - \dfrac{1}{10R_3C_1}v_{C_2}v_{C_3} \\ \dot{v}_{C_2} = -\dfrac{1}{R_4C_2}v_{C_1} + \dfrac{1}{10R_5C_2}v_{C_1}v_{C_3} - \dfrac{1}{R_7C_2}V_b + \dfrac{1}{R_6C_2}v_{C_2}\left(\dfrac{R_{11}}{R_{12}} + \dfrac{R_{11}}{100R_{13}}v_{C_4}^2\right) \\ \dot{v}_{C_3} = -\dfrac{1}{R_8C_3}v_{C_3} - \dfrac{1}{10R_9C_3}v_{C_1}v_{C_2} \\ \dot{v}_{C_4} = \dfrac{1}{R_{10}C_4}v_{C_2} \end{cases} \tag{49}$$

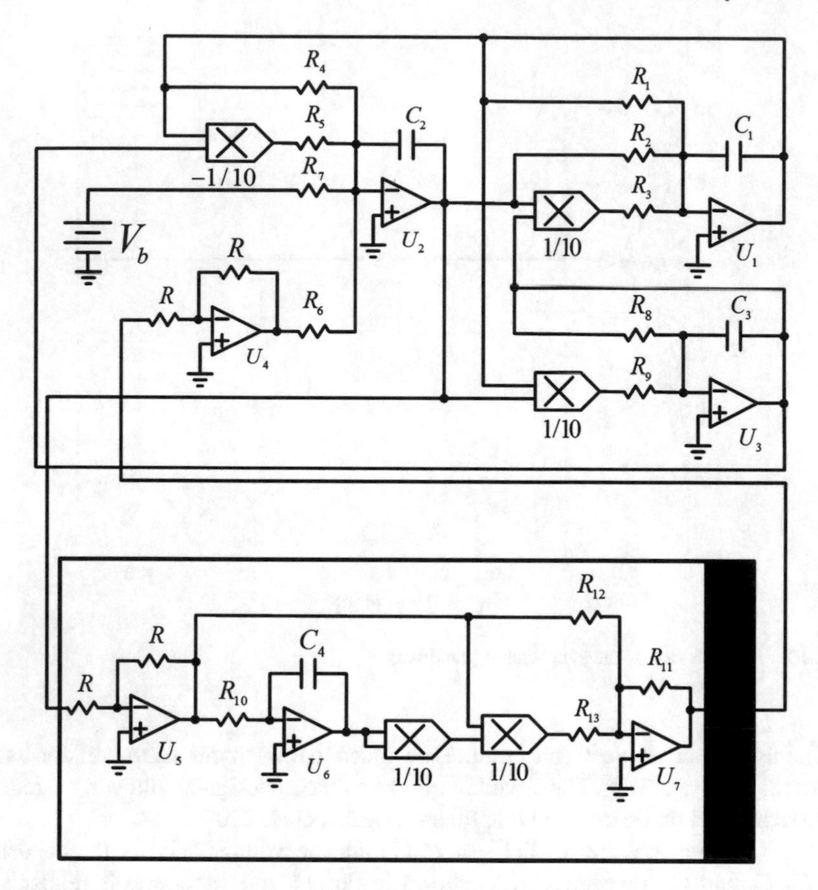

Fig. 14 Schematic of the circuit design for the memristor-based system

where

$$a = \frac{1}{R_6 C_2} \text{ and } b = \frac{1}{R_7 C_2} V_b \quad (50)$$

The operational amplifiers in the electronic circuit are TL084 ones. The power supplies are ±15 V.

We set the values of the circuit components as follows:

$$R_1 = R_3 = 1.8\,\text{k}\Omega,\ R_2 = 3.6\,\text{k}\Omega,\ R_4 = 3\,\text{k}\Omega,\ R_5 = R_9 = 1.5\,\text{k}\Omega \quad (51)$$

$$R_6 = 180\,\text{k}\Omega,\ R_7 = 90\,\text{k}\Omega,\ R_8 = R_{10} = R_{11} = R_{12} = R = 18\,\text{k}\Omega \quad (52)$$

$$R_{13} = 0.75\,\text{k}\Omega,\ C_1 = C_2 = C_3 = C_4 = 10\,\text{nF, and } V_b = 1\text{m}\,V_{DC} \quad (53)$$

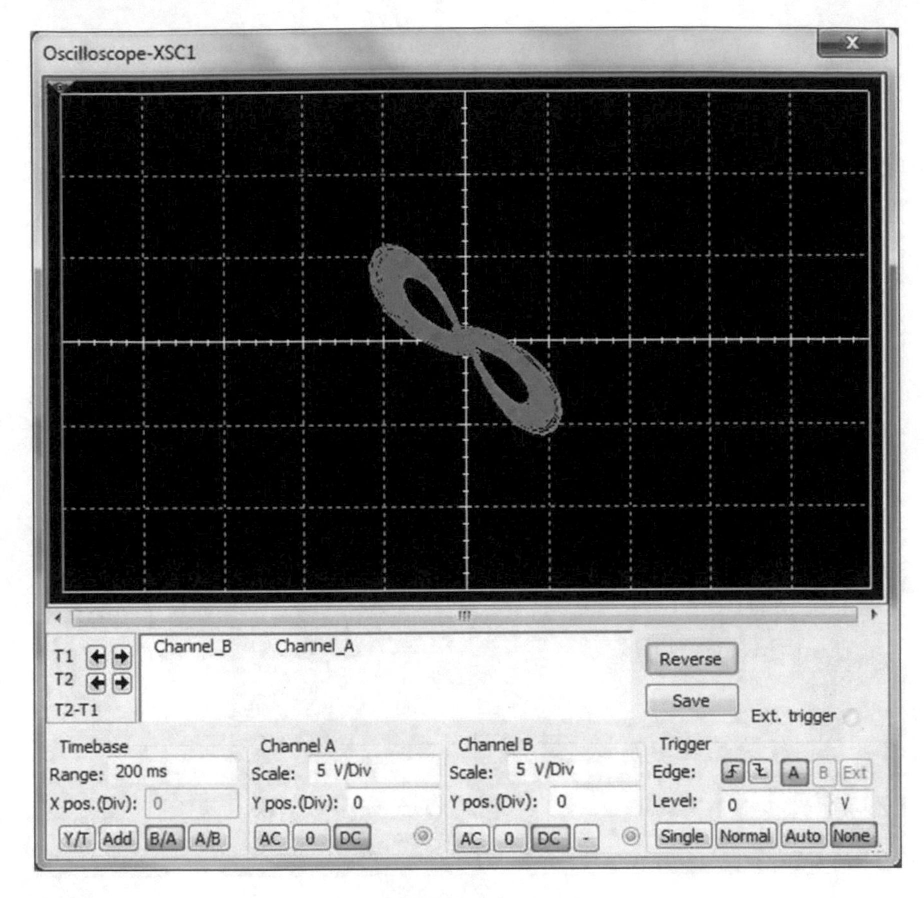

Fig. 15 Multisim plot of the hyperchaotic attractor in the $v_{C_1} - v_{C_2}$ plane

The circuit design of Fig. 14 is implemented in the electronic simulation package Multisim. Figures 15, 16 and 17 show the 2-D plots of the hyperchaotic attractor of the designed circuit obtained from Multisim package.

7 Conclusions

In this work, a memristor-based hyperchaotic system has been studied. This system displays rich dynamical behavior as confirmed by qualitative properties, numerical simulations and circuit implementation. Moreover, the possibility of adaptive control and synchronization schemes of memristor-based hyperchaotic systems have been

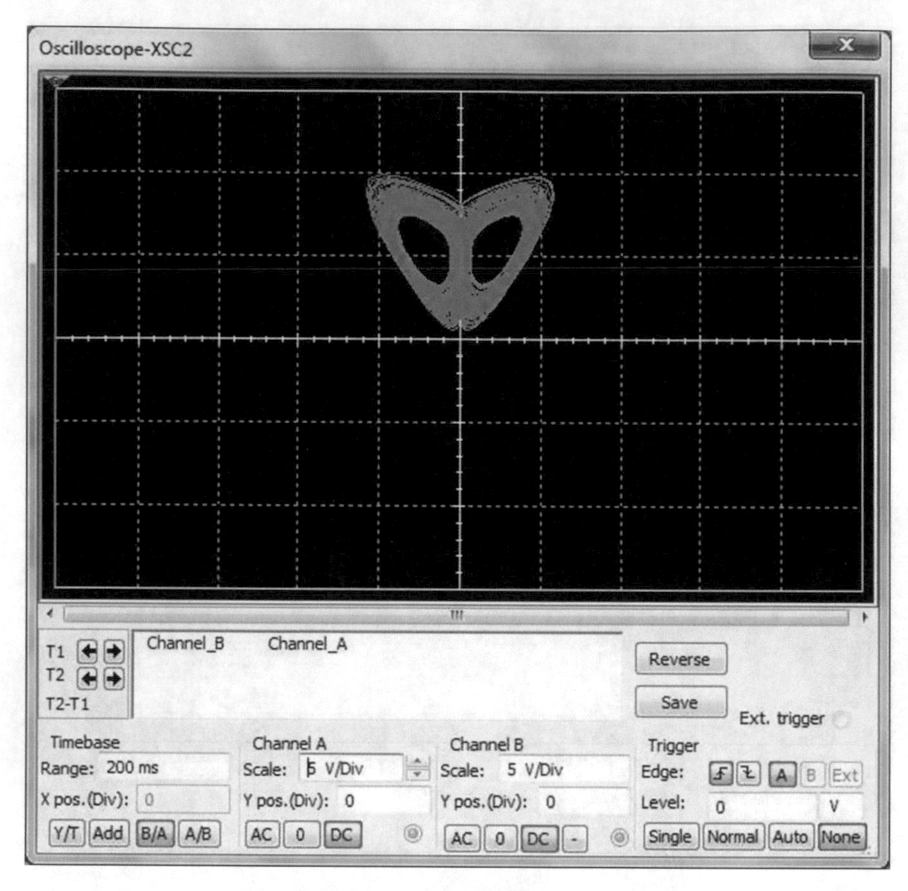

Fig. 16 Multisim plot of the hyperchaotic attractor in the $v_{C_1} - v_{C_3}$ plane

designed via adaptive control method and Lyapunov stability theory. MATLAB simulations are shown to illustrate the phase portraits, adaptive control and synchronization results for the memristor-based hyperchaotic system. It is worth noting that the presence of the memristor creates some special and unusual features. For example, such memristor-based systems can exhibit hyperchaos although it has no equilibrium points. Also, it is well-known that hyperchaotic system, which is characterized by more than one positive Lyapunov exponent, exhibits a higher level of complexity than a conventional chaotic system. Hence, we can apply this memristor-based hyperchaotic system in practical applications like encryption, cryptosystems, neural networks and secure communications.

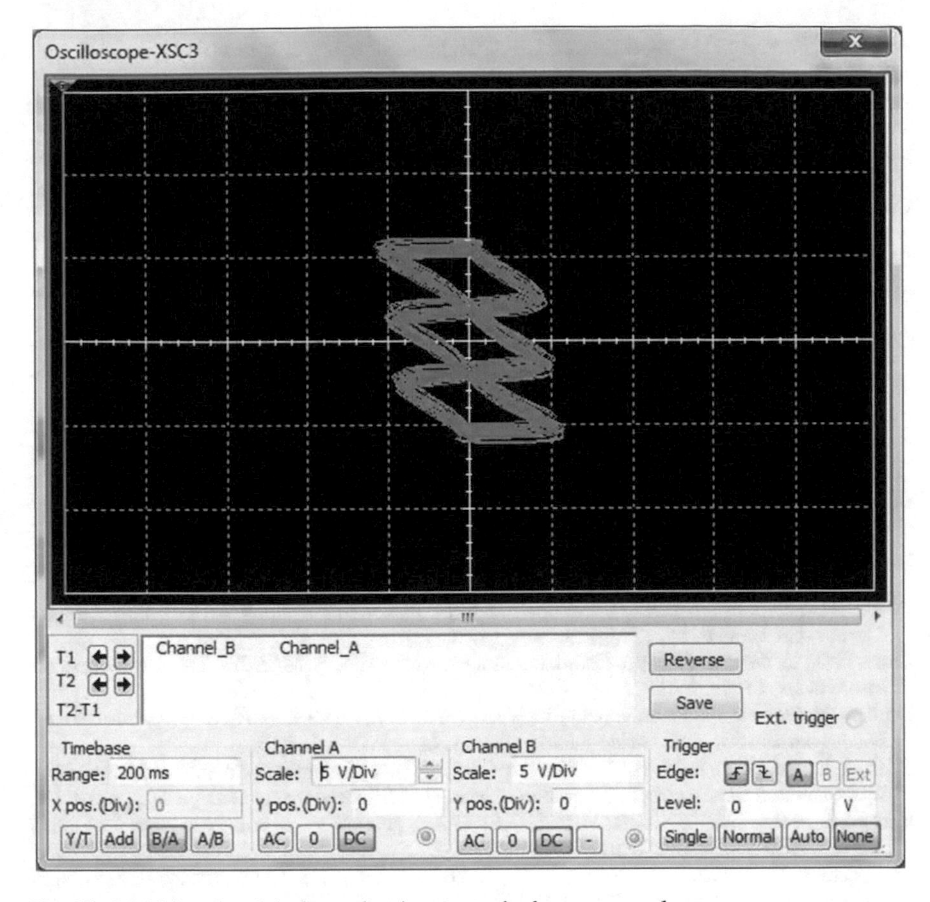

Fig. 17 Multisim plot of the hyperchaotic attractor in the $v_{C_1} - v_{C_4}$ plane

References

Abdurrahman, A., Jiang, H., & Teng, Z. (2015). Finite-time synchronization for memristor-based neural networks with time-varying delays. *Neural Networks*, *69*, 20–28.

Adhikari, S. P., Yang, C., Kim, H., & Chua, L. O. (2012). Memristor bridge synapse-based neural network and its learning. *IEEE Transactions on Neural Networks and Learning Systems*, *23*, 1426–1435.

Adhikari, S. P., Sad, M. P., Kim, H., & Chua, L. O. (2013). Three fingerprints of memristor. *IEEE Transactions on Circuits and Systems I: Regular Papers*, *60*(11), 3008–3021.

Albuquerque, H. A., Rubinger, R. M., & Rech, P. C. (2008). Self-similar structures in a 2D parameter-space of an inductorless Chua's circuit. *Physics Letters A*, *372*, 4793–4798.

Arneodo, A., Coullet, P., & Tresser, C. (1981). Possible new strange attractors with spiral structure. *Commonications in Mathematical Physics*, *79*(4), 573–576.

Azar, A. T., & Vaidyanathan, S. (2015). *Chaos modeling and control systems design* (Vol. 581). Germany: Springer.

Azar, A. T., & Vaidyanathan, S. (2016). *Advances in chaos theory and intelligent control*. Berlin, Germany: Springer.

Azar, A. T., Vaidyanathan, S., & Ouannas, A. (2017). *Fractional order control and synchronization of chaotic systems*. Berlin, Germany: Springer.
Bao, B. C., Liu, Z., & Xu, B. P. (2010). Dynamical analysis of memristor chaotic oscillator. *Acta Physica Sinica*, *59*(6), 3785–3793.
Cai, G., & Tan, Z. (2007). Chaos synchronization of a new chaotic system via nonlinear control. *Journal of Uncertain Systems*, *1*(3), 235–240.
Carroll, T. L., & Pecora, L. M. (1991). Synchronizing chaotic circuits. *IEEE Transactions on Circuits and Systems*, *38*(4), 453–456.
Chen, G., & Ueda, T. (2002). *Chaos in circuits and systems*. Singapore: World Scientific.
Chen, G., & Ueta, T. (1999). Yet another chaotic attractor. *International Journal of Bifurcation and Chaos*, *9*(7), 1465–1466.
Chen, W. H., Wei, D., & Lu, X. (2014). Global exponential synchronization of nonlinear time-delay Lure systems via delayed impulsive control. *Communications in Nonlinear Science and Numerical Simulation*, *19*(9), 3298–3312.
Cheng, C. J., & Cheng, C. B. (2013). An asymmetric image cryptosystem based on the adaptive synchronization of an uncertain unified chaotic system and a cellular neural network. *Communications in Nonlinear Science and Numerical Simulation*, *18*(10), 2825–2837.
Chua, L. O. (1971). Memristor-the missing circuit element. *IEEE Transactions on Circuit Theory*, *18*(5), 507–519.
Chua, L. O. (1994). Chua's circuit: An overview ten years later. *Journal of Circuits, Systems and Computers*, *04*, 117–159.
Chua, L. O., & Yang, L. (1988a). Cellular neural networks: Applications. *IEEE Transactions on Circuits and Systems*, *35*, 1273–1290.
Chua, L. O., & Yang, L. (1988b). Cellular neural networks: Theory. *IEEE Transactions on Circuits and Systems*, *35*, 1257–1272.
Dudkowski, D., Jafari, S., Kapitaniaka, T., Kuznetsov, N. V., Leonov, G. A., & Prasad, A. (2016). Hidden attractors in dynamical systems. *Physics Reports*, *637*, 1–50.
Fitch, A. L., Yu, D. S., Iu, H. H. C., & Sreeram, V. (2012). Hyperchaos in a memristor-based modified canonical Chua's circuit. International Journal of Bifurcation and Chaos, *22*(6), 1250133
Fortuna, L., Frasca, M., & Xibilia, M. G. (2009). *Chua's circuit Implementations: Yesterday, today and tomorrow*. Singapore: World Scientific.
Frederickson, P., Kaplan, J. L., Yorke, E. D., & York, J. A. (1983). The Lyapunov dimension of strange attractors. *Journal of Differential Equations*, *49*, 185–207.
Gan, Q., & Liang, Y. (2012). Synchronization of chaotic neural networks with time delay in the leakage term and parametric uncertainties based on sampled-data control. *Journal of the Franklin Institute*, *349*(6), 1955–1971.
Itoh, M., & Chua, L. O. (2008). Memristor oscillators. *International Journal of Bifurcation and Chaos*, *18*(11), 3183–3206.
Jiang, G. P., Zheng, W. X., & Chen, G. (2004). Global chaos synchronization with channel time-delay. *Chaos, Solitons & Fractals*, *20*(2), 267–275.
Joglekar, Y. N., & Wolf, S. J. (2009). The elusive memristor: Properties of basic electrical circuits. *European Journal of Physics*, *30*(4), 661–675.
Karthikeyan, R., & Sundarapandian, V. (2014). Hybrid chaos synchronization of four-scroll systems via active control. *Journal of Electrical Engineering*, *65*(2), 97–103.
Khalil, H. K. (2001). *Nonlinear systems* (3rd ed.). New Jersey, USA: Prentice Hall.
Kuznetsov, N. V., & Leonov, G. A. (2014). Hidden attractors in dynamical systems: systems with no equilibria, multistability and coexisting attractors. *IFAC Proceedings Volumes*, *47*(3), 5445–5454.
Leonov, G. A., Kuznetsov, N. V., & Vagaitsev, V. I. (2011). Localization of hidden Chua's attractors. *Physics Letters A*, *375*(23), 2230–2233.
Li, D. (2008). A three-scroll chaotic attractor. *Physics Letters A*, *372*(4), 387–393.
Li, G. H., Zhou, S. P., & Yang, K. (2007). Controlling chaos in Colpitts oscillator. *Chaos, Solitons and Fractals*, *33*, 582–587.

Li, H., Wang, L., & Duan, S. (2014). A memristor-mased scroll chaotic system—Design, analysis and circuit implementation. International Journal of Bifurcation and Chaos, *24*(07), 1450099
Li, Q., Hu, S., Tang, S., & Zeng, G. (2013). Hyperchaos and horseshoe in a 4D memristive system with a line of equilibria and its implementation. *International Journal of Circuit Theory and Applications*, *42*(11), 1172–1188.
Liu, L., Wu, X., & Hu, H. (2004). Estimating system parameters of Chua's circuit from synchronizing signal. *Physics Letters A*, *324*(1), 36–41.
Lorenz, E. N. (1963). Deterministic periodic flow. *Journal of the Atmospheric Sciences*, *20*(2), 130–141.
Lü, J., & Chen, G. (2002). A new chaotic attractor coined. *International Journal of Bifurcation and Chaos*, *12*(3), 659–661.
Matsumoto, T. (1984). A chaotic attractor from Chua's circuit. *IEEE Transactions on Circuits and Systems*, *31*, 1055–1058.
Muthuswamy, B. (2010). Implementing memristor based chaotic circuits. *International Journal of Bifurcation and Chaos*, *20*(5), 1335–1350.
Muthuswamy, B., & Chua, L. O. (2010). Simplest chaotic circuit. *International Journal of Bifurcation and Chaos*, *20*(5), 1567–1580.
Muthuswamy, B., & Kokate, P. (2009). Memristor based chaotic circuits. *IETE Technical Review*, *26*(6), 417–429.
Pecora, L. M., & Carroll, T. L. (1990). Synchronization in chaotic systems. *Physical Review Letters*, *64*(8), 821–824.
Pehlivan, I., Moroz, I. M., & Vaidyanathan, S. (2014). Analysis, synchronization and circuit design of a novel butterfly attractor. *Journal of Sound and Vibration*, *333*(20), 5077–5096.
Pham, V. T., Volos, C., Jafari, S., & Wang, X. (2014). Generating a novel hyperchaotic system out of equilibrium. *Optoelectronics and Advanced Materials-Rapid Communications*, *8*, 535–539.
Pham, V. T., Volos, C. K., Vaidyanathan, S., Le, T. P., & Vu, V. Y. (2015). A memristor-based hyperchaotic system with hidden attractors: Dynamics, synchronization and circuital emulating. *Journal of Engineering Science and Technology Review*, *8*(2), 205–214.
Rasappan, S., & Vaidyanathan, S. (2012a). Global chaos synchronization of WINDMI and Coullet chaotic systems by backstepping control. *Far East Journal of Mathematical Sciences*, *67*(2), 265–287.
Rasappan, S., & Vaidyanathan, S. (2012b). Hybrid synchronization of n-scroll Chua and Lur'e chaotic systems via backstepping control with novel feedback. *Archives of Control Sciences*, *22*(3), 343–365.
Rasappan, S., & Vaidyanathan, S. (2012c). Synchronization of hyperchaotic Liu system via backstepping control with recursive feedback. *Communications in Computer and Information Science*, *305*, 212–221.
Rasappan, S., & Vaidyanathan, S. (2013). Hybrid synchronization of *n*-scroll chaotic Chua circuits using adaptive backstepping control design with recursive feedback. *Malaysian Journal of Mathematical Sciences*, *7*(2), 219–246.
Rasappan, S., & Vaidyanathan, S. (2014). Global chaos synchronization of WINDMI and Coullet chaotic systems using adaptive backstepping control design. *Kyungpook Mathematical Journal*, *54*(1), 293–320.
Rössler, O. E. (1976). An equation for continuous chaos. *Physics Letters A*, *57*(5), 397–398.
Sampath, S., Vaidyanathan, S., Volos, C. K., & Pham, V. T. (2015). An eight-term novel four-scroll chaotic system with cubic nonlinearity and its circuit simulation. *Journal of Engineering Science and Technology Review*, *8*(2), 1–6.
Sarasu, P., & Sundarapandian, V. (2011a). Active controller design for the generalized projective synchronization of four-scroll chaotic systems. *International Journal of Systems Signal Control and Engineering Application*, *4*(2), 26–33.
Sarasu, P., & Sundarapandian, V. (2011b). The generalized projective synchronization of hyperchaotic Lorenz and hyperchaotic Qi systems via active control. *International Journal of Soft Computing*, *6*(5), 216–223.

Sarasu, P., & Sundarapandian, V. (2012). Adaptive controller design for the generalized projective synchronization of 4-scroll systems. *International Journal of Systems Signal Control and Engineering Application*, *5*(2), 21–30.
Sarasu, P., & Sundarapandian, V. (2012b). Generalized projective synchronization of three-scroll chaotic systems via adaptive control. *European Journal of Scientific Research*, *72*(4), 504–522.
Sarasu, P., & Sundarapandian, V. (2012c). Generalized projective synchronization of two-scroll systems via adaptive control. *International Journal of Soft Computing*, *7*(4), 146–156.
Shang, Y., Fei, W., & Yu, H. (2012). Analysis and modeling of internal state variables for dynamic effects of nonvolatile memory devices. *IEEE Transactions on Circuits and Systems I: Regular Papers*, *59*, 1906–1918.
Shin, S., Kim, K., & Kang, S. M. (2011). Memristor applications for programmable analog ICs. *IEEE Transactions on Nanotechnology*, *410*, 266–274.
Sprott, J. C. (1994). Some simple chaotic flows. *Physical Review E*, *50*(2), 647–650.
Sprott, J. C. (2011). A proposed standard for the publication of new chaotic systems. *International Journal of Bifurcation and Chaos*, *21*(9), 2391–2394.
Strukov, D., Snider, G., Stewart, G., & Williams, R. (2008). The missing memristor found. *Nature*, *453*, 80–83.
Sundarapandian, V. (2010). Output regulation of the Lorenz attractor. *Far East Journal of Mathematical Sciences*, *42*(2), 289–299.
Sundarapandian, V. (2011). Output regulation of the Arneodo-Coullet chaotic system. *Communications in Computer and Information Science*, *133*, 98–107.
Sundarapandian, V. (2013). Analysis and anti-synchronization of a novel chaotic system via active and adaptive controllers. *Journal of Engineering Science and Technology Review*, *6*(4), 45–52.
Sundarapandian, V., & Karthikeyan, R. (2011a). Anti-synchronization of hyperchaotic Lorenz and hyperchaotic Chen systems by adaptive control. *International Journal of Systmes Signal Control and Engineering Application*, *4*(2), 18–25.
Sundarapandian, V., & Karthikeyan, R. (2011b). Anti-synchronization of Lü and Pan chaotic systems by adaptive nonlinear control. *European Journal of Scientific Research*, *64*(1), 94–106.
Sundarapandian, V., & Karthikeyan, R. (2012a). Adaptive anti-synchronization of uncertain Tigan and Li systems. *Journal of Engineering and Applied Sciences*, *7*(1), 45–52.
Sundarapandian, V., & Karthikeyan, R. (2012b). Hybrid synchronization of hyperchaotic Lorenz and hyperchaotic Chen systems via active control. *Journal of Engineering and Applied Sciences*, *7*(3), 254–264.
Sundarapandian, V., & Pehlivan, I. (2012). Analysis, control, synchronization, and circuit design of a novel chaotic system. *Mathematical and Computer Modelling*, *55*(7–8), 1904–1915.
Sundarapandian, V., & Sivaperumal, S. (2011). Sliding controller design of hybrid synchronization of four-wing chaotic systems. *International Journal of Soft Computing*, *6*(5), 224–231.
Suresh, R., & Sundarapandian, V. (2013). Global chaos synchronization of a family of *n*-scroll hyperchaotic Chua circuits using backstepping control with recursive feedback. *Far East Journal of Mathematical Sciences*, *73*(1), 73–95.
Tang, F., & Wang, L. (2005). An adaptive active control for the modified Chua's circuit. *Physics Letters A*, *346*, 342–346.
Tetzlaff, R. (2014). *Memristors and memristive systems*. Berlin, Germany: Springer.
Tigan, G., & Opris, D. (2008). Analysis of a 3D chaotic system. *Chaos, Solitons and Fractals*, *36*, 1315–1319.
Vaidyanathan, S. (2011a). Hybrid chaos synchronization of Liu and Lü systems by active nonlinear control. *Communications in Computer and Information Science*, *204*, 1–10.
Vaidyanathan, S. (2011b). Output regulation of the unified chaotic system. *Communications in Computer and Information Science*, *204*, 84–93.
Vaidyanathan, S. (2012a). Analysis and synchronization of the hyperchaotic Yujun systems via sliding mode control. *Advances in Intelligent Systems and Computing*, *176*, 329–337.
Vaidyanathan, S. (2012b). Anti-synchronization of Sprott-L and Sprott-M chaotic systems via adaptive control. *International Journal of Control Theory and Applications*, *5*(1), 41–59.

Vaidyanathan, S. (2012c). Global chaos control of hyperchaotic Liu system via sliding control method. *International Journal of Control Theory and Applications*, *5*(2), 117–123.
Vaidyanathan, S. (2012d). Output regulation of the Liu chaotic system. *Applied Mechanics and Materials*, *110–116*, 3982–3989.
Vaidyanathan, S. (2012e). Sliding mode control based global chaos control of Liu-Liu-Liu-Su chaotic system. *International Journal of Control Theory and Applications*, *5*(1), 15–20.
Vaidyanathan, S. (2013a). A new six-term 3-D chaotic system with an exponential nonlinearity. *Far East Journal of Mathematical Sciences*, *79*(1), 135–143.
Vaidyanathan, S. (2013b). Analysis and adaptive synchronization of two novel chaotic systems with hyperbolic sinusoidal and cosinusoidal nonlinearity and unknown parameters. *Journal of Engineering Science and Technology Review*, *6*(4), 53–65.
Vaidyanathan, S. (2013c). Analysis, control and synchronization of hyperchaotic Zhou system via adaptive control. *Advances in Intelligent Systems and Computing*, *177*, 1–10.
Vaidyanathan, S. (2014a). A new eight-term 3-D polynomial chaotic system with three quadratic nonlinearities. *Far East Journal of Mathematical Sciences*, *84*(2), 219–226.
Vaidyanathan, S. (2014b). Analysis and adaptive synchronization of eight-term 3-D polynomial chaotic systems with three quadratic nonlinearities. *European Physical Journal: Special Topics*, *223*(8), 1519–1529.
Vaidyanathan, S. (2014c). Analysis, control and synchronisation of a six-term novel chaotic system with three quadratic nonlinearities. *International Journal of Modelling, Identification and Control*, *22*(1), 41–53.
Vaidyanathan, S. (2014d). Generalized projective synchronisation of novel 3-D chaotic systems with an exponential non-linearity via active and adaptive control. *International Journal of Modelling, Identification and Control*, *22*(3), 207–217.
Vaidyanathan, S. (2014e). Global chaos synchronization of identical Li-Wu chaotic systems via sliding mode control. *International Journal of Modelling, Identification and Control*, *22*(2), 170–177.
Vaidyanathan, S. (2015a). 3-cells Cellular Neural Network (CNN) attractor and its adaptive biological control. *International Journal of PharmTech Research*, *8*(4), 632–640.
Vaidyanathan, S. (2015b). A 3-D novel highly chaotic system with four quadratic nonlinearities, its adaptive control and anti-synchronization with unknown parameters. *Journal of Engineering Science and Technology Review*, *8*(2), 106–115.
Vaidyanathan, S. (2015c). A novel chemical chaotic reactor system and its adaptive control. *International Journal of ChemTech Research*, *8*(7), 146–158.
Vaidyanathan, S. (2015d). Adaptive backstepping control of enzymes-substrates system with ferroelectric behaviour in brain waves. *International Journal of PharmTech Research*, *8*(2), 256–261.
Vaidyanathan, S. (2015e). Adaptive biological control of generalized Lotka-Volterra three-species biological system. *International Journal of PharmTech Research*, *8*(4), 622–631.
Vaidyanathan, S. (2015f). Adaptive chaotic synchronization of enzymes-substrates system with ferroelectric behaviour in brain waves. *International Journal of PharmTech Research*, *8*(5), 964–973.
Vaidyanathan, S. (2015g). Adaptive control of a chemical chaotic reactor. *International Journal of PharmTech Research*, *8*(3), 377–382.
Vaidyanathan, S. (2015h). Adaptive control of the FitzHugh-Nagumo chaotic neuron model. *International Journal of PharmTech Research*, *8*(6), 117–127.
Vaidyanathan, S. (2015i). Adaptive synchronization of chemical chaotic reactors. *International Journal of ChemTech Research*, *8*(2), 612–621.
Vaidyanathan, S. (2015j). Adaptive synchronization of generalized Lotka-Volterra three-species biological systems. *International Journal of PharmTech Research*, *8*(5), 928–937.
Vaidyanathan, S. (2015k). Adaptive synchronization of novel 3-D chemical chaotic reactor systems. *International Journal of ChemTech Research*, *8*(7), 159–171.
Vaidyanathan, S. (2015l). Adaptive synchronization of the identical FitzHugh-Nagumo chaotic neuron models. *International Journal of PharmTech Research*, *8*(6), 167–177.

Vaidyanathan, S. (2015m). Analysis, control and synchronization of a 3-D novel jerk chaotic system with two quadratic nonlinearities. *Kyungpook Mathematical Journal*, *55*, 563–586.
Vaidyanathan, S. (2015n). Analysis, properties and control of an eight-term 3-D chaotic system with an exponential nonlinearity. *International Journal of Modelling, Identification and Control*, *23*(2), 164–172.
Vaidyanathan, S. (2015o). Anti-synchronization of brusselator chemical reaction systems via adaptive control. *International Journal of ChemTech Research*, *8*(6), 759–768.
Vaidyanathan, S. (2015p). Chaos in neurons and adaptive control of Birkhoff-Shaw strange chaotic attractor. *International Journal of PharmTech Research*, *8*(5), 956–963.
Vaidyanathan, S. (2015q). Chaos in neurons and synchronization of Birkhoff-Shaw strange chaotic attractors via adaptive control. *International Journal of PharmTech Research*, *8*(6), 1–11.
Vaidyanathan, S. (2015r). Coleman-Gomatam logarithmic competitive biology models and their ecological monitoring. *International Journal of PharmTech Research*, *8*(6), 94–105.
Vaidyanathan, S. (2015s). Dynamics and control of brusselator chemical reaction. *International Journal of ChemTech Research*, *8*(6), 740–749.
Vaidyanathan, S. (2015t). Dynamics and control of tokamak system with symmetric and magnetically confined plasma. *International Journal of ChemTech Research*, *8*(6), 795–803.
Vaidyanathan, S. (2015u). Global chaos synchronization of chemical chaotic reactors via novel sliding mode control method. *International Journal of ChemTech Research*, *8*(7), 209–221.
Vaidyanathan, S. (2015v). Global chaos synchronization of the forced Van der Pol chaotic oscillators via adaptive control method. *International Journal of PharmTech Research*, *8*(6), 156–166.
Vaidyanathan, S. (2015w). Global chaos synchronization of the Lotka-Volterra biological systems with four competitive species via active control. *International Journal of PharmTech Research*, *8*(6), 206–217.
Vaidyanathan, S. (2015x). Lotka-Volterra population biology models with negative feedback and their ecological monitoring. *International Journal of PharmTech Research*, *8*(5), 974–981.
Vaidyanathan, S. (2015y). Lotka-Volterra two species competitive biology models and their ecological monitoring. *International Journal of PharmTech Research*, *8*(6), 32–44.
Vaidyanathan, S. (2015z). Output regulation of the forced Van der Pol chaotic oscillator via adaptive control method. *International Journal of PharmTech Research*, *8*(6), 106–116.
Vaidyanathan, S. (2016). Anti-synchronization of 3-cells Cellular Neural Network attractors via integral sliding mode control. *International Journal of PharmTech Research*, *9*(1), 193–205.
Vaidyanathan, S., & Azar, A. T. (2015a). Analysis and control of a 4-D novel hyperchaotic system. In A. T. Azar & S. Vaidyanathan (Eds.), *Chaos modeling and control systems design*. Studies in computational intelligence (Vol. 581, pp. 19–38). Germany: Springer.
Vaidyanathan, S., & Azar, A. T. (2015b). Analysis, control and synchronization of a nine-term 3-D novel chaotic system. In A. T. Azar & S. Vaidyanathan (Eds.), *Chaos modelling and control systems design*. Studies in computational intelligence (Vol. 581, pp. 19–38). Germany: Springer.
Vaidyanathan, S., & Azar, A. T. (2015c). Anti-synchronization of identical chaotic systems using sliding mode control and an application to Vaidhyanathan-Madhavan chaotic systems. *Studies in Computational Intelligence*, *576*, 527–547.
Vaidyanathan, S., & Azar, A. T. (2015d). Hybrid synchronization of identical chaotic systems using sliding mode control and an application to Vaidhyanathan chaotic systems. *Studies in Computational Intelligence*, *576*, 549–569.
Vaidyanathan, S., & Madhavan, K. (2013). Analysis, adaptive control and synchronization of a seven-term novel 3-D chaotic system. *International Journal of Control Theory and Applications*, *6*(2), 121–137.
Vaidyanathan, S., & Pakiriswamy, S. (2013). Generalized projective synchronization of six-term Sundarapandian chaotic systems by adaptive control. *International Journal of Control Theory and Applications*, *6*(2), 153–163.
Vaidyanathan, S., & Pakiriswamy, S. (2015). A 3-D novel conservative chaotic system and its generalized projective synchronization via adaptive control. *Journal of Engineering Science and Technology Review*, *8*(2), 52–60.

Vaidyanathan, S., & Rajagopal, K. (2011a). Anti-synchronization of Li and T chaotic systems by active nonlinear control. *Communications in Computer and Information Science*, *198*, 175–184.

Vaidyanathan, S., & Rajagopal, K. (2011b). Global chaos synchronization of hyperchaotic Pang and Wang systems by active nonlinear control. *Communications in Computer and Information Science*, *204*, 84–93.

Vaidyanathan, S., & Rajagopal, K. (2011c). Global chaos synchronization of Lü and Pan systems by adaptive nonlinear control. *Communications in Computer and Information Science*, *205*, 193–202.

Vaidyanathan, S., & Rasappan, S. (2011). Global chaos synchronization of hyperchaotic Bao and Xu systems by active nonlinear control. *Communications in Computer and Information Science*, *198*, 10–17.

Vaidyanathan, S., & Pakiriswamy, S. (2015). A 3-D novel conservative chaotic system and its generalized projective synchronization via adaptive control. *Journal of Engineering Science and Technology Review*, *8*(2), 52–60.

Vaidyanathan, S., & Sampath, S. (2011). Global chaos synchronization of hyperchaotic Lorenz systems by sliding mode control. *Communications in Computer and Information Science*, *205*, 156–164.

Vaidyanathan, S., & Sampath, S. (2012). Anti-synchronization of four-wing chaotic systems via sliding mode control. *International Journal of Automation and Computing*, *9*(3), 274–279.

Vaidyanathan, S., & Volos, C. (2015). Analysis and adaptive control of a novel 3-D conservative no-equilibrium chaotic system. *Archives of Control Sciences*, *25*(3), 333–353.

Vaidyanathan, S., & Volos, C. (2016a). *Advances and Applications in Chaotic Systems*. Berlin, Germany: Springer.

Vaidyanathan, S., & Volos, C. (2016b). *Advances and applications in nonlinear control systems*. Berlin, Germany: Springer.

Vaidyanathan, S., Volos, C., & Pham, V. T. (2014a). Hyperchaos, adaptive control and synchronization of a novel 5-D hyperchaotic system with three positive Lyapunov exponents and its SPICE implementation. *Archies of Control Sciences*, *24*(4), 409–446.

Vaidyanathan, S., Volos, C., & Pham, V. T. (2014b). Hyperchaos, adaptive control and synchronization of a novel 5-D hyperchaotic system with three positive Lyapunov exponents and its SPICE implementation. *Archives of Control Sciences*, *24*(4), 409–446.

Vaidyanathan, S., Volos, C., Pham, V. T., Madhavan, K., & Idowu, B. A. (2014c). Adaptive backstepping control, synchronization and circuit simulation of a 3-D novel jerk chaotic system with two hyperbolic sinusoidal nonlinearities. *Archives of Control Sciences*, *24*(3), 375–403.

Vaidyanathan, S., Idowu, B. A., & Azar, A. T. (2015a). Backstepping controller design for the global chaos synchronization of Sprott's jerk systems. *Studies in Computational Intelligence*, *581*, 39–58.

Vaidyanathan, S., Rajagopal, K., Volos, C. K., Kyprianidis, I. M., & Stouboulos, I. N. (2015b). Analysis, adaptive control and synchronization of a seven-term novel 3-D chaotic system with three quadratic nonlinearities and its digital implementation in LabVIEW. *Journal of Engineering Science and Technology Review*, *8*(2), 130–141.

Vaidyanathan, S., Volos, C., Pham, V. T., & Madhavan, K. (2015c). Analysis, adaptive control and synchronization of a novel 4-D hyperchaotic hyperjerk system and its SPICE implementation. *Archives of Control Sciences*, *25*(1), 5–28.

Vaidyanathan, S., Volos, C. K., Kyprianidis, I. M., Stouboulos, I. N., & Pham, V. T. (2015d). Analysis, adaptive control and anti-synchronization of a six-term novel jerk chaotic system with two exponential nonlinearities and its circuit simulation. *Journal of Engineering Science and Technology Review*, *8*(2), 24–36.

Vaidyanathan, S., Volos, C. K., & Madhavan, K. (2015e). Analysis, control, synchronization and SPICE implementation of a novel 4-D hyperchaotic Rikitake dynamo System without equilibrium. *Journal of Engineering Science and Technology Review*, *8*(2), 232–244.

Vaidyanathan, S., Volos, C. K., & Pham, V. T. (2015f). Analysis, adaptive control and adaptive synchronization of a nine-term novel 3-D chaotic system with four quadratic nonlinearities and its circuit simulation. *Journal of Engineering Science and Technology Review*, *8*(2), 181–191.

Vaidyanathan, S., Volos, C. K., & Pham, V. T. (2015g). Global chaos control of a novel nine-term chaotic system via sliding mode control. In A. T. Azar & Q. Zhu (Eds.), *Advances and applications in sliding mode control systems*. Studies in computational intelligence (Vol. 576, pp. 571–590). Germany: Springer.

Vaidyanathan, S., Volos, C. K., Pham, V. T., & Madhavan, K. (2015h). Analysis, adaptive control and synchronization of a novel 4-D hyperchaotic hyperjerk system and its SPICE implementation. *Archives of Control Sciences*, *25*(1), 135–158.

Vaidyanathan, S., Volos, C. K., Rajagopal, K., Kyprianidis, I. M., & Stouboulos, I. N. (2015i). Adaptive backstepping controller design for the anti-synchronization of identical WINDMI chaotic systems with unknown parameters and its SPICE implementation. *Journal of Engineering Science and Technology Review*, *8*(2), 74–82.

Volos, C. K., Kyprianidis, I. M., & Stouboulos, I. N. (2012). A chaotic path planning generator for autonomous mobile robots. *Robotics and Autonomous Systems*, *60*(4), 651–656.

Volos, C. K., Kyprianidis, I. M., & Stouboulos, I. N. (2013). Image encryption process based on chaotic synchronization phenomena. *Signal Processing*, *93*(5), 1328–1340.

Volos, C. K., Kyprianidis, I. M., Stouboulos, I. N., Tlelo-Cuautle, E., & Vaidyanathan, S. (2015). Memristor: A new concept in synchronization of coupled neuromorphic circuits. *Journal of Engineering Science and Technology Review*, *8*(2), 157–173.

Wang, L., Zhang, C., Chen, L., Lai, J., & Tong, J. (2012a). A novel memristor-based rSRAM structure for multiple-bit upsets immunity. *IEICE Electronics Express*, *9*, 861–867.

Wang, X., & Ge, C. (2008). Controlling and tracking of Newton-Leipnik system via backstepping design. *International Journal of Nonlinear Science*, *5*(2), 133–139.

Wang, X., Xu, B., & Luo, C. (2012b). An asynchronous communication system based on the hyperchaotic system of 6th-order cellular neural network. *Optics Communications*, *285*(24), 5401–5405.

Wei, Z., & Yang, Q. (2010). Anti-control of Hopf bifurcation in the new chaotic system with two stable node-foci. *Applied Mathematics and Computation*, *217*(1), 422–429.

Xiao, X., Zhou, L., & Zhang, Z. (2014). Synchronization of chaotic Lure systems with quantized sampled-data controller. *Communications in Nonlinear Science and Numerical Simulation*, *19*(6), 2039–2047.

Yalcin, M. E., Suykens, J. A. K., & Vandewalle, J. (2004). True random bit generator from a double-scroll attractor. *IEEE Transactions on Circuits and Systems I: Regular Papers*, *51*(7), 1395–1404.

Yang, J. J., Strukov, D. B., & Stewart, D. R. (2013). Memristive devices for computing. *Nature Nanotechnology*, *8*, 13–24.

Zhou, W., Xu, Y., Lu, H., & Pan, L. (2008). On dynamics analysis of a new chaotic attractor. *Physics Letters A*, *372*(36), 5773–5777.

Zhu, C., Liu, Y., & Guo, Y. (2010). Theoretic and numerical study of a new chaotic system. *Intelligent Information Management*, *2*, 104–109.

Modern System Design Using Memristors

Lauren Guckert and Earl Swartzlander Jr.

Abstract Since memristors have recently come to the forefront of the computer architecture field, the majority of the research is still in its infancy. The most popular application for memristors is memories, namely crossbars, but preliminary work has shown their use in logic circuits too. This work explores two approaches to memristor logic, IMPLY operations and MAD gates. While IMPLY has been successfully demonstrated and popularized in previous works, it suffers from long latencies and destructive operations. MAD gates have been shown to overcome these issues, offering a lower area and lower latency alternative. These two approaches are described, implemented, and analyzed against each other and other proposed approaches to memristor logic. Both methodologies are then presented in the context of a crossbar, showing how IMPLY and MAD operations can be performed on memory cells. It is shown that they offer improved logic-in-memory implementations over alternative proposed works. Lastly, general considerations when designing memristor-based circuits are discussed and future directions of research are motivated.

Keywords Memristors · IMPLY · MAD gates · Crossbar · Logic-in-memory

1 Introduction

Memristors were first introduced in 1971 (Chua 1971; Strukov 2008). Since memristors are a relatively new concept in the field of computer architecture, there is not yet a standardized approach to incorporating them into modern system

L. Guckert (✉) · E. Swartzlander Jr.
Department of Electrical and Computer Engineering,
The University of Texas at Austin, Austin, TX 78712, USA
e-mail: lguckert@utexas.edu

E. Swartzlander Jr.
e-mail: eswartzla@aol.com

© Springer International Publishing AG 2017

S. Vaidyanathan and C. Volos (eds.), *Advances in Memristors, Memristive Devices and Systems*, Studies in Computational Intelligence 701,
DOI 10.1007/978-3-319-51724-7_6

design. Current research focuses on exploring various ways to harness the characteristics and strengths of memristors to improve future designs in terms of performance, area, and energy. This research falls into 3 broad categories: memristor design, logic applications for memristors, and memory applications for memristors. This chapter will focus on the latter two, exploring ways to apply current memristor models to redesign traditional CMOS circuits for the nanoscale domain.

Even within the bounds of memristor applications, there is a multitude of parallel and orthogonal research approaches to implement and optimize designs. Much of the research focuses on memory applications because memristor characteristics lend themselves particularly well to this domain. Thus, research into logic applications for memristors is still in its infancy and has great potential to improve. Although there have been many approaches proposed, it is not clear that a realistic, optimal, fabricate-able implementation of logic gates with memristors has been discovered.

There is another area of research at the intersection of memory and logic designs —logic-in-memory memory designs. Logic-in-memory is the idea of moving away from traditional Von Neumann designs where the data is brought from memory to the computational circuit and instead the computation is brought from the CPU to the memory. This avoids the overhead of reading and writing the data to and from memory, improving performance and power consumption. This prospect has spurred a subset of memristor research that focuses on how to bridge the gap between memory applications and logic applications such that a single standardized approach can be applied to both. The research attempts to find logic designs using memristors that can also be applied to a crossbar memory context. In this way, both traditional Von-Neumann and progressive logic-in-memory system designs can be pursued with memristors.

This chapter covers the current state-of-the-art in memristor-based logic-in-memory circuit design. First, the most common approach to memristor logic, the IMPLY operation, is presented. Many works have shown that the IMPLY operation can be used to implement common logic circuits (Mahajan 2014; Teimoory 2014, 2015; Rose 2012). Although this approach offers an effective, popular way to perform Boolean operations, it has shortcomings in terms of high latency, serialized operations (Corinto 2012; Deng 2013; Devolder 2008), and a high number of drivers. Second, MAD gates, or Memristors-As-Drivers gates, are introduced as a way to overcome some of the issues with the IMPLY approach (Guckert 2016). Both of the approaches stray from traditional Von-Neumann paradigm by bringing the drivers (computation) to the memristors (data). The implementations are analyzed and compared against other approaches to logic in terms of latency, area, and power. Then, applications of the IMPLY and MAD gates are shown, both in the realm of logic and memory. A multiplexer and a 4 × 4 crossbar memory are used as motivating examples. Descriptions for optimizations, limitations, as well as general considerations when designing with memristors are also discussed.

2 IMPLY Memristor Logic

IMPLY based Boolean operations are built from the basic IMPLY circuitry shown in Fig. 1 (Bickerstaff 2010).

The circuitry requires two memristors (one for each operand), a pull-down resistor, and driver circuitry to apply the appropriate voltages. The operation is performed by applying a voltage V_{cond} to memristor P and a voltage V_{set} to memristor Q concurrently where $V_{cond} < V_{set}$. It is assumed that the value of P and Q have been set in the memristors before execution. The corresponding truth table for the IMPLY operation is given in Table 1.

The value of the pull-down resistor R_g, $V_{cond,}$ and V_{set} are determined by the internal parameters of the memristors themselves. For example, depending on the threshold voltages or currents and the doping width of the memristors, the correct value of R_g must be set to ensure that the correct current flows through memristors P and Q when V_{cond} and V_{set} are applied.

Recall that memristors can be written if a high-magnitude current is driven through the memristor, else it can be read. In the context of an IMPLY operation, the Q memristor is written with the result while the P memristor remains unchanged. Thus, the current in the circuit during the IMPLY operation must be controlled such that the voltage drop across P is not enough to change its value, but the voltage drop across Q is enough to change its value. This is ensured through the selection of V_{cond}, V_{set}, and R_g. Throughout this chapter $V_{cond} = 1.6$ V, $V_{set} = 2.5$ V, and

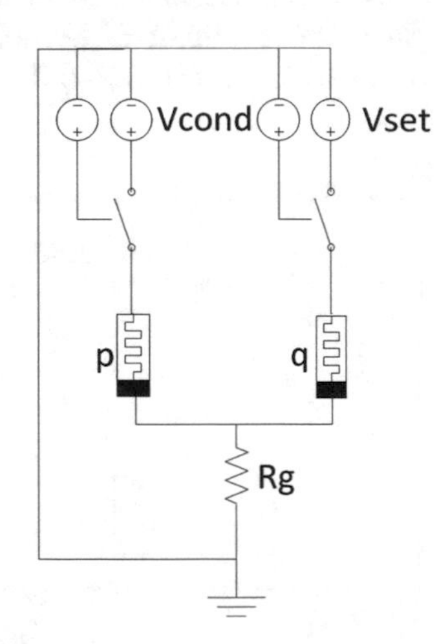

Fig. 1 Circuitry for the IMPLY operation p → q

Table 1 Truth table for the IMPLY operation

Case	*p*	*q*	*p* → *q*
1	0	0	1
2	0	1	1
3	1	0	0
4	1	1	1

Table 2 IMPLY implementations of Boolean operations (Bickerstaff 2010)

Operation	Implementations	#Steps	#Memristors
P NAND Q	P IMP (Q IMP 0)	2	3
P AND Q	(P IMP (Q IMP 0)) IMP 0	3	4
P NOR Q	((P IMP 0) IMP Q) IMP 0	5	6
P OR Q	(P IMP 0) IMP Q	4	5
P XOR Q	(P IMP Q) IMP ((Q IMP P) IMP 0)	8	7
NOT P	P IMP 0	1	2

$R_g = 2$ K are used. Also, high resistance and low resistance are selected to be $R_{high} = 100$ K and $R_{low} = 1$ K respectively (Kvatinsky 2014b). The VTeam model (Kvatinsky 2015) is used to represent these values in simulation.

From Table 1, it can be seen that the NOT operation can be computed by setting the value in memristor P and the value 0 in memristor Q and performing an IMPLY. Thus, by adding the ability to perform a FALSE operation, essentially a memristor reset, the IMPLY operation becomes functionally complete. The reset operation can be performed by applying a V_{reset} voltage to the memristor where $|V_{reset}| > |V_{set}|$. For this chapter, assume $V_{reset} = -5$ V. In order to implement more complex operations, the IMPLY circuitry can be extended by introducing more memristors in parallel and performing consecutive serialized IMPLY operations on pairs at a time. The resultant Boolean operations and their costs are given in Table 2.

A '0' represents a reset memristor with a high-resistance, or a logical '0' value. An example circuit of extending the IMPLY circuitry to implement a NAND operation is shown in Fig. 2.

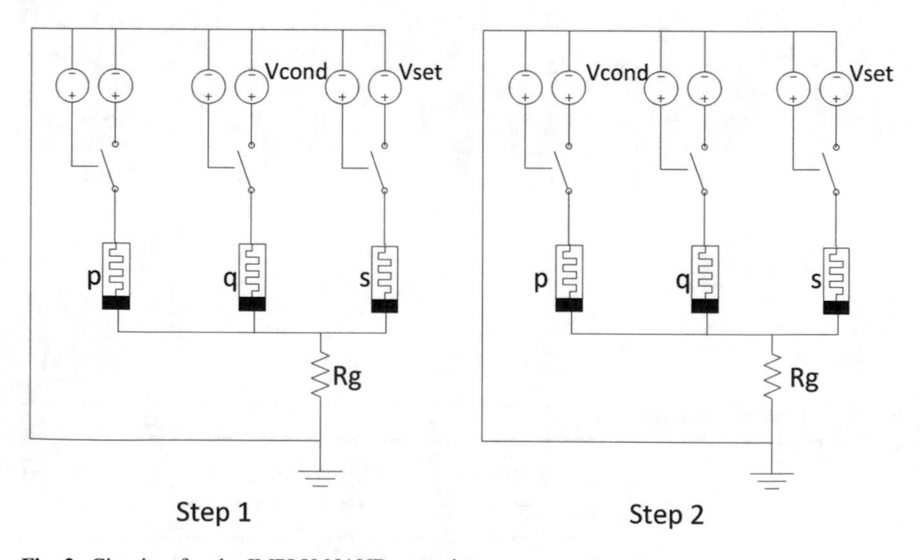

Fig. 2 Circuitry for the IMPLY NAND operation

A third memristor, s, is added which is cleared, or initialized to 0, or high resistance. In the first step, q IMP 0 is performed by applying the read voltage V_{cond} to the q memristor and the write voltage V_{set} to the s memristor. No voltage is applied to the p memristor. At the end of this step, the result of q IMP 0 lies in the s memristor. In the second step, the IMPLY operation is performed on the p memristor and the result from step 1. p IMP (q IMP 0) is performed by applying V_{cond} to the p memristor and V_{set} to the s memristor (where the result from step 1 is). Again, the result of Step 2 lies in the s memristor. p NAND q has been completed. A similar explanation exists for the remaining Boolean operations.

In addition to performing the IMPLY operations themselves, it is necessary to perform a copy operation on the operands which are overwritten for some operations. For example, in the XOR operation, after performing p IMP q, the result of this operation lies in the q memristor rather than the original value of q. Thus, the operation q IMP p required for the XOR operation is no longer possible. A copy of q must be created before performing p IMP q. This is done by performing two consecutive NOT operations, p IMP 0, and (p IMP 0) IMP 0, incurring 2 steps and 2 memristors. For the same reason, a copy of p must also be created. Thus the total cost is 8 steps and 7 memristors.

2.1 *IMPLY Implementation of a Multiplexer*

Once it is understood how to perform basic Boolean logic with the IMPLY operation, the concept can be extended to implement more complex building blocks to expand the portfolio of memristor structures. As an example, consider a 1-bit multiplexer. The execution steps for an implementation of this IMPLY-based multiplexer are shown in Table 3.

S represents the select line memristor and $\overline{S}$ represents the memristor holding the inverse of the select line. Note that this series of steps assumes that the memristors have been initialized ahead of time with a single write operation (as described in the previous section). First, the operations from Table 2 are used to compute A NAND S. Next, B AND $\overline{S}$ is computed in the same manner. Finally, the two

Table 3 Steps for a 1-bit multiplexer (A AND S) OR (B AND $\overline{S}$) = Out

Step	Mux functionality	Mux goal
1	S IMP 0	NOT S
2	A IMP (S IMP 0)	A NAND S
3	$\overline{S}$ IMP 0	NOT ($\overline{S}$)
4	B IMP ($\overline{S}$ IMP 0)	B NAND $\overline{S}$
5	(A IMP (S IMP 0)) IMP 0	A AND S
6	((A IMP (S IMP 0)) IMP 0) IMP (B IMP ($\overline{S}$ IMP 0))	(A AND S) OR (B AND $\overline{S}$) == Out

outputs are OR'ed together to produce the output. In total, the multiplexer requires 6 IMPLY steps to produce the result. There are four input memristors for the A, B, S, and $\overline{S}$ inputs and a single memristor for the output Out. In addition to these 5 memristors, there are two more memristors required in steps 1 and 5 as initialized, reset memristors to hold the IMPLY results. The initialized reset memristor used in Step 3 does not contribute an additional memristor because it is used as the Out memristor as the execution continues. Thus this implementation requires 6 IMPLY steps and 7 memristors.

However, in the context of the IMPLY operation, it is often possible to optimize out many of the steps and thus reduce the number of memristors by rearranging the Boolean logic. The equation for a multiplexer is optimized for the context of traditional CMOS Boolean gates and their transistor complexity. However, the same tradeoffs do not exist in the memristor domain and IMPLY Boolean operations. Thus, the multiplexer equation is rewritten to an equivalent form.

$$(\mathrm{A\ AND\ S})\mathrm{OR}(\mathrm{B\ AND\ }\overline{\mathrm{S}}) == \overline{(\mathrm{B\ NAND\ }\overline{\mathrm{S}})}\mathrm{OR}\overline{(\mathrm{A\ NAND\ S})} = \mathrm{Out}$$

By interpreting the equation for the multiplexer as such, the number of steps is reduced. The execution steps for the optimized multiplexer are given in Table 4.

No internal working memristors are necessary, only the input and output memristors. Thus, the design is maximally optimized in terms of area.

Note that if the multiplexer were extended to wider than a single bit, the select line will need to be replicated for each bit since the results of the IMPLY operations overwrite the S and $\overline{S}$ memristors.

The number of memristors is now just 5 memristors. The equivalent schematic is shown in Fig. 3. Thus this implementation requires 4 IMPLY steps and 5 memristors.

Optimizations can be performed on this multiplexer once it has contextual information. For example, consider a circuit where the multiplexer uses outputs from a separate circuit as its inputs A and B. Rather than having dedicated input memristors for the A and B inputs, the output memristors from the prior circuit can be sensed instead. In this way, the A and B memristors can be removed entirely, reducing the complexity of the multiplexer to just 3 memristors. The delay remains the same, however now instead of applying the V_{cond} signal to the A and B memristors in their respective steps, the V_{cond} signal is applied to the output memristors in the other circuit which contain A and B. Note that the memristors in

Table 4 Steps for an optimized 1-bit multiplexer $\overline{(\mathrm{B\ NAND\ }\overline{\mathrm{S}})}$ OR $\overline{(\mathrm{A\ NAND\ S})} = \mathrm{Out}$

Step	Mux functionality	Mux goal
1	A IMP $\overline{S}$	A NAND S
2	B IMP S	B NAND $\overline{S}$
3	(A IMP $\overline{S}$) IMP Out	A AND S
4	(B IMP S) IMP Out	(A AND S) OR (B AND $\overline{S}$) = Out

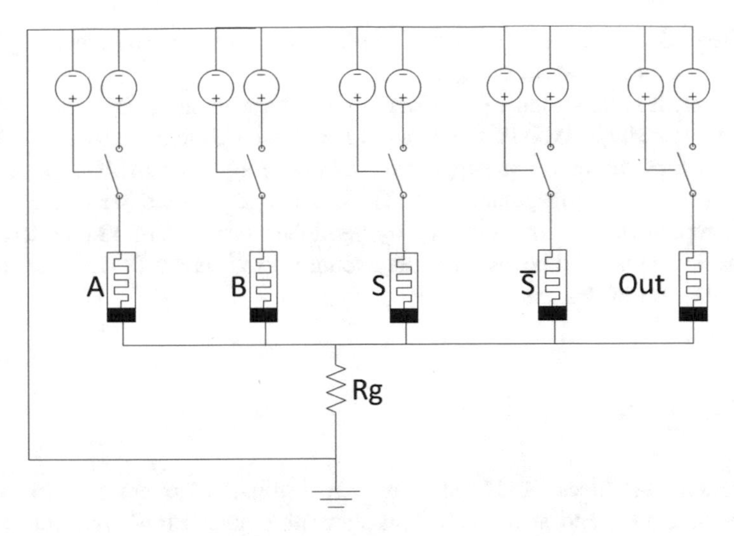

Fig. 3 Optimized multiplexer implemented with memristors using IMPLY logic

the first circuit and the second circuit must be connected in the IMPLY circuit form in order to properly execute the IMPLY operation.

Similarly, the output memristor can be reused as the input to the circuit which the multiplexer feeds into. This further reduces the overhead of the design by 1 memristor and maintains the same delay.

Optimizations such as these are common in the context of memristor based design. Since IMPLY is based on drivers applying voltage signals, it doesn't matter where the data lies, only where the drivers are asserted. Thus, many copy or move operations and duplicate memristors can be removed in favor of memristor reuse without incurring any delay penalty. This is a key characteristic that enables IMPLY to be used for logic-in-memory applications. The core of logic-in-memory is the idea of bringing the computation to the data, just as the IMPLY operation brings the voltage signals and drivers (computation) to the memristors (data).

2.2 *Considerations*

Although the IMPLY operation has come to the forefront of memristor logic, it has limitations in terms of latency optimizations. Strategic Boolean simplification and expression adjustment can be performed to reduce the number of operations, but there is still a fundamental limitation in the serialization of IMPLY steps. For example, to perform an AND operation, 3 IMPLY steps are required as compared to a single gate delay in traditional CMOS. Other operations, such as the XOR operation, are more costly, requiring 8 serialized steps as compared to 1. In most if not all designs, this causes the IMPLY implementation to have a higher step latency than the equivalent CMOS design.

However, it is important to consider what a "step" is in each of these domains. The delay of a single IMPLY pulse is not necessarily equivalent to a gate delay in CMOS. In fact, research has suggested that the delay of an IMPLY pulse can be shorter than a gate propagation time. Thus, although the number of steps for the IMPLY implementation of a circuit may be higher than its CMOS counterpart, the real-time delay may still be less. The step count is used here to facilitate the analysis and comparison of designs.

3 MAD Gates

This section introduces MAD gates, or Memristors-As-Drivers gates, to achieve lower power, area, and step count than alternative memristor-based gate designs. MAD gates were previously presented (Guckert 2016) as an approach to overcome some of the limitations of the IMPLY operation, but they have been shown to trump all other memristor logic approaches too (Kvatinsky 2012, 2014a; Zhang 2015a, b; Goto 1960; Zhu 2013).

MAD gates combine the driver methodology of the IMPLY circuits with threshold logic to achieve the benefits of the IMPLY operation without the high latency of serialized operations. The driver signals, V_{cond}, V_{set}, and V_{reset} are selected as read, write, and reset signals using the same design principles as the IMPLY operation to be $V_{cond} = 1.6$ V, $V_{set} = 2.5$ V and $V_{reset} = -5$ V.

Let's start with an example circuit for the AND operation in Fig. 4.

Fig. 4 MAD AND gate

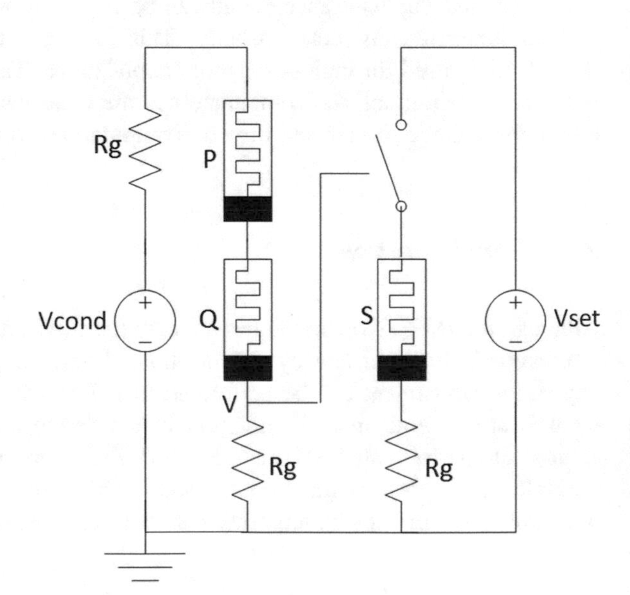

Similar to the IMPLY circuit, this implementation requires two memristors for the inputs, P and Q. However, in the MAD gate, these memristors are now connected in series and a third memristor, labeled S, is introduced for the result. For consistency, this example also assumes that the values of P and Q are preloaded into the input memristors as described in the IMPLY description. The pull down resistor R_g is selected as was done for the IMPLY operation. However, now two additional resistors are required in series with the input memristors and one in series with the output memristor.

To perform the AND operation, the read voltage V_{cond} is applied to the input memristors in series. This is similar to the application of the V_{cond} voltage in standard IMPLY operations. The application of the V_{set} voltage is slightly altered; the V_{set} voltage on the output memristor is now gated by the voltage of the input circuit at node V. In other words, depending on the voltage sensed at node V, the V_{set} voltage will or will not be applied to the output memristor. Specifically, if the voltage at node V is greater than the threshold of the switch on the output memristor, the switch will close and the V_{set} signal will be applied to the output memristor. This effectively sets the memristor to a logical '1'. If the voltage at node V is not greater than the threshold of the switch, the gate will remain open and the output memristor will remain unchanged.

The remaining design parameter is the correct threshold for the switch. Call this V_{apply}. To understand how V_{apply} is computed, consider the example of Fig. 4 where R_g has been assigned 10 K ohms and V_{cond} remains 1.6 V. Note this is an exception to the rule that $R_g = 2$ K ohms throughout this work and is done purely for example purposes. In this circuit, when both input memristors are logical '1', the voltage at node V is 16/22 V according to voltage division. If memristors P and Q are both logical '0', the voltage is 16/220 V and if one of the memristors is '1' and the other is '0', the voltage of node V is 16/121 V.

For the AND operation, the output memristor should only be set to '1' if both the inputs P and Q are '1'. This only happens if the voltage at node V is greater than the threshold of the switch, V_{apply}. Since the voltage at node V is 16/22 V when both inputs are '1', V_{apply} must be less than 16/22 V. This will ensure that the V_{set} signal is properly applied to the output memristor when both inputs are '1'. Similarly, node V must be less than V_{apply} for the remaining cases. Since the other possible values of node V are 16/220 and 16/121 V, V_{apply} must be greater than 16/121 V. Thus, V_{apply} should be selected in the range 16/121 V $< V_{apply} <$ 16/22 V.

The same circuit can be reused to implement the remaining Boolean operations by varying the value of V_{apply}. For example, for the OR operation, V_{set} is to be applied to the output memristor when either or both of the input memristors are a '1'. Thus, if node V is equal to 16/121 or 16/22 V, the switch should close. If node V is equal to 16/220 V, the gate should remain open. Thus, for the OR operations, V_{apply} should satisfy 16/220 V $< V_{apply} <$ 16/121 V.

The circuit can also be used to accomplish a COPY operation. The NOT and COPY operations only require a single input memristor, but take the same form otherwise. The respective circuits for the OR, XOR, NOT, and COPY operations are shown in Fig. 5.

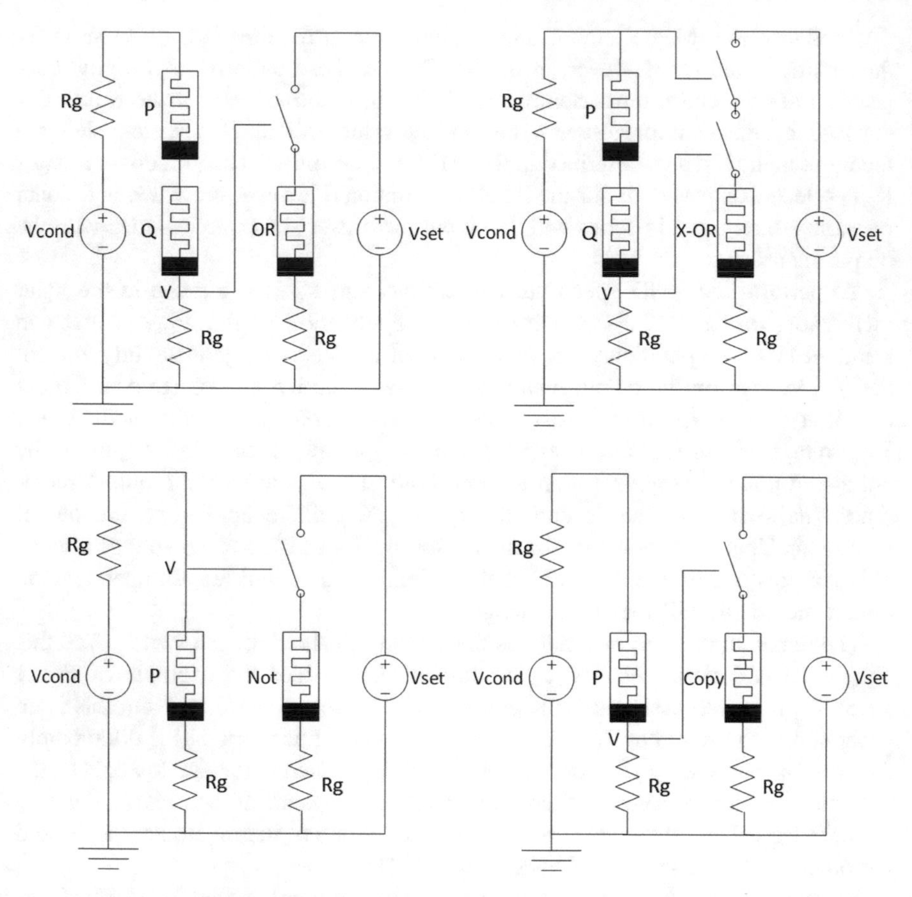

Fig. 5 MAD OR, XOR, NOT, and COPY gates

In this manner, only a single step (one application of the V_{cond} and V_{set} signals) is required for all Boolean operations and the COPY operation. This improves over not only the latency of IMPLY operations (Kvatinsky 2014b), but all other previous proposals to memristor logic too—hybrid-CMOS (Kvatinsky 2012), MAGIC gates (Kvatinsky 2014a), Zhang et al. (2015a, b), threshold gate (Goto 1960; Huisman 1995) implementations and others (Zhu 2013). A full comparison of the step counts for the various gates is given in Table 5.

The number of steps to complete the operations is not necessarily equivalent to the delay across the various approaches. For example, a hybrid-CMOS step is a gate delay whereas an IMPLY, MAGIC, or MAD step is the application of a drive signal. However, MAD gates offer fast switching times of about 0.4 ns which is 3 times faster than the transition times of 1.25 ns in other works on logic-in-memory. Taking these considerations into account, MAD gates require fewer steps than all previous approaches to memristor logic. They also offer a more complete set of

Table 5 Step count comparison for memristor-based gates

Op	IMPLY	Hybrid-CMOS	MAGIC	Zhang et al.	Threshold	MAD
NAND	2	2	1	3	2	1
AND	3	1	1	1	1	1
NOR	5	2	1	3	2	1
OR	4	1	1	1	1	1
XOR	8	3	N/A	N/A	3	1
NOT	1	1	1	2	1	1

Table 6 Area comparisons for memristor-based Boolean gates

Op	IMPLY	Hybrid-CMOS	MAGIC	Zhang et al.	Threshold	MAD
NAND	3 memristors 3 drivers	2 memristors 2 MOSFETs	3 memristors	2 memristors	2 memristors 1 GOTO pair 2 MOSFETs	3 memristors 2 drivers
AND	4 memristors 4 drivers	2 memristors	3 memristors	2 memristors	2 memristors 1 GOTO pair	3 memristors 2 drivers
NOR	6 memristors 6 drivers	2 memristors 2 MOSFETs	3 memristors	2 memristors	2 memristors 1 GOTO pair 2 MOSFETs	3 memristors 2 drivers
OR	6 memristors 6 drivers	2 memristors	3 memristors	2 memristors	2 memristors 1 GOTO pair	3 memristors 2 drivers
XOR	7 memristors 7 drivers	6 memristors 2 MOSFETs	N/A	N/A	5 memristors 3 GOTO pairs 2 MOSFETs	3 memristors 2 drivers
NOT	2 memristors 2 drivers	2 MOSFETs	2 memristors	2 memristors	2 MOSFETs	2 memristors 2 drivers

operations. Although MAGIC gates also offer a single step, operations on the same inputs cannot be performed in parallel and the XOR operation is not implemented.

MAD gates also reduce the necessary area for memristor logic as compared to previous proposals. A complete breakdown of the area comparisons for the various Boolean operations is given in Table 6.

In general, the MAD gates offer improved area for all of the Boolean operations, especially in comparison to the IMPLY operation (Kvatinsky 2014b). This is partially due to the fact that no intermediate memristors are required for the MAD operation. Only memristors for the inputs and outputs are necessary. Also, there is no need for signal restoration in circuits built from the proposed gate structure as required in some alternative approaches like hybrid-CMOS. This is because signals do not propagate through the circuits, but rather serve as sense voltages. For the gates which MAGIC (Kvatinsky 2014a) and Zhang et al. (2015a, b) have implemented, the proposed MAD gates have slightly higher area. However, the area measurements for these approaches do not take into consideration the additional circuitry required for resetting, writing, and reading the memristors such as switches or comparators. They also do not report on the logic required for concatenating the gates. Thus, it is likely that MAD gates require the least area.

Table 7 Energy comparisons (J) for memristor-based Boolean gates

Operation	IMPLY	Hybrid-CMOS	Zhang et al.	MAD
NAND	7.1e-13	N/A	2.5e-19	3e-14
AND	1.044e-12	1.75e-13	2.5e-19	3e-14
NOR	1.044e-12	N/A	2.5e-19	3.3e-14
OR	7.1e-13	1.75e-13	2.5e-19	3.3e-14
XOR	3.03e-12	N/A	2.5e-19	3.3e-14
NOT	3.4e-13	1.75e-13	2.5e-19	3.3e-14

MAD designs offer further area savings when the same inputs are used for multiple gates by reusing the input circuitry. For example, performing P AND Q and P OR Q in parallel would require the input memristors P and Q and two output memristors. Thus, for N gates using the same inputs P and Q, the design only requires N + 2 memristors rather than 3 N. Additionally, since MAD gate inputs are not overwritten during operation, the gates can be used repeatedly without having to reread the inputs or save/copy them.

In terms of power, MAD gates also improve in energy consumption over most prior work as shown in Table 7.

Energy is reported for all prior works where it was given. For the MAD gates, the energy was calculated by integrating V * I characteristics across the execution of the operation. The MAD gates improve energy by an order of magnitude over the IMPLY approach, mostly because of the large number of steps required for IMPLY operations. Although the measurements suggest that Zhang et al. requires consumes less energy than MAD gates, these calculations represent the mean power per bit, and it is unclear exactly how this value is calculated.

A final benefit of MAD gates over all alternative approaches to memristor logic is that all MAD gates are constructed from the same uniform standardized cell that can be configured with a threshold that depends on the gate and application. Additionally, these designs do not suffer from the concatenation, parallelizability, and fanout challenges that hybrid-CMOS, logic-in-memory, and MAGIC gates approaches have. Lastly, the MAD gates are currently able to be fabricated and modeled, rendering them a more practical option than some progressive implementations.

3.1 MAD Implementation of a Multiplexer

To show how MAD memristor concepts can be used beyond individual gates, again consider the multiplexer example. An optimized implementation for a 1-bit wide MAD multiplexer is shown in Fig. 6.

The MAD multiplexer is essentially a combination of two singular MAD AND gates. For the multiplexer equation (A AND S) OR (B AND $\overline{S}$) = Out, an AND gate is needed to compute A AND S and an AND gate is needed to compute B

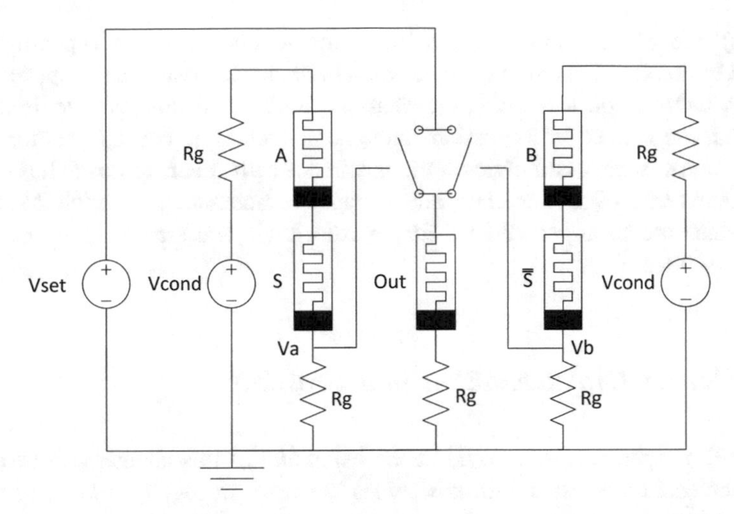

Fig. 6 MAD multiplexer implementation

AND $\overline{S}$. S and $\overline{S}$ represent the select line of the multiplexer and it's inverse. Also an output memristor Out is used to hold the result. As was done for the basic AND gate, the V_{set} signal is gated to the output memristor. However, now there are two switches gating the signal, one for each AND gate. The thresholds of the two switches are both selected to correspond to an AND gate as described in the previous section.

The multiplexer executes by applying V_{cond} to both of the AND gates in parallel, while applying V_{set} to the output memristor Out. If either V_a or V_b is above the threshold voltage of a MAD AND operation, its corresponding switch will close and allow V_{set} to drive the output memristor, setting it to '1'. If neither of the AND operations corresponds to a '1', both switches will remain open and the output memristor will remain at '0'. In other words if (A AND S) OR (B AND $\overline{S}$) is a 1, the output will be 1. Thus, the multiplexer exhibits correct functionality.

The multiplexer requires one memristor per input and output for a total of 5 memristors. This is equivalent to the fully optimized IMPLY multiplexer. However, the latency is reduced to just a single step for the entire multiplexer operation as compared to 4. This directly corresponds to lower energy consumption as well.

4 IMPLY and MAD Implementations in a Crossbar

Since the discovery and exhibition of memristors, their most prevalent application has been crossbar memories. The low area, high density, and 2-terminal properties of memristors make them a prime candidate for memory applications. Memristors can be used to construct a crossbar memory with area $4F^2$ per bit, where F is the feature size. This replaces DRAM cells which require 6 transistors (3-terminal devices) and exhibit poor density. Also, memristors have the capability to be stacked to create 3-D memory structures. These characteristics have spurred much research focusing on leveraging crossbars for efficient memory and logic designs.

Within the context of a crossbar, there have been various proposals for how to perform reads, writes, and logic on the memristor cells. Each approach has different tradeoffs in terms of latency, area, and voltage characteristics. Both the IMPLY and MAD gates are strong candidates for crossbar logic and are discussed in the following sections.

4.1 IMPLY Implementation in a Crossbar

The IMPLY operation introduced at the beginning of this chapter can be directly implemented in a crossbar (Strukov 2008) as shown in Fig. 7.

Pull down resistors are placed on all rows to allow operations to perform on any row in the crossbar in a given cycle. The p and q memristors lie on a single row of the crossbar while the pull down resistor R_g serves as a resistor for the entire row. It is not necessary for the memristors p and q to lie on consecutive bit cells, but it is

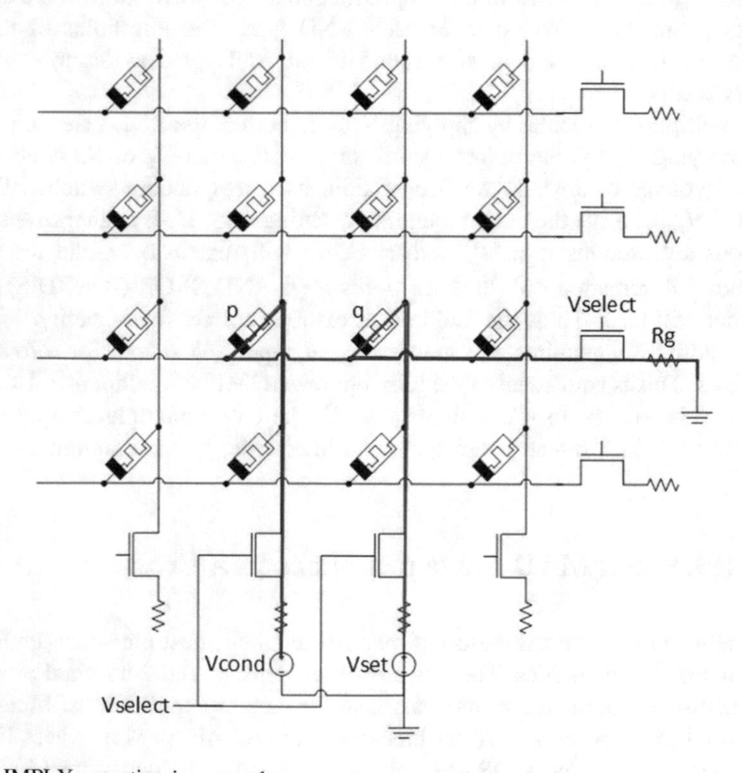

Fig. 7 IMPLY operation in a crossbar

necessary they lie on the same row so that their terminals are connected properly as per the IMPLY circuit. To perform the IMPLY operation, V_{select} drives both bit lines and the row line of the input memristors high. V_{cond} is applied to the p memristor (highlighted in red) while V_{set} is applied to the q memristor (highlighted in purple) and the row is connected to GND (highlighted in blue). This directly matches the traditional IMPLY circuitry. After this step, the result of p IMP q lies in the q memristor.

4.2 *MAD Implementation in a Crossbar*

MAD gates can also be transposed into a crossbar structure to allow for logic-in-memory. However, the traditional MAD gate circuitry must be slightly altered. First, in the standard MAD Boolean gate, the input memristors are connected in series by connecting their opposite-polarity terminals, i.e. one input memristor's p-terminal is connected to the second input memristor's n-terminal. This is required in order to be able to maintain the values of the input memristors when V_{cond} is applied. Because of memristor fundamentals, if their common-polarity terminals are connected in series, the values of the input memristors can change when V_{cond} is applied. However, in a crossbar structure, the p-terminals of both input memristors necessarily share a common row line, connecting their common-polarity. Because of this, the way the MAD input memristors are connected must be slightly altered when translated into a crossbar. This can be seen in Fig. 8.

The red highlighted path follows the V_{cond} signal, the purple highlighted path follows the V_{set} signal, and the blue highlighted path follows the voltage division threshold signal. The only change is performed on the input memristors and the application of V_{cond}. The other signals remain unchanged. The input memristors are

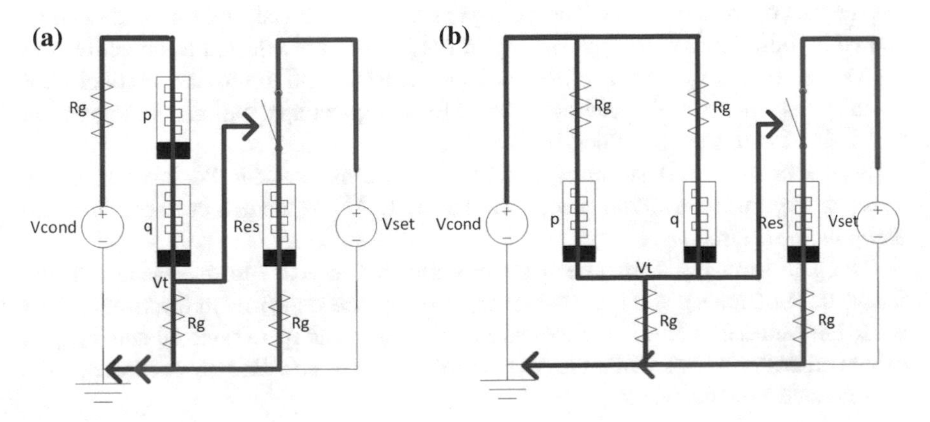

Fig. 8 Translation of **a** a MAD AND gate into **b** a crossbar form

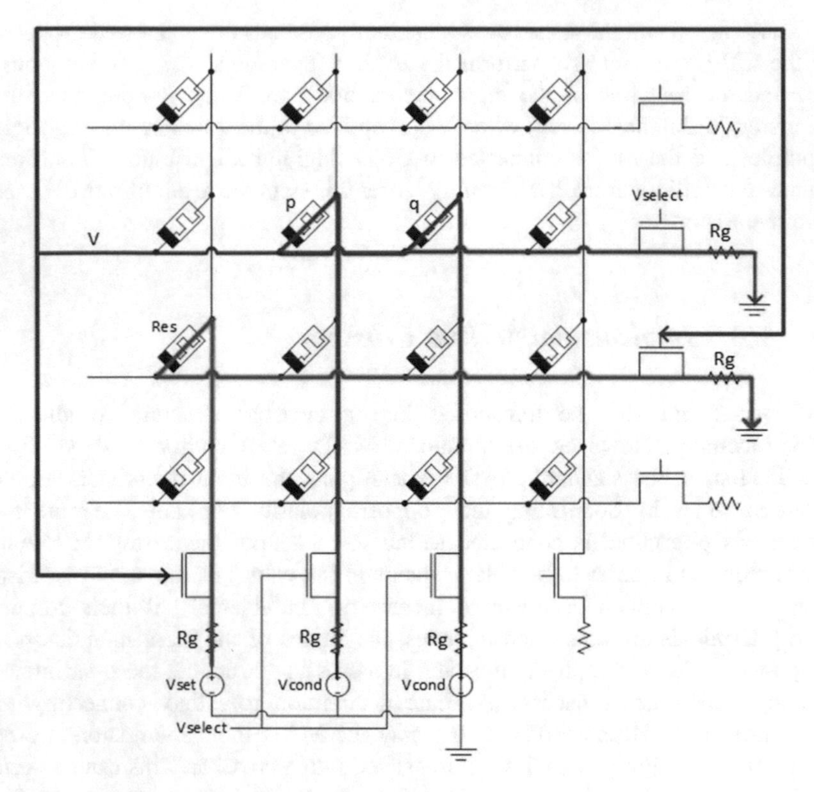

Fig. 9 MAD AND gate in a crossbar structure

still used in conjunction with the pull down resistors to function as a voltage divider circuit; however it takes a different form. Instead of connecting the input memristors in series and applying V_{cond} and GND to the terminals, V_{cond} is applied to the n-terminal of both input memristors and their shared row line is connected to GND via the pull down memristor. The voltage at V is still sensed, but this node is now shared by both input resistors. The resistor R_g values are selected to coincide with the values commonly used in other works in crossbars and to provide simplicity for calculations. However, R_g can be selected for the given application, staying within the design parameters specified in this chapter.

Now that the MAD structure has been transformed, it can be inserted into a crossbar. Figure 9 shows the execution of a MAD AND gate in a crossbar structure using the new circuitry.

The gate from Fig. 8 has been directly mapped in a one-to-one fashion to the crossbar. The inputs p and q are two memristors on the same row in the crossbar. To sense the values of p and q, the row and bit lines of p and q are selected and V_{cond} is driven on both bit lines while the row is grounded. The row line now represents the sensed node voltage, V.

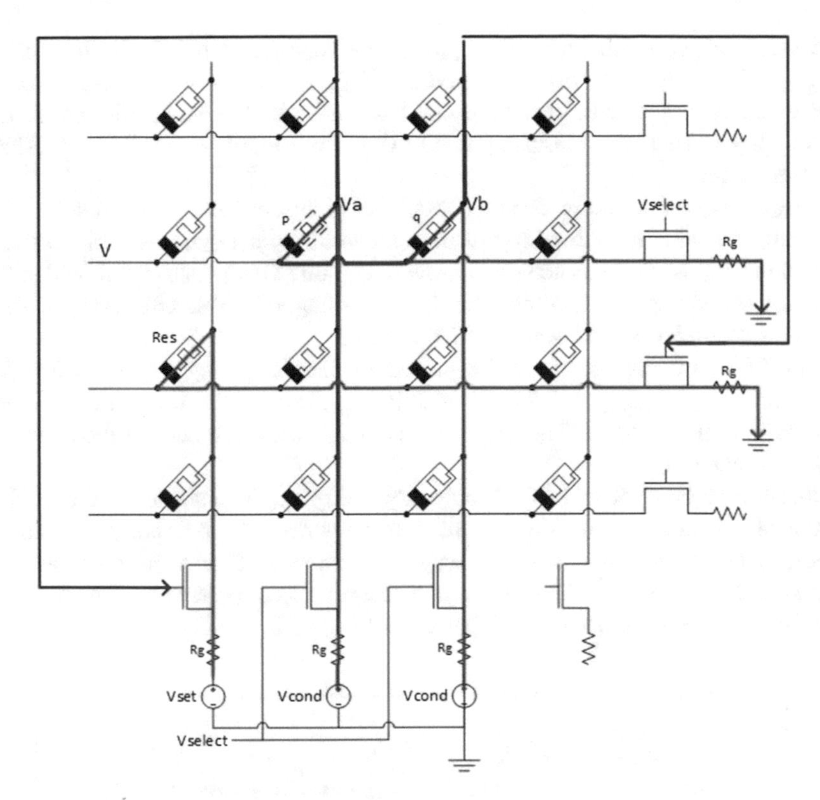

Fig. 10 MAD XNOR gate in a crossbar structure

In the second step, the sensed value at V is used to drive the select lines on the result memristor's row and bit lines. If the value of V is greater than the threshold of the select line gates, then V_{set} and GND will be driven to the result memristor Res and it will be set to 1. Otherwise, the row and bit lines will not be selected and Res will remain at 0. V_{apply} is calculated using the same methodology as the standard MAD gates.

For the XOR and XNOR operation, the same crossbar circuit is used but now V_a and V_b are sensed rather than V. Figure 10 shows this for the XNOR operation.

For example purposes, Let $R_g = 2$ K ohms, low resistance = 1 K, and high resistance = 100 K. When both input memristors are 0, the voltages at V_a and V_b are 0.975 V. When input memristor p is a 1 and q is a 0, the voltage at V_a is 0.6 V and the voltage at V_b is 0.982 V. If the values of the inputs are swapped the voltages at V_a and V_b swap accordingly. If both inputs are a 1, the voltages at V_a and V_b are 5/7 V. Thus, together, V_a and V_b can be used to differentiate the four input scenarios.

In the original XNOR MAD gate, two separate conditions on two switches must both be true to gate the V_{set} signal to the result memristor. This was implemented by placing two switches in parallel between the V_{set} signal and the result memristor, one gated by V_a (with threshold V_{applyA}) and one gated by V_b (with threshold V_{applyB}). To correctly perform an XNOR, $V_{applyA} > 0.6$ V and $V_{applyB} > 0.6$ V. The result memristor will only be set to 1 when both of these conditions are true, which only occurs when both inputs are 0 or both inputs are 1. Thus, a XNOR operation occurs.

However, this structure is not conducive to the context of a crossbar. In a crossbar, the same functionality can be achieved by placing one switch on the bit line gating V_{set} and one switch on the row line gating GND. One switch will still be gated by the voltage V_a and the other by the voltage V_b. This effectively achieves the same operation as the original XNOR gate.

For XOR, two switches will be placed in parallel gating V_{set} such that $V_{applyA} > 0.975$ V and $V_{applyB} > 0.975$ V. If either of these conditions is true, the result memristor will be set to 1. This corresponds to the cases when one of the inputs is 0 and the other is 1.

Recall that the NOT and COPY operations only require a single input memristor in the MAD gate design. Since these two gates do not use input memristors in series, there are no issues with the way the terminals of the memristors are connected in the gate design. Thus, their gate designs do not need to be altered at all to translate into the crossbar context. This can be seen in Fig. 11.

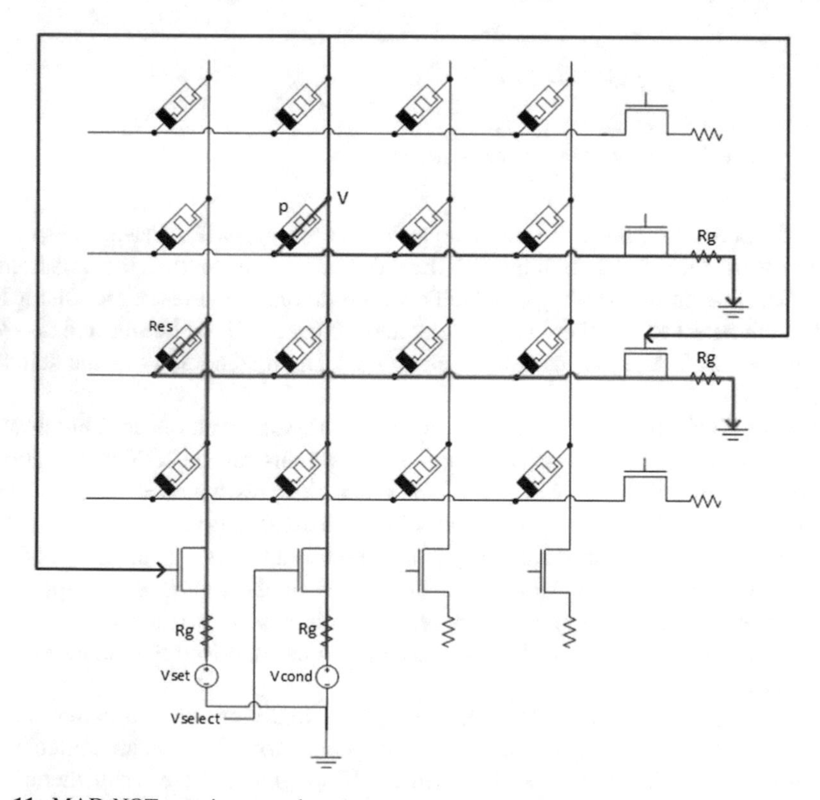

Fig. 11 MAD NOT gate in a crossbar structure

V_{select} is driven high for the row and bit line of the input memristor to create a circuit from V_{cond}, through the input memristor and pull down resistor, to GND. In the next step, the voltage sensed at V is used to determine if V_{set} should be gated to the result memristor Res. This is identical to the original MAD NOT gate execution.

4.3 Crossbar Analysis and Comparison

The IMPLY operation is the most commonly used approach to logic-in-memory. It has been shown to function correctly on physical memristors and also has the benefit of implementing a logically complete set of Boolean operations. Thirdly, since it can easily be mapped to the crossbar, it is a natural first step in memristor memory. However, the IMPLY operation suffers from long-latency, serialized operations and overwrites inputs during operation. Many other approaches have been proposed to improve over these limitations. Other approaches include MAGIC gates (Kvatinsky 2014a), CRS cells (Yang 2016), and more (Zhang et al. 2015a, b), but none are as established as the IMPLY operation.

None of these proposals for memristor crossbar logic, including IMPLY, can perform multiple operations in parallel and each has their own tradeoffs and shortcomings. MAD gates overcome all the issues associated with the other proposals and offer a complete set of logic-in-memory operations with lower area and step counts. They also offer a logically complete set of operations that handles high fanout, does not overwrite operands, and can perform the COPY operation.

MAD gates require only 1 memristor per operand and 2 steps for every Boolean operation. Although it only takes a single step to perform an IMPLY, it takes multiple IMPLY steps to achieve most Boolean operations. For example, the XOR operation requires as many as 6 steps and 5 memristors in a crossbar. MAGIC gates have the same complexity and latency as MAD gates but only the NOR gate has been successfully mapped into a crossbar. Since every other Boolean operation must be performed via a series of NOR operations, their delays are much higher than their MAD counterparts. Zhang, et al. propose a novel OR gate that can exist in a crossbar and pair it with another AND gate design and the IMPLY NOT operation to create a logically complete set for logic-in-memory operations. However, these gates destroy one of the operands during operation. MAD gates use an additional memristor to store the result, leaving the inputs intact for subsequent use. Zhang et al. also takes more steps for the NOR and NAND operations and does not offer implementations for the XOR or XNOR operations.

CRS cells have been proposed as an alternative to the commonly used memristor cells in a crossbar to achieve lower delay and area. However, each CRS cell uses two memristors rather than one and requires initialization on all operands before Boolean computation. This incurs an extra delay for each operation. This design also overwrites one of the input operands to store the result. CRS cells are only capable of performing NAND-AND and NOR-OR operations, each of which requires 3 steps as compared to 2 in MAD. All other Boolean operations must be

Table 8 Latency comparisons for the MAD crossbar and prior crossbar approaches

Op	IMPLY	MAGIC	CRS	Zhang et al.	MAD
p NAND q	2	N/A	3	3	2
p AND q	3	N/A	6	2	2
p NOR q	5	1	3	3	2
p OR q	4	N/A	6	2	2
p XOR q	8	N/A	6	N/A	2
NOT p	1	N/A	3	2	2

Table 9 Area comparisons for the MAD crossbar and prior crossbar approaches

Op	IMPLY	MAGIC	CRS	Zhang et al.	MAD
p NAND q	3	N/A	6	2	3
p AND q	4	N/A	8	2	3
p NOR q	6	3	6	2	3
p OR q	6	N/A	8	2	3
p XOR q	7	N/A	8	N/A	3
NOT p	2	N/A	6	2	2

constructed from a series of these operations, requiring a total of N + 2 steps where N is the number of NAND and NOR operations required for the Boolean operation.

A full breakdown of the comparison between the proposed MAD crossbar (Guckert 2016) operations with prior work is given in Tables 8 and 9.

One consequence of the MAD approach is the use of multiple voltages. The V_{cond}, V_{set}, and V_{reset} are all necessary for execution. Although this is equivalent to the number of voltage sources used by the IMPLY operation, other proposed approaches such as MAGIC gates only require a single voltage source. However, this is likely because this approach is only capable of implementing a single operation, namely the NOR, in a crossbar. If additional Boolean operations were tackled, especially the XOR operation, it is likely that the number of voltages required would increase to match or exceed those required by MAD. The CRS and Zhang, et al. approaches both require 4 voltage sources, more than MAD and IMPLY. In Zhang, et al., the first three voltages are similar to those in the IMPLY and MAD approaches, and the fourth voltage has the same magnitude as V_{cond} but opposite polarity.

In addition to low latency and low area, MAD gates also offer increased flexibility over previous approaches. In the IMPLY circuitry, the result memristor inherently must lie on the same row as the input memristors since one of the inputs is used for the output. In a MAD operation in a crossbar, the result can lie on any row line and bit line. This is important for full system designs if logic-in-memory is going to be harnessed for performance gains. If data is frequently being moved around in the memory in order to read, write, and execute logic, the advantages of performing the logic in the memory rather than the CPU is diminished or even eliminated. Operations in the other proposed approaches suffer from this risk, but MAD operations in memory do not.

Also, because the voltages associated with the inputs are sensed in a single step and used as drivers in a latter step, these voltages can be used multiple times for operations, which use the same inputs. This can result in increased latency and energy savings by removing the sense step on the input memristors for all additional operations using the same inputs. For example, if it is desired to execute A AND B followed by A OR B, the first step will send the voltage division circuitry on the input memristors A and B, the second step will drive the write operation for the A AND B result memristor, and the third step will drive the write operation for the A OR B result memristor. In a naïve approach, the sense operation would occur again before writing the A OR B result memristor. Using this knowledge, logic-in-memory operations that use the same inputs should be prioritized to execute consecutively.

Another benefit of MAD gates, and perhaps the most critical, is the fact that the input memristors are not overwritten at any time during execution. This allows for reuse of the operands for later operations if needed. It also retains this data in memory to be read or written later by the system. Since all of the other proposed approaches to crossbar logic overwrite one or more of the input operands during execution, the original data is lost. Thus, if the system were to need this data for another operation or a simple write, it would need to execute additional instructions to reproduce this value. This would involve reading the data out of memory and storing it in auxiliary space, then performing logic-in-memory operations, moving the result of the operations to another location, and finally writing back the original value from auxiliary space to the memory.

Lastly, MAD gates do not require any memristors or steps for holding intermediate values in the calculation of a Boolean operation. All memristors in a MAD gate, whether in a logic or memory context, are either inputs or outputs of the Boolean operation. In other approaches, such as CRS and IMPLY, numerous memristors are required during intermediate steps. This leads to extra memory cells that do not hold any "real" valuable data. This has long-reaching effects that degrade performance, area, and power in full system designs. MAD gates overcome all of these deficiencies.

5 Memristor Design Considerations

There are a few design principles that should be discussed and analyzed in order to design successful memristor designs. This section will discuss properties and advantages of memristors as well as issues and considerations that arise specific to the IMPLY and MAD contexts.

5.1 Pipelining Logic

One of the most important benefits of the presented approaches to memristor logic is the ability to parallelize and pipeline operations to improve bandwidth and reduce latency and area. Recall that both the IMPLY and MAD methodologies harness the idea of bringing the computation to the data through the use of voltage drivers. Because of this, the data is not "flowing" through the system as it is in traditional Von Neumann systems. Thus, it is possible to reuse the memristor computational elements as soon as they finish their operation.

Recall the multiplexer example. Let's assume this multiplexer lies in a larger system and is used to produce an input to another logic block which is also constructed from memristors. As soon as the latter logic block applies V_{cond} to the output of the mux and senses its value into its memristor circuitry, the multiplexer has effectively completed its operation. Now, the multiplexer can accept new inputs and begin a second multiplexer operation while the latter logic block continues to process the previous input. This can be done in parallel because the data is contained in disjoint memristors in separate circuits and only the computations, or the driver signals, are moving. In this way, the effective latency of the entire circuit is reduced to the largest latency of the individual blocks. As a result, IMPLY and MAD approaches to logic offer vastly improved bandwidth and latency as compared to traditional CMOS logic circuits.

5.2 General Considerations

First, it is always good practice to revisit the parameterized values assigned to components each time a new application, variation, or optimization is introduced. For example, the doping width and low resistance and high resistance values could be changed for the memristors. Similarly, the values of the pull down resistors may need to change. Depending on the functionality, sensitivity, and context of the circuit and the processing technology, different values should be selected. Also, it is important to note that many of these principles work in simulation, but extra care must be in taken on fabricated memristors. One of the major drawbacks of current state-of-the-art memristors is their subpar endurance and their high variability. Thus, memristors should be studied thoroughly before being used for critical, highly-sensitive, or frequently executed circuit designs.

5.3 Interference and Sneak Paths

One issue with memristor crossbars is sneak paths (Snider 2004; David 1968; Lynch 1969; Shin 2012; Cassuto 2013). This refers to the ability for current to traverse through other memristor cells when the V_{cond} and V_{set} signals are applied. This can cause two issues. First, if the current is high enough, the value of unintended memristors can be affected. Second, it can weaken the currents through the intended cells, causing the operation to function incorrectly. There are multiple ways to reconcile this issue; however each comes with a different cost tradeoff. For example, one approach to eliminate this issue is to apply a voltage V_p to every inactive line in the crossbar (Zhu 2013).

The MAD implementations in the crossbar have another possibility for interference between the input memristors and the result memristor. Activating multiple row lines will lead to interference and false currents in the crossbar. Because of this, the sense and write stages must occur in successive cycles rather than in the same cycle. The voltage sensed in the first stage can be buffered or stored in a standard latch for use in the subsequent set cycle. This is not an issue for the IMPLY operation since one of the input memristors is also used as the result memristor.

5.4 Boolean Operation and Threshold Gate Selection

Depending on the Boolean operation being performed on a MAD gate or MAD crossbar, the threshold voltage, or V_{apply}, of the gated drivers will vary. However, the hardware cannot be reconfigured dynamically to alter this value. This is not an issue for logic circuits since their operations are determined by their functionality, but the crossbar must adapt to this complexity.

In a real implementation, a single switch cannot be used to drive a given line in the crossbar. Instead, every row and bit line in the crossbar must have the ability to be driven by multiple inputs, each corresponding to a different Boolean operation. The encoding for the Boolean operation will be used to select which one of the input lines is used to drive the given row and column of the result memristor. A visualization of this can be seen in Fig. 12.

Now, the V_{set} signal has multiple paths to the crossbar, one for each Boolean operation. Let the first path correspond to the AND operation. Thus, a single switch, S_{and} will lie on this path and have a V_{apply} corresponding to the voltage calculated in prior sections. Let the second path correspond to the OR operation. A single switch, S_{or}, will lie on this path with its calculated V_{apply} value. Both will be driven by the V_s signal. The third path will have two parallel switches for the XOR operation, S_{xora} and S_{xorb}, driven by the voltages V_a and V_b. The process will continue for each of the remaining operations. The path corresponding to the encoded Boolean operation will be selected and gated to the relevant bit line in the crossbar. Similar logic exists for the row lines.

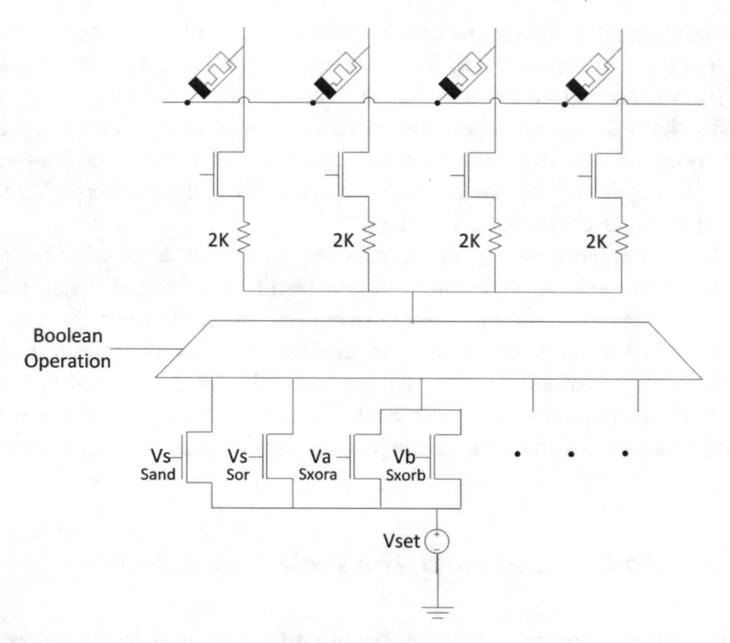

Fig. 12 Driver logic for selecting Boolean operations

5.5 *Threshold Sensitivity*

The final concern of this design is the sensitivity of the circuitry in correctly differentiating the voltages at the sense nodes to determine the threshold satisfaction on the row and bit lines. For example, for the XOR operation, the switches must be sensitive enough to differentiate 0.975 versus 0.982 V in the example.

There are two solutions to this issue. The first solution is to reselect the parameterized values of the system, namely the values of low-resistance, high-resistance, and the pull down resistor R_g. For example, by changing the value of the resistors on the row and bit lines, the value of V_{cond}, and the low and high resistances of the memristors, the threshold voltages can be changed. The values selected in this paper were done for simplicity and serve as motivation and do not necessarily represent ideal values. Note that care must be taken when these parameters are chosen to abide by the rules of IMPLY design as specified in prior work (Kvatinsky 2014b).

The second solution is to add additional circuitry to the logic for the XOR operation. For example, an additional switch driven by V_s can be placed in series on the path to the result memristor. Let this switch have threshold voltage V_{applyS} such that $V_{applyS} > 0.25$ V. This switch will be used to protect against the case that the XOR logic does not correctly differentiate 0.975 and 0.982 V when both input memristors are 0. Since the value of Vs is about 50 mV in this scenario, the threshold condition for this switch will not be satisfied and the result memristor will not be set. Thus, even if the initial two switches cannot correctly resolve 0.975 versus 0.982 V, the result will be correct. The consequence of this approach is additional complexity.

6 Conclusion

This chapter introduced the concept of memristor logic and its applications to standard logic circuits, crossbars, and non-traditional logic-in-memory. Two approaches to memristor logic were presented: IMPLY and MAD gates. A multiplexer is used as an example to introduce how logic is performed in each methodology and compare the costs and benefits of each.

The IMPLY operation is the most popularized approach due to its ability to implement a logically complete set of Boolean operations and its conformity to a crossbar structure. However, IMPLY operations suffer from high latencies and a large number of memristors. Overall, this can lead to poor efficiency.

MAD gates have been shown to offer a low latency, low area alternative to prior approaches to memristor-based logic. MAD gates require a single cycle regardless of the Boolean operation, including the XOR, XNOR, and COPY operations. They also do not overwrite input memristors and only require 1 memristor per operand. This improves over all previously proposed approaches to memristor logic in terms of both area and step count. MAD gates can also be transformed into the context of a crossbar, but the latency of each operation increases to two steps. This is in part to allow flexibility on the crossbar structure, allowing the result to lie in any row and allowing the input memristors to be reused. Overall, MAD gates provide an optimally low-area and low-latency implementation for memristor circuit designs in both the logic and memory settings.

There are many design parameters and practices that should be considered when designing memristor-based circuits. Depending on the context of the circuit, these considerations can change. For example, appropriate values of driver voltages, threshold voltages, and pull down resistors all depend heavily on the approach to memristor logic, where it's being used, and the memristor properties themselves. In addition to these design decisions, memristors can exhibit some complications that traditional CMOS circuits do not. For example, issues such as sneak paths and interference can arise and should be handled or addressed to ensure correct functionality. Finally, benefits of memristors can be harnessed, such as their ability to pipeline execution, improving their latency and resource utilization over CMOS designs.

Future research in this area can follow many different avenues. There is still a multitude of exploration to be done in exploring the implementation of more complex circuits using proposed approaches to memristor logic. Secondly, both memristor logic circuits and crossbar memories capable of performing logic-in-memory can be further optimized and demonstrated in real, fabricated systems. Specifically, exploring how non-Von Neumann systems and pipelined logic designs can be constructed from memristors to create modern, low area, low delay systems will prove very interesting. Lastly, research should explore better models and simulation tools for the memristor realm are key to enabling this field to progress.

References

Bickerstaff, K., & Swartzlander, E., Jr. (2010). Memristor-based arithmetic. In *Forty-Fourth Asilomar Conference on Signals, Systems and Computers* (pp. 1173–1177).

Cassuto, Y., Kvatinsky, S., & Yaakobi, E. (2013). Sneak-path constraints in memristor crossbar arrays. In *IEEE International Symposium on Information Theory* (pp. 156–160).

Chua, L. (1971). Memristor-the missing circuit element. *IEEE Transactions on Circuit Theory, 18,* 507–519.

Corinto, F., & Ascoli, A. (2012). A boundary condition-based approach to the modeling of memristor nanostructures. *IEEE Transactions on Circuits and Systems I: Regular Papers, 59,* 2713–2726.

David, C., & Feldman, B. (1968). High-speed fixed memories using large-scale integrated resistor matrices. *IEEE Transactions on Computers, C-17*, 721–728.

Deng, Y., Huang, P., Chen, B., Yang, X., Gao, B., Wang, J., et al. (2013). Rram crossbar array with cell selection device: A device and circuit interaction study. *IEEE Transactions on Electron Devices, 60,* 719–726.

Devolder, T., Hayakawa, J., Ito, K., Takahashi, H., Ikeda, S., Crozat, P., et al. (2008). Single-shottime-resolved measurements of nanosecond-scale spin-transfer induced switching: Stochastic versus deterministic aspects. *Physical Review Letters, 100,* 057206.

Goto, E., Murata, K., Nakazawa, K., Nakagawa, K., Moto-Oka, T., Ishibashi, Y., et al. (1960). Esaki diode high-speed logical circuits. *IRE Transactions on Electronic Computers, EC-9*, 25–29.

Guckert, L., & Swartzlander, E., Jr. (2016). MAD gates—memristor logic design using driver circuitry. *IEEE Transactions on Circuits and Systems II: Express Briefs*. doi:10.1109/TCSII.2016.2551554.

Huisman, T., & Vassiliadis, S. (1995). Counters and multipliers with threshold logic.

Kvatinsky, S., Wald, N., Satat, G., Kolodny, A., Weiser, U., & Friedman, E. (2012). MRL—memristor ratioed logic. In *13th International Workshop* on *Cellular Nanoscale Networks and Their Applications* (pp. 1–6).

Kvatinsky, S., Satat, G., Wald, N., Friedman, E., Kolodny, A., & Weiser, U. (2014a). Memristor-based material implication (IMPLY) logic: design principles and methodologies. *IEEE Transactions on Very Large Scale Integration (VLSI) Systems, 22,* 2054–2066.

Kvatinsky, S., Belousov, D., Liman, S., Satat, G., Wald, N., Friedman, E., et al. (2014b). MAGIC memristor aided LoGIC. *IEEE Transactions on Circuits and Systems II: Express Briefs, 61,* 1–5.

Kvatinsky, S., Ramadan, M., Friedman, E., & Kolodny, A. (2015). VTEAM—a general model for voltage controlled memristors. *IEEE Transactions on Circuits and Systems II: Express Briefs, 62,* 786–790.

Lynch, W. (1969). Worst-case analysis of a resistor memory matrix. *IEEE Transactions on Computers, C-18*, 940–942.

Mahajan, D., Musaddiq, M., & Swartzlander, E., Jr. (2014). Memristor based adders. In *Forty-Eighth Asilomar Conference on Signals, Systems and Computers* (pp. 1256–1260).

Rose, G., & Stan, M. (2007). A programmable majority logic array using molecular scale electronics. *IEEE Transactions on Circuits and Systems I: Regular Papers, 54,* 2380–2390.

Rose, G., Rajendran, J., Manem, H., Karri, R., & Pino, R. (2012). Leveraging memristive systems in the construction of digital logic circuits. *Proceedings of the IEEE, 100,* 2033–2049.

Shin, S., Kim, K., & Kang, S. (2012). Analysis of passive memristive devices array: data-dependent statistical model and self-adaptable sense resistance for RRAMs. *Proceedings of the IEEE, 100,* 2021–2032.

Snider, G., Kuekes, P., & Williams, R. (2004). CMOS-like logic in defective nanoscale crossbars. *Nanotechnology, 15,* 881.

Strukov, D., Snider, G., Stewart, D., & Williams, R. (2008). The missing memristor found. *Nature, 453,* 80–83.

Teimoory, M., Amirsoleimani, A., Shamsi, J., Ahmadi, A., Alirezaee, S., & Ahmadi, M. (2014). Optimized implementation of memristor-based full adder by material implication logic. In *21st IEEE International Conference on Electronics, Circuits and Systems* (pp. 562–565).

Teimoory, M., Amirsoleimani, A., Ahmadi, A., Alirezaee, S., Salimpour, S., & Ahmadi, M. (2015). Memristor-based linear feedback shift register based on material implication logic. In *Proceedings of 22nd European Conference on Circuit Theory and Design* (pp. 24–26).

Yang, Y., Mathew, J., Pontarelli, S., Ottavi, M., & Pradhan, D. (2016). Complementary resistive switch-based arithmetic logic implementations using material implication. *IEEE Transactions on Nanotechnology, 15,* 94–108.

Zhang, Y., Shen, Y., Wang, X., & Guo, Y. (2015a). A novel design for memristor-based logic switch and crossbar circuits. *IEEE Transactions on Circuits and Systems I: Regular Papers, 62,* 1402–1411.

Zhang, Y., Shen, Y., Wang, X., & Guo, Y. (2015b). A novel design for a memristor-based OR gate. *IEEE Transactions on Circuits and Systems II: Express Briefs, 62,* 781–785.

Zhu, X., Yang, X., Wu, C., Xiao, N., Wu, J., & Yi, X. (2013). Performing stateful logic on memristor memory. *IEEE Transactions on Circuits and Systems II: Express Briefs, 60,* 682–686.

RF/Microwave Applications of Memristors

Milka Potrebić, Dejan Tošić and Dalibor Biolek

Abstract Memristor-based technology could be utilized, potentially, to enhance performance of many RF/microwave subsystems. Application of memristors in RF/microwave circuits, and in a broader context in electromagnetic systems, is another challenging field for researchers and engineers. In this application frontier, the research efforts might be divided, for example, into the following important classes of applications: (1) frequency selective surface, reconfigurable planar absorber, (2) reconfigurable antenna, direct antenna modulation, (3) RF/microwave filter, split-ring resonator filter, hairpin-line filter, capacitively coupled resonator filter, quasi-Gaussian lossy filter, (4) Wilkinson power divider. Memristors could be exploited as linear resistors with programmable resistance, which can be accurately adjusted to a desired or specified value. Precise controllability of the memristance value might be important for tuning microwave circuits and optimizing their performance. In several applications, such as filters, the high-frequency range of the operation enforces the memristor into the role of a linear resistor whose resistance can be adjusted electronically. On the other hand, some applications, such as reconfigurable electromagnetic absorbers, benefit from memristors as electromagnetic switches. Due to the unavailability of commercial memristors, it is necessary to use accurate circuit-level simulations for experimenting with the memristor-based RF/microwave circuits and for studying their performance. RF/microwave circuit simulators, which use the HSPICE engine for the time-domain transient simulation,

M. Potrebić (✉) · D. Tošić
School of Electrical Engineering, University of Belgrade, Bulevar kralja Aleksandra 73,
PO Box 35-54, 11120 Belgrade, Serbia
e-mail: milka_potrebic@etf.rs

D. Tošić
e-mail: tosic@etf.rs

D. Biolek
Department of Electrical Engineering, University of Defense, Kounicova 65,
662 10 Brno, Czech Republic
e-mail: dalibor.biolek@unob.cz

© Springer International Publishing AG 2017

S. Vaidyanathan and C. Volos (eds.), *Advances in Memristors, Memristive Devices and Systems*, Studies in Computational Intelligence 701,
DOI 10.1007/978-3-319-51724-7_7

such as NI AWR Microwave Office, can be used to verify the expected functionality of the considered memristor-based circuits.

Keywords Memristor · Reconfigurable planar absorber · Reconfigurable antenna · RF/microwave filter · Wilkinson power divider

1 Introduction

The memristor is a two-terminal one-port electric circuit element, envisioned, postulated and conceptualized by Chua, characterized by a constitutive relation between the time integral of the element's current and the time integral of the element's voltage (Chua 1971). The element is detailed by the inventor in many recent papers, e.g. (Chua 2011, 2012, 2015). Several scientific books and monographs were dedicated to memristors, e.g. (Adamatzky and Chua 2014; Tetzlaff 2014; Radwan and Fouda 2015; Vaidyanathan and Volos 2016a, b).

Memristor symbol and the constitutive relation are shown in Fig. 1.

Conventionally, q is called the charge and φ is called the flux of the memristor but these quantities need not have any physical interpretations.

The memristor is said to be charge-controlled if its constitutive relation can be expressed by

$$\varphi = \Phi(q) \tag{1}$$

where $\Phi(q)$ is a continuous and piecewise-differentiable function with bounded slopes. Differentiating (1) with respect to time t, the memristor port equation is obtained

$$v = \frac{\mathrm{d}\varphi}{\mathrm{d}t} = \frac{\mathrm{d}\Phi(q)}{\mathrm{d}q}\frac{\mathrm{d}q}{\mathrm{d}t} = M(q)\, i \tag{2}$$

where

$$M(q) = \frac{\mathrm{d}\Phi(q)}{\mathrm{d}q} \tag{3}$$

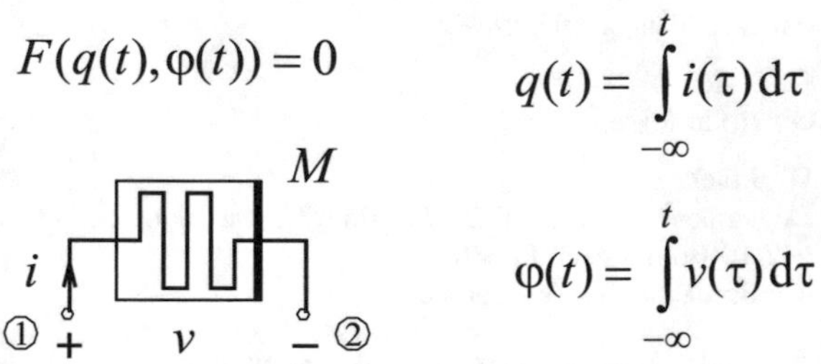

Fig. 1 Memristor symbol and constitutive relation

$$F(q(t), \varphi(t)) = 0$$

$$q(t) = \int_{-\infty}^{t} i(\tau)\,\mathrm{d}\tau$$

$$\varphi(t) = \int_{-\infty}^{t} v(\tau)\,\mathrm{d}\tau$$

is called the memristance at q. Just as memristor is an acronym for memory resistor, memristance is an acronym for memory resistance. It should be noted that the memristance at any time depends on the entire past history of the element's current. Equations (2) and (3) define an ideal memristor.

The memristor exhibits a distinctive "fingerprint" characterized by a pinched hysteresis loop, a double-valued Lissajous figure passing through the origin, confined to the first and the third quadrants of the v-i plane. Consequently, Chua establishes the following identification of memristors (Chua 2011): "Any two-terminal device which exhibits a pinched hysteresis loop in the v-i plane when driven by any bipolar periodic voltage or current waveform, for any initial conditions, is a memristor. The loop shrinks to a straight line whose slope depends on the excitation waveform, as the excitation frequency tends to infinity".

An important and salient feature of the memristor is that it exhibits non-volatile memory (Chua 2012): when a memristor is opened or short-circuited, or when the excitation is switched off, the memristor holds its charge and the flux and "memorizes" its memristance—it is a resistor with memory.

In addition to the three traditional fundamental passive circuit elements, the resistor, the capacitor and the inductor, the memristor is the fourth basic ideal (pure) element of electric circuits characterized by a state-dependent Ohm's law. In a broader sense, the memristor begins a subclass of memristive systems introduced by Chua and Kang (1976).

Extensive analysis of the memristor salient properties and the detailed memristor fingerprints summary, from the simulation and modeling viewpoint, are presented in Tetzlaff (2014) along with the generalization to memristive systems and non-electrical applications.

The successful implementation of memristor is a titanium-dioxide nano device fabricated at Hewlett-Packard Laboratories (Strukov 2008). The memristive behavior was observed by the Hewlett Packard researchers during their experimental work with nanoscale crossbar memory arrays. This pure solid-state implementation of memristor, without an internal power supply, is sometimes referred to as the HP memristor. Nanoscale RF memristive switch was reported by Pi et al. (2015).

Recently, some companies (Bio Inspired Technologies 2016, Knowm) have announced memristor implementations, for example as 44-pin PLCC and 16-pin DIP packages. Commercial memristors for RF/microwave applications are still not available as off-the-shelf components, so reliable circuit models are needed to explore and simulate application circuits which exploit memristor's potential (Pickett et al. 2009; Biolek et al. 2013, 2015; Biolek and Biolek 2014; Ascoli et al. 2013a, b, c; Bayat et al. 2015).

The direct physical realization of memristor as the fourth basic circuit element opens new vistas and research interests in many application fields ranging from digital memories to analog devices. Moreover, special issues of the eminent IEEE publications have been dedicated to this emerging technology (IEEE 2012, 2013).

Despite an immense interest among researchers and engineers on the memristor, commercial memristors are still not available, so various memristor models have been reported to help simulate application circuits. SPICE (Simulation Program with Integrated Circuit Emphasis) is a general-purpose simulation program which allows the testing of complex circuits before they are actually implemented experimentally, so it can be useful for simulating memristor-based circuits (Abdalla and Pickett 2011; Batas and Fiedler 2011; Prodromakis et al. 2011; Corinto and Ascoli 2012; Eshraghian et al. 2012; Kolka et al. 2012; Ascoli et al. 2013a, b, c; Kvatinsky et al. 2013a, b; Pershin and Di Ventra 2013; Yakopcic et al. 2013).

Memristors hold promise for use in diverse applications ranging from digital memories and logic to analog circuits and systems. In analog circuits, the resistance may require a continuous value, so memristors might be used as configurable components and the desired resistance could be initialized by a specific procedure, different from the expected circuit operation (Pershin and Di Ventra 2010; Kvatinsky et al. 2013a, b; Ascoli et al. 2013a, b, c).

Memristor-based technology could be utilized, potentially, to enhance performance of many RF/microwave subsystems, as well. Application of memristors in RF/microwave circuits, and in a broader context in electromagnetic systems, is another challenging field for researchers and engineers. In this application frontier, the research efforts might be divided into several important classes of applications, such as frequency selective surfaces, antennas, filters, and other RF/microwave circuits (Xu et al. 2014a, b; Potrebić and Tošić 2015; Gregory and Werner 2015).

Bray and Werner utilize memristors as electromagnetic switches to implement a frequency selective surface (Bray and Werner 2009, 2010). Werner and Gregory analyze a memristor-based electromagnetic absorber (Werner and Gregory 2012). Wang et al. report the broadband radiation properties of a microstrip patch L-band antenna directly modulated by dual high-frequency resistive memristors (Wang et al. 2011). Gregory details on finite-difference time-domain (FDTD) modeling of memristive devices and some configurable memristor-based electromagnetic devices (Gregory 2013). Sombrin et al. use the ideal memristor as a behavioral model for passive non-linearity in filters, antennas and connections (Sombrin et al. 2014). Gregory and Werner analyze a polarization-switchable patch antenna with memristors as microwave switches (Gregory and Werner 2014). Xu et al. analyze a planar ultra-wideband (UWB) monopole antenna with memristor-based reconfigurable notched band (Xu et al. 2014a, b). Wu et al. explore the feasibility of fabrication transient photonic memristor at microwave frequencies with metamaterials (Wu et al. 2014). Xu et al. incorporate a memristor in a microstrip transmission line as a load, analyze single memristor-loaded split-ring resonator filter, and utilize a memristor as a carrier-wave modulator connecting a microstrip patch antenna to the ground (Xu et al. 2014a, b). Potrebić and Tošić utilize memristors as configurable linear resistors in a power divider, coupled-resonator bandpass filters, and a low-reflection quasi-Gaussian lowpass filter with lossy elements, and propose memristor-based bandpass filters that feature suppression of parasitic frequency pass bands and widening of the desired rejection band (Potrebić and Tošić 2015).

For RF/microwave applications memristors could be exploited as linear resistors with programmable resistance, which can be accurately adjusted to a desired or specified value. Precise controllability of the memristance value might be important for tuning microwave circuits and optimizing their performance. In several applications, such as filters, the high-frequency range of the operation enforces the memristor into the role of a linear resistor whose resistance can be adjusted electronically. On the other hand, some applications, such as reconfigurable electromagnetic absorbers, benefit from memristors as electromagnetic switches.

Due to the unavailability of commercial memristors, it is necessary to use accurate circuit-level simulations for experimenting with the memristor-based RF/microwave circuits and for studying their performance. RF/microwave circuit simulators, which use the HSPICE engine (HSPICE 2016) for the time-domain transient simulation, such as NI AWR Microwave Office (NI 2016), can be used to verify the expected functionality of the considered memristor-based circuits.

The rest of the chapter is organized as follows. The next Section provides a brief description of transient simulations of RF/microwave circuits with memristors. Section 3 presents memristor-based reconfigurable frequency selective surfaces and planar absorbers. Antennas with memristors are presented in Sect. 4. Memristor-based RF/microwave passive circuits are dealt with in Sect. 5. Finally, Sect. 6 outlines the conclusions of this study and some thoughts for future work.

2 Simulation of RF/Microwave Circuits with Memristors

SPICE (Simulation Program with Integrated Circuit Emphasis) can be used for the simulation of RF/microwave circuits from DC to microwave frequencies higher than 100 GHz. Transmission lines are often encountered in these circuits, so a versatile transmission line model is needed for simulations from a simple lossless line to complex frequency-dependent lossy coupled lines. In general, transmission line simulation is challenging and time-consuming since extracting the transmission line parameters from physical geometry takes a significant effort.

Transient simulation is also important for predicting the time-domain behavior of RF/microwave circuits, especially when they include memristors. An advanced SPICE-based simulator is needed to achieve accurate transient simulations of high frequency circuits containing substrate-specific distributed components, coupled transmission lines, discontinuities, and components characterized by frequency-dependent multiport parameters obtained from numerical electromagnetic (EM) analysis. In addition, the skin effect, proximity effect, edge effect, dielectric and conductor losses, and surface roughness should be taken into account for realistic models of distributed RF/microwave devices. Furthermore, all models should preserve the interdependencies of frequency-domain data, the correct asymptotic behavior of the data at high frequencies, and ensure causality in the time-domain simulation (Wedge et al. 2005). It should be emphasized that all models intended for use in transient simulations must be causal (having a response that does not appear before the stimulus).

NI AWR Microwave Office (MWO) is an example of an environment in which the model generation is performed automatically for various RF/microwave components. Moreover, the native HSPICE netlist can be incorporated as a model of a multiport device, e.g. the memristor. As a result, microwave circuits with memristors can be simulated in the time domain with HSPICE, which is included in the environment (Wasserman et al. 2005). The W-element and the S-element are the crucial HSPICE elements for the transient simulation of RF/microwave circuits. Generally, multiconductor transmission line structures are represented by the W-element and the frequency dependant N-port parameters are represented by the S-element.

Idealized simple RF/microwave circuits with memristors can be simulated by LTspice, a free SPICE implementation (Engelhardt 2015), as well (Potrebić and Tošić 2015). The LTspice model of the ideal memristor, Fig. 2, is based on the memristorR1 model and subcircuit proposed by Biolek et al. (2013) and Tetzlaff (2014). The corresponding fingerprint, a pinched hysteresis loop passing through the origin, is shown in Fig. 3.

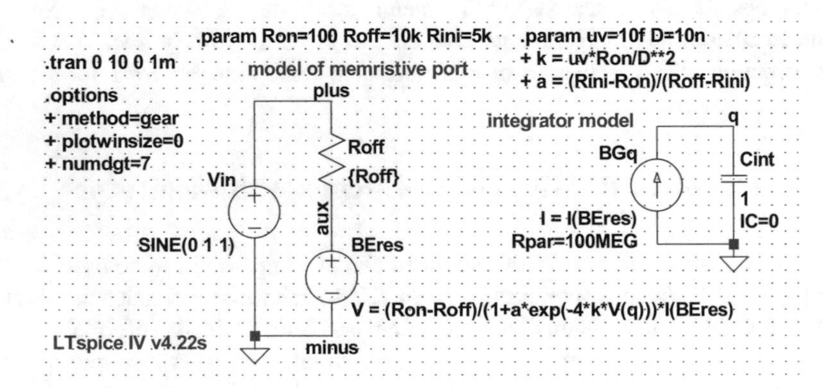

Fig. 2 LTspice model of the ideal memristor based on the memristor R1 model and subcircuit proposed by Biolek et al.

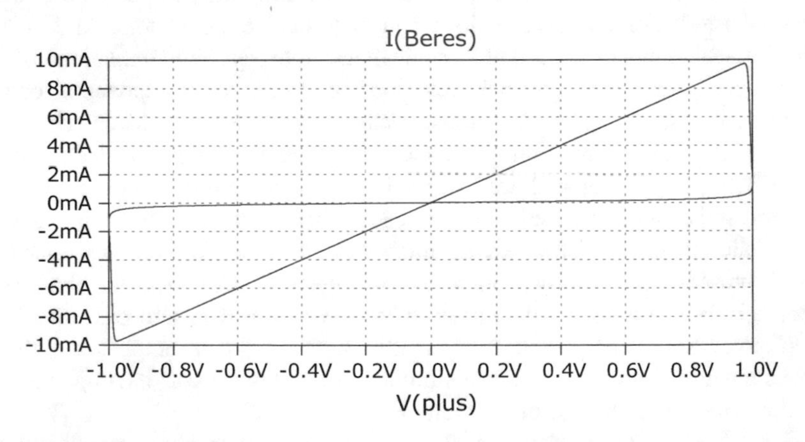

Fig. 3 Fingerprint of the ideal memristor of Fig. 2: Pinched hysteresis loop, a double-valued Lissajous figure passing through the origin, confined to the first and the third quadrants of the v-i plane

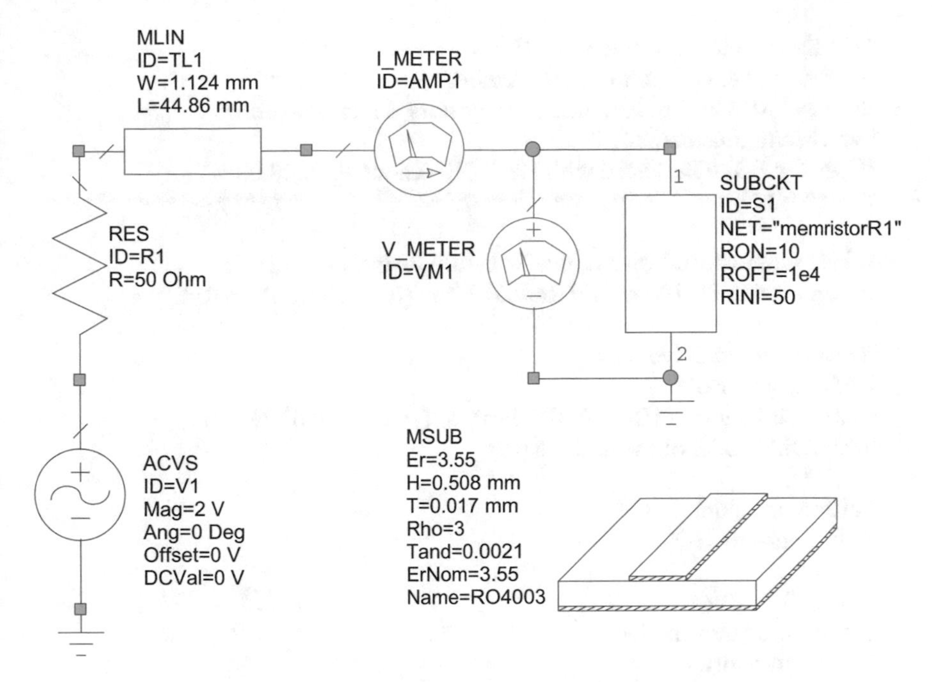

Fig. 4 RF/microwave circuit with memristor in NI AWR Microwave Office

Transient simulation of an RF/microwave circuit with frequency dependant parameters is exemplified by the circuit of Fig. 4, in NI AWR Microwave Office (MWO), and the corresponding response is shown in Fig. 5. The circuit consists of a microstrip transmission line section terminated by a memristor and driven by a sinusoidal voltage source: amplitude 2 V, frequency 1 GHz.

The microstrip line is implemented on the standard Rogers RO4003 substrate (Rogers 2015) characterized by the following physical parameters: relative permittivity 3.55, loss tangent 0.0021, thickness 0.508 mm, copper cladding 17 μm, width 1.124 mm, length 44.86 mm. The characteristic impedance of the line is 50 Ω. Conductor losses are set via the Rho parameter (the bulk resistivity of conductor metal normalized to gold) and the empirical value of 3 takes into account the surface roughness and the skin effect.

Memristor is represented by the two-terminal element SUBCKT NET = "memristorR1" specified as a native HSPICE netlist containing the ideal memristor model R1 proposed by Biolek et al. The initial resistance of the memristor is set to 50 Ω. The netlist is an HSPICE subcircuit as follows (Biolek et al. 2013):

```
**** Ideal memristor model R1 ****
*D. Biolek, M. Di Ventra, Y. V. Pershin*
*Reliable SPICE Simulations of Memristors, Memcapacitors*
* and Meminductors, 2013*
*Code for HSPICE; tested with HSPICE Version A-2008.03*
**********************************************************

.subckt memristorR1 plus minus Ron=100 Roff=10k Rini=5k
.param uv=10f D=10n k='uv*Ron/D**2' a='(Rini-Ron)/(Roff-Rini)'

*model of memristive port
Roff plus aux 'Roff'
Eres aux minus vol='(Ron-Roff)/(1+a*exp(-4*k*V(q)))*I(Eres)'
*end of the model of memristive port

*integrator model
Gx 0 Q cur='i(Eres)'
Cint Q 0 1
Raux Q 0 100meg
*end of integrator model
.ends memristorR1
```

The HSPICE time-domain simulation is performed from 0 to 8 ns with the time step of 0.005 ns. NI AWR Microwave Office computes the frequency-dependant parameters of the dispersive lossy microstrip transmission line and generates the corresponding causal model of the circuit for HSPICE. Next, HSPICE is

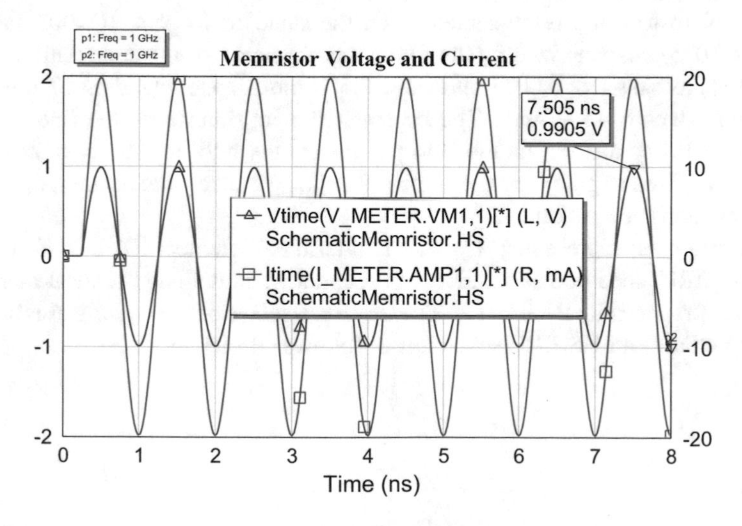

Fig. 5 Time-domain response of the circuit shown in Fig. 4

automatically invoked to carry out the time-domain simulation. The simulation results are available for presentation, e.g., via the V_METER and I_METER elements.

Initial RF/microwave circuit design often starts with idealized components. Therefore, a memristor-based microwave circuit composed of ideal elements is expected to give a good insight into the circuit operation and performance. This work considers only circuits with ideal memristors and is meant to be a proof-of-concept on the potential application of memristors for RF/microwave circuits. Our model of the memristor is a simplified version of a real operating threshold-type memristive device.

We can use the ideal memristor as a linear resistor with a programmable resistance, which can be adjusted with accuracy and reproducibility by auxiliary programming circuitry. It is assumed that the signal to be processed, e.g. filtered, should under no circumstance affect the value of the (programmed) memristance. The resistance value is equal to Rini, shown in Fig. 2, in our modeling.

It is expected that at frequencies higher than 100 MHz, which are typical for RF/microwave circuit operation, the memristor has behavior similar to that of a linear resistor. Consequently, it can be assumed that the distortions of the signals processed by memristor-based microwave circuits should be negligible. We increase the excitation frequency of Fig. 2 to analyze the memristor dynamics with regard to the required frequency behavior. For frequencies over 100 kHz the pinched hysteresis loop of Fig. 3 degenerates to straight line implying a limiting frequency for a linear operation and the FFT analysis shows that the higher harmonics are more that 80 dB below the fundamental.

Precision variable resistors are important for RF/microwave circuits, e.g. in impedance matching and for tuning the frequency characteristics. Therefore, memristors might be promising elements for this application field.

3 Reconfigurable Frequency Selective Surfaces

Investigation of potential applications of RF memristive switches to electromagnetic devices is initiated by Bray and Werner (2009). These researchers present a passive reconfigurable frequency selective surface (FSS) with memristors (Bray and Werner 2010). FSS can be switched at long standoff distances by changing its reflectivity via a low frequency control pulse sent through incident electromagnetic radiation. Since the resistance of the memristor is practically unchanged at high frequencies, it can function as a variable high frequency resistor which is controlled by a low frequency charge delivery. Consequently, a memristor can be used as a passive electromagnetic switch.

The FSS is implemented as a periodic array of square metallic patches located in the center of a wire grid, Fig. 6. A memristor is placed at the center of each wire on the grid. The metallic pattern and memristors are placed on top of a dielectric substrate.

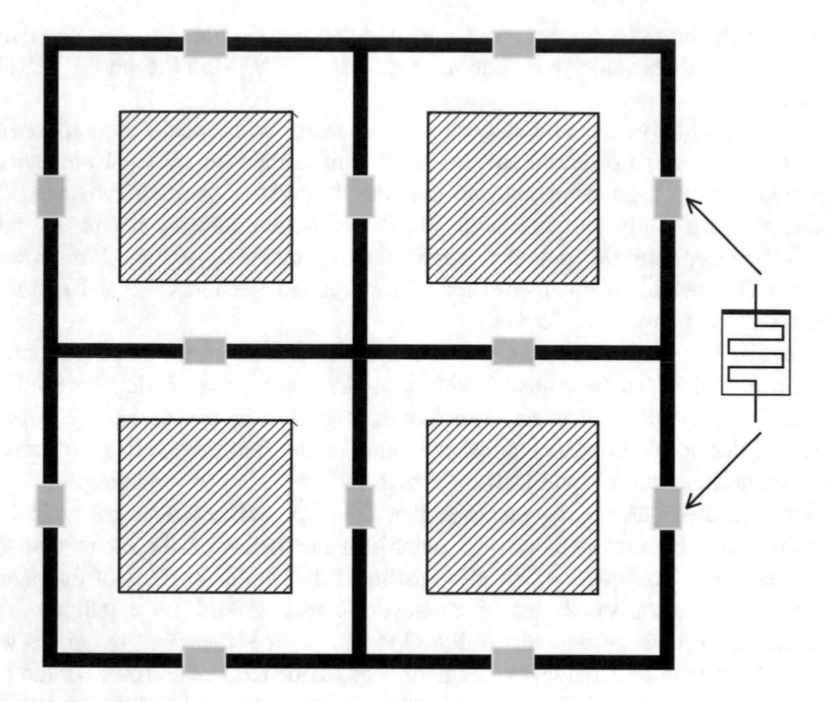

Fig. 6 Memristor-based reconfigurable bandpass frequency selective surface

When the memristor resistance is very high, say greater than 5000 Ω, the FSS is reflective at some frequency band. When the memristor resistance is very low, say less than 100 Ω, the FSS is transmissive. Memristors are electromagnetically switched by a low frequency control signal, e.g. a positive or negative Gaussian pulse that is superimposed over the RF signal. The control signal should be able to deliver enough positive or negative charge to change the resistance of memristors. No auxiliary electronic circuitry is required on the FSS to control the memristance of memristors—it is accomplished by electromagnetic waves incident on the FSS. It should be noted that the reconfigurable FSS is polarization sensitive.

Memristor-based radio frequency devices are further investigated by Werner and Gregory who present a reconfigurable planar absorber with memristors, Fig. 7 (Werner and Gregory 2012).

Unlike the bandpass frequency selective surface of Fig. 6, this planar absorber has an electrically conducting backplane (PEC, perfect electric conductor). Reconfigurability can be achieved by applying a voltage source at the edges of the structure. No power or signal wires need be routed throughout the structure, as the memristors form a voltage divider for reconfiguration. The absorber exhibits peak absorption at some frequency of the incident electromagnetic wave.

The above work is elaborated in the doctoral dissertation (Gregory 2013) and revisited in the recent paper (Gregory and Werner 2015).

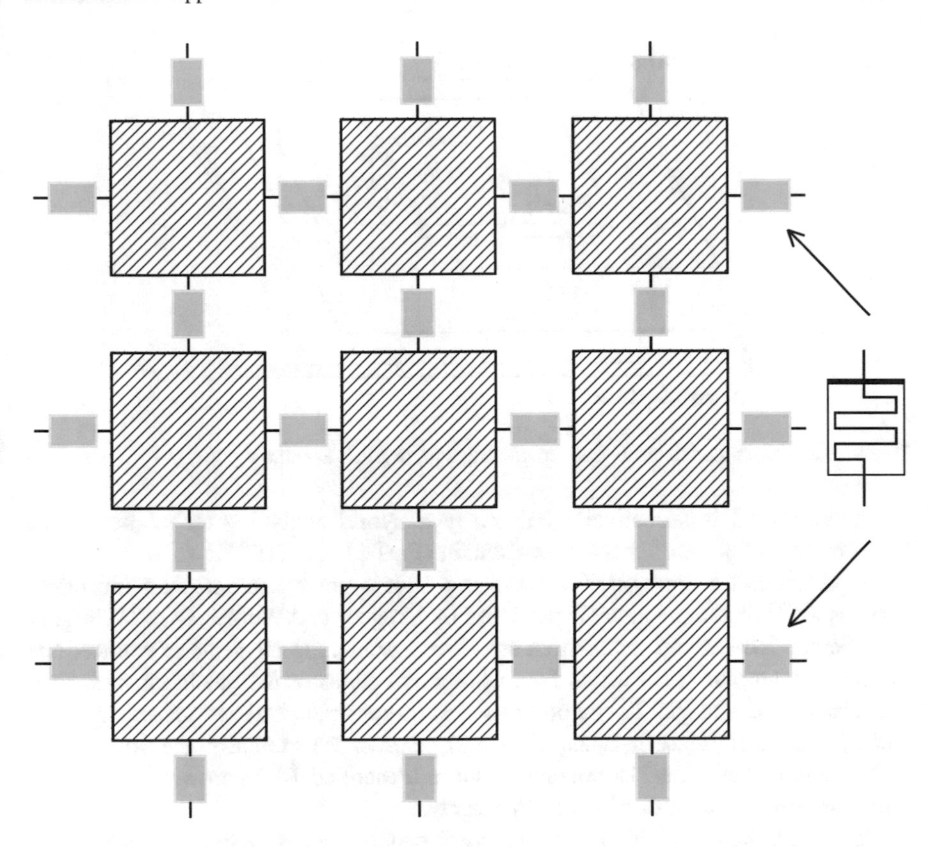

Fig. 7 Memristor-based reconfigurable planar absorber

4 Antennas with Memristors

Antennas and antenna systems might be another field for potential application of memristive devices. Wang et al. report the first broadband radiation properties of a microstrip patch antenna modulated by dual high-frequency resistive memristors (Wang et al. 2011). The authors investigate an L-band microstrip square patch antenna which is directly modulated by two memristors, Fig. 8. Memristors are connected between the patch and the ground plane. The antenna resonates at some RF/microwave frequency and is narrowband in the absence of memristors. When the memristor resistance is high, say about 15 kΩ, the antenna produces the electromagnetic (EM) radiation. The radiation changes the memristance, say to about 50 Ω, which in turn corresponds to the "OFF" state of the antenna, generating a tiny radiation: the antenna is periodically switched "OFF" and "ON". The effect of embedding memristors in the antenna is the broadband electromagnetic radiation. The authors develop an EM-field-circuit model of the dual memristor structure first.

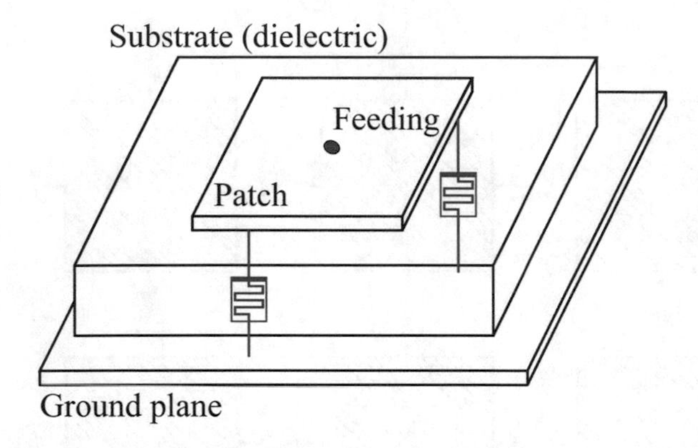

Fig. 8 Memristors-modulated L-band microstrip square patch antenna

Next, the model is numerically solved by a hybrid analyzing technique, which incorporates a finite-difference time-domain (FDTD) and SPICE3 solver.

A polarization-reconfigurable microstrip square patch antenna with the memristor as the switching element is proposed by Gregory and Werner, Fig. 9 (Gregory and Werner 2014). Four memristors are used as switching elements to connect one of two transmission lines which feed orthogonal edges of the patch antenna. Two memristors, M1 and M3, lie flat on the dielectric substrate, and two memristors, M2 and M4, are connected from the top metal layer to the ground plane underneath. The antenna is designed to operate at some prescribed RF/microwave frequency with a narrow bandwidth of several percents.

The non-volatility of the memristor as a switch is an attractive property, eliminating any need for a constant voltage or current bias as with PIN diodes and other RF/microwave switching devices.

The antenna is reconfigured by applying low-frequency signals to the same transmission line that feeds the antenna. Applying a positive voltage turns on two

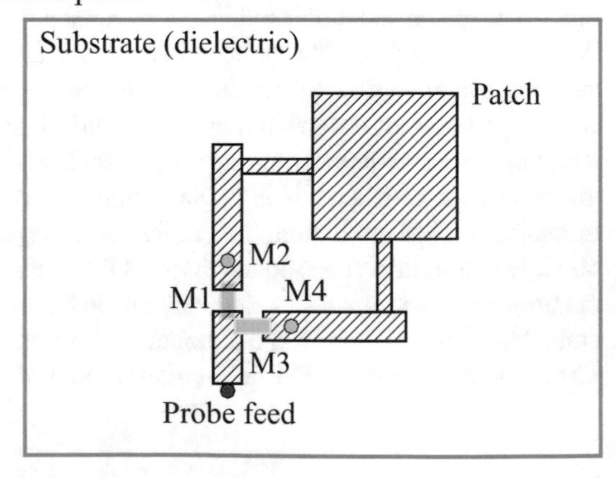

Fig. 9 Memristor-based polarization-reconfigurable patch antenna (*top view*)

horizontal memristors M1 and M3 and turns off two vertical memristors M2 and M4; applying a negative voltage has the opposite effect. The antenna can switch between two orthogonal linear polarizations.

The authors use a finite difference time-domain (FDTD) simulation tool to accurately model the devices under switching conditions, including all of the effects of the accompanying electromagnetic structure, while radio frequency (RF) signals are applied.

The above work is elaborated in the doctoral dissertation (Gregory 2013) and revisited in the recent paper (Gregory and Werner 2015).

A band-switching microstrip patch antenna with the memristor as the switching element is proposed by Gregory and Werner, Fig. 10 (Gregory and Werner 2015). The central rectangular metallic patch is connected to two metallic rectangular extensions with six memristors acting as resistive switches. Two additional inductors are placed between the patch extensions and the ground plane to form a DC path for controlling the memristors.

Memristors connect patch extensions to the central patch and change (lower) the resonant frequency of the antenna when the memristances are low. Consequently, the proposed reconfigurable design achieves switching between two bands of operation.

A finite difference time-domain (FDTD) software tool is used to simulate the structure and to obtain the performance of the dual-band patch antenna.

The above work is elaborated in the doctoral dissertation (Gregory 2013).

Direct antenna modulation (DAM) technique using a memristor as a carrier-wave modulator when connected across the microstrip patch antenna to the ground is presented by Xu et al. Fig. 11 (Xu et al. 2014a, b). An RF carrier wave signal is directly modulated by the memristor with bias controlled by a low-frequency baseband information signal.

A square half-wavelength microstrip patch antenna is designed to have a prescribed resonant frequency. The probe feed point is set along the diagonal of the

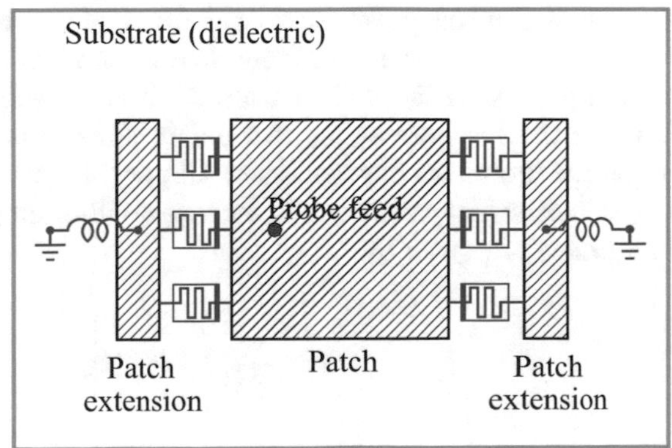

Fig. 10 Memristor-based band-switching patch antenna (*top view*)

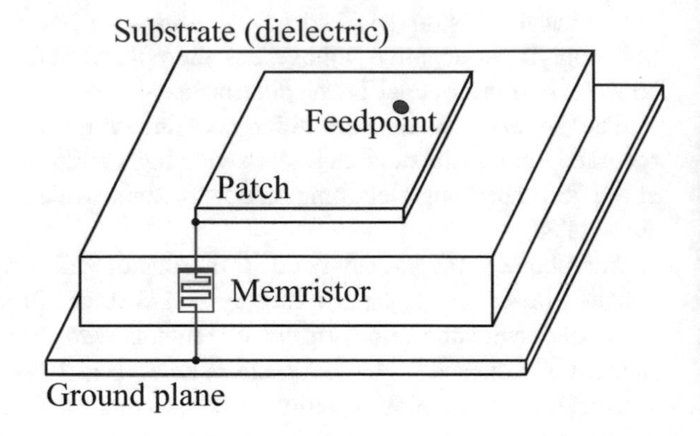

Fig. 11 Memristor-based directly modulated patch antenna

patch plane, where the feed point impedance has a prescribed value, say 50 Ω. The memristor is placed at the corner of the square patch.

The authors propose two SPICE circuit models of the memristor and utilize a hybrid circuit-field analysis technique in the finite difference time-domain (FDTD) simulator (Xu et al. 2014a, b).

5 RF/Microwave Passive Circuits with Memristors

RF/microwave passive circuits might benefit from the memristor specific features. Some examples according to the recent paper (Potrebić and Tošić 2015) are presented in this section.

The Wilkinson power divider (Pozar 2012) is a passive three-port linear time-invariant microwave network, matched at all ports, which is used for power division or power combining. In power division, an input signal is divided into two output signals of lesser power. The divider is usually implemented in planar technologies, such as microstrip and stripline. It can be designed with an arbitrary power division ratio, but we shall consider the equal-split (3 dB) case only.

The Wilkinson power divider 3 dB, Fig. 12, consists of a resistor and two quarter-wave lossless transmission-line sections. We can replace the resistor with a memristor, excite the divider with a 2 V amplitude 1 GHz sinusoidal signal, and observe the response at the two output ports terminated by matched loads. The reference (nominal) impedances of all ports are 50 Ω.

The scattering parameters of the ideal Wilkinson power divider, at the operating frequency, are given by

$$\mathbf{S} = \frac{-\mathrm{j}}{\sqrt{2}}\begin{vmatrix} 0 & 1 & 1 \\ 1 & 0 & 0 \\ 1 & 0 & 0 \end{vmatrix}, \mathrm{j} = \sqrt{-1}. \tag{4}$$

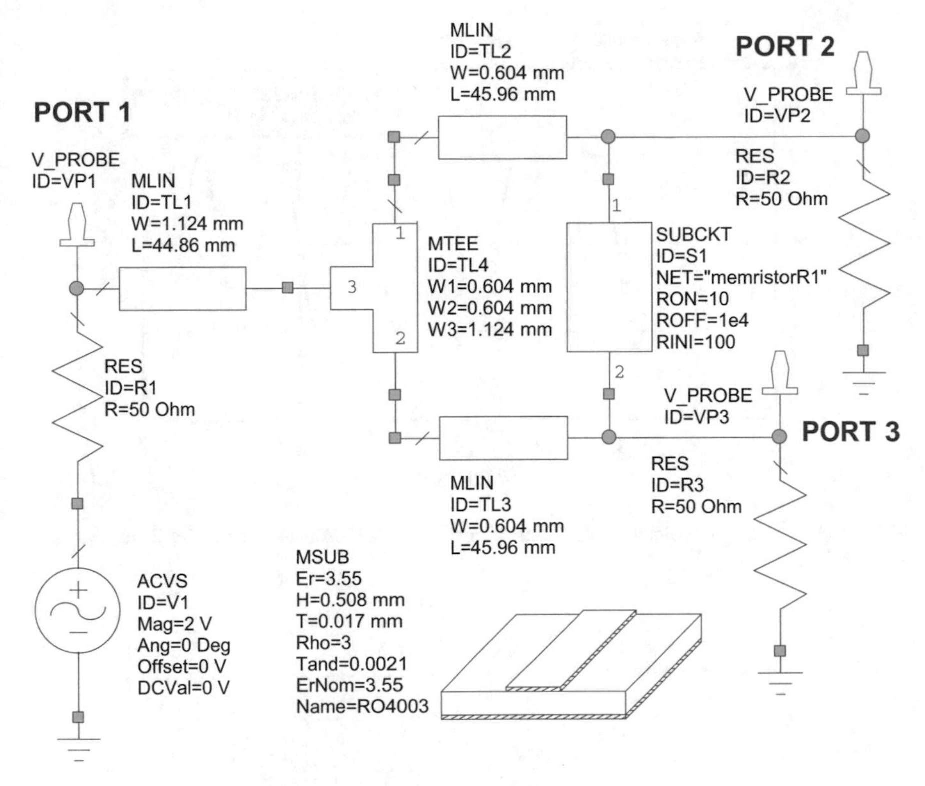

Fig. 12 NI AWR Microwave Office (MWO) circuit model of the microstrip Wilkinson power divider 3 dB with memristor

The instantaneous voltages at the divider ports are shown in Figs. 13, 14, and 15. The amplitude at the input port is 1 V and the amplitude at the output ports is about 0.7 V that verifies the equal-split power division. It should be noted that the output port amplitude is slightly smaller than expected due to the losses in the transmission line sections.

From the practical viewpoint it is important to assume/provide the following:

(1) signal to be processed by the divider should under no circumstance affect the value of the programmed memristance,
(2) distortions of the signals processed by memristor-based divider circuits should be negligible,
(3) input signal power should under no circumstance exceed the memristor power rating and power-handling capability,
(4) memristor chip packaging should be suitable for the microstrip technology implementation.

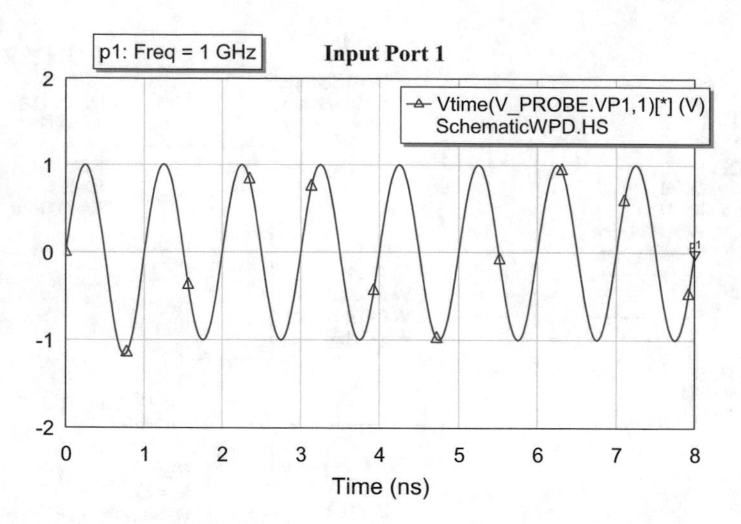

Fig. 13 Instantaneous voltage at the input port of the Wilkinson power divider 3 dB shown in Fig. 12

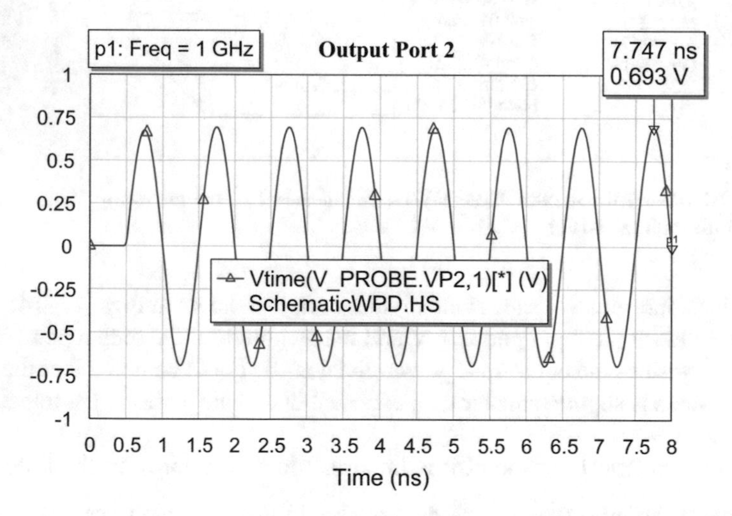

Fig. 14 Instantaneous voltage at the first output port of the Wilkinson power divider 3 dB shown in Fig. 12

We hope that memristors might be useful in the design of the Wilkinson power divider because the memristance can be adjusted—programmed—precisely to a desired value which is not the case with ordinary microwave resistors.

Low-reflection transmission-line quasi-Gaussian lowpass filter with lossy elements is a microwave circuit that might be suitable for the memristor-based design. Gaussian-like frequency-domain transfer functions are often desirable in digital signal transmission because they do not yield overshoots and ringing in the time

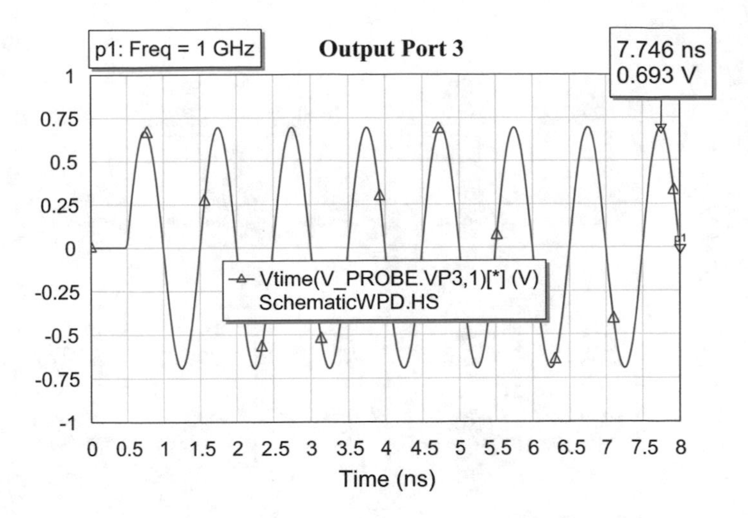

Fig. 15 Instantaneous voltage at the second output port of the Wilkinson power divider 3 dB shown in Fig. 12

domain, so a special class of low-reflection filters is required. Lossy elements are introduced in the filter realization in order to achieve a good matching.

NI AWR Microwave Office (MWO) circuit model of the lossy quasi-Gaussian lowpass filter proposed by Djordjević et al. (2003) is shown in Fig. 16. The voltage source amplitude is set to 2 V in order to generate the transmission scattering parameter. The corresponding frequency response is shown in Fig. 17. The phase characteristic is linear in a wide frequency range. The amplitude characteristic is not very selective.

Potrebić and Tošić propose a modified lossy filter, Fig. 18, in which the resistors are replaced by memristors (Potrebić and Tošić 2015). The filter is excited by a sum of three sinusoidal signals of amplitudes 2.5 V, 0.8 V and 0.5 V, at frequencies 1 GHz, 3 GHz and 5 GHz, respectively. This excitation approximates a bipolar rectangular pulse train with the period of 1 ns and the amplitude of 2 V. The nominal impedances of the ports are 50 Ω, i.e. the source and load impedances are 50 Ω, so the signals at the input and output ports swing from –1 to +1 V.

The expected benefit of this approach is easier and precise tunability of the required resistances due to the inherent tunability feature of the memristor by programming its memristance. The corresponding response is shown in Fig. 19.

The hairpin-line bandpass filter (Hong 2011) is a coupled-resonator filter realized with a cascade of pairs of parallel-coupled open-circuited transmission lines. It is suitable for planar implementations, such as the microstrip or stripline technology, as it is easy to fabricate due to the absence of short circuits. Practically, this filter is obtained by folding the planar half-wavelength resonators into a "U" shape. For an accurate design of the hairpin-line bandpass filter full-wave electromagnetic simulations are required.

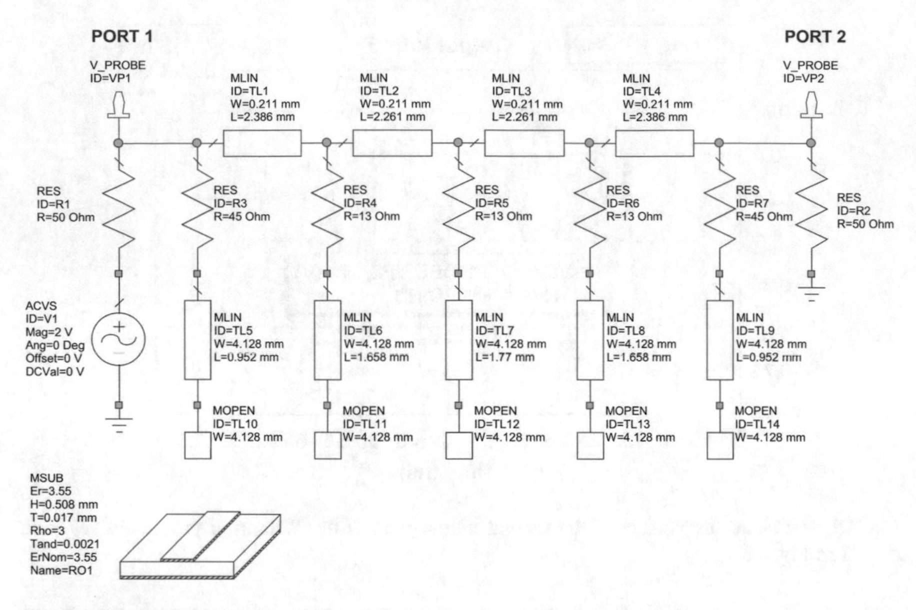

Fig. 16 NI AWR Microwave Office (MWO) circuit model of the low-reflection transmission-line quasi-Gaussian lowpass filter with resistors as lossy elements

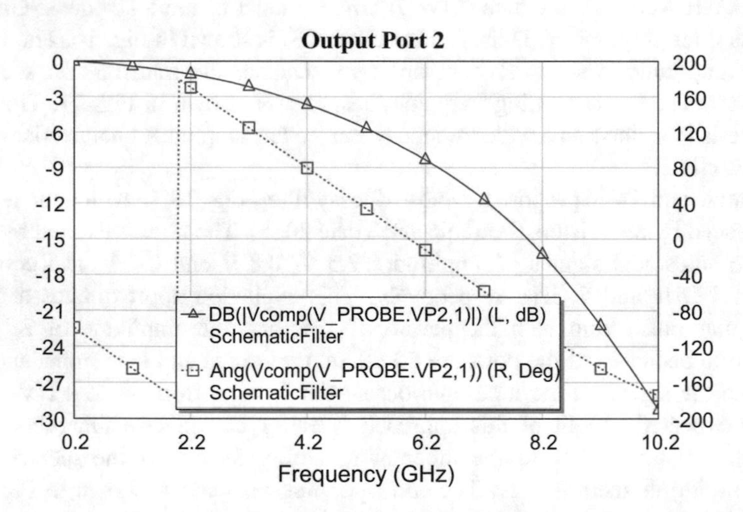

Fig. 17 Magnitude and phase of the transmission scattering parameter of the lossy lowpass filter. The phase characteristic (*dotted*) is very linear in a wide frequency range, thus implying the flat group delay characteristics

The magnitude of the transmission scattering parameter of a typical hairpin-line filter is presented in Fig. 20. The filter is designed as a bandpass filter with the center frequency at 1 GHz and only one pass band is desired. Evidently, undesired pass bands exist, which is a known side effect of all-transmission-line filter realizations.

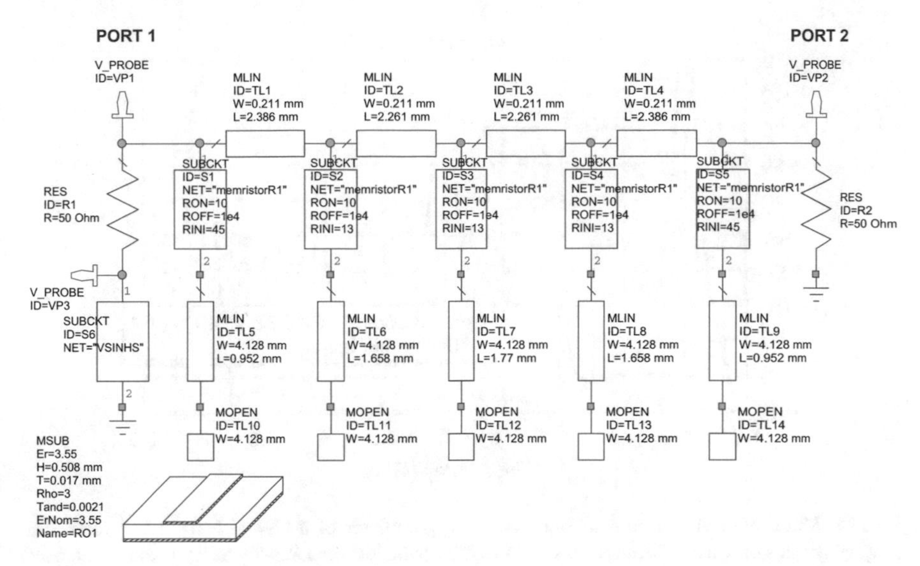

Fig. 18 NI AWR Microwave Office (MWO) circuit model of the low-reflection transmission-line quasi-Gaussian lowpass filter with memristors

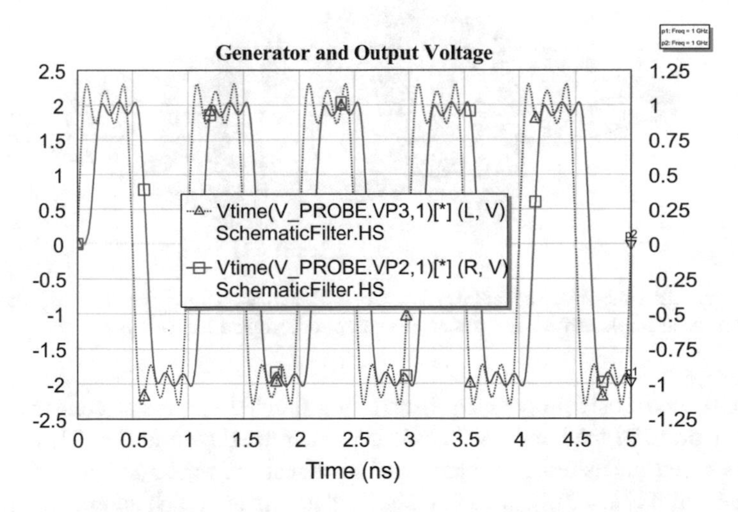

Fig. 19 Time-domain response of the lossy lowpass filter with memristors. The output signal is a very good replica of the generator signal due to the linear phase response, i.e. because of the nearly constant group delay

The hairpin-line filter is an electrically symmetrical network. Symmetrical passive two-port networks can be conveniently analyzed by using Bartlett's bisection theorem, Bartlett and Brune's theorem, and the even- and odd-mode analysis (Wanhammar 2009).

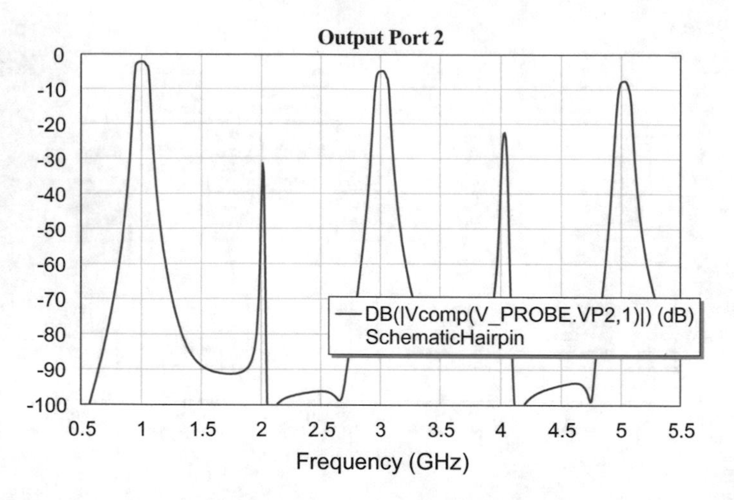

Fig. 20 Magnitude of the transmission scattering parameter of a typical hairpin-line filter. The frequency response has undesired pass bands. The parasitic pass bands occur at the frequencies which are even multiples of the desired pass-band center frequency

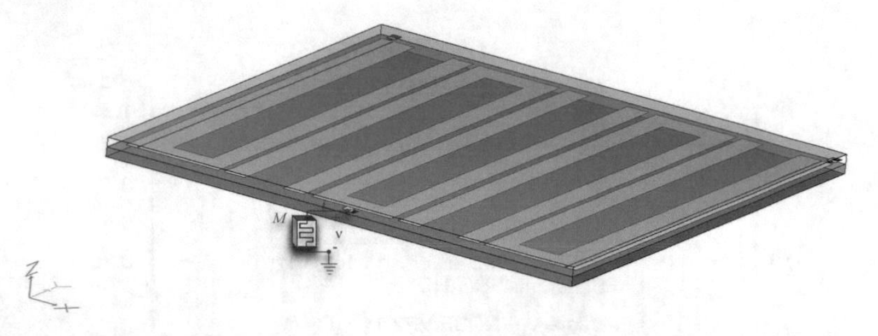

Fig. 21 Physical three-dimensional (3D) model of the bandpass hairpin-line filter with memristor. The filter can be implemented in a planar microstrip, or stripline technology

Potrebić and Tošić propose insertion of a grounded memristor connected to the symmetry point of the bandpass hairpin-line filter, as shown in Fig. 21, in order to suppress some undesired pass bands and to widen the rejection band of the filter (Potrebić and Tošić 2015). It can be shown that this approach suppresses parasitic pass bands at the frequencies which are even multiples of the desired pass-band center frequency.

NI AWR Microwave Office (MWO) circuit model of the stripline bandpass hairpin-line filter with memristor is shown in Fig. 22. It suppresses some undesirable pass bands and widens the rejection band. The filter is excited by two sinusoidal RF signals with amplitudes of 2 V. The frequency of the first signal is at 1 GHz, which is the center frequency of the pass band, so the signal passes through the filter. The second signal at 2 GHz, which is the frequency within the first undesired pass band of the filter, is suppressed by insertion of the memristor.

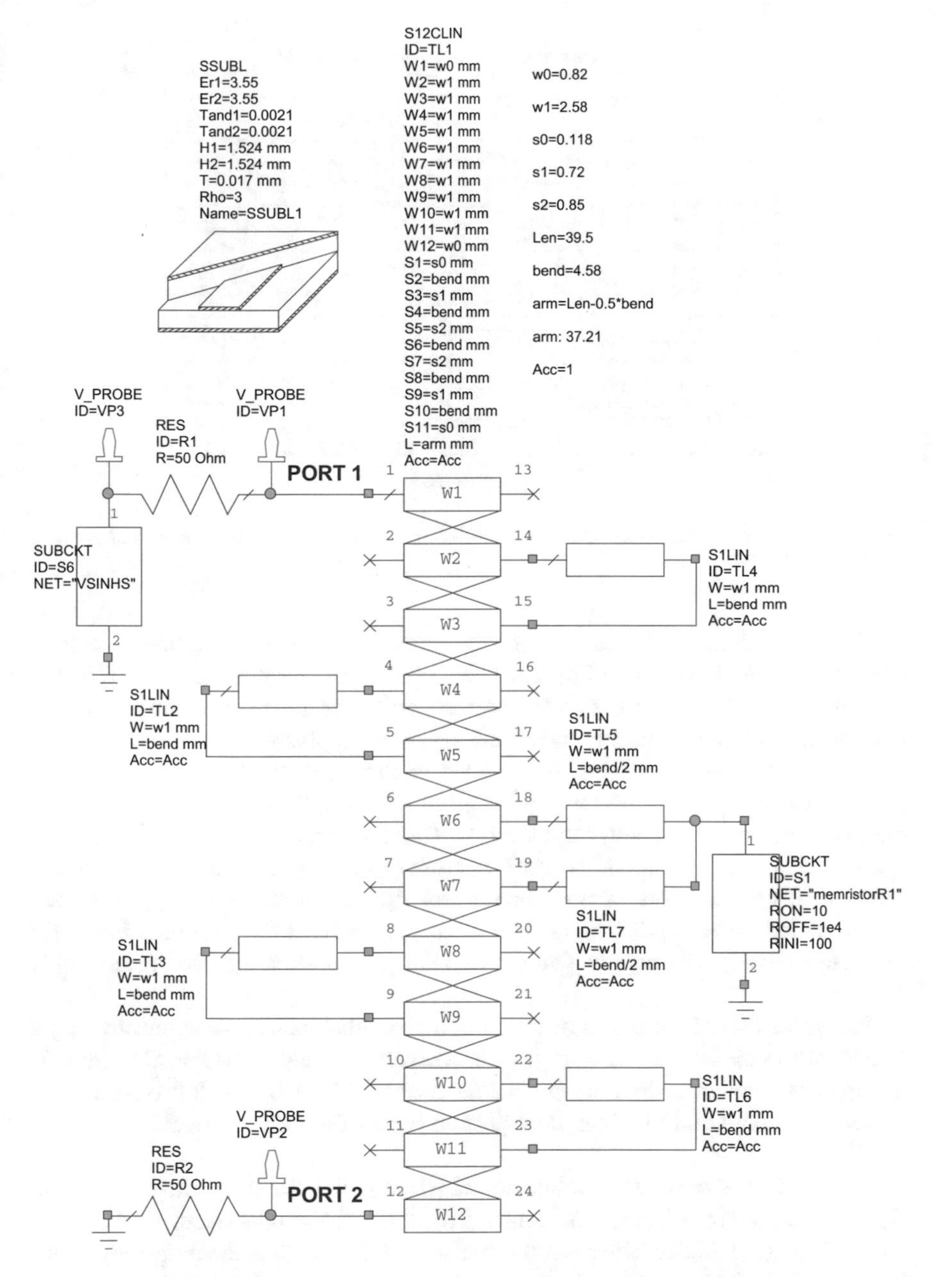

Fig. 22 NI AWR MWO model of the bandpass hairpin-line filter with memristor, which suppresses some undesirable pass bands and widens the rejection band

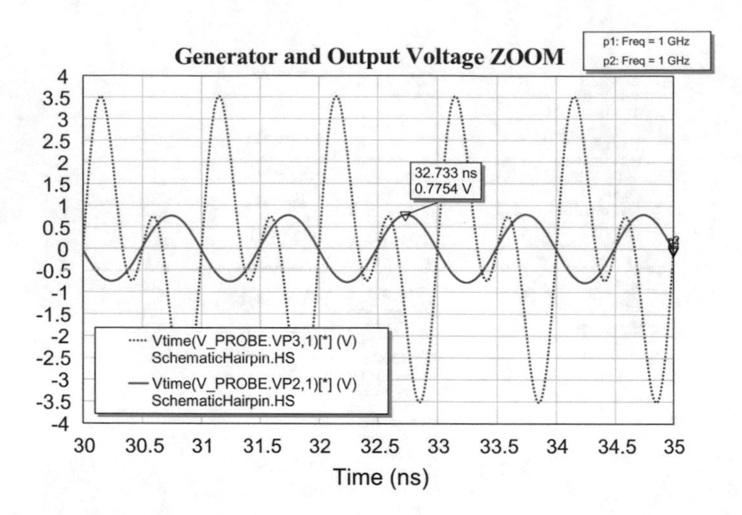

Fig. 23 Time-domain response of the hairpin-line filter with memristors. The output signal is sinusoidal at 1 GHz

The time-domain response of the memristor-based stripline bandpass hairpin-line filter is shown in Fig. 23. The output signal is sinusoidal at 1 GHz. It is slightly attenuated due to the dielectric and conductor losses. The second component of the excitation, the RF signal at 2 GHz, is suppressed.

The above concept can be also applied to the capacitively coupled resonator filter, which is useful for narrowband applications, usually with bandwidths of less than 10% of center frequency (Hong 2011). This filter can be implemented in planar technologies, e.g. as the end-coupled microstrip half-wavelength resonator bandpass filter. The resonators are open-end microstrip resonators that are approximately a half guided wavelength long at the center frequency of the bandpass filter. The resonators are capacitively coupled through the gap between the two adjacent open ends.

Potrebić and Tošić propose insertion of a grounded memristor connected to the middle of the central resonator of the capacitively coupled resonator filter in order to suppress some undesired pass bands (at even multiples of the desired pass-band center frequency) and to widen the rejection band of the filter (Potrebić and Tošić 2015).

The filter is excited by two sinusoidal RF signals. The frequency of the first signal is at 1 GHz, which is the center frequency of the pass band, so the signal passes through the filter. The second signal at 2 GHz, which is the frequency within the first undesired pass band of the filter, is suppressed by insertion of the memristor.

Memristors might be potentially utilized for the design of reconfigurable RF/microwave planar filters, as well. Xu et al. design a reconfigurable microstrip bandpass filter based on a memristor-loaded resonator (Xu et al. 2014a, b).

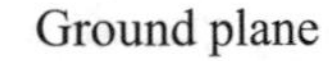

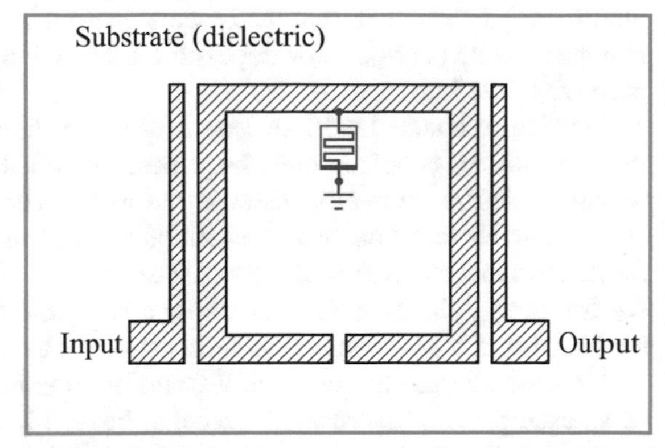

Fig. 24 Reconfigurable microstrip bandpass filter based on a memristor-loaded resonator (*top view*)

The structure of the conventional single-mode open-loop bandpass filter with memristor is shown in Fig. 24. The second resonant frequency can be generated or suppressed depending on the memristance while the fundamental resonant frequency is not affected. When the memristance is high, the second passband is generated and the filter could become a dual-band bandpass filter, which is the same as the conventional single-mode open-loop filter without a memristor. As the memristance decreases, the second passband disappears.

In contrast to the conventional RF/microwave switches, the memristor-based reconfigurable bandpass filter does not require a continuous power supply and consumes little energy (Xu et al. 2014a, b).

6 Conclusion

Memory circuit elements are gaining significant attention owing to their ubiquity and potential use in miscellaneous areas of engineering. In this work we have presented prospective utilization of memristors in microwave passive circuits. Memristors are exploited as linear resistors with programmable resistance, which can be accurately adjusted to a desired or specified value. Precise controllability of the memristance value might be important for tuning microwave circuits and optimizing their performance.

The signals processed by microwave circuits are typically sinusoidal with very high frequencies with respect to the memristor characteristics. Consequently, we expect that the memristor should keep its memristance at the initial value, which has been setup by some other control circuitry. Therefore, we may possibly replace resistors by memristors in the traditional designs of microwave circuits. It should be noted that the way of modeling the nonlinear behavior of the memristor in this case is not so much important because at such frequency range the memristor behaves as a pure linear resistor.

The Wilkinson power divider has been presented in this work as an example of a microwave device that inherently comprises a resistor in its realization. The resistance value is critical for the expected operation, so the memristor might be a promising solution.

Coupled-resonator half-wavelength bandpass filters, the hairpin-line filter and the capacitively coupled filter, have been modified in this work by inserting a memristor at the symmetry plane of the filters. The memristor-based filters have shown a wider rejection band and suppression of the unwanted pass bands at the frequencies that are even multiples of the center frequency of the desired pass band. By fine tuning the memristance value, a compromise between attenuation in the pass band and the amount of suppression could be achieved.

Lowpass transmission-line quasi-Gaussian filter has been presented in this work as an example of a microwave filter that inherently comprises lossy resistive elements. The filter design requires optimization of the resistances to achieve the target performance—linear phase characteristic. Accordingly, the memristor might be beneficial as a replacement for resistors and the memristance electronic adjustment might be a solution to the problem.

Due to the unavailability of memristors, it has been necessary to use models that would allow (1) experimenting with the memristor-based RF/microwave circuits via simulation programs, LTspice and MWO-HSPICE in this work, and (2) studying the performance of the circuits.

RF/microwave applications of memristors might also include implementations of switching components for various types of reconfigurable electromagnetic devices, such as frequency selective surfaces, planar absorbers, band-switching patch antennas, polarization-switching patch antennas, reconfigurable planar bandpass filters with memristor-loaded resonators, and directly modulated patch antennas. A specialized finite-difference time-domain (FDTD) simulation code has been employed to design reconfigurable electromagnetic devices with embedded memristors. Hybrid circuit-field analysis techniques in the FDTD simulator software have been found suitable for exploring these devices.

Simulation of the considered memristor-based microwave circuits and devices have verified the expected functionality and encouraged us to further explore the memristor deployment in the field of the RF/microwave engineering.

Acknowledgements This work was supported in part by the Ministry of Education, Science and Technological Development of the Republic of Serbia under Grant TR 32005. The authors would like to acknowledge the contribution of the EU COST Action IC1401.

References

Abdalla, H., & Pickett, M.D. (2011). SPICE modeling of memristors. In *Proceedings of the IEEE International Symposium on Circuits and Systems (ISCAS)*, Rio de Janeiro (Brasil), pp. 1832–1835.

Adamatzky, A., & Chua, L. O. (2014). *Memristor networks*. New York: Springer.

Ascoli, A., Tetzlaff, R., Corinto, F., Mirchev, M., & Gilli, M. (2013a). Memristor-based filtering applications. In *Proceedings of the 14th Latin American Test Workshop (LATW)*, Cordoba (Argentina), 1–6.

Ascoli, A., Corinto, F., Senger, V., & Tetzlaff, R. (2013b). Memristor model comparison. *IEEE Circuits and Systems Magazine, 13*(2), 89–105.

Ascoli, A., Tetzlaff, R., Corinto, F., & Gilli, M. (2013c). PSpice switch-based versatile memristor model. In *Proceedings of the IEEE International Symposium on Circuits and Systems (ISCAS)*, Beijing (China), pp. 205–208.

Batas, D., & Fiedler, H. (2011). A memristor SPICE implementation and a new approach for magnetic flux-controlled memristor modeling. *IEEE Transactions on Nanotechnology, 10*(2), 250–255.

Bayat, F. M., Hoskins, B., & Strukov, D. B. (2015). Phenomenological modeling of memristive devices. *Applied Physics A, 118,* 779–786.

Biolek, D., Di Ventra, M., & Pershin, Y. V. (2013). Reliable SPICE simulations of memristors, memcapacitors and meminductors. *Radioengineering, 22*(4), 945–968.

Biolek, D., & Biolek, Z. (2014). Fourth fundamental circuit element: SPICE modeling and simulation. Chapter 4 in R. Tetzlaff (Ed.), Memristors and Memristive Systems. New York: Springer.

Biolek, D., Biolek, Z., Biolková, V., & Kolka, Z. (2015). Reliable modeling of ideal generic memristors via state-space transformation. *Radioengineering, 24*(2), 393–407.

Bio Inspired Technologies, LLC, Boise, Idaho, USA. Retrieved June 2016, from http://www.bioinspired.net/.

Bray, M.G., & Werner, D.H. (2009). Passive electromagnetic switching with memristors. In *Proceedings of the 2009 IEEE International Symposium on Antennas and Propagation*, Charleston, SC, USA, June 1–5.

Bray, M.G., & Werner, D.H. (2010). Passive switching of electromagnetic devices with memristors. *Applied Physics Letters, 96*(7), 073504 1–3.

Chua, L.O. (1971). Memristor—The missing circuit element. *IEEE Transactions on Circuit Theory, CT-18*(5), 507–519.

Chua, L. O., & Kang, S. M. (1976). Memristive devices and systems. *Proceedings of the IEEE, 64*(2), 209–223.

Chua, L. O. (2011). Resistance switching memories are memristors. *Applied Physics A, 102,* 765–783.

Chua, L. O. (2012). The fourth element. *Proceedings of the IEEE, 100*(6), 1920–1927.

Chua, L. O. (2015). Everything You wish to know about memristors but are afraid to ask. *Radioengineering, 24*(2), 319–368.

Corinto, F., & Ascoli, A. (2012). A boundary condition-based approach to the modeling of memristor nanostructures. *IEEE Transactions on Circuits and Systems I, 59*(11), 2713–2726.

Djordjević, A. R., Zajić, A. G., Steković, A. S., Nikolić, M. M., Marićević, Z. A., & Schemmann, M. F. C. (2003). On a class of low-reflection transmission-line quasi-Gaussian low-pass filters and their lumped-element approximations. *IEEE Transactions on Microwave Theory and Techniques, 51*(7), 1871–1877.

Engelhardt, M. (2015). SPICE differentiation. *LT Journal of Analog Innovation*, January 10–16.

Eshraghian, K., Kavehei, O., Cho, K.-R., Chappell, J. M., Iqbal, A., Al-Sarawi, S. F., et al. (2012). Memristive device fundamentals and modeling: Applications to circuits and systems simulation. *Proceedings of the IEEE, 100*(6), 1991–2007.

Gregory, M.D. (2013). New methods in ultra-wideband array design and finite-difference time-domain modeling of memristive devices. Doctoral dissertation, *Pennsylvania State University*.

Gregory, M.D., & Werner, D.H. (2014). Reconfigurable electromagnetics devices enabled by a non-linear dopant drift memristor. In *Proceedings of the IEEE Antennas and Propagation Society International Symposium (APSURSI)*, Memphis (USA), pp. 563–564.

Gregory, M. D., & Werner, D. H. (2015). Application of the memristor in reconfigurable electromagnetic devices. *IEEE Antennas and Propagation Magazine, 57*(1), 239–248.

HSPICE, Synopsys, Inc., Mountain View, CA 94043, USA. Retrieved June, 2016, from http://www.synopsys.com/.

Hong, J.-S. (2011). *Microstrip filters for RF/microwave applications* (2nd ed.). Hoboken: Wiley.

IEEE. (2012). Memristors: Devices, models and applications. *Proceedings of the IEEE, 100*(6).

IEEE. (2013). Special issue on memristors: theory and applications. *IEEE Circuits and Systems Magazine, 13*(2).

Knowm, Inc., PO Box 4698, Santa Fe, NM, 87502-4698, USA. Retrieved June, 2016, from http://knowm.org/.

Kolka, Z., Biolek, D., & Biolková, V. (2012). Hybrid modelling and emulation of mem-systems. *International Journal of Numerical Modeling: Electronic Networks, Devices and Fields, 25*(3), 216–225.

Kvatinsky, S., Friedman, E. G., Kolodny, A., & Weiser, U. C. (2013a). TEAM: ThrEshold adaptive memristor model. *IEEE Transactions on Circuits and Systems I, 60*(1), 211–221.

Kvatinsky, S., Friedman, E. G., Kolodny, A., & Weiser, U. C. (2013b). The desired memristor for circuit designers. *IEEE Circuits and Systems Magazine, 13*(2), 17–22.

NI AWR Design Environment, National Instruments, Inc., El Segundo, CA 90245, USA. Retrieved June, 2016, from http://ni.com/awr.

Pershin, Y. V., & Di Ventra, M. (2010). Practical approach to programmable analog circuits with memristors. *IEEE Transaction on Circuits and Systems I, 57*(8), 1857–1864.

Pershin, Y. V., & Di Ventra, M. (2013). Spice model of memristive devices with threshold. *Radioengineering, 22*(2), 485–489.

Pi, S., Ghadiri-Sadrabadi, M., Bardin, J. C., & Xia, Q. (2015). Nanoscale memristive radiofrequency switches. *Nature Communications, 7519*(6), 1–9. doi:10.1038/ncomms8519.

Pickett, M. D., Strukov, D. B., Borghetti, J. L., Yang, J. J., Snider, G. S., Stewart, D. R., et al. (2009). Switching dynamics in titanium dioxide memristive devices. *Journal of Applied Physics, 106*(074508), 1–6.

Potrebić, M., & Tošić, D. (2015). Application of memristors in microwave passive circuits. *Radioengineering, 24*(2), 408–419.

Pozar, D. M. (2012). *Microwave engineering* (4th ed.). Hoboken: Wiley.

Prodromakis, T., Peh, B. P., Papavassiliou, C., & Toumazou, C. (2011). A versatile memristor model with nonlinear dopant kinetics. *IEEE Transactions on Electron Devices, 58*(9), 3099–3105.

Radwan, A. G., & Fouda, M. E. (2015). *On the mathematical modeling of memristor, memcapacitor, and meminductor*. New York: Springer.

Rogers Corporation, USA. (2015). RO4000 series high frequency circuit materials. Retrieved June, 2016, from http://www.rogerscorp.com/acs/products/54/ro4003c-laminates.aspx.

Sombrin, J., Michel, P., Soubercaze-Pun, G., & Albert, I. (2014). Memristors as non-linear behavioral models for passive inter-modulation simulation. In *Proceedings of the 9th European Microwave Integrated Circuit Conference (EuMIC)*, Rome (Italy), pp. 385–388.

Strukov, D. B., Snider, G. S., Stewart, D. R., & Williams, R. S. (2008). The missing memristor found. *Nature, 453*(7191), 80–83.

Tetzlaff, R. (2014). *Memristors and memristive systems*. New York: Springer.

Vaidyanathan, S., & Volos, C. (2016a). *Advances and applications in nonlinear control systems*. Berlin: Springer.

Vaidyanathan, S., & Volos, C. (2016b). *Advances and applications in chaotic systems*. Berlin: Springer.

Wang, L., Yuan, M., Xiao, T., Joines, W. T., & Liu, Q. H. (2011). Broadband electromagnetic radiation modulated by dual memristors. *IEEE Antennas and Wireless Propagation Letters, 10*, 623–626.

Wanhammar, L. (2009). *Analog filters using MATLAB*. New York: Springer.

Wasserman, E., Neilson, D., & Mido, T. (2005). Applied wave research's analog office extends HSPICE transient simulations to RF frequencies. [online] NI AWR Design Environment, National Instruments, Inc., El Segundo, CA 90245, USA. Retrieved June, 2016, from http://ni.com/awr.

Werner, D.H., & Gregory, M.D. (2012). The memristor in reconfigurable radio frequency devices. In *Proceedings of the IEEE Antennas and Propagation Society International Symposium (APSURSI)*, Chicago (USA), pp. 1–2.

Wedge, S., Wasserman, E., & Neilson, D. (2005). Transient simulations at RF frequencies. *Microwave Journal* 1–4.

Wu, H., Zhou, J., Lan, C., Guo, Y., & Bi, K. (2014). Microwave memristive-like nonlinearity in a dielectric metamaterial. *Scientific Reports, 4*(5499), 1–6.

Xu, K., Zhang, Y., Spiegel, R.J., Joines, W.T., & Liu, Q.H. (2014a). Memristor-based UWB antenna with reconfigurable notched band. In *Proceedings of Abstracts of the Progress in Electromagnetics Research Symposium*, Guangzhou (China), p. 1656.

Xu, K. D., Zhang, Y. H., Wang, L., Yuan, M. Q., Fan, Y., Joines, W. T., et al. (2014b). Two memristor SPICE models and their applications in microwave devices. *IEEE Transactions on Nanotechnology, 13*(3), 607–616.

Yakopcic, C., Taha, T. M., Subramanyam, G., & Pino, R. E. (2013). Generalized memristive device SPICE model and its application in circuit design. *IEEE Transactions on Computer-Aided Design of Integrated Circuits and Systems, 32*(8), 1201–1214.

Theory, Modeling and Design of Memristor-Based Min-Max Circuits

S.H. Amer, A.H. Madian, Hany ElSayed, A.S. Emara and H.H. Amer

Abstract Neuromorphic systems have recently emerged as promising candidates for future computing paradigms. Min-Max circuits are indispensable building blocks in Artificial Neural Networks and Fuzzy systems. For instance, the inference engine in Fuzzy controllers that constitutes the decision-making unit in such systems is a Min-Max circuit. Conventionally, transistor-based architectures were adopted in the design of Min-Max circuits. Several designs have been reported that primarily focus on reducing the area consumption (some were voltage mode and others were current mode). However, the miniaturized features of the memristor and the peculiar characteristics it exhibits have driven researchers to use it in state-of-the-art Min-Max circuits. This work addresses the theory, design and modeling of memristor-based Min-Max circuits. Basics of memristor-based Min-Max circuits are addressed through an elaborate explanation of 2-input Min-Max circuits. First, the working principle is explained based on Ohm's and Kirchhoff's Laws. Then, the theory is generalized to an arbitrary number 'N' of inputs (N-ary memristor-based Min-Max circuits) via a formal mathematical proof. An important feature of the memristor is the existence of a threshold below which no change in the state variable (no switching in the case of Min-Max circuits) occurs. Although some existing models overlook the threshold behavior of memristors, most experimental data does confirm the existence of a threshold and, accordingly, it is essential to incorporate its effect in Min-Max circuits. Furthermore, failure to abide by the threshold restrictions results in a circuit malfunction not just a parametric failure (i.e. increased power consumption, increased delay.... etc.) which further necessitates a careful and thorough modeling of the effect of the threshold on the circuit's behavior. Modeling of the threshold will be approached in

---

S.H. Amer (✉) · A.S. Emara · H.H. Amer
Electronics and Communications Engineering Department, The American University in Cairo, New Cairo, Cairo, Egypt
e-mail: sherif_amer@aucegypt.edu

A.H. Madian
Radiation Engineering Department, Egyptian Atomic Energy Authority, Cairo, Egypt

H. ElSayed
Electronics and Communications Department, Cairo University, Giza, Egypt

© Springer International Publishing AG 2017
S. Vaidyanathan and C. Volos (eds.), *Advances in Memristors, Memristive Devices and Systems*, Studies in Computational Intelligence 701,
DOI 10.1007/978-3-319-51724-7_8

two ways. First, an analytical approach is adopted to derive a closed form expression for the effect of the threshold on the circuit. Then, an algorithm is developed (implemented in MATLAB) that emulates the circuit operation. The algorithm runs exhaustive simulations on memristor states and input voltage vectors for different circuit sizes (number of inputs) to verify the derived model. The implications of the derived model are twofold: (1) it provides a closed formula for designers who wish to design memristor-based Min-Max circuits (2) it demonstrates a clear trade-off between the size and the resolution of the circuit.

Keywords Memristor • Fuzzy logic • Min-Max circuits • Beyond von neumann • Computational intelligence

1 Introduction

Complementary Metal Oxide Semiconductor (CMOS) technology has for so long stood as the cornerstone of all Very Large Scale Integration (VLSI) systems. However, other technologies such as the memristor (Chua 1971) have been recently proposed that promise to push the semiconductor industry into new paradigms. The perpetual down scaling of CMOS technology has provided an ever-enhanced performance of electronic circuits over the past decades. With each technology node (a technology node is defined as the channel length of the transistor), average power consumption has decreased, device speed has been boosted and more integration has become achievable. This enhancement was predicted in 1965 by Gordon Moore and has ever since been known as “Moore’s Law”.

The sustainability of Moore’s Law, however, cannot last much longer due to two major challenges which are (1) the CMOS science has reached the fundamental physical limits (Thompson and Parathasarathy 2006) which prohibits further miniaturization and (2) process variations have skyrocketed in such Nano-scale regime.

These issues have instigated significant research trying to provide novel and innovative solutions to assuage the aforementioned challenges. Several endeavors have been proposed on both circuit and device levels. On the circuit level, new structures and circuit architectures have been proposed such as multilayered Integrated Circuits where an extra spatial dimension is exploited to provide a higher functionality per chip area ratio (i.e. 3D ICs) (Thompson and Parathasarathy 2006). Also, on the device level, researchers have sought out new devices such as carbon nanotubes, spintronics and FinFETs. Amongst the new devices tackled by the research community, a novel device known as “memristor” stands as a powerful candidate that has the potential to push the microelectronics industry into new paradigms.

Memristors are newly characterized devices that were first theoretically predicted by Leon Chua in 1971 but had not been physically realized until 2008 when HP announced the first manufactured memristor based on the Titanium dioxide TiO2 process (Sturkov et al. 2008). Ever since that date, a significant research has been undertaken in the area of memristors and memristor-based systems. Leon Chua postulated that the memristor constitutes the missing link between the electric charge and the magnetic flux. He published a paper in 1971 (Chua 1971) in which he provided a merely theoretical treatment for the memristor element. Later in 1976 (Chua and Kang 1976), he published another paper generalizing the concept of the memristor from an electrical element to a whole system theory. Recently, in 2015, Chua published an article summarizing his work on memristors (Chua 2015).

In order for the memristor element to be integrated into commercial CAD tools, several mathematical models were developed. The published models vary from simple linear models (Sturkov et al. 2008) to complex nonlinear models (Pickett et al. 2009). Linear models are simple both analytically and computationally. Yet, experimental data showed a noticeable deviation from those models. On the other hand, nonlinear models are accurate but complex (Kvatinsky et al. 2013).

Despite the immaturity of the memristor science that calls for the need for more theoretical work regarding the memristor element and, accordingly, the continuous refinement of the models, the peculiar behavior of memristors and their miniaturized size have instigated a surge in memristor-based applications in which those features are leveraged to deliver new functions such as resistive RAMs (ReRAMs) and Neuromorphic and Fuzzy circuits, or improve the performance of transistor-based architectures such as in the case of digital and analog circuits.

Memristors have been primarily used in four applications: analog circuits, digital circuits, memories and Neuromorphic (sometimes referred to as biologically-inspired systems or beyond Von Neumann architectures) and Fuzzy systems. Also, several researchers have leveraged memristors and memristive systems in unconventional computing applications such as in chaotic systems (Vaidyanathan and Volos 2016a, b) and nonlinear control systems (Vaidyanathan and Volos 2016a, b).

The analog programmability of the memristor has inspired many researchers to utilize it in several applications other than memories and digital design. One potential application is programmable analog circuits. In Pershin and Di Ventra (2010), the authors reinvented a number of currently existing and extensively used analog blocks by employing memristors in the design. Their idea is hinged upon the threshold behavior of the memristor. The authors utilized this fact by building a memristor-based analog circuit that operates in two phases. In phase '1', the programming phase, high voltages (voltages higher than the threshold of the memristors), are used to program the memristor to the desired resistive value. In phase '2', the analog operation, low voltages (lower than the threshold of the memristor) are applied to perform the analog functionality of the circuit. For further details, the reader is referred to Pershin and Di Ventra (2010).

The non-volatility of memristors and their miniaturized features have instigated their use in state of the art memories known as Resistive RAMs. Unlike in regular CMOS designs where the logic is stored as a voltage, logic in Resistive RAMs is stored as resistive value whereby the information stored is not lost when the power supply is switched off owing to the peculiar nature of the memristive behavior (hence, non-volatile). Also, their miniaturized size has enabled building denser memory arrays. Three major challenges have been encountered by researchers in designing resistive RAMs which are: (1) non-destructive reading operation in which the reading circuitry and the applied read voltages should be designed in such a way as not to corrupt the data stored in the memory cell as in Elshamy et al. (2014) (2) process variations and their effect on read/write operations as in Niu et al. (2010) (3) sneak paths testing in memory arrays as in Kannan et al. (2015).

Another interesting application, in which the compact size of the memrsitor is leveraged, is digital applications. In essence, there are two types of digital applications: logic in memory and conventional logic. In Kvatinsky et al. (2014), material implication logic family was presented in which memristors are used as memory elements as well as perform logic operations. On the other hand, other logic families such as Memristor Ratioed Logic (MRL) described in Kvatinsky et al. (2012) uses memristors as computational elements such as in the case of standard CMOS architectures.

Amongst the very promising emerging computer architectures is the concept of Neuromorphic computing and Fuzzy systems. While such architectures can be implemented using transistors, memristors exhibit peculiar characteristics that are well suited to such systems, which enable the implementation of high density and power efficient systems (Rose et al. 2012).

Memristors were employed in the design of two major building blocks in Fuzzy systems, which are: the Fuzzifier and the Defuzzifier. In Merrikh-Bayat et al. (2011), a memristor-based Fuzzifier was proposed. Memristors were implemented in the crossbar structure. The values of the fuzzy sets were stored in the memristances and OPAMPs were used to convert the membership degrees of the system variable(s) in the fuzzy sets into voltage values. The same structure was later employed in the design of Neuro-Fuzzy systems (Merrikh-Bayat et al. 2013). In Amer et al. (2015a, b), a memristor-based Center-Of-Gravity Defuzzifier was proposed. Four approaches were historically adopted for the hardware implementation of COG defuzzifier circuits. First, the fully digital technique was proposed in Watanabe et al. (1990) in which multiplication/division were performed via iterative addition/subtraction. However, this brings about significant speed limitations and occupies a relatively large chip area. Second, the voltage follower aggregator structure was utilized (Hoseini et al. 2010). This structure does not contain divider circuits, which makes it advantageous. However, it uses several operational amplifiers which makes it area consuming. Third, a current mode approach was proposed. It has the advantage of simple addition/subtraction of signals (Farshidi 2008). Yet, the voltage mode approach is often preferred since most sensors and auxiliary devices communicate with fuzzy systems in voltage mode (Hoseini et al. 2010).

Amongst the major building blocks found in Neuromorphic and Fuzzy systems are Min-Max circuits. For instance, in Fuzzy controllers, Min-Max circuits are used in the Fuzzy inference engine. While transistor-based architectures were commonly adopted in the design of Min-Max circuits, memristor-based Min-Max circuits have been proven to outperform their transistor-based counterparts, primarily, in area occupancy.

To this end, this chapter will focus on developing the theory of memristor-based Min-Max circuits. First, 2-input memristor-based Min-Max circuits are thoroughly investigated. The working principle is discussed and design equations are developed based on Kirchhoff's and Ohm's Laws. Furthermore, the effect of the memristor threshold is analyzed and a closed form design constrain is developed. Second, the theory is generalized to arbitrary number 'N' of inputs whereby the working principle of the generalized N-input circuits is demonstrated via a formal proof. Also, the effect of the memristor threshold on N-input circuits is developed. Finally, in order to put things into perspective, memristor-based min-max circuit are compared to commonly used transistor-based architectures in order to show the advantages gained from incorporating memristors in the design of Min-Max circuits.

2 2-Input Memristor-Based Min-Max Circuits

In general, the governing equation for the Min-Max operation is presented as follows:

$$X_{min} = \text{Min}\ (X_1, \ldots, X_n) \tag{1}$$

$$X_{max} = \text{Max}\ (X_1, \ldots, X_n) \tag{2}$$

If both v_1 and v_2 are equal, no current flows through the circuit and $v_0 = v_1 = v_2$. If $v_1 > v_2$, where $v_1 = v_{max}$ and $v_2 = v_{min}$, a current flows from the upper memristor to the lower one whereby it flows outside the thick line in the upper memristor and inside the thick line in the lower one. According to the definition of the memristor, the upper memristor will switch to 'OFF' acquiring the maximum resistance R_{off} while the lower one will switch to 'ON' acquiring the minimum resistance R_{on}. Since, by definition, $R_{on} \ll R_{off}$ or, equivalently, $G_{off} \ll G_{on}$ ($'G'$ is the memductance and defined as the reciprocal of the memristance such that $G = \frac{1}{R}$) and from Kirchhoff's law, it can be shown that the output voltage is computed as follows (Amer et al. 2015a, b):

$$V_0 = \frac{V_{max} G_{off} + V_{min} G_{on}}{G_{off} + G_{on}} \approx V_{min} \tag{3}$$

It can be readily shown that reversing the polarity of the memristors in Fig. 1 implements a maximum operation. Therefore, the forthcoming analysis will be only concerned with the minimum circuit.

An important characteristic of the memristor is the existence of a threshold below which no change in the memristance occurs. Hence, it is important to model

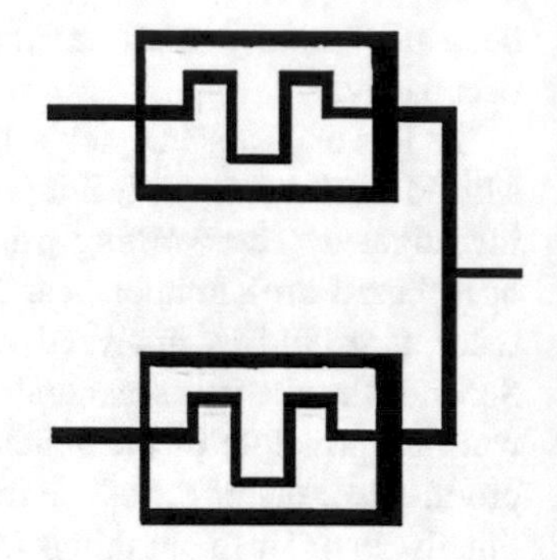

Fig. 1 2-input minimum circuits

the effect of the threshold on the operation of Min-Max circuits. Assuming a current controlled memristor (Kvatinsky et al. 2013), there are two thresholds I_{on} and I_{off} that correspond to the memristor switching from 'ON' to 'OFF' or from 'OFF' to 'ON', respectively. I_{on} has a negative value corresponding to current flowing to the left while I_{off} is positive corresponding to current flowing to the right in Fig. 1. Also, in general, I_{on} and I_{off} do not have to be equal. However, for the purpose of this work, in order to simplify the analysis, I_{on} and I_{off} are assumed equal which is known as symmetric switching. This can be formalized as follows:

$$|I_{on}| = |I_{off}| = |I_t| \tag{4}$$

From now on, the absolute sign will be dropped and $|I_t|$ will be expressed as I_t.

In order to ensure the proper operation for the circuit, both memristors in Fig. 1 must be able to switch under all states. By inspection, since the circuit is a simple one, it can be shown that the lowest current occurs when both memristors are 'OFF' and the voltage difference between both inputs is minimal.

$$\frac{\partial V}{2R_{off}} > I_t \tag{5}$$

∂V is the minimum allowable difference between both inputs which reflects the resolution of the circuit, R_{off} is the maximum resistance of the memristor and I_t is the threshold current of the memristor.

3 N-Ary Memristor-Based Min-Max Circuit

This section will generalize the theory of memristor-based Min-Max circuits to N-input circuits. The first subsection will provide a proof that, similar to 2-input structure presented in Fig. 1, an N-input structure does implement Min-Max operation. The next subsection will derive a closed form expression for the effect of the threshold on the circuit.

3.1 Proof of N-Input Memristor-Based Min-Max Circuit

Definitions and Assumptions:

(a) $G = 1/R$
(b) If I_i is negative (left), then $G_i = G_{on}$
(c) If I_i is positive (right), then $G_i = G_{off}$
(d) Switching time of the memristors is ignored (i.e., only steady state conditions are considered)
(e) $G_{on} \gg G_{off}$ irrespective of the number of memristors 'n'
(f) $(|v_i - v_O|).G_i > I_t$.

(a), (b), and (c) are inherent properties in the memristor device which were explained in previous sections. In essence, memristors possess a certain switching time (i.e. time taken to switch from G_{on} to G_{off} or vice versa) depending on the material characteristics. However, only steady state conditions are considered since the aim is to prove the viability of the structure not its switching dynamics. (e) is considered valid throughout the development of the proof. However, it will be clear later in this section that (e) imposes design restrictions on N-ary Min-Max circuits. Finally, (f) is of a prime importance since it places restrictions on the applied voltages as a function of the memristor Current threshold. The effect of the threshold will be ignored in the proof. Yet, its impact will be studied in detail in the next section.

Suppose $v_{min} \le v_i \le v_{max} \forall v_i$ and $G_i \in \{G_{on}, G_{off}\}$, then from Kirchhoff's Current Law (KCL) and Ohm's Law applied at v_O:

$$v_O = \frac{\sum_{i=1}^{n} v_i.G_i}{\sum_{i=1}^{n} G_i} \tag{6}$$

Suppose, without loss of generality, that $v_1 \le v_2 \le \cdots \le v_n$, where $v_1 = v_{min}$ and $v_n = v_{max}$, then:

$$v_1 \le v_O \le v_n \tag{7}$$

Suppose initially at the start of the operation that v_o assumes an arbitrary value between v_1 and v_n such that:

$$v_1 \ldots v_k < v_o \tag{8}$$

$$v_{k+1} \ldots v_n > v_o \tag{9}$$

Then:

$$G_1 = \cdots = G_k = G_{on} \tag{10}$$

$$G_{k+1} = \cdots = G_n = G_{off} \tag{11}$$

Substituting back in (6):

$$v_{ON} = \frac{G_{on} \sum_{i=1}^{k} v_i + G_{off} \sum_{i=k+1}^{n} v_i}{kG_{on} + (n-k)G_{off}} \tag{12}$$

where v_{ON} is the new output voltage. From (e), $G_{on} \gg G_{off}$ and (12) is reduced to:

$$v_{ON} = \frac{\sum_{i=1}^{k} v_i}{k} \tag{13}$$

Hence, $v_1 < v_{oN} < v_k$ and from (8), $v_k < v_o$. Therefore:

$$v_{ON} < v_o \tag{14}$$

Therefore, it is concluded that the process is recursive since for any arbitrary output voltage v_o, the new output voltage is v_{ON} and is less than v_o.

Also, note that when the output voltage changes from a value v_o to a new value v_{ON}, since $v_{oN} < v_k$, some memristors (Δ) switch from ON ($G_i = G_{on}$) to OFF ($G_i = G_{off}$). Therefore the change in the output voltage can be modeled as:

$$\Delta v = \frac{G_{on} \sum_{i=1}^{k} v_i + G_{off} \sum_{i=k+1}^{n} v_i}{kG_{on} + (n-k)G_{off}} - \frac{G_{on} \sum_{i=1}^{k-\Delta} v_i + G_{off} \sum_{i=k-\Delta+1}^{n} v_i}{(k-\Delta)G_{on} + (n-k+\Delta)G_{off}} \tag{15}$$

Such that $\Delta v = v_o - v_{oN}$. After some mathematical manipulation, (15) reduces to:

$$\Delta v = \frac{(k-\Delta) \sum_{i=k-\Delta+1}^{k} v_i - \Delta \sum_{i=1}^{k-\Delta} v_i}{k(k-\Delta)} \tag{16}$$

It was shown in (14) that v_{ON} is always less than v_0. However, it must be shown that the decrease in the output voltage Δv has a minimal finite value throughout the operation to ensure that the output voltage will eventually gravitate to v_{min} in a finite time. Since finding the minimal Δv might be mathematically tedious, especially, that the variables k and Δ in (16) can only assume integers, it is enough to show that Δv is always finite for all k and Δ and is always positive since we define $\Delta v = v_o - v_{oN}$ where $v_{oN} < v_o$. Since, by definition, $v_1 \le v_2 \le \cdots \le v_n$, each v_i in $\sum_{i=k-\Delta+1}^{n} v_i$ is, individually, larger than each v_i in $\sum_{i=1}^{k-\Delta} v_i$. Hence, assuming $v_{MIN} = \text{minimum}(v_{k-\Delta+1} \ldots . v_n) \forall v_i$ and $v_{MAX} = \text{maximum}(v_1 \ldots . v_{k-\Delta}) \forall v_i$, then from (16):

$$\Delta v \geq \frac{\Delta (v_{MIN} - v_{MAX})}{k} \tag{17}$$

where $v_{MIN} > v_{MAX}$. Since, by definition, k and Δ are finite integers between '1' and 'n' and $k > \Delta$, then Δv is a finite positive number. Note that v_{MIN}/v_{MAX} should not be confused with v_{min}/v_{max} which are the global minimum/maximum voltages in the system. Therefore, it can be shown that since:

- The N-ary minimum circuit has a minimum voltage (boundary) "V_{min}"
- The new output voltage is always smaller than the old output voltage "$V_{0N} < V_o$" and the process is recursive
- Δv is a finite positive number

It can be concluded that $v_0 = v_{min}$.

(f) is crucially important for the proper functioning of the circuit. In essence, this assumption is what allows the cancellation of the G_{off} term throughout the proof. However, this cancellation is not always valid but constrained by the values of applied voltages 'v_i' and the number of memristors 'n'. These constraints can be derived from (12) yielding:

$$\frac{n-k}{k} \ll \frac{G_{on}}{G_{off}} \tag{18}$$

$$\frac{\sum_{i=k+1}^{n} v_i}{\sum_{j=1}^{k} v_j} \ll \frac{G_{on}}{G_{off}} \tag{19}$$

To ensure the proper functioning of the system, (18) and (19) have to hold true under worst case state which is when $k = 1$, $v_j = v_{min}$ and $v_i = v_{max}$ (largest possible left hand side in (18) and (19)). Substituting back in (18) and (19):

$$n - 1 \ll \frac{G_{on}}{G_{off}} \tag{20}$$

$$(n-1)\left(\frac{v_{max}}{v_{min}}\right) \ll \frac{G_{on}}{G_{off}}. \tag{21}$$

3.2 *Effect of the Memristor Threshold*

As mentioned earlier in the introduction and the previous section, the threshold behavior of the memristor poses a crucial challenge in the design of memristor-based Min-Max circuits. Given a specific Current controlled memrsitor with threshold current 'I_t', ON resistance 'R_{on}', OFF resistance 'R_{off}' and arbitrary size 'n', design constraints on the values of the input voltages are derived. This problem will be approached, first, analytically. Then, a MATLAB code is developed to validate the results computationally.

It is important to provide first an intuitive explanation to the problem at hand. While most memrsitors applications are mainly constrained with the type of memristor that can be programmed in an incremental fashion, another type of memrsitors exists which can only assume two values: Low Resistive State (LRS) and High Resistive State (HRS). In essence, memristors can be classified into two types: analog memrsitors and binary memristors (Truang et al. 2014). In order to simplify the analysis and make it feasible to arrive at a closed form constraint, the forth coming discussion will be restricted to binary memristors that can only assume two values: $'R'_{on}$ representing (LRS) and $'R'_{off}$ representing (HRS).

The idea behind the analysis is to ensure that any memristor is able to switch at any state out of the 2^N states that the system might take (a total of 'N' memristors each can take two values G_{on}/G_{off}). Although it was proven earlier that the system can only assume 'N' states (only one memristor out of the 'N' is ON), this only holds for steady state response of the system. Stated differently, the voltages have already settled at the memristors and the switching is going to take place in order to output the minimum voltage at the output. However, in the transient state (i.e. while the voltages are transitioning from one set of input voltage to another set), the system can assume any state out of the 2^N possible states. Having said that, the following analysis will derive generic analytical formulae that present the restriction on the input voltages as a function of the memristance and the threshold current $'I'_t$.

3.2.1 Analysis of the Effect of the Memristor Threshold

In order to ensure the proper functioning for the circuit, two conditions have to be met: (i) at least one memristor $G_i = G_{on}$ where $v_i = v_{min}$ and (ii) $G_j = G_{off}\ \forall j$ where $v_j \neq v_{min}$. Intuitively, if a group of memrsitors have the minimum voltage applied to them, from Kirchhoff's law, they become parallel and if only one memristor is ON, the effective resistance of the whole group is G_{on}. However, if only one memristor that has a voltage higher than the minimum applied to it is ON, it will contribute to the output voltage by pulling up the output node and, subsequently, the output voltage deviates from the minimum and the circuit malfunctions.

In general, based on Ohm's law and writing v_0 as a weighted average of all inputs, any current, for instance I_1, can be presented as:

$$I_1 = G_1.(v_1 - \frac{\sum_{j=1}^{n} v_j.G_j}{\sum_{j=1}^{n} G_j}) \tag{22}$$

Rearranging the terms in (22):

$$I_1 = \frac{G_1}{G_1 + \cdots + G_n}[G_2(v_1 - v_2) + \cdots + G_n(v_1 - v_n)] \tag{23}$$

Let $v_i = v_{min} + m_i.\partial V = v_{max} - x_i.\partial V$ where m_i and x_i are integers and $0 \leq m_i, x_i \leq \frac{V_{max} - V_{min}}{\partial V}$ such that for a given memrsitor G_i, $m_i = 0$ corresponds to $v_i = v_{min}$ and $m_i = \frac{V_{max} - V_{min}}{\partial V}$ corresponds to $v_i = v_{max}$ while $x_i = 0$ corresponds to $v_i = v_{max}$ and $x_i = \frac{V_{max} - V_{min}}{\partial V}$ corresponds to $v_i = v_{min}$. For example, assume a three input circuit with input vector **V** = (0.1, 0.2, 0.3 V). Then, $v_{min=0.1}$, $\partial V = 0.1$, $m_2 = 1$ and $m_3 = 2$. By the same token, v_{max} and x_i can be deduced accordingly. It is important to note that v_{min}/v_{max}, in this particular analysis, are not necessarily the global minimum/maximum for the circuit. However, v_{min}/v_{max} represent the minimum/maximum values of the input vector applied to the circuit at a particular instant in time. For instance, while the global minimum/maximum voltages for the circuit might be 0/1 V, in this example, $v_{min}/v_{max} = 0.1/0.3$ V.

Assume, without loss of generality, that $v_1 = v_{min}$ and $v_n = v_{max}$

$$I_1 = \frac{-G_1 \partial V}{G_1 + \cdots + G_n}[G_2 m_2 + \cdots + G_n m_n] \tag{24}$$

$$I_n = \frac{G_n \partial V}{G_1 + \cdots + G_n}[G_1 x_1 + \cdots + G_{n-1} x_{n-1}] \tag{25}$$

Note that the negative sign in (24) indicates, for example, that the current is flowing to the left in the upper most memrsitor in Fig. 1, since, by definition, $v_1 = v_{min}$. Therefore, henceforth, the negative sign will be dropped since it indicates no more than the direction of the current. Also, note that $m_1 = 0$ since $v_1 = v_{min}$. (25) can be inferred accordingly for the case of $v_n = v_{max}$ based on the earlier discussion.

Condition (i):

Let an arbitrary number of memristor 'k' be 'ON' Such that (24) can be rewritten as:

$$I_1 = \frac{G_{off} \partial V}{kG_{on} + (n-k)G_{off}}[G_{on} \sum_{k} m_i + G_{off} \sum_{n-k-1} m_j] \tag{26}$$

Notice that the goal is to minimize (26) in order to find the worst case current (lowest current that results in circuit malfunctioning) and ensure that it is higher than the 'I_t'. This will ensure that condition (i) is satisfied. There are two ways to achieve this which are (1) vary 'k' and 'n' such that the number of 'ON' and 'OFF' memrsitors change (2) vary the values of m_i and m_j which, essentially, change the values of the input voltages. Also, notice that both ways are independent and can be treated separately since the combination of 'ON' and 'OFF' memristors during the transition state is independent from the steady state voltage values at the inputs.

Case (1): $1 \le k \le n-1$:

$$I_1 \approx \frac{G_{off}\partial V}{kG_{on}}[G_{on}\sum_{k} m_i + G_{off}\sum_{n-k-1} m_j] \tag{27}$$

Notice that, from (i), only one memristor is required to be ON. Hence, if for any of the 'k' memristors that are 'ON', $m_i = 0$, (i) will be automatically satisfied and there would be no need to worry about switching of the memristors. Conversely, the analysis is concerned with the case were none of the memrsitors for which $v_i = v_{min}$ is 'ON' which means that minimum $(\sum_k m_i) = k$ (i.e. $m_i \ge 1\, \forall i \in k$). Hence, (27) can be written as:

$$I_1 \approx \frac{G_{off}\partial V}{kG_{on}}[kG_{on}] \approx G_{off}\partial V \tag{28}$$

Case (2): $k = 0$:
Equation (26) is reduced to:

$$I_1 = \frac{\partial V}{n}[G_{off}\sum_{n-1} m_j] \tag{29}$$

If $\text{minimum}\left(\sum_{n-1} m_j\right) = 0$, this would mean that all voltages are equal to each other and equal to v_{min}, in which case no current flows and the voltage is transmitted normally to the output. Hence, $\text{minimum}\left(\sum_{n-1} m_j\right) = 1$ and (29) can be written as

$$I_1 = \frac{G_{off}\partial V}{n} \tag{30}$$

Condition (ii):
Using the same argument in (26), (25) can be described as:

$$I_n = \frac{G_{on}\partial V}{kG_{on} + (n-k)G_{off}}[G_{on}\sum_{k-1} x_i + G_{off}\sum_{n-k} x_j] \tag{31}$$

Case (1): $k = 1$:

$$I_n \approx \partial V[G_{off}\sum_{n-1} x_j] \tag{32}$$

Following the same reasoning as before, $\text{minimum}(\sum_{n-1} x_j) = 1$ and, accordingly,

$$I_n \approx G_{off} \partial V \tag{33}$$

Case (2): $1 < k \leq n-1$:

$$I_n = \frac{\partial V}{k}[G_{on} \sum_{k-1} x_i + G_{off} \sum_{n-k} x_j] \tag{34}$$

n order to minimize (34), as mentioned earlier, only one of the two summations can be zero, not both. Hence, since $G_{on} > G_{off}$, minimum($\sum_{k-1} x_i$) = 0 and minimum ($\sum_{n-k} x_j$) = 1.

$$I_n \approx \frac{G_{off} \partial V}{k} \tag{35}$$

As mentioned earlier in condition (ii), $G_j = G_{off} \forall j$ where $v_j \neq v_{min}$. While (34) was only concerned with the memristor with the maximum voltage applied to it, once this memristor switches to OFF, the memristor with lower voltage than the maximum becomes the new maximum voltage (i.e. $k \to k-1$). Since the minimum value for the current 'I_n' is what is sought in this analysis, (34) can be further minimized to (largest possible denominator). Formally, $\frac{G_{off}\partial V}{n-1} \leq I_{v_j \neq v_{min}} \leq \frac{G_{off}\partial V}{2}$. Hence, minimum ($I_{v_j \neq v_{min}}$) = $\frac{G_{off}\partial V}{n-1}$.

$$I_n \approx \frac{G_{off} \partial V}{n-1} \tag{36}$$

Case (3): k = n:

$$I_n \approx \frac{G_{on} \partial V}{n} \tag{37}$$

Hence, the worst case current can be computed as minimum ($G_{off}\partial V$, $\frac{G_{off}\partial V}{n}$, $\frac{\partial V G_{off}}{n-1}$, $\frac{\partial V G_{on}}{n}$) which is obviously $\frac{G_{off}\partial V}{n}$ and therefore:

$$\frac{\partial V}{nR_{off}} > I_t \tag{38}$$

Therefore, given a memristor with OFF resistance 'R_{off}' and threshold current 'I_t', there exists a trade-off between the minimum allowed voltage difference '∂V' and the size of the circuit 'n'. Also, note that substituting n = 2, we arrive at (5) which is the case for two input circuits.

3.2.2 Simulative Analysis of the Memristor Threshold

In order to validate the model in (38), an algorithm is developed that emulates the circuit operation (Amer et al. 2016). The algorithm initializes a current threshold 'I_t' and runs an exhaustive simulation over all circuit pictures where a picture is defined as a combination of memristor states and applied input voltages. For every picture, the memristors are allowed to switch until the output voltage stabilizes at its final value (i.e. no more switching of the memristors is taking place). If the final output voltage is the minimum voltage, this particular picture is said to have succeeded. A picture succeeds when the output voltage gravitates to v_{min}. In contrast, a picture fails when the system is stuck at an output voltage that is not v_{min}. For example, for input voltages V1 = 0.1, V2 = 0.2 and V3 = 0.3, the memristances should switch to M1 = R_{on}, M2 = R_{off} and M3 = R_{off}. If for instance the system stabilizes at M1 = M2 = M3 = R_{off} and the system cannot switch any further, the output voltage is not the minimum voltage and the picture is said to have failed. This failure occurs because a memristor or more are not able to switch because their currents are below the threshold current 'I_t'. Algorithm 1 presents a pseudo code for this procedure.

Algorithm 1. Modeling the effect of the threshold

1. Define G_{on} and G_{off}
2. Define threshold current: I_t
3. Loop1: $(v_1, v_2, v_3, G_1, G_2, G_3)$
4. Calculate $v_{min} = \text{minimum}(v_i)$
5. Compute $v_0 = \frac{\sum v_i G_i}{\sum G_i}$
6. Flag UP
7. Loop2: While Flag UP
8. Compute $I_i = (v_i - v_0) \times G_i$
9. Update G_i
10. Compute $v_{0n} = \frac{\sum v_i G_i}{\sum G_i}$ (updated G_i)
11. If $v_{0n} == v_0$ (Check if the output stabilized)
12. Flag DOWN
13. If $v_{0n} - v_{min} > \varepsilon$ (Check for failing states)
14. Output Failing state
15. End If
16. Else
17. $v_{0n} = v_0$
18. End If
19. End Loop2
20. End Loop1

Algorithm 1 generates the output failing picture(s) for every choice of I_t. Hence, I_t is decreased until no failing pictures occur as mentioned above. The choice of 'ε' is critical since it is the failing criterion (i.e. the criterion that decides whether failing pictures exist or not). Moreover, its value is a function in the applied voltages and the size of the circuit. Therefore, the next section is devoted to finding the proper ε. Note that although, theoretically speaking, $v_O = v_{min}$, this is never precisely true since the 'ON' resistance of the memristor R_{on} has some finite value. Yet, the difference is extremely small such that $v_O \approx v_{min}$.

In order to develop a model for the effect of the threshold on the system, two curves are plotted. Figure 2 plots I_t against $1/R_{off}$ for some specific size of the min-max circuit (n = 3) and a constant voltage step ($\partial V = 0.1$ V), where the voltage step is the minimum voltage difference between any two allowed voltage levels. Figure 3 plots I_t against ∂V at constant $R_{off} = 200\,K\Omega$ (Kvatinsky et al. 2013) and different sizes for the Min-Max circuit namely: n = 2, 3, 4, 5, 6, 10.

Note that Figs. 2 and 3 depict a strictly linear relation with almost perfect correlation ($R^2 = 1$ in Fig. 2 and $R^2 > 0.99$ in Fig. 3 due to minor numerical errors) which is in agreement with (38). It is no surprise, however, that the relation is perfectly linear. In essence, there is a specific combination of memristor states and applied voltages that results in the worst case state for all sizes of the circuit. This combination is when v_{min} is applied to 'n−1'memristors, $v_{min} + \partial V$ is applied to the *n*th memristor and all memristors have the maximum resistance R_{off}. Intuitively, the worst case picture occurs when the circuit has the maximum resistive state, all memristors are 'OFF', minimum potential difference applied to the circuit, only one memristor has higher voltage than the rest of the memristors and the difference is minimal (∂V).

Note that Epsilon 'ε' is crucial for it is not only the failing criterion, as mentioned earlier, but also its value changes based on the size of the circuit and the applied input voltages. Hence, in order to properly model it, another algorithm is developed that, similar to Algorithm 1, emulates the circuit operation but does not account for the effect of the threshold. Ignoring the effect of the threshold ensures the proper switching for all memristors and, accordingly, avoids any circuit

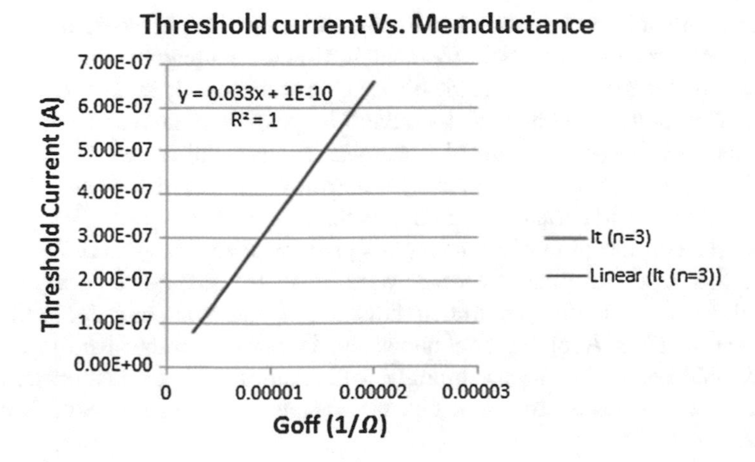

Fig. 2 Threshold current versus memductance

malfunction. Exhaustive simulations are run as before and all deviations of the output voltage from the minimum voltage are recorded. Then, 'ε' is assigned the maximum deviation. Thus, 'ε' can be interpreted as the maximum deviation of the output voltage under which the circuit functions properly. Therefore, if the deviation is more than 'ε', the circuit is considered to have malfunctioned as in step 13 in Algorithm 1.

Algorithm.2. Epsilon '$\boldsymbol{\varepsilon}$' generation

1. Define Array E
2. Loop1: (v_i, G_i)
3. Calculate v_{min} = minimum(v_i)
4. Compute $v_0 = \frac{\sum v_i G_i}{\sum G_i}$
5. Flag UP
6. Loop2: While Flag UP
7. Compute $I_i = (v_i - v_0) \times G_i$
8. Update G_i
9. Compute $v_{0n} = \frac{\sum v_i G_i}{\sum G_i}$ (updated G_i)
10. If$v_{0n} == v_0$
11. Flag DOWN
12. deviation $= v_{0n} - v_{min}$
13. Store deviation in Array E: E= [E deviation]
14. End If
15. $v_{0n} = v_0$
16. End Loop2
17. End Loop1
18. Output ε = maximum (E)

Algorithm 2 was validated against Spice simulations to validate the results as shown in Table 1:

It is important to comment on two important assumptions in the derivation of the threshold constraints. First, both the analytical and the computational approaches assumed that the memrsitor is of the binary type that can only assume two resistive values. Without this assumption, it wouldn't be possible to arrive at a closed form expression and, hence, the derived constraint is only valid for binary memristors. Thus, it is recommended to use binary memristors for memristor-based Min-Max circuits. The other important assumption is the discrete nature of the inputs. While some works considered continuous input signals such as sinusoids in Biolek et al. (2014), this work, inspired by early work in mermristor-based Fuzzy systems (Merrikh-Bayat et al. 2011), considered discrete signals again in order to make the derivation possible. Applying continuous signals makes it rather hard to abide by the threshold condition and, accordingly, will result in a circuit malfunction. It is, therefore, recommended to adopt discrete signals in memristor-based Min-Max circuits.

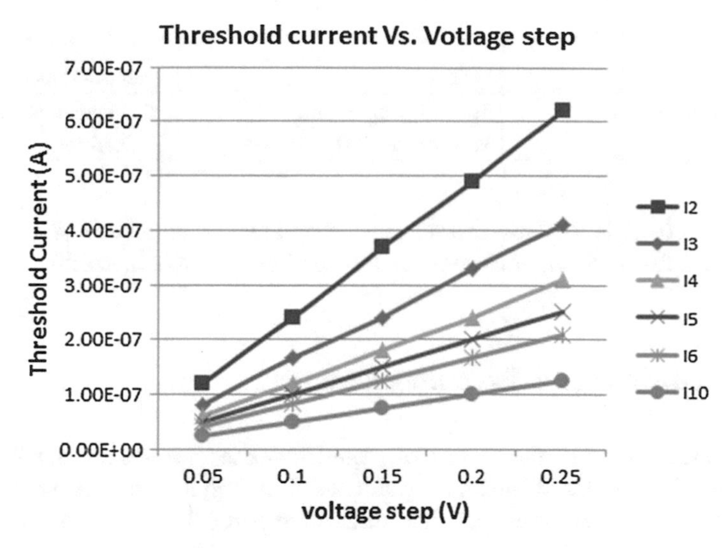

Fig. 3 Threshold current versus memductance. I2 is the curve associated with n = 2

Table 1 Validation of ε

# of memristors	Algorithm 2.	Spice
2	0.0005	0.00049
3	0.0099	0.00099
4	0.0015	0.00149
5	0.002	0.00199
6	0.0025	0.00249
10	0.0045	0.00447

4 Comparison with Conventional Transistor-Based Architectures

In order to demonstrate the advantages rendered by the memristor-based implementation of Min-Max circuits over their transistor based counterparts, the proposed circuit is compared against the most commonly and widely used transistor based architecture known as WTA-LTA (Padash et al. 2011) structure where WTA stands for Winner-Takes-All and LTA stand for Looser-Takes-All. The base of comparison is the area occupancy since, according to Yosefi et al. (2009), it is the primary metric for hardware-based Fuzzy systems. Table 2 provides the comparison in terms of transistor/memristor count per input (note that memristors are smaller than transistors). Also, note that the memristor/transistor count rises linearly with the size of the circuit (number of inputs). For example, in the case of 2 inputs, the transistor count is 30/20 for Min/Max functions, respectively, while the

Table 2 Comparison between memristor-based and transistor-based Min-Max circuits

	WTA-LTA	Memristor-based
Transistor/memristor count	Min circuit/Max circuit	Min circuit/Max circuit
	15 Transistors/10 Transistors	1 Memristor

memristor count is 2. This demonstrates that area savings become even more pronounced for multi-input circuits (as the number of inputs increase).

5 Conclusions and Recommendations

This chapter discussed the theory of memristor-based Min-Max circuits. While conventional transistor technologies can be used in such systems, memristors have exhibited peculiar characteristics that make them particularly well suited for such structures. First, the working principle was introduced through investigating simple 2-input circuits. Also, the effect of the memristor threshold was carefully modeled. Later, the theory of memristor-based Min-Max circuits was extended to N-inputs. It was demonstrated through a formal proof that the same structure proposed for 2-input circuits can be extrapolated to N-input circuits while abiding by the proper design constraints. Unlike in the case of the 2-input circuits, deriving a closed form expression for N-input circuits is highly complex. Thus, two assumptions where employed, namely: Binary memristors and discrete input signals in order to make the derivation of closed form expression feasible. Finally, it was shown that given a specific memristor with OFF resistance R_{off} and threshold current I_t, there exists a trade-off between the minimum allowed voltage difference and the size of the circuit.

References

Amer, S., et al. (2015a). Memristor-based center-of-gravity (COG) defuzzifier circuit. In *The European Conference on Circuit Theory and Design, Trondheim, August 2015.*

Amer, S., et al. (2015b). Design and analysis of memristor-based min-max circuit. In *The International Conference on Electronics, Circuits and Systems, Cairo, Dec 2015.*

Amer, S., et al. (2016). Effect of the Memristor threshold current on memristor-based min-max circuits. In *The International Conference on Modern Circuits and Systems Technologies, Thessaloniki, May 2016.*

Chua, L. (1971). Memristor—The missing element. *IEEE Transactions on Circuit Theory, 18,* 507–519.

Chua, L. (2015). Everything you wish to know about memristors but are afraid to ask. *Radioengineering, 24,* 319–368.

Chua, L., & Kang, M. (1976). Memristive devices and systems. *Proceedings of the IEEE, 64,* 209–223.

Elshamy, M., et al. (2014). A novel nondestructive readout circuit for memristor-based memory arrays. In *The Canadian Conference on Electrical and Computer Engineering, Toronto, May 2014*.

Farshidi, E. (2008). A low-power current-mode defuzzifier for fuzzy logic controllers. In *The International Conference on Signals, Circuits and Systems, Hammamet, November 2008*.

Hoseini, P., et al. (2010). Circuit design of voltage mode center of gravity defuzzifier in CMOS process. In *The International Conference on Electronic Devices, Systems and Applications, Kuala Lumpur, April 2010*.

Kannan, S., et al. (2015). Modeling, detection, and diagnosis of faults in multilevel memristor memories. *IEEE Transactions on Computer Aided Design of Integrated Circuits and Systems, 34,* 822–834.

Kvatinsky, S., et al. (2012). MRL—memristor ratioed logic. In *The International Workshop on Cellular Nanoscale Networks and Their Applications, Turin, August 2012*.

Kvatinsky, S., et al. (2013). TEAM: Threshold adaptive memristor model. *IEEE Transaction on circuits and Systems I: Regular Papers, 60,* 211–221.

Kvatinsky, S., et al. (2014). Memristor-based implication (IMPLY) logic: Design principles and methodologies. *IEEE Transaction on Very Large Scale Integration Systems, 22,* 2065–2066.

Merrikh-Bayat, F., et al. (2011). Memristor crossbar-based hardware implementation of fuzzy membership functions. In *The International Conference of Fuzzy Systems and Knowledge Discovery, Shanghai, July 2011*.

Merrikh-Bayat, F., & Shouraki, S. (2013). Memristive neuro-fuzzy system. *IEEE Transactions on Cybernetics, 43,* 269–285.

Niu, D., et al. (2010). Impact of process variations on emerging memristor. In *ACM/IEEE Design Automation Conference, Anaheim, June 2010*.

Padash, M., et al. (2011). A high precision high frequency VLSI multi-input min-max circuit based on WTA-LTA cells. In *The International Conference on Electronic Devices, Systems and Applications, Kuala Lampur, April 2011*.

Pershin, Y., & Di Ventra, M. (2010). Practical approach to programmable analog circuits with memristors. *IEEE Transactions on Circuits and Systems I: Regular Papers, 57,* 1857–1864.

Pickett, M., et al. (2009). Switching dynamics in titanium dioxide memristive devices. *Journal of Applied Physics, 106,* 1–6.

Rose, G., et al. (2012). Leveraging memristive systems in the construction of digital logic circuits. *Proceedings of the IEEE, 100,* 2033–2249.

Sturkov, D., et al. (2008). The missing memristor found. *Nature, 453,* 80–83.

Thompson, S., & Parathasarathy, S. (2006). Moore's law: the future of Si microelectronics. *Materials Today, 9,* 20–25.

Truang, S., et al. (2014). Neuromorphic crossbar circuit with nanoscale filamentary-switching binary memristors for speech recognition. *Nanoscale Research Letters, 9,* 629.

Vaidyanathan, S., & Volos, C. (2016a). *Advances and applications in nonlinear control systems*. Berlin: Springer.

Vaidyanathan, S., & Volos, C. (2016b). *Advances and applications in chaotic systems*. Berlin: Springer.

Watanbe, H., & Dettloff, D. (1990). A VLSI fuzzy logic controller with reconfigurable, cascadable architecture. *IEEE Journal Solid state Circuits, 25,* 376–382.

Yosefi, G., et al. (2009). Design of new CMOS current mode min and max circuits for FLC chip applications. In *The European Conference on Circuit Theory and Design, Anatalya, August 2009*.

Analysis of a 4-D Hyperchaotic Fractional-Order Memristive System with Hidden Attractors

Christos Volos, V.-T. Pham, E. Zambrano-Serrano, J.M. Munoz-Pacheco, Sundarapandian Vaidyanathan and E. Tlelo-Cuautle

Abstract In 1695, G. Leibniz laid the foundations of fractional calculus, but mathematicians revived it only 300 years later. In 1971, L.O. Chua postulated the existence of a fourth circuit element, called memristor, but Williams's group of HP Labs realized it only 37 years later. In recent years, few unusual dynamical systems, such as those with a line of equilibriums, with stable equilibria or without equilibrium, which belong to chaotic systems with hidden attractors, have been reported. By looking at these interdisciplinary and promising research areas, in this chapter, a fractional-order 4-D memristive system with a line of equilibria is introduced. In particular, a hyperchaotic behavior in a simple fractional-order memristor-based system is presented. Systematic studies of the hyperchaotic

C. Volos (✉)
Physics Department, Aristotle University of Thessaloniki, 54124 Thessaloniki GR, Greece
e-mail: volos@physics.auth.gr; chvolos@gmail.com

V.-T. Pham
School of Electronics and Telecommunications, Hanoi University of Science and Technology, 01 Dai Co Viet, Hanoi, Vietnam
e-mail: pvt3010@gmail.com

E. Zambrano-Serrano
Department of Applied Mathematics, IPICYT, San Luis Potosi, SLP 78216 San Luis Potosí, Mexico
e-mail: ernesto.zambrano@ipicyt.edu.mx

J.M. Munoz-Pacheco
Electronics Department, Autonomous University of Puebla (BUAP), 72570 Puebla, Mexico
e-mail: jesusm.pacheco@correo.buap.mx

S. Vaidyanathan
Research and Development Centre, Vel Tech University, Avadi, Chennai 600062, Tamil Nadu, India
e-mail: sundarvtu@gmail.com

E. Tlelo-Cuautle
Department of Electronics, INAOE, Tonantzintla Puebla, 72840 Cholula, Mexico
e-mail: etlelo@inaoep.mx

© Springer International Publishing AG 2017

S. Vaidyanathan and C. Volos (eds.), *Advances in Memristors, Memristive Devices and Systems*, Studies in Computational Intelligence 701,
DOI 10.1007/978-3-319-51724-7_9

behavior in the integer and fractional-order form of the system are performed using phase portraits, Poincaré maps, bifurcation diagrams and Lyapunov exponents. Simulation results show that both integer-order and fractional-order system exhibit hyperchaotic behavior over a wide range of control parameter. Finally, the electronic circuits for the evaluation of the theoretical model of the proposed integer and fractional-order systems are presented.

Keywords Memristive system • Hyperchaos • Fractional order • Hidden attractors • Nonlinear circuit

1 Introduction

The announcement of the realization of a solid-state thin film two terminal memristor at Hewlett-Packard Labs in 2008 (Strukov et al. 2008), brought a revolution in various scientific fields. Many phenomena in systems, such as in thermistors, which internal state depends on the temperature (Sapoff and Oppenheim 1963), spintronic devices which resistance varies according to their spin polarization (Pershin and Di Ventra 2008) and molecules which resistance changes according to their atomic configuration (Chen et al. 2003), could be explained now with the use of the memristor. Also, electronic circuits with memory circuit elements could simulate processes typical of biological systems, such as the adaptive behavior of unicellular organisms (Pershin et al. 2009) and the learning and associative memory (Pershin and Di Ventra 2010).

Also, a considerable number of potential memristor-based applications have been reported like adaptive filter (Driscoll et al. 2010), high-speed low-power processors (Yang et al. 2013), pattern recognition systems (Corinto et al. 2012), neural networks (Adhikari et al. 2012; Ascoli and Corinto 2013), programmable analog integrated circuits (Shin et al. 2011), and so on (Ascoli et al. 2013; Tetzlaff 2014). Interestingly, the intrinsic nonlinear characteristic of memristor has been exploited in implementing novel chaotic oscillators with complex dynamics (Itoh and Chua 2008; Muthuswamy 2010; Bo-Cheng et al. 2011; Driscoll et al. 2011; Buscarino et al. 2012a, b; Corinto and Ascoli 2012).

It is very interesting to ask naturally whether there exists a memristor-based system that is hyperchaotic. Some authors have recently answered this question by introducing some memristor-based hyperchaotic systems, motivated by the complex dynamical behaviors of hyperchaotic systems and the special features of memristor. Hyperchaos was generated by combining a memristor with cubic nonlinear characteristics and a modified canonical Chua's circuit (Fitch et al. 2012). However, this memristor-based modified canonical Chua's circuit is a five-dimensional hyperchaotic system. Also, by extending the HP memristor-based canonical Chua's oscillator, a six-dimensional hyperchaotic oscillator was designed (Buscarino et al. 2012c). The authors of this work used a configuration based on two HP memristors in antiparallel. In other interesting works, four-dimensional hyperchaotic memristive systems were discovered by Li et al. (Li et al. 2014; Li et al. 2015). The last examples belong to a new category of chaotic systems with hidden attractors.

According to a new classification of chaotic dynamics (Leonov et al. 2011a; Leonov et al. 2011b; Jafari and Sprott 2013; Leonov and Kuznetsov 2013), there are two types of attractors: self-excited attractors and hidden attractors. A self-excited attractor has a basin of attraction that is excited from unstable equilibria. In contrast, hidden attractor cannot be found by using a numerical method in which a trajectory started from a point on the unstable manifold in the neighborhood of an unstable equilibrium. Furthermore, in contrary to self-excited attractors, hidden attractors cannot be computed by the standard procedure because its basin of attraction does not contain neighborhoods of any equilibria.

So, the discovery of dynamical systems with hidden attractors is a great challenge due to their appearance in many research fields such as in mechanics, secure communication and electronics (Leonov and Kuznetsov 2011c; Kuznetsov et al. 2011; Pham et al. 2014e; Sharma et al. 2015). For example, hidden attractor in smooth Chua's system was reported in Leonov et al. (2012). Hidden oscillations in mathematical model of drilling system (Leonov et al. 2014) and hidden oscillations in nonlinear control systems (Leonov and Kuznetsov 2011a, b; Vaidyanathan and Volos 2016a, b) were witnessed. Various examples of hidden attractors were also summarized in Brezetskyi et al. (2015; Jafari et al. 2015; Shahzad et al. 2015; Sprott 2015). Hidden attractors were observed in a 4-D Rikitake dynamo system (Vaidyanathan et al. 2015a) or 5-D hyperchaotic Rikitake dynamo system (Vaidyanathan et al. 2015b). Hidden attractors in a chaotic system with an exponential nonlinear term were introduced in Pham et al. (2015a). In addition, algorithms for searching for hidden oscillations were presented in Leonov and Kuznetsov (2011b), Leonov et al. (2011c).

In the last five years there has been an increasing interest in chaotic and especially in hyperchaotic systems with the presence of hidden attractors (Jafari and Sprott 2013; Pham et al. 2014a; Pham et al. 2014b; Wei et al. 2014; Vaidyanathan et al. 2015b). Also, it is worth noting that memristors have been applied in the investigation of chaotic and hyperchaotic systems with hidden attractors. Pham et al. introduced a simple neural network having a memristive synaptic weight, which can exhibit hyperchaos although it possesses no equilibrium points (Pham et al. 2014c). A new memristive system, which does not display any equilibria but exhibits periodic, chaotic, and also hyperchaotic dynamics in a particular range of the parameters space, was presented in Pham et al. (2014d). Also, a novel 4-D memristor-based hyperchaotic system with hidden attractor was studied in Pham et al. (2015b). However, motivated by complex dynamical behaviors, especially of hyperchaotic systems, noticeable characteristics of memristor, and unknown features of hidden attractors, the study of memristive hyperchaotic systems with hidden attractors is still an open research subject.

In this chapter, a 4-D hyperchaotic fractional order memristive system with hidden attractors is introduced. Owing the presence of a memristive device, this particular hyperchaotic system has no equilibrium points. The rest of the chapter is organized as follows. Related works are summarized in Sect. 2. Section 3 provides the mathematical model of the memristive hyperchaotic system while the dynamics and properties of the system are presented in Sect. 4. The electronic circuit for the evaluation of the theoretical model of the proposed memristive hyperchaotic system is introduced in Sect. 5. The fractional order form of the proposed system and its analysis are provided in Sect. 6, while Sect. 7 presents its electronic circuit realization. Finally, conclusions are drawn in Sect. 8.

2 Related Work

Fractional calculus is a very old branch of mathematics, which mainly deals with derivatives and integrals of arbitrary non-integer order. It was firstly introduced 300 years ago, but it only developed as a pure mathematical branch (Oldhamm and Spanier 1974; Kenneth 1993; Podlubny 1999; Butzer and Westphal 2000). In the last few decades, it is found that the fractional-order derivatives are extremely useful to describe many real-world phenomena in fields such as in control theory, material and mechanical systems, acoustics and thermal systems, signal processing and system identification, reconfigurable hardware and so on (Concepcion et al. 2014; Dumitru et al. 2014; Ghasemi et al. 2014; Richard 2014; Santanu 2015).

The main aim of the researchers, who work on this field, is to find chaotic behavior in fractional-order systems. Usually, chaotic attractors can not be observed in nonlinear systems whose order is less than three, so it is highly interesting to analyze the routes to chaos of fractional systems with low orders. Recently, there has been a trend to transform integer-order in fractional-order forms. In this direction, it was proven that many fractional-order nonlinear differential systems behave chaotically, for instance, fractional-order Lorenz system (Grigorenko and Grigorenko 2003), fractional-order Rössler system (Li and Chen 2004), fractional-order Duffing oscillators (Gao and Yu 2005), fractional-order Chua circuit (Cafagna and Grassi 2008), fractional-order Chen system (Li and Peng 2004), fractional-order Lü system (Lu 2006), fractional-order Liu system (Wang and Wang 2007) and so on. Due to the non local properties of fractional differential operators, topological structures of fractional-order systems are different from the traditional classical differential ones. Recently, control, synchronization and circuit implementation of fractional-order chaotic systems have received much attention.

Compared to integer-order, the fractional chaotic systems have the following advantages:

1. the fractional derivatives have complex geometrical interpretation because of their nonlocal character and high nonlinearity,
2. the power spectrum of fractional-order chaotic systems fluctuates complexly increasing the chaoticity in frequency domain and
3. the computational complexity goal is also achieved.

Therefore, new fractional chaotic systems are crucial to enhance the performance of several integer-order chaos-based applications.

The last few years the research has been expanded into the design of fractional-order hyperchaotic systems. Dadras et al. developed a four-wing fractional-order hyperchaotic attractor generated from a 4-D system with one equilibrium (Dadras et al. 2012). Wu et al. studied the case of synchronization between two coupled identical new fractional-order hyperchaotic systems (Wu et al. 2009). Also, Matouk et al. presented the stability conditions, hyperchaos and control in a novel fractional-order hyperchaotic system. (Matouk 2009). Other novel fractional-order hyperchaotic systems and the cases of their synchronization by using various methods, were presented in Yu and Li (2008), Deng et al. (2009), Ping et al. (2009), Bai et al. (2012).

It is worth noting that all the above mentioned fractional-order chaotic and hyperchaotic systems are characterized by one or more equilibrium points. However, a very challenging topic is about the study of fractional-order systems without equilibrium points. In this regard, referring to the presence of chaos or hyperchaos in fractional-order systems with no equilibria, only very few works have been published (Li et al. 2011; Cafagna and Grassi 2013; Zhou and Huang 2014; Cafagna and Grassi 2015). On the other hand, referring to the presence of hyperchaos in fractional-order memristive systems without equilibria, to the best of our knowledge, no paper has been published in the literature so far.

3 The Memristive Hyperchaotic System Without Equilibria

In this section a new memristive hyperchaotic system without equilibria is presented in details. Firstly, the model of the memristive device will be analyzed, while next the mathematical description of the 4-D hyperchaotic system will be introduced.

3.1 Model of the Memristive Device

In 1976 Chua and Kang introduced a memristive system by generalizing the original definition of a memristor (Chua and Kang 1976). In general, a memristive system is described by

$$\begin{aligned} \dot{\omega}_m &= F(\omega_m, u_m, t) \\ y_m &= G(\omega_m, u_m, t) u_m \end{aligned} \tag{1}$$

where u_m, y_m and ω_m denote the input, output, and state of the memristive system, respectively. The function F is a continuous n-dimensional vector function and G is a continuous scalar function. Based on the definition of memristive system, a memristive device is proposed by the following form:

$$\begin{aligned} \dot{\omega}_m &= u_m \\ y_m &= (1 + 0.24\omega_m^2 - 0.0016\omega_m^4)u_m \end{aligned} \tag{2}$$

Hence the function G is a fourth degree polynomial function. In order to investigate the fingerprints of memristive device (2), an external bipolar period signal is applied across its terminals.

The external sinusoidal stimulus is given by

$$u_m = \mathrm{A}\sin(2\pi ft) \tag{3}$$

where A is the amplitude and f is the frequency. From the first equation of (2), the state variable of the memristive device is described by

$$\begin{aligned} \omega_m(t) &= \int_{-\infty}^{t} u_m(\tau)d\tau = \omega_m(0) + \int_{-\infty}^{t} \mathrm{A}\sin(2\pi ft)d\tau \\ &= \omega_m(0) + \frac{A}{2\pi f}(1 - \cos(2\pi ft)) \end{aligned} \tag{4}$$

Substituting (3) and (4) into (2), it is easy to derive the output of the memristive device y_m. Therefore, the output y_m depends on the frequency and amplitude of the applied input stimulus. Figure 1 shows the hysteresis loop of the memristive device (2) when it is driven by the periodic signal (3) with different frequencies.

Obviously, the proposed memristive device exhibits a “pinched hysteresis loop” in the input-output plane (Biolek et al. 2011, 2012). In addition, when the excitation frequency increases, the hysteresis lobe area decreases monotonically. Moreover, when the frequency is adequately large, the pinched hysteresis loop shrinks to a single-valued function. It is worth noting that the hysteresis loop of the memristive device (2) pinched at different input amplitudes (see Fig. 2). Additionally, the output y_m also depends on the initial state of memristive device, as depicted in Fig. 3. Thus, according to Adhikari et al. (2013), Biolek et al. (2013) the three main fingerprints of memristive system have been observed in the proposed memristive device (2) (Figs. 1–3).

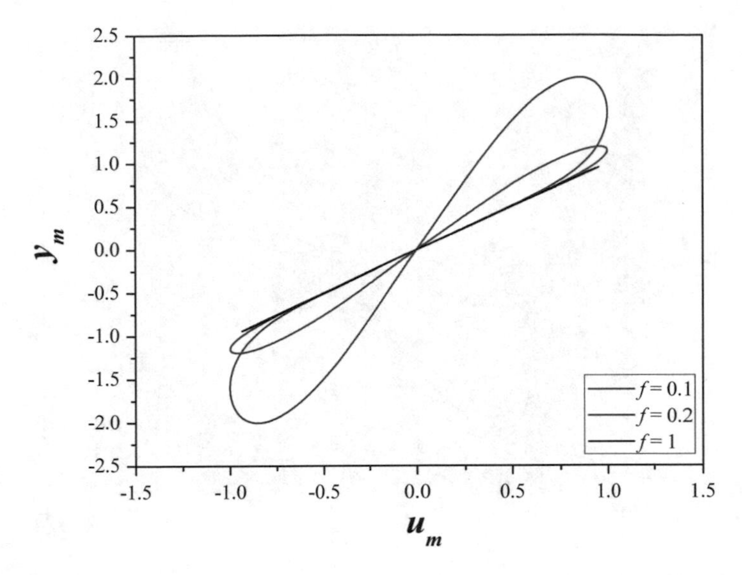

Fig. 1 Hysteresis loops of the proposed memristive device (2) driven by a sinusoidal stimulus (3), when $A = 1$, $\omega_m(0) = 0$ and varying frequency f

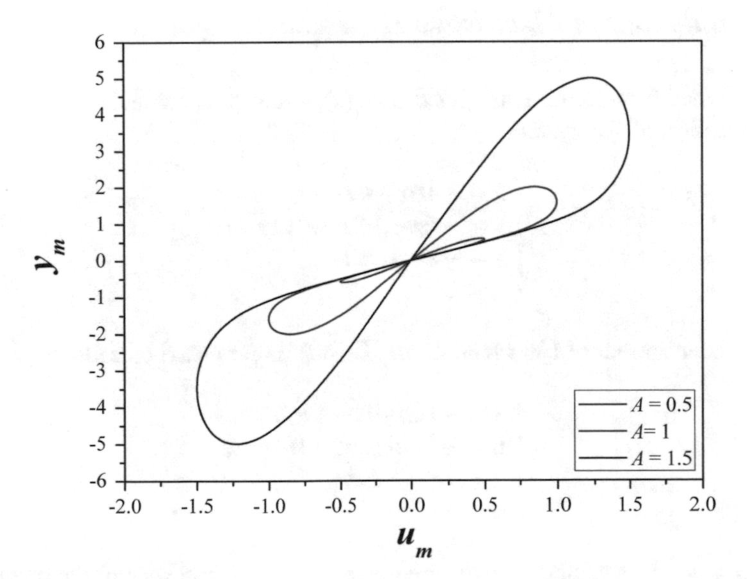

Fig. 2 Hysteresis loops of the proposed memristive device (2) driven by a sinusoidal stimulus (3), when $f = 0.1$, $\omega_m(0) = 0$ and varying amplitude A

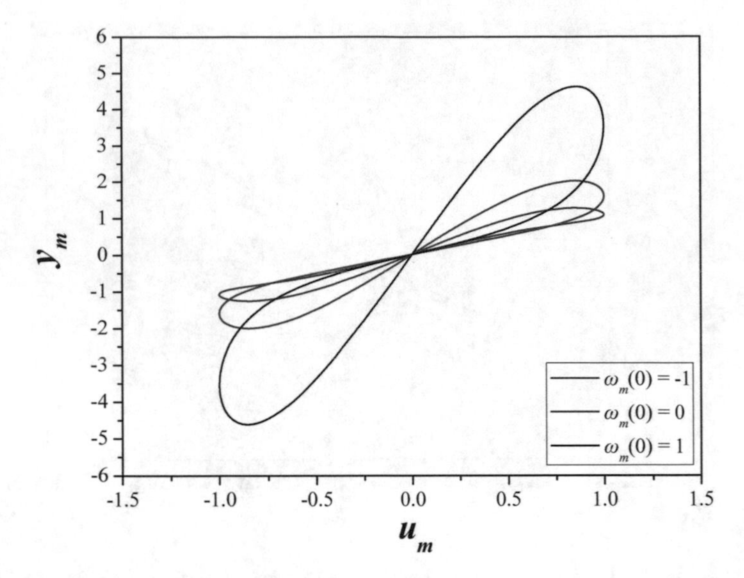

Fig. 3 Hysteresis loops of the proposed memristive device (2) driven by a sinusoidal stimulus (3), when $A = 1, f = 0.1$ and using different initial states $\omega_m(0)$

3.2 *Model of the Memristive Hyperchaotic System*

Based on the introduced memristive device (2), a four-dimensional system, which is a modification of the system

$$\begin{cases} \dot{x} = -10x - bu - uz \\ \dot{u} = -6x + 1.2xz + 0.1y - c \\ \dot{z} = -z - 1.2xu \\ \dot{\omega} = u \end{cases} \tag{5}$$

that has been introduced in Pham et al. (2014d), is proposed as follows:

$$\begin{cases} \dot{x} = -10x - bu - uz \\ \dot{u} = -6x + 1.2xz + 0.1y - c \\ \dot{z} = -z - 1.2u^2 \\ \dot{\omega} = u \end{cases} \tag{6}$$

where $u = u_m$, $y = y_m$ and $\omega = \omega_m$ are the input, output, and state of the memristive system, while b, c are two parameters. The difference between the two systems is that the term xu in the third equation of (5) has been changed with u^2 in (6).

The equilibrium points of the 4-D system (6) are obtained by solving the equations

$$-10x - bu - uz = 0 \tag{7}$$

$$-6x + 1.2xz + 0.1y - c = 0 \tag{8}$$

$$-z - 1.2u^2 = 0 \tag{9}$$

$$u = 0 \tag{10}$$

From Eqs. (7), (9), (10) we have $x = u = z = 0$. Then, the Eq. (8) becomes

$$c = 0 \tag{11}$$

Equation (11) is inconsistent, thus there is no equilibrium in system (6).
The divergence of system (6) is defined as:

$$\nabla V = \frac{\partial \dot{x}}{\partial x} + \frac{\partial \dot{u}}{\partial u} + \frac{\partial \dot{z}}{\partial z} + \frac{\partial \dot{\omega}}{\partial \omega} = -10 + 0.24\omega^2 - 0.0016\omega^4 \tag{12}$$

So, system (6) is not conservative, and, furthermore, the contraction/expansion of volumes is not uniform in the four dimensional space.

For the parameter values $b = 5$ and $c = 0.001$ and for the initial conditions ($x(0)$, $u(0)$, $z(0)$, $\omega(0)$) = (0, 1, 0.2, 0,3, 0,4), the calculated Lyapunov exponents (LE_i) of the 4-D memristive system (6) are:

$$LE_1 = 0.14779,\ LE_2 = 0.01571,\ LE_3 = 0,\ LE_4 = -15.66248 \tag{13}$$

There is more than one positive Lyapunov exponents in the LE spectrum (13), thus the 4-D memristive system (6) exhibits hyperchaotic behavior (see its hyperchaotic attractor in Fig. 4).

The Kaplan–Yorke fractional dimension, that presents the complexity of attractor (Sprott 2003; Strogatz 1994), is defined by

$$D_{KY} = j + \frac{1}{|L_{j+1}|} \sum_{i=1}^{j} L_i \tag{14}$$

where j is the largest integer satisfying $\sum_{i=1}^{j} L_i \geq 0$ and $\sum_{i=1}^{j+1} L_i < 0$.Therefore, the Kaplan–Yorke dimension of the novel memristive hyperjerk system (6) is calculated as: $D_{KY} = 3.0104$, which is fractional.

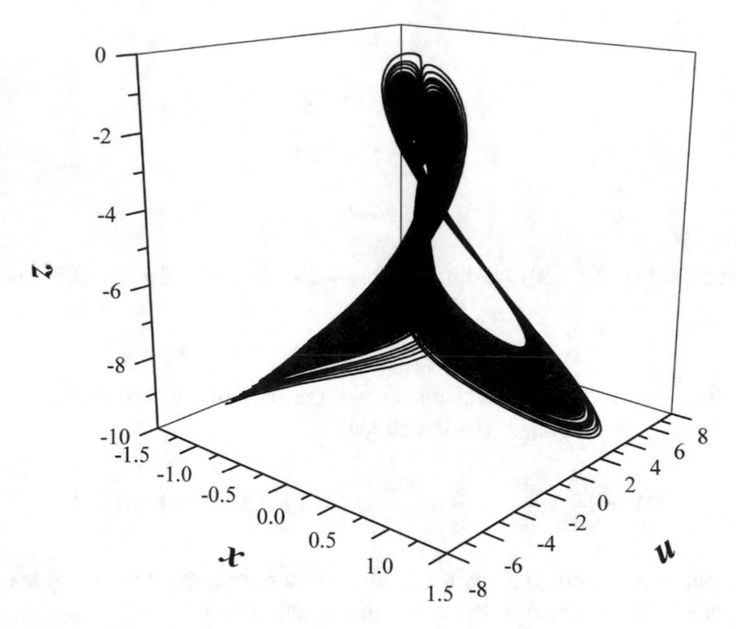

Fig. 4 Hyperchaotic attractor of system (6), for $b = 5$, and $c = 0.001$, in the (x, u, z)-space

4 Dynamics of the Memristive Hyperchaotic System

In this section the system's dynamical behavior is explored by using well-known tools of nonlinear theory, such as the bifurcation diagram, the diagram of Lyapunov exponents, the phase portraits and the Poincaré maps.

The bifurcation diagram provides a useful tool in nonlinear science because it shows the change of system's dynamical behavior. We investigate the dynamics of system (6) further by using this tool for the bifurcation parameter b. In more details, Fig. 5a presents the bifurcation diagram of the variable x versus the parameters b. The system's complexity has also been verified by the corresponding diagram of Lyapunov exponents (LE) versus the parameter b (see Fig. 5b). For system dynamical behavior's better observation, from the diagram of Lyapunov exponents, only the three of the four Lyapunov exponents are depicted in Fig. 5b.

It is well known that Lyapunov exponents measure the exponential rates of the divergence and convergence of nearby trajectories in the phase space of the chaotic system (Strogatz 1994). So, in order to have detailed view of the memristive system (6), the Lyapunov exponents have been calculated using the algorithm in Wolf et al. (1985) and are predicted in Fig. 5b. Obviously, Lyapunov spectrum indicates more clearly than the bifurcation diagram that there are some narrow windows of limit cycles [$b \in (5.46, 5.49)$, $b \in (5.61, 5.82)$, $b \in (6.53, 7.00)$] and of chaotic behavior [$b \in (4.50, 4.55]$, $b \in (5.82, 5.94)$], two windows of quasiperiodic behavior through which the system is driven to chaos [$b \in (5.55, 5.61]$, $b \in (5.94, 6.53)$] and

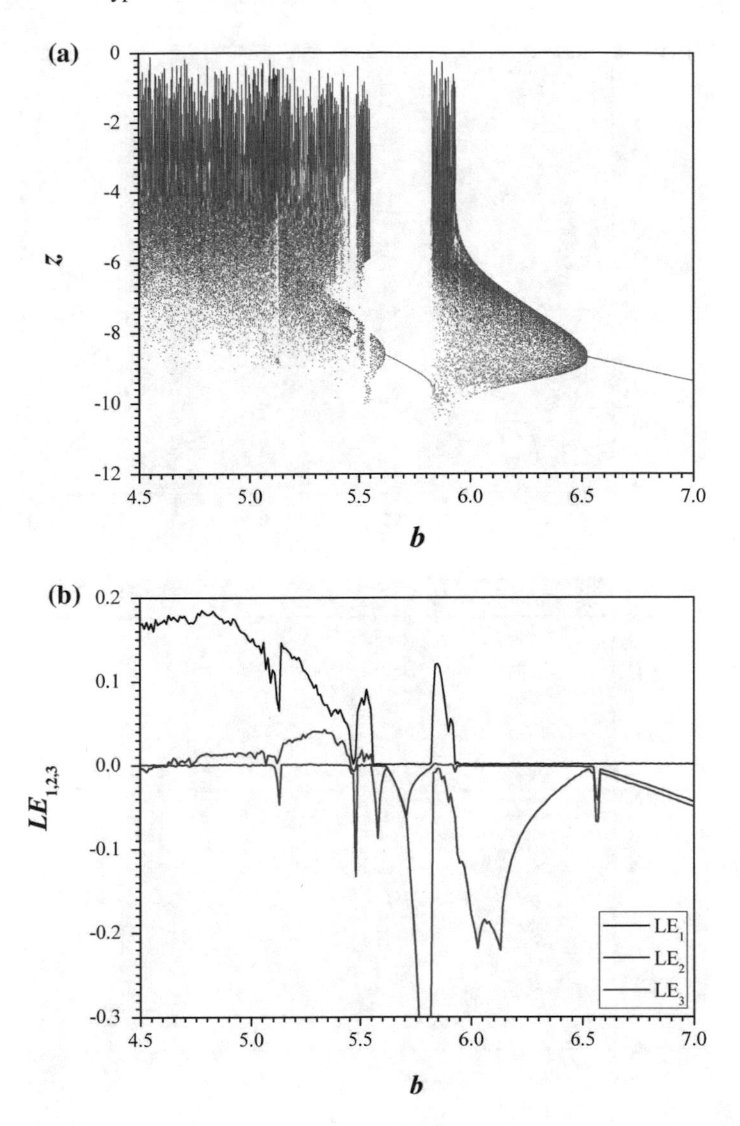

Fig. 5 **a** Bifurcation diagram of z versus the parameter b and **b** the respective diagram of Lyapunov exponents versus the parameter b, for $c = 0.001$, with initial conditions ($x(0)$, $u(0)$, $z(0)$, $\omega(0)$) = (0, 1, 0.2, 0,3, 0,4)

of hyperchaotic behavior [$b \in (4.55, 5.46]$, $b \in (5.49, 5.55)$]. In Figs. 6, 7, 8 and 9 the phase portraits in (x, z)-plane and the corresponding Poincaré maps in (z, ω)-plane are displayed, in order to present each one of system's dynamical behavior (limit cycle, quasiperiodic, chaotic and hyperchaotic).

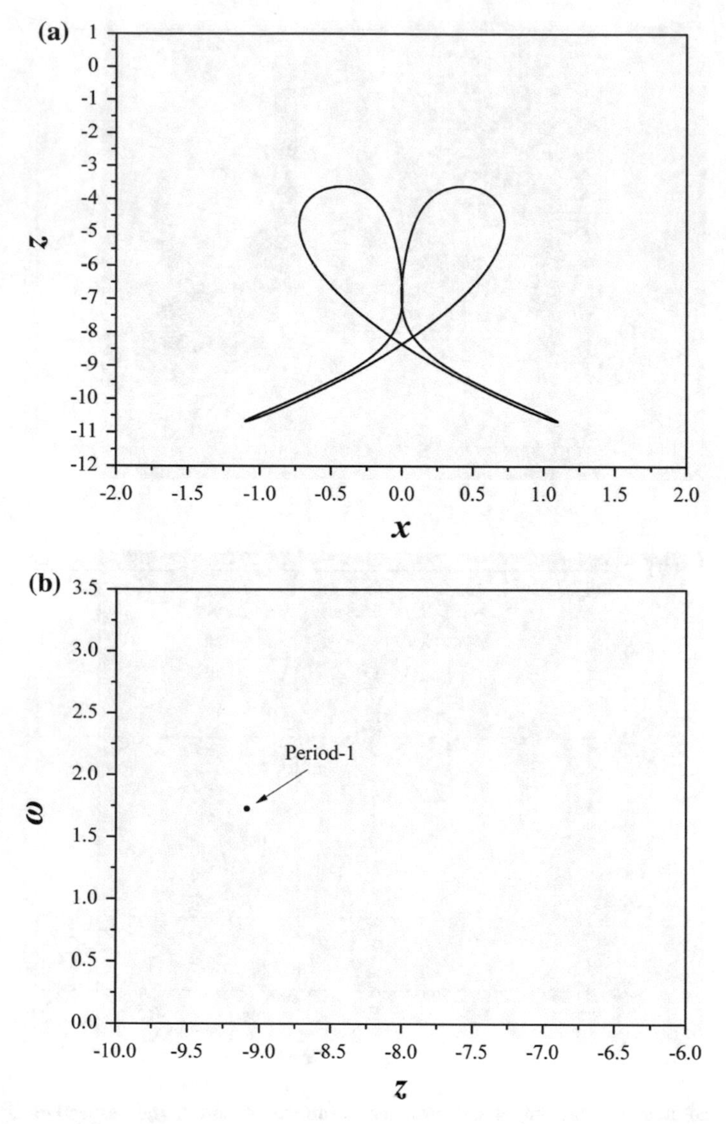

Fig. 6 **a** Phase portrait of x versus z and **b** the respective Poincaré map of z versus ω, for $b = 6.8$ and $c = 0.001$, with initial conditions $(x(0), u(0), z(0), \omega(0)) = (0, 1, 0.2, 0{,}3, 0{,}4)$. The discrete point in the Poincaré map indicates the system's periodic behavior

5 Circuit Realization of the Memristive Hyperchaotic System

In this section, the electronic circuit, which emulates the mathematical model of the proposed memristive hyperchaotic system (6) is presented in order to show its feasibility. The designed circuit, which is built by using off-the-shelf electronic

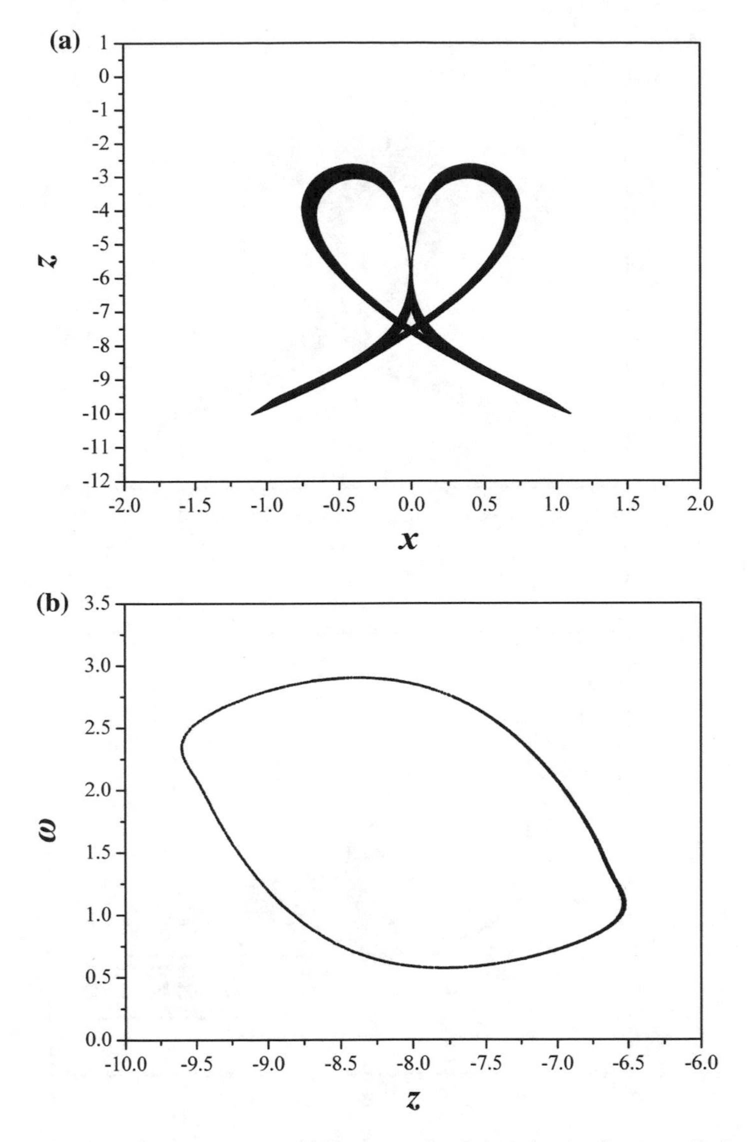

Fig. 7 **a** Phase portrait of x versus z and **b** the respective Poincaré map of z versus ω, for $b = 6.1$ and $c = 0.001$, with initial conditions $(x(0), u(0), z(0), \omega(0)) = (0, 1, 0.2, 0{,}3, 0{,}4)$. The closed curve in the Poincaré map indicates the system's quasiperiodic behavior

components, is an effective way for discovering dynamics of the theoretical model practically.

Circuital design, especially of hyperchaotic systems plays an important role on the field of nonlinear science due to its applications in many other fields, such as in secure communication, signal processing, random bit generator, or path planning for autonomous mobile robot etc. (Barakat et al. 2013; Gamez-Guzman et al. 2009;

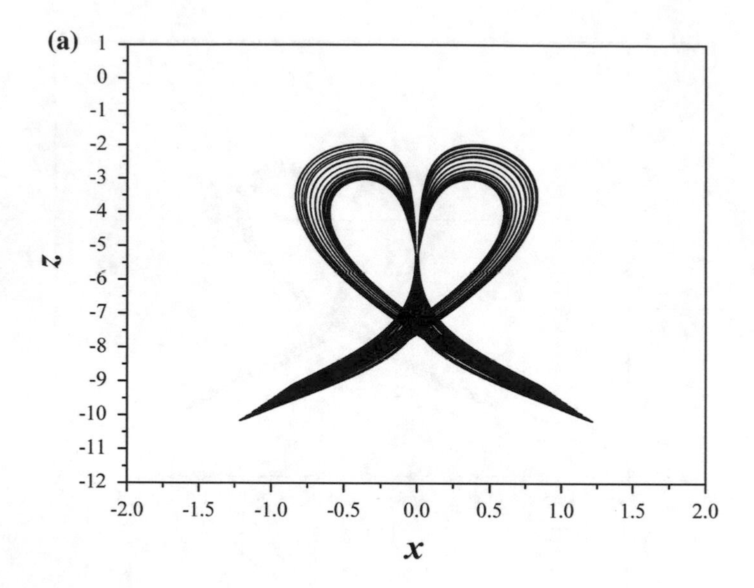

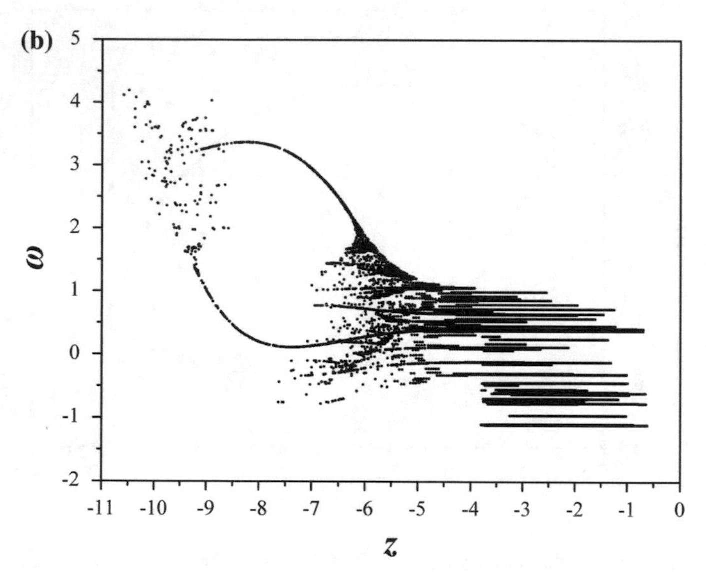

Fig. 8 **a** Phase portrait of x versus z and **b** the respective Poincaré map of z versus ω, for $b = 5.9$ and $c = 0.001$, with initial conditions ($x(0)$, $u(0)$, $z(0)$, $\omega(0)$) = (0, 1, 0.2, 0,3, 0,4). The strange attractor in the Poincaré map indicates the system's chaotic behavior

Sadoudi et al. 2013; Volos et al. 2012, 2013; Yalcin et al. 2004). In addition, circuital implementation of chaotic/hyperchaotic systems is also provide an effective approach for investigating dynamics of such theoretical models (Buscarino et al. 2009; Sundarapandian and Pehlivan 2012). For example, hyperchaotic attractors can be observed on the oscilloscope easily or experimental bifurcation

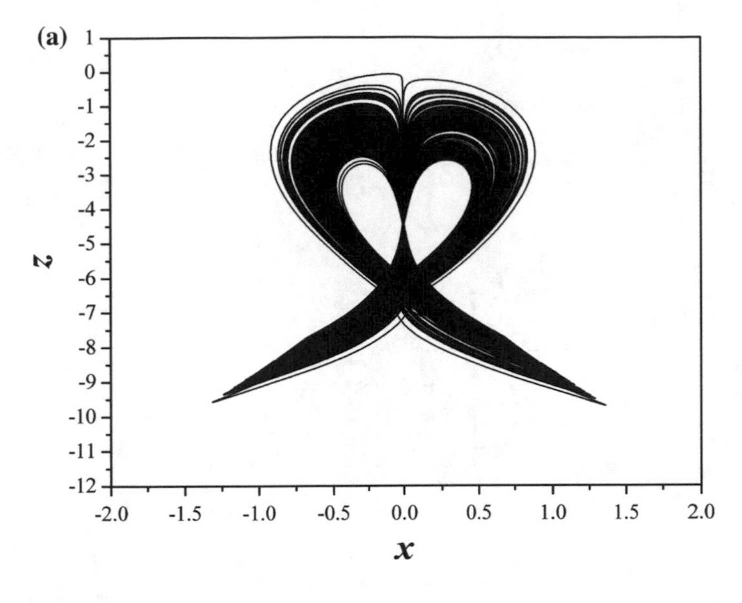

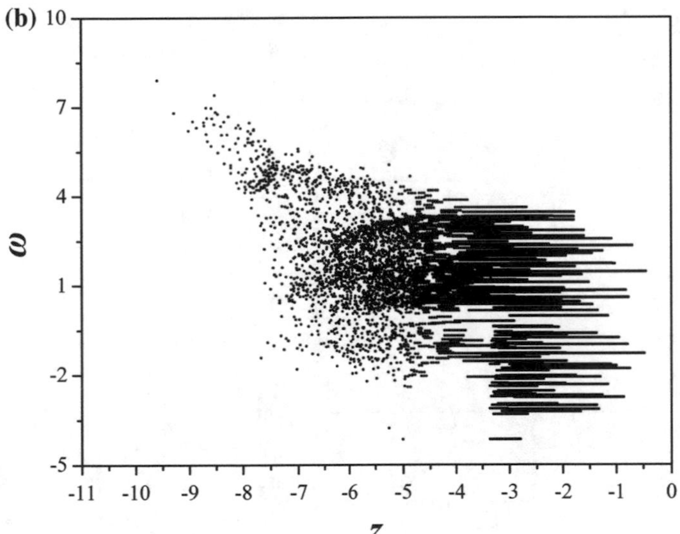

Fig. 9 **a** Phase portrait of x versus z and **b** the respective Poincaré map of z versus ω, for $b = 5$ and c = 0.001, with initial conditions ($x(0)$, $u(0)$, $z(0)$, $\omega(0)$) = (0, 1, 0.2, 0,3, 0,4). The strange attractor in the Poincaré map in combination with the two positive Lyapunov exponents indicates the system's hyperchaotic behavior

diagrams can be obtained by varying the values of variable resistors (Bouali et al. 2012; Fortuna et al. 2009).

Figure 10 depicts the schematic of the circuit. The main circuit that realizes the system (6), has four integrators (U1–U4) and two inverting amplifiers (U5, U6),

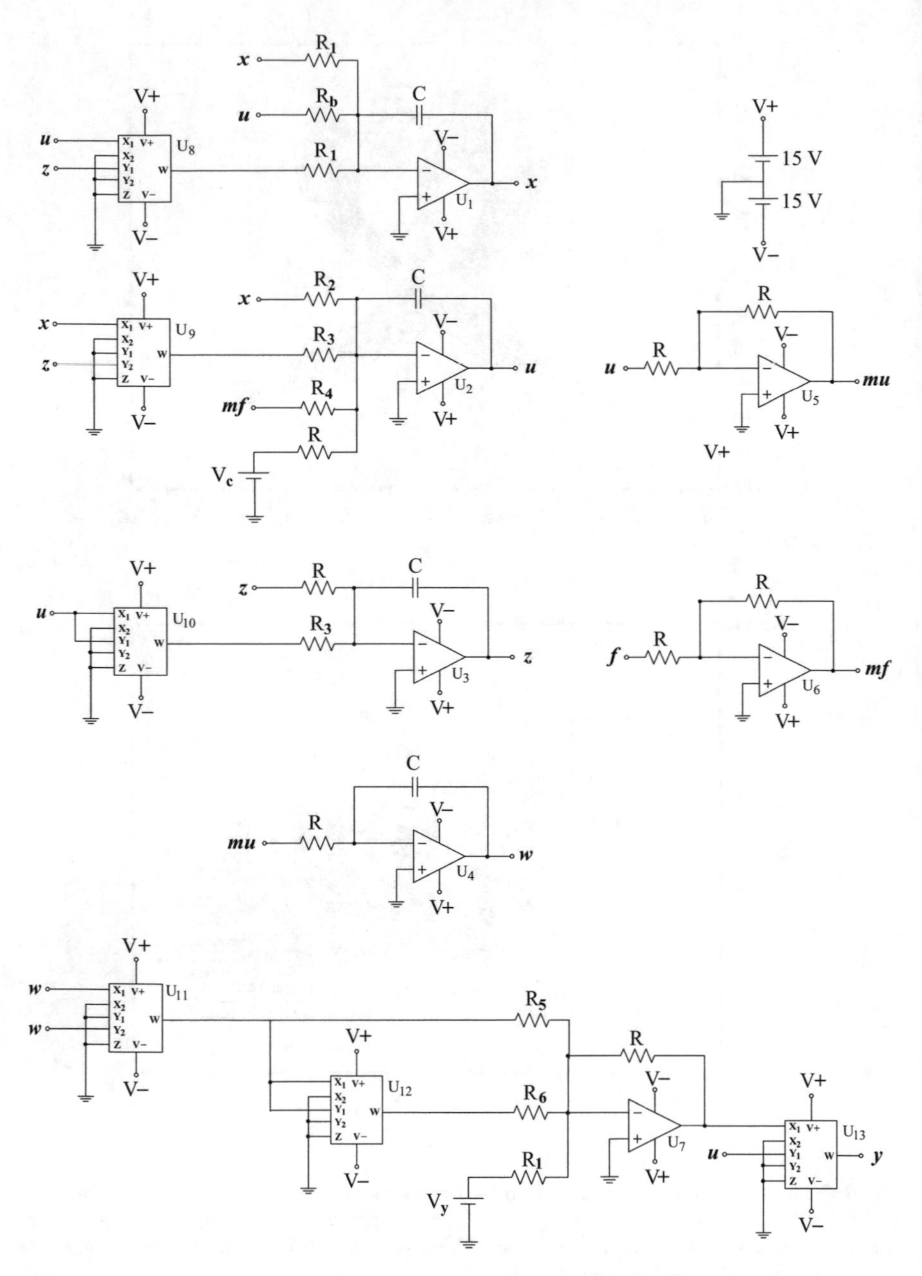

Fig. 10 Schematic of the circuit, which emulates the memristive hyperchaotic system

which are implemented with the operational amplifier TL084, as well as three signals multipliers (U8–U10) by using the analog multiplier AD633. Also, the circuital realization of the memristor's function y of Eq. (2), is also depicted in

Fig. 10. Indeed the sub-circuit of the proposed memristive system only emulates the memristive function because there is not any commercial off-the-shelf memristive device in the market yet. So, the memristor's function is realized by using an inverting adder amplifier (U7) and three signals multipliers (U11–U13).

By applying Kirchhoff's circuit laws, the corresponding circuital equations of the designed circuit can be written as:

$$\begin{cases} \dot{x} = \frac{1}{RC}\left(- \frac{R}{R_1}x - \frac{R}{R_b}u - \frac{R}{10V \cdot R_1}uz \right) \\ \dot{u} = \frac{1}{RC}\left(- \frac{R}{R_2}x + \frac{R}{10V \cdot R_3}xz + \frac{R}{R_5}y - V_c \right) \\ \dot{z} = \frac{1}{RC}\left(-z - \frac{R}{10V \cdot R_3}u^2 \right) \\ \dot{\omega} = \frac{1}{RC}u \end{cases} \tag{15}$$

where

$$y = \left(\frac{R}{10V \cdot R_1}V_y + \frac{R}{(10V)^2 \cdot R_5}\omega^2 - \frac{R}{(10V)^4 \cdot R_6}\omega^4 \right)u \tag{16}$$

In system (15), the variables x, u, z and ω correspond to the voltages in the outputs of the integrators U1–U4. Normalizing the differential equations of system (15), by using $\tau = t/RC$, we can see that this system is equivalent to the system (6), with $b = R/R_b$. The circuit components have been selected as: $R = 10$ kΩ, $R_1 = 1$ kΩ, $R_2 = 2$ kΩ, $R_3 = 0.866$ kΩ, $R_4 = 100$ kΩ, $R_5 = 0.416$ kΩ, $R_6 = 0.625$ kΩ, $C = 10$ nF, $V_c = 1$ V_{DC}, $V_y = 0.001$ V_{DC}, while the power supplies of all active devices are ± 15 V_{DC}. For the chosen set of components the system's (6) parameter is $c = 0.001$, while the value of the parameter b is adjusted via the resistor R_b.

The designed circuit is implemented in the electronic simulation package MULTISIM and the obtained results are displayed in Figs. 11 and 12. Theoretical attractors (see Figs. 6a and 9a) are similar with the circuital ones (see Figs. 11 and 12).

6 The Fractional-Order Memristive Hyperchaotic System

Fractional calculus is a generalization of integration and differentiation to non-integer-order fundamental operator. Different definitions of fractional order integration and differentiation have emerged during the development of fractional order theory. Some definitions are the Grünwald-Letnikov definition, the Cauchy integral formula, the Riemann-Liouville definition and the Caputo definition (Podlubny 1999).

Let $L^1 = L^1[a, b]$, $0 \leq a < b < \infty$, be a class of Lebesgue integrable function on [a, b]. Then, the Riemman-Liouville definition of a fractional integral for the function $f(t) \in L^1$, of order $a > 0$, and $t > 0$, is given by

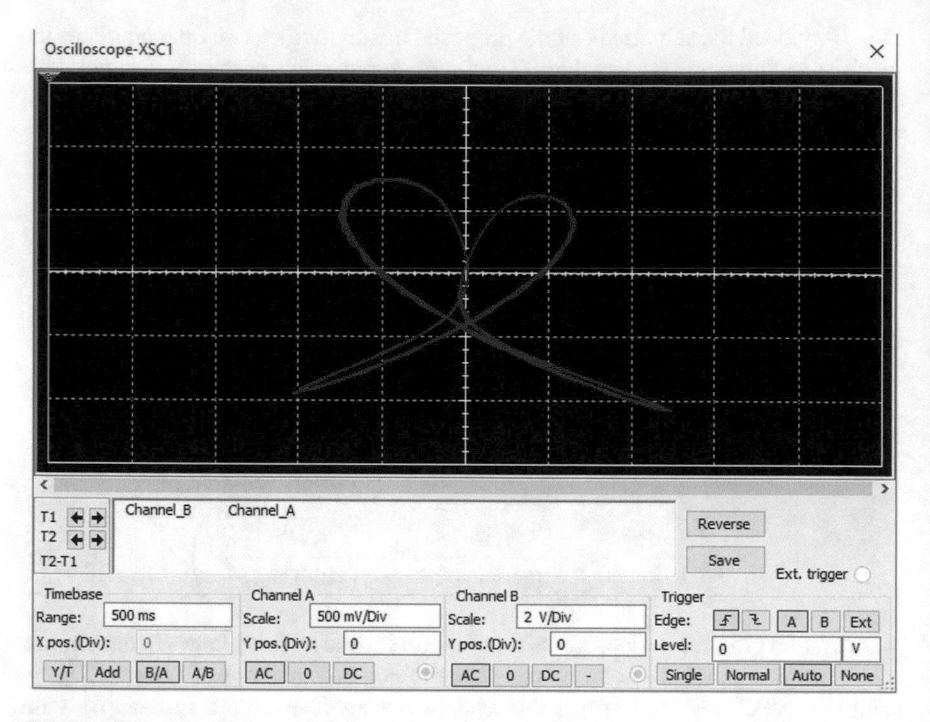

Fig. 11 Limit cycle of the designed circuit obtained from MULTISIM in the (x, z)-phase plane, for $b = 6.8$, $c = 0.001$

$$I^{\alpha}f(t) = \frac{1}{\Gamma(\alpha)} \int_{0}^{t} \frac{f(\tau)}{(t-\tau)^{1-\alpha}} d\tau \tag{17}$$

From the practical point of view the Riemman-Liouville definition requires the knowledge of the non-integer order derivatives of the function at $t = 0$. But the problem does not exist in the Caputo definition of the fractional derivative, which is called a smooth fractional derivative, and it is described by

$$D^{\alpha} = I^{m-\alpha} y^{(m)}(x), \quad \alpha > 0 \tag{18}$$

where $m = \lceil \alpha \rceil$, $y^{(m)}$ is the general m-order derivative, and I^{α} is the Riemman-Liouville integral operator. In general the operator D^{α} is called α-order Caputo differential operator, and is widely used in engineering field. The main advantage of using the Caputo definition is that the initial conditions of the fractional order differential equations are in the same form as the initial conditions of integer order differential equations, and there are clear interpretations of the initial conditions for integer orders. Moreover, it has the benefit of possessing a value of zero when it is applied to a constant.

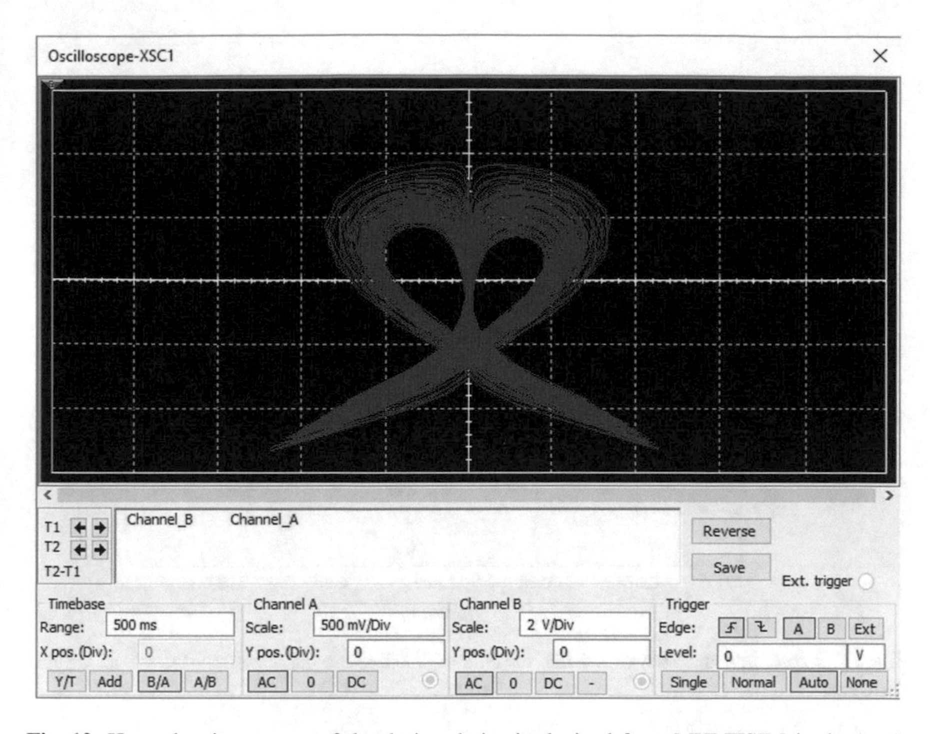

Fig. 12 Hyperchaotic attractor of the designed circuit obtained from MULTISIM in the (x, z)-phase plane, for $b = 5$, $c = 0.001$

The Laplace transform of the fractional integration operator is

$$I^{\alpha}(s) = L\left\{\frac{t^{\alpha-1}}{\Gamma(\alpha)}\right\} = \frac{1}{s^{\alpha}} \tag{19}$$

the implicit fractional differentiation is defined as the dual operation of the fractional integration. If $y(t) = I^{\alpha}(x(t))$ or $Y(s) = \frac{1}{s^{\alpha}}X(s)$, then $x(t)$ is the αth fractional order derivative of $y(t)$ defined as:

$$x(t) = D^{\alpha}(y(t)) \quad \text{or} \quad X(s) = s^{\alpha}Y(s) \tag{20}$$

In this scenario and by taking into account Refs. (Deng and Lu 2007; Petras 2011; Zambrano-Serrano et al. 2016), the fractional order memristive hypechaotic system of (6) can be described by

$$\begin{cases} D^{\alpha}x = -10x - bu - uz \\ D^{\alpha}u = -6x + 1.2xz + 0.1y - c \\ D^{\alpha}z = -z - 1.2u^{2} \\ D^{\alpha}\omega = u \end{cases} \tag{21}$$

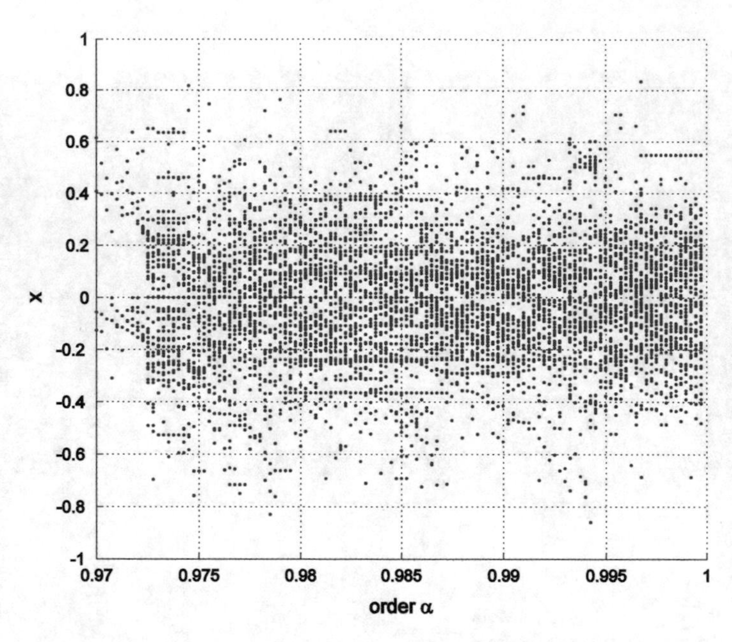

Fig. 13 Bifurcation diagram of x with $b = 5$, $c = 0.001$ and commensurate order α as varying parameter

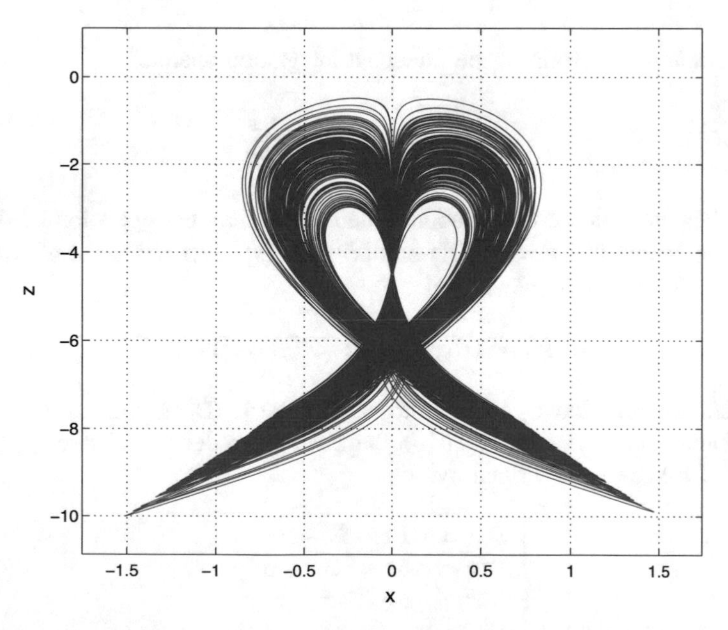

Fig. 14 Hyperchaotic attractor without equilibrium points from the system (21) in the (x, z)-phase plane, for $b = 5$, $c = 0.001$ and $\alpha = 0.98$

where α is the fractional order satisfying $0 < \alpha < 1$. By preserving the same values of system's parameters for integer case, the chaotic behavior of (21) is observed when $0.9725 \leq \alpha < 1$ as demonstrated by the bifurcation diagram in Fig. 13. The resulting hyperchaotic attractor by setting $b = 5$, $c = 0.001$ and order $\alpha = 0.98$ is given in Fig. 14. We have calculated numerically the spectrum of Lyapunov exponents in order to show the chaotic behavior of system (21). TISEAN software has been considered as a tool to obtain the spectrum of Lyapunov exponents (Hegger et al. 1998), from the numerical time series of the state variable x produced from system (21). The calculated Lyapunov exponents (LE_i) of the 4-D fractional-order hyperchaotic memristive system (6) are $LE_1 = 0.2715$, $LE_2 = 0.0171$, $LE_3 = 0$, $LE_4 = -0.2889$.

7 Circuit Realization of the Fractional-Order Memristive Hyperchaotic System

The electronic circuit synthesis of the fractional-order memristive hyperchaotic system (21) is shown in this section. The basic idea is similar to integer order case. It means that analog computing approach can be also used to design four integration channels. The main difference lies on the integration, which must be performed for fractional orders. As previously reported in Krishna and Reddy (2008), the fractional impedances have been pointed out as a solution to design fractional order operators. The fractional impedance, or fractance for short, is an electrical element which exhibits fractional order impedance properties. The impedance of the fractance device in the complex frequency domain is given by

$$Z(s) = as^{\alpha} \Rightarrow Z(j\omega) = a\omega^{\alpha} e^{j(\pi\alpha/2)} \tag{22}$$

where ω is the angular frequency, and the parameter α, for the special case of $\alpha = 1$ this element represents an inductor, for $\alpha = -1$ represents a capacitor while for $\alpha = 0$ represents a resistance. In the range $-2 < \alpha < 0$ this element generally can be considered to represent a fractional order capacitor. Alike, in the range $0 < \alpha < 2$ this element can be considered to represent a fractional order inductor. At $\alpha = -2$ the element represents the frequency-dependent-negative-resistor (Radwan et al. 2008). A physical fractance device does not yet available in the form of a single commercial device. So, the fractance device can be emulated via higher order passive RC or RLC trees, chains or even a net grid type networks (Krishna and Reddy 2008).

The fractance has many interesting properties. The phase angle is constant independent of the frequency, its magnitude versus frequency is nonlinear which can increase or decrease the effect of frequency, which means that it depends only on the value of fractional order α. Furthermore by using an operational amplifier, a fractional order integration can be attained. The main issue to design a fractance

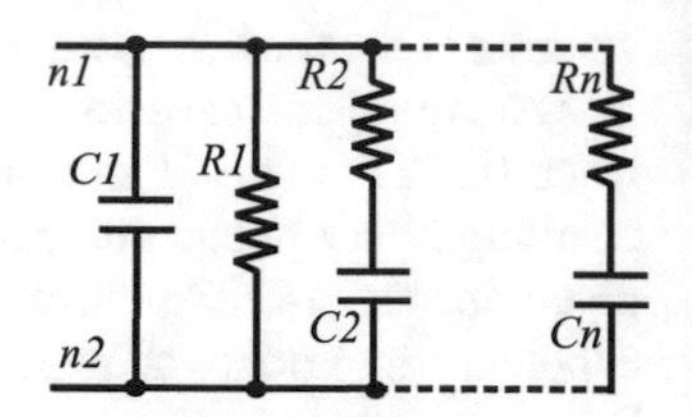

Fig. 15 Fractional impedance with n-resistors and n-capacitors

device is to find the rational approximation of the fractional order operator. This means, that the design of fractance with order α can be done considering the rational approximations. A general network to design a fractance for any order is shown in Fig. 15.

To compute a solution of a fractional order system considering rational approximations, first the fractional order equations of the system is considered in the frequency domain, and then Laplace transform of the fractional integral operator is replaced by its integer order approximation. One of the most common methods to find a rational approximation of the fractional operators is Charef method (Charef et al. 1992). Therein, the goal is to find zeros and poles of a transfer function that has similar amplitude diagram as $1/s^{\alpha}$ in a given frequency range. The fractional operator $1/s^{\alpha}$ has a Bode amplitude diagram characterized by a slope of I (-20α) dB/decade. Therefore in this method, the slope is approximated by a number of zigzag straight lines connected together with individuals slopes of (-20α) and 0 dB/decade. The order of this linear approximation system depends on the desired bandwidth and accuracy.

By considering $\alpha = 0.98$, the approximation of $1/s^{0.98}$ with error of approximately 0.5 dB and bandwidth of 10^{-2} – 10^{2} rad/s, and considering the method presented in Charef et al. (1992) is given by

$$\frac{1}{s^{0.98}} \approx \frac{1.14s^2 + 136.3s + 45.6}{s^3 + 134.5s^2 + 50.73s + 0.05365} \tag{23}$$

Then, the high-order rational approximation in (23) is designed by using three capacitors and three resistors from Fig. 15.

The values of circuit elements are obtained by determining the transfer function $H(s)$ between n_1 and n_2 as follows:

$$H(s) = R_1 a || \frac{1}{sC_1 a} || R_2 a + \frac{1}{sC_2 a} || R_3 a + \frac{1}{sC_3 a} \tag{24}$$

where C_0 is a unit parameter. In this manner, Fig. 16 depicts the schematic of the electronic circuit of (21) by using commercial devices such as operational amplifiers TL084, analog multipliers AD633, and passive components. Herein, devices U1, U2, U3, and U4 implement the fractance device to get a fractional order integrator for $\alpha = 0.98$. Circuit equations are analogous to integer-order case in (15) and (16). Then, by using the same values for circuit components as given in

Sect. 5, and let $G(s) = H(s)C_0 = \frac{1}{s^{0.98}}$ and $C_0 = 1\mu F$ in (24), the values for fractance device are: R_1a = 849.9 MΩ, R_2a = 24.38 MΩ, R_3a = 76.55 MΩ, C_1a = 8.771 μF, C_2a = 1.222 μF and C_3a = 1.095 μF.

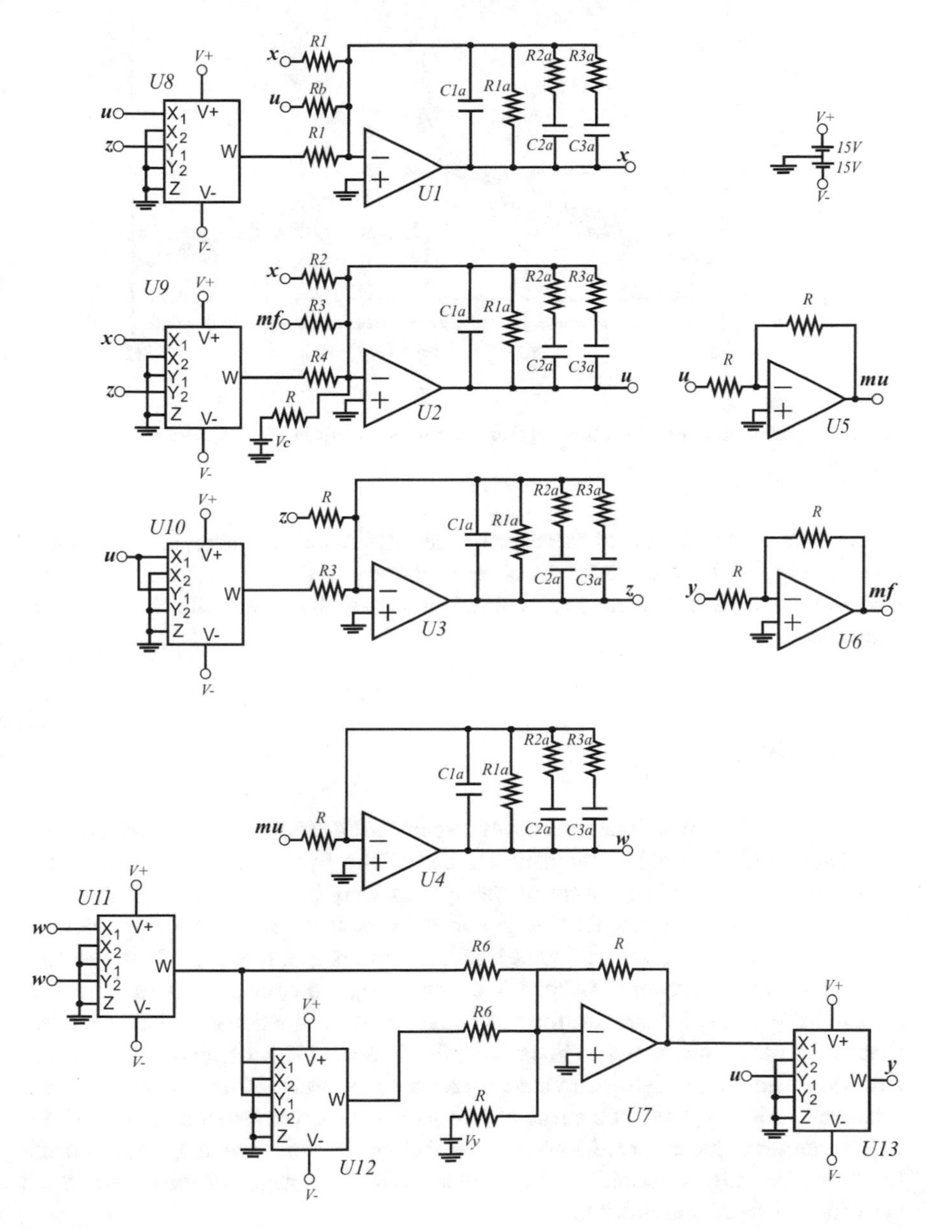

Fig. 16 Circuit implementation of fractional order hyperchaotic memristive system in (21) with fractional order $\alpha = 0.98$

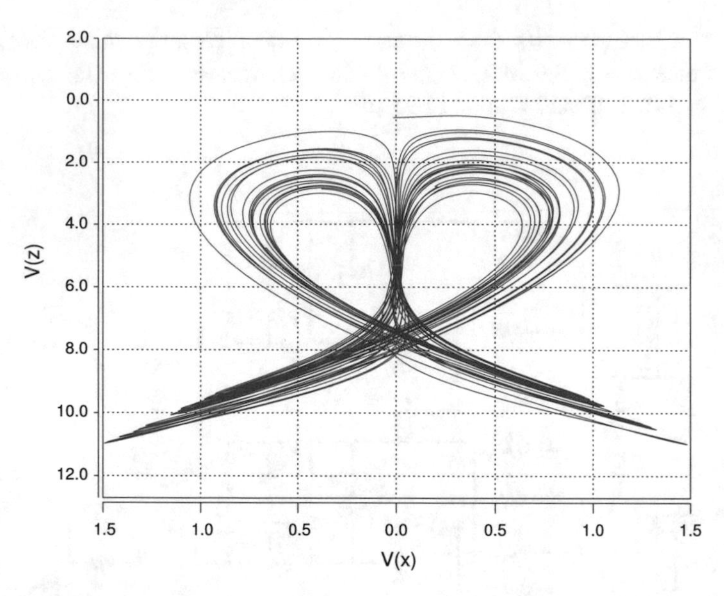

Fig. 17 Hyperchaotic attractor observed from electronic circuit in Fig. 16, with fractional order $\alpha = 0.98$

The circuit simulation of fractional order hyperchaotic memristive system is given in Fig. 17. By comparing the chaotic attractor in Fig. 17 with that in Fig. 14, it can be concluded that the circuit simulations are consistent with the numerical simulations.

8 Conclusion

In this chapter the first fractional-order hyperchaotic memristive system has been introduced and analyzed. Systematic studies of the hyperchaotic behavior in the integer and fractional-order form of the system were performed using phase portraits, Poincaré maps, bifurcation diagrams and Lyapunov exponents. Hyperchaotic behavior was observed with different fractional orders as a function of the system's parameters. The spectrum of Lyapunov exponent was computed to give a validation of the hyperchaotic behavior for the attractor with fractional order $\alpha = 0.98$. Finally, an electronic circuit to implement the fractional order hyperchaotic system has been also introduced. A fractance device was designed in order to get the fractional order hyperchaotic system from its integer-order version. Circuit simulations confirm the observed hyperchaotic behavior. This research would enable future engineering applications by considering the advantages of memristor-based systems and fractional order theory.

References

Adhikari, S. P., Yang, C., Kim, H., & Chua, L. O. (2012). Memristor bridge synapse-based neural network and its learning. *IEEE Transactions on Neural Networks and Learning Systems, 23*(9), 1426–1435.

Adhikari, S. P., Sah, M. P., Kim, H., & Chua, L. O. (2013). Three fingerprints of memristor. *IEEE Transactions on Circuits and Systems., 60*(11), 3008–3021.

Ascoli, A., & Corinto, F. (2013). Memristor models in a chaotic neural circuit. *International Journal of Bifurcation and Chaos in Applied Sciences and Engineering, 23*(3), 1350052.

Ascoli, A., Corinto, F., Senger, V., & Tetzlaff, R. (2013). Memristor model comparison. *IEEE Circuits and Systems Magazine, 13*(2), 89–105.

Bai, J., Yu, Y., Wang, S., & Song, Y. (2012). Modified projective synchronization of uncertain fractional order hyperchaotic systems. *Communications in Nonlinear Science and Numerical Simulation, 17*(4), 1921–1928.

Barakat, M., Mansingka, A., Radwan, A. G., & Salama, K. N. (2013). Generalized hardware post processing technique for chaos–based pseudorandom number generators. *ETRI Journal, 35,* 448–458.

Biolek, D., Biolek, Z., & Biolková, V. (2011). Pinched hysteretic loops of ideal memristors, memcapacitors and meminductors must be "self-crossing". *Electronics Letters, 47*(25), 1385–1387.

Biolek, Z., Biolek, D., & Biolková, V. (2012). Computation of the area of memristor pinched hysteresis loop. *IEEE Transactions on Circuits and Systems II: Express Briefs, 59*(9), 607–611.

Biolek, Z., Biolek, D., & Biolková, V. (2013). Analytical computation of the area of pinched hysteresis loops of ideal mem-elements. *Radioengineering, 22*(1), 132–135.

Bo-Cheng, B., Jian-Ping, X., Guo-Hua, Z., Zheng-Hua, M., & Ling, Z. (2011). Chaotic memristive circuit: equivalent circuit realization and dynamical analysis. *Chinese Physics B, 20*(12), 120502.

Bouali, S., Buscarino, A., Fortuna, L., Frasca, M., & Gambuzza, L. V. (2012). Emulating complex business cycles by using an electronic analogue. *Nonlinear Analysis: Real World Applications, 13*, 2459–2465.

Brezetskyi, S., Dudkowski, D., & Kapitaniak, T. (2015). Rare and hidden attractors in Van der Pol-Duffing oscillators. *The European Physical Journal Special Topics, 224,* 1459–1467.

Buscarino, A., Fortuna, L., & Frasca, M. (2009). Experimental robust synchronization of hyperchaotic circuits. *Physica D: Nonlinear Phenomena, 238,* 19171922.

Buscarino, A., Fortuna, L., Frasca, M., Gambuzza, L. V., & Sciuto, G. (2012a). Memristive chaotic circuits based on cellular nonlinear networks. *International Journal of Bifurcation and Chaos in Applied Sciences and Engineering, 22*(03), 1250070.

Buscarino, A., Fortuna, L., Frasca, M., & Gambuzza, L. V. (2012b). A chaotic circuit based on Hewlett-Packard memristor. Chaos: An Interdisciplinary. *Journal of Nonlinear Science, 22*(2), 023136.

Buscarino, A., Fortuna, L., Frasca, M., & Gambuzza, L. V. (2012c). A gallery of chaotic oscillators based on HP memristor. *International Journal of Bifurcation and Chaos in Applied Sciences and Engineering, 22,* 1330014–1330015.

Butzer, P., & Westphal, U. (2000). *An introduction to fractional calculus*. Singapore: World Scientific.

Cafagna, D., & Grassi, G. (2008). Fractional-order Chua's circuit: Time-domain analysis, bifurcation, chaotic behavior and test for chaos. *International Journal of Bifurcation and Chaos in Applied Sciences and Engineering, 18*(3), 615–639.

Cafagna, D., & Grassi, G. (2013). Elegant chaos in fractional-order system without equilibria. *Mathematical Problems in Engineering*, 2013.

Cafagna, D., & Grassi, G. (2015). Fractional-order systems without equillibria: The first example of hyperchaos and its application to synchronization. *Chinese Physics B, 24*(8), 080502.

Charef, A., Sun, H. H., Tsao, Y. Y., & Onaral, B. (1992). Fractal system as represented by singularity function. *IEEE Transactions on Automatic Control, 37*(9), 1465–1470.

Chen, Y., Jung, G. Y., Ohlberg, D. A. A., Li, X. M., Stewart, D. R., Jeppesen, J. O., et al. (2003). Nanoscale molecular-switch crossbar circuits. *Nanotechology, 14,* 462–468.

Chua, L. O., & Kang, S. M. (1976). Memristive devices and systems. *Proceedings of the IEEE, 64,* 209–223.

Concepcion, A., Chen, Y., Vinagre, B., Xue, D., & Feliu-Batlle, V. (2014). *Fractional-order systems and controls: Fundamentals and applications*. London, UK: Springer.

Corinto, F., & Ascoli, A. (2012). Memristor based elements for chaotic circuits. *IEICE Nonlinear Theory and Its Applications, 3*(3), 336–356.

Corinto, F., Ascoli, A., & Gilli, M. (2012). Analysis of current-voltage characteristics for memristive elements in pattern recognition systems. *International Journal of Circuit Theory and Applications, 40*(12), 1277–1320.

Dadras, S., Momeni, H. R., Qi, G., & Wang, Z. (2012). Four-wing hyperchaotic attractor generated from a new 4D system with one equilibrium and its fractional-order form. *Nonlinear Dynamics, 67,* 1161–1173.

Deng, H., Li, T., Wang, Q., & Li, H. (2009). A fractional-order hyperchaotic system and its synchronization. *Chaos, Solitons and Fractals, 41*(2), 962–969.

Deng, W., & Lu, J. (2007). Generating multi-directional multi-scroll chaotic attractors via a fractional differential hysteresis system. *Physics Letters A, 369*(5–6), 438–443.

Driscoll, T., Quinn, J., & Klein, S. (2010). Memristive adaptive filters. *Applied Physics Letters, 97* (9), 093502.

Driscoll, T., Pershin, Y. V., Basov, D. N., & Di Ventra, M. (2011). Chaotic memristor. *Applied Physics A, 102*(4), 885–889.

Dumitru, B., Guvenc, Z., & Tenreiro Machado, J. (2014). *New trends in nanotechnology and fractional calculus applications*. Germany: Springer.

Fitch, A. L., Yu, D., Iu, H. H. C., & Sreeram, V. (2012). Hyperchaos in an memristor-based modified canonical Chua's circuit. *International Journal of Bifurcation and Chaos in Applied Sciences and Engineering, 22,* 1250133–1250138.

Fortuna, L., Frasca, M., & Xibilia, M. G. (2009). *Chua's circuit implementation: Yesterday, today and tomorrow*. Singapore: World Scientific.

Gamez-Guzman, L., Cruz-Hernandez, C., Lopez-Gutierrez, R., & Garcia-Guerrero, E. E. (2009). Synchronization of chua's circuits with multiscroll attractors: Application to communication. *Communications in Nonlinear Science and Numerical Simulation, 14,* 2765–2775.

Gao, X., & Yu, J. (2005). Chaos in the fractional order periodically forced complex duffing's oscillators. *Chaos, Solitons and Fractals, 24*(4), 1097–1104.

Ghasemi, S., Tabesh, A., & Askari-Marnani, J. (2014). Application of fractional calculus theory to robust controller design for wind turbine generators. *IEEE Transactions on Energy Conversion, 29,* 780–787.

Grigorenko, I., & Grigorenko, E. (2003). Chaotic dynamics of the fractional Lorenz system. *Physical Review Letters, 91,* 34101–34104.

Hegger, R., Kantz, H., & Schereiber, T. (1998). Practical implementation of nonlinear time series methods: The TISEAN package. *Chaos, 9*(1999), 1–27.

Itoh, M., & Chua, L. O. (2008). Memristor oscillators. *International Journal of Bifurcation and Chaos in Applied Sciences and Engineering, 18*(11), 3183–3206.

Jafari, S., & Sprott, J. C. (2013). Simple chaotic flows with a line equilibrium. *Chaos, Solitons and Fractals, 57,* 79–84.

Jafari, S., Sprott, J. C., & Nazarimehr, F. (2015). Recent new examples of hidden attractors. *The European Physical Journal Special Topics, 224,* 1469–1476.

Kenneth, S. (1993). *An introduction to fractional calculus and fractional differential equations*. USA: Wiley-Interscience.

Krishna, B. T., & Reddy, K. V. V. S. (2008). Active and passive realization of fractance device of order 1/2. *Active and Pasive Electronic Components, 2008*(369421), 1–5.

Kuznetsov, N. V., Leonov, G. A., & Seledzhi, S. M. (2011). Hidden oscillations in nonlinear control systems. *IFAC Proceedings, 18,* 2506–2510.

Leonov, G. A., Kuznetsov, N. V., Kuznetsova, O. A., Seledzhi, S. M., & Vagaitsev, V. I. (2011a). Hidden oscillations in dynamical systems. *WSEAS Transactions on Systems and Control, 6*(2), 54–67.

Leonov, G. A., Kuznetsov, N. V., Kuznetsova, O. A., Seldedzhi, S. M., & Vagaitsev, V. I. (2011b). Hidden oscillations in dynamical systems. *Transactions on Systems and Control, 6,* 54–67.

Leonov, G. A., & Kuznetsov, N. V. (2011a). Analytical-numerical methods for investigation of hidden oscillations in nonlinear control systems. *IFAC Proceedings, 18,* 2494–2505.

Leonov, G. A., & Kuznetsov, N. V. (2011b). Algorithms for searching for hidden oscillations in the Aizerman and Kalman problems. *Doklady Mathematics, 84,* 475–481.

Leonov, G. A., Kuznetsov, N. V., & Vagaitsev, V. I. (2011c). Localization of hidden Chua's attractors. *Physics Letters A, 375,* 2230–2233.

Leonov, G. A., Kuznetsov, N. V., & Vagaitsev, V. I. (2012). Hidden attractor in smooth Chua system. *Phys. D, 241,* 1482–1486.

Leonov, G. A., & Kuznetsov, N. V. (2013). Hidden attractors in dynamical systems: from hidden oscillations in Hilbert-Kolmogorov, Aizerman, and Kalman problems to hidden chaotic attractor in Chua circuits. *International Journal of Bifurcation and Chaos in Applied Sciences and Engineering, 23*(1), 1330002.

Leonov, G. A., Kuznetsov, N. V., Kiseleva, M. A., Solovyeva, E. P., & Zaretskiy, A. M. (2014). Hidden oscillations in mathematical model of drilling system actuated by induction motor with a wound rotor. *Nonlinear Dynamics, 77,* 277–288.

Li, C., & Chen, G. (2004). Chaos and hyperchaos in the fractional-order Rössler equations. *Physica A: Statistical Mechanics and its Applications, 341*, 55–61.

Li, C., & Peng, G. (2004). Chaos in Chen's system with a fractional order. Chaos, *Solitons & Fractals, 22*(2), 443–450.

Li, H., Liao, X., & Luo, M. (2011). A novel non-equilibrium fractional-order chaotic system and its complete synchronization by circuit implementation. *Nonlinear Dynamics, 68*(1), 137–149.

Li, Q., Hu, S., Tang, S., & Zeng, G. (2014). Hyperchaos and horseshoe in a 4D memristive system with a line of equilibria and its implementation. *International Journal of Circuit Theory and Applications, 42,* 1172–1188.

Li, Q., Zeng, H., & Li, J. (2015). Hyperchaos in a 4D memristive circuit with infinitely many stable equilibria. *Nonlinear Dynamics, 79,* 2295–2308.

Lu, J. (2006). Chaotic dynamics of the fractional-order Lü system and its synchronization. *Physics Letters A, 354*(4), 305–311.

Matouk, A. E. (2009). Stability conditions, hyperchaos and control in a novel fractional order hyperchaotic system. *Physics Letters A, 373*(25), 2166–2173.

Muthuswamy, B. (2010). Implementing memristor based chaotic circuits. *International Journal of Bifurcation and Chaos in Applied Sciences and Engineering, 20*(05), 1335–1350.

Oldhamm, K., & Spanier, J. (1974). *The fractional calculus*. NewYork, USA: Academic-Press.

Pershin, Y. V., & Di Ventra, M. (2008). Spin memristive systems: Spin memory effects in semiconductor spintronics. *Physics Review B, 78,* 113309/1–113309/4.

Pershin, Y. V., La Fontaine, S., & Di Ventra, M. (2009). Memristive model of amoeba learning. *Physics Review E, 80*, 021926/1–021926/6.

Pershin, Y. V., & Di Ventra, M. (2010). Experimental demonstration of associative memory with memristive neural networks. *Neural Networks, 23,* 881.

Petras, I. (2011). *Fractional-order nonlinear system: modeling, analysis and simulation.* Berlin Heidelberg: Springer.

Pham, V. T., Rahma, F., Frasca, M., & Fortuna, L. (2014a). Dynamics and synchronization of a novel hyperchaotic system without equilibrium. *International Journal of Bifurcation and Chaos in Applied Sciences and Engineering, 24*(06), 1450087.

Pham, V. T., Volos, C. K., Jafari, S., & Wang, X. (2014b). Generating a novel hyperchaotic system out of equilibrium. *Optoelectronics and Advanced Materials, Rapid Communications, 8* (5–6), 535–539.

Pham, V. T., Volos, C. K., Jafari, S., Wang, X., & Vaidyanathan, S. (2014c). Hidden hyperchaotic attractor in a novel simple memristive neural network. *Optoelectronics and Advanced Materials, Rapid Communications, 8,* 1157–1163.

Pham, V.T., Volos, C.K., and Gambuzza, L.V. (2014d). A memristive hyperchaotic system without equilibrium. *The Scientific World Journal*, 2014.

Pham, V.-T., Jafari, S., Volos, C. K., Wang, X., & Golpayegani, S. M. R. H. (2014e). Is that really hidden? The presence of complex fixed-points in chaotic flows with no equilibria. *International Journal of Bifurcation and Chaos in Applied Sciences and Engineering, 24,* 1450146.

Pham, V. T., Vaidyanathan, S., Volos, C. K., & Jafari, S. (2015a). Hidden attractors in a chaotic system with an exponential nonlinear term. *The European Physical Journal Special Topics, 224,* 1507–1517.

Pham, V. T., Volos, C. K., Vaidyanathan, S., Le, T. P., & Vu, V. Y. (2015b). A memristor-based hyperchaotic system with hidden attractors: Dynamics, synchronization and circuital emulating. *Journal of Engineering Science and Technology Review, 8*(2), 205–214.

Ping, Z., Li-Jia, W., & Xue-Feng, C. (2009). A novel fractional-order hyperchaotic system and its synchronization. *Chinese Physics B, 18*(7), 2674.

Podlubny, I. (1999). *Fractional differential equations*. NewYork, USA: Academic Press.

Radwan, A. G., Soliman, A. S., & Elwakil, A. S. (2008). First-order generalized to the fractional domain. *Journal of Circuits, Systems, and Computers, 17*(1), 55–66.

Richard, H. (2014). *Fractional calculus: An introduction for physicists*. Singapore: World Scientific Publishing.

Sadoudi, S., Tanougast, C., Azzaz, M. S., & Dandache, A. (2013). Design and FPGA implementation of a wireless hyperchaotic communication system for secure realtime image transmission. *EURASIP Journal on Image and Video Processing, 943,* 1–18.

Santanu, S. (2015). *Fractional calculus with applications for nuclear reactor dynamics*. London, UK: Productivity Press.

Sapoff, M., & Oppenheim, R. M. (1963). Theory and application of self-heated thermistors. *Proceedings of the IEEE, 51,* 1292–1305.

Shahzad, M., Pham, V. T., Ahmad, M. A., Jafari, S., & Hadaeghi, F. (2015). Synchronization and circuit design of a chaotic system with coexisting hidden attractors. *The European Physical Journal Special Topics, 224,* 1637–1652.

Sharma, P. R., Shrimali, M. D., Prasad, A., Kuznetsov, N. V., & Leonov, G. A. (2015). Control of multistability in hidden attractors. *The European Physical Journal Special Topics, 224,* 1485–1491.

Shin, S., Kim, K., & Kang, S. M. (2011). Memristor applications for programmable analog ICs. *IEEE Transactions on Nanotechnology, 10*(2), 266–274.

Sprott, J. C. (2003). *Chaos and times-series analysis*. Oxford, UK: Oxford University Press.

Sprott, J. C. (2015). Strange attractors with various equilibrium types. *The European Physical Journal Special Topics, 224,* 1409–1419.

Strogatz, S. H. (1994). *Nonlinear dynamics and chaos: With applications to physics, biology, chemistry, and engineering*. Massachusetts, USA: Perseus Books.

Strukov, D., Snider, G., Stewart, G., & Williams, R. (2008). The missing memristor found. *Nature, 453,* 80–83.

Sundarapandian, V., & Pehlivan, I. (2012). Analysis, control, synchronization, and circuit design of a novel chaotic system. *Mathematical and Computer Modelling, 55*, 1904–1915.

Tetzlaff, R. (2014). *Memristors and Memristive Systems*. New York, USA: Springer.

Vaidyanathan, S., Volos, C. K., & Pham, V. T. (2015a). Analysis, control, synchronization and spice implementation of a novel 4-D hyperchaotic Rikitake dynamo system without equilibrium. *Journal of Engineering Science and Technology Review, 8,* 232–244.

Vaidyanathan, S., Pham, V. T., & Volos, C. K. (2015b). A 5-D hyperchaotic Rikitake dynamo system with hidden attractors. *The European Physical Journal Special Topics, 224*(8), 1575–1592.

Vaidyanathan, S., & Volos, C. K. (2016a). *Advances and applications in nonlinear control systems*. Berlin, Germany: Springer.

Vaidyanathan, S., & Volos, C. K. (2016b). *Advances and applications in chaotic systems*. Berlin, Germany: Springer.

Volos, C. K., Kyprianidis, I. M., & Stouboulos, I. N. (2012). A chaotic path planning generator for autonomous mobile robots. *Robotics and Autonomous Systems, 60,* 651–656.

Volos, C. K., Kyprianidis, I. M., & Stouboulos, I. N. (2013). Image encryption process based on chaotic synchronization phenomena. *Signal Processing, 93,* 1328–1340.

Wang, X., & Wang, M. (2007). Dynamic analysis of the fractional-order Liu system and its synchronization. *Chaos, 17*(3), 033106.

Wei, Z., Wang, R., & Liu, A. (2014). A new finding of the existence of hidden hyperchaotic attractors with no equilibria. *Mathematics and Computers in Simulation, 100,* 13–23.

Wolf, A., Swift, J. B., Swinney, H. L., & Vastano, J. (1985). Determining Lyapunov exponents from a time series. *Physica D: Nonlinear Phenomena, 16*(3), 285–317.

Wu, X., Lu, H., & Shen, S. (2009). Synchronization of a new fractional-order hyperchaotic system. *Physics Letters A, 373*(27), 2329–2337.

Yalcin, M. E., Suykens, J. A. K., & Vandewalle, J. (2004). True random bit generation from a double-scroll attractor. *IEEE Transactions on Circuits and Systems I: Regular Papers, 51,* 1395–1404.

Yang, J. J., Strukov, D. B., & Stewart, D. R. (2013). Memristive devices for computing. *Nature Nanotechnology, 8*(1), 13–24.

Yu, Y., & Li, H. X. (2008). The synchronization of fractional-order Rössler hyperchaotic systems. *Physica A: Statistical Mechanics and its Applications, 387*(5), 1393–1403.

Zambrano-Serrano, E., Campos-Canton, E., & Munoz-Pacheco, J. M. (2016). Strange attractors generated by a fractional order switching system and its topological horseshoe. *Nonlinear Dynamics, 83*(3), 1629–1641.

Zhou, P., & Huang, K. (2014). A new 4-D non-equilibrium fractional-order chaotic system and its circuit implementation. *Communications in Nonlinear Science and Numerical Simulation, 19* (6), 2005–2011.

Adaptive Control and Synchronization of a Memristor-Based Shinriki's System

Christos Volos, Sundarapandian Vaidyanathan, V.-T. Pham, H.E. Nistazakis, I.N. Stouboulos, I.M. Kyprianidis and G.S. Tombras

Abstract The recent discovery of memristor has aroused great interest in the scientific community about this new fourth circuit element and its applications in spintronic devices, ultra-dense information storage, neuromorphic circuits and programmable electronics. Also, the intrinsic nonlinear characteristic of memristor has been exploited in implementing novel chaotic oscillators with complex dynamics, by replacing their nonlinear elements with memristors. However, the increased systems' complexity, due to the use of memristor, have been raised significantly the interest for studying the cases of control of such systems as well as the synchronization of coupled memristive systems. So, to this direction, this chapter presents an adaptive controller, which is designed to stabilize a

C. Volos (✉) · I.N. Stouboulos · I.M. Kyprianidis
Physics Department, Aristotle University of Thessaloniki, 54124 Thessaloniki GR, Greece
e-mail: chvolos@gmail.com; volos@physics.auth.gr

I.N. Stouboulos
e-mail: stouboulos@physics.auth.gr

I.M. Kyprianidis
e-mail: imkypr@auth.gr; imkypr@physics.auth.gr

S. Vaidyanathan
Research and Development Centre, Vel Tech University, Avadi, Chennai 600062 Tamil Nadu, India
e-mail: sundarvtu@gmail.com

V.-T. Pham
School of Electronics and Telecommunications, Hanoi University of Science and Technology, 01 Dai Co Viet, Hanoi, Vietnam
e-mail: pvt3010@gmail.com

H.E. Nistazakis · G.S. Tombras
Department of Electronics, Computers, Telecommunications and Control Faculty of Physics, National and Kapodistrian University of Athens, 15784 Athens, GR, Greece
e-mail: enistaz@phys.uoa.gr; henistaz@gmail.com

G.S. Tombras
e-mail: gtombras@phys.uoa.gr

© Springer International Publishing AG 2017

S. Vaidyanathan and C. Volos (eds.), *Advances in Memristors, Memristive Devices and Systems*, Studies in Computational Intelligence 701,
DOI 10.1007/978-3-319-51724-7_10

memristor-based chaotic system with unknown memristor's parameters. Moreover, an adaptive controller is designed to achieve global chaos synchronization of the memristor-based chaotic systems with unknown memristor's parameters. The proposed chaotic system is a modified Shinriki nonlinear circuit, in which its nonlinear positive conductance has been replaced with a first order memristive diode bridge. All the main adaptive results in this chapter are proved using Lyapunov stability theory. The simulation results confirm the effectiveness of the proposed control and synchronization schemes.

Keywords Memristor · Shinriki system · Chaos · Control · Synchronization

1 Introduction

Three attractive inventions of Professor Leon Chua: the Chua's circuit (Matsumoto 1984), the Cellular Neural/Nonlinear Networks (CNNs) (Chua and Yang 1988a, b), and the memristor (Chua 1971; Chua and Kang 1976) are considered as the major breakthroughs in the literature of the nonlinear science. While Chua's circuit and CNNs have been studied and applied in various areas, such as secure communications, random generators, signal processing, pattern formation of modelling of complex systems (Arena 2005; Chua 1994, 1998), studies on memristor have only received significant attention recently after the realization of a solid-state thin film two-terminal memristor at Hewlett-Packard Laboratories (Strukov et al. 2008).

After this realization, a considerable number of potential memristor-based applications have been reported because memristor can be applied in different potential areas. The more important of them are related with spiking neural network, high-speed computing, synapses of biological systems, flexible circuits, nonvolatile memory, adaptive filter, pattern recognition systems, artificial intelligence, modeling of complex systems or low power devices and sensing (Adhikari et al. 2012; Ascoli et al. 2013; Ascoli and Corinto 2013; Corinto et al. 2012; Driscoll et al. 2010; Joglekar and Wolf 2009; Shin et al. 2011; Tetzlaff 2014). Interestingly, the intrinsic nonlinear characteristic of memristor has been exploited in implementing novel chaotic oscillators with complex dynamics (Bo-Cheng et al. 2011; Buscarino et al. 2012a, b; Corinto et al. 2012; Corinto and Ascoli 2012; Driscoll et al. 2011; Itoh and Chua 2008; Muthuswamy 2010).

Furthermore, the study of control of a chaotic system investigates methods for designing feedback control laws that globally or locally asymptotically stabilize or regulate the outputs of a chaotic system. Many methods have been developed for the control and tracking of chaotic systems such as active control (Chen 1999; Mahmoud et al. 2007; Nbendjo et al. 2003; Nbendjo and Woafo 2007), adaptive control (Chen 2011; Zheng 2011; Lin 2008; Luo et al. 2010; Vaidyanathan and Volos 2016a, b), backstepping control (Laoye et al. 2009; Lin 2010; Yassen 2006) and sliding mode control (Bartoszewicz and Patton 2007; Edwards and Spurgeon 1998; Utkin 1993, 2004; Utkin et al. 2009; Young et al. 1999).

Furthermore, chaos synchronization problem deals with the synchronization of a couple of systems called the master or drive system and the slave or response system. To solve this problem, control laws are designed so that the output of the slave system tracks the output of the master system asymptotically with time. The study of chaos in the last decades had a tremendous impact on the foundations of science and engineering and one of the most recent exciting developments is the discovery of chaos synchronization, which possibility was first reported by Fujisaka and Yamada and later by Pecora and Carroll (Fujisaka and Yamada 1983; Pecora and Carroll 1990). Because of the "butterfly" effect, the synchronization of chaotic systems is a challenging problem in the chaos literature even when the initial conditions of the master and slave systems are nearly identical because of the exponential divergence of the outputs of the two systems in the absence of any control. Different types of synchronization such as complete synchronization (Landsman and Schwartz 2007; Lin and He 2005; Liu 2002; Mahmoud and Mahmoud 2010; Pecora and Carroll 1990), antisynchronization (Kim et al. 2003; Li and Zhou 2007; Wang 2009; Wedekind and Parlitz 2002; Zhang and Sun 2004), hybrid synchronization (Barajas-Ramírez et al. 2003; Karthikeyan and Sundarapandian 2014; Xie 2002), lag synchronization (Li et al. 2005; Rosenblum et al. 1997; Shahverdiev et al. 2002; Taherion and Lai 1999), phase synchronization (Pikovsky et al. 1997; Rosenblum et al. 1996, 2001), anti-phase synchronization (Astakhov et al. 2000; Cao and Lai 1998; Liu et al. 2006), generalized synchronization (Kocarev and Parlitz 1996; Rulkov et al. 1995; Wang and Guan 2006; Yang and Duan 1998), projective synchronization (Li and Xu 2004; Mainieri and Rehacek 1999), generalized projective synchronization (Li 2007; Sarasu and Sundarapandian 2011; Yan and Li 2005), have been studied in the chaos literature.

Since the discovery of chaos synchronization, different approaches have been proposed to achieve it, such as PC method (Pecora and Carroll 1990), active control method (Agiza and Yassen 2001; Idowu et al. 2009; Vaidyanathan and Rajagopal 2011; Vincent 2008), adaptive control method (Chen and Lü 2002a, b; Vaidyanathan and Rajagopal 2012), backstepping control method (Huang 2005; Tan 2003; Yassen 2007) and sliding mode control method (Tavazoei and Haeri 2008; Yau 2004; Zhang and Xu 2010).

In this chapter, adaptive control and synchronization schemes for a memristor-based chaotic system have been developed. The proposed system is a modified Shinriki nonlinear circuit, in which its nonlinear positive conductance has been replaced with a first order memristive diode bridge. All the main adaptive results in this chapter are proved using Lyapunov stability theory. Simulation results prove the effectiveness of the proposed control and synchronization schemes.

The rest of this chapter is organized as follows. Related works are summarized in Sect. 2. Section 3 provides the mathematical model of the memristor-based Shinriki system, while the dynamics and properties of the system are presented in Sect. 4. The adaptive control scheme of the memristor-based Shinriki system is introduced in Sect. 5, while the adaptive synchronization scheme between two coupled memristor-based Shinriki system is presented in Sect. 6. Finally, conclusions are drawn in Sect. 7.

2 Related Works

Based on the complex dynamical behavior that memristive systems present, in the last five years many interesting control and synchronization schemes in those systems have been proposed. These schemes are presented in details in this section.

Firstly, in 2012, Wu et al. proposed some sufficient conditions for guarantying the exponential synchronization of the coupled memristor-based recurrent neural on drive-response concept (Wu et al. 2012).

Two different types of anti-synchronization algorithms are presented by Wu and Zeng in order to achieve the exponential anti-synchronization of coupled memristive recurrent neural networks (Wu and Zeng 2013). Huang and his co-workers investigated the problem of intermittent control of a memristor-based Chua's oscillator and presented the oscillator as the T-S fuzzy model system (Huang et al. 2013). Also, in 2013, a novel kind of compound synchronization between four memristor chaotic oscillator systems was investigated, where the drive systems have been conceptually divided into two categories: scaling drive systems and base drive systems (Sun et al. 2013).

In 2014, Zhang and Shen have investigated the exponential synchronization of coupled memristor-based chaotic neural networks with both time-varying delays and general activation functions (Zhang and Shen 2014). In the same year, the problem of global exponential synchronization for a class of memristor-based Cohen–Grossberg neural networks with time-varying discrete delays and unbounded distributed delays was studied (Yang et al. 2014). The problem of exponential lag synchronization control of memristive neural networks via the fuzzy method and its application in pseudorandom number generators has been presented in Wen et al. (2014a). In Wang et al. (2014) the synchronization control of memristor-based recurrent neural networks with impulsive perturbations or boundary perturbations is studied. Also, in 2014, the synchronization problem of memristive systems with multiple networked input and output delays via observer-based control has been investigated (Wen et al. 2014b).

Pham et al. 2015, studied the case of anti-synchronization between coupled memristor-based hyperchaotic systems with hidden attractors (Pham et al. 2015). The global robust synchronization of multiple memristive neural networks with nonidentical uncertain parameters is presented in Yang et al. (2015). Wen and his co-workers studied the problem of circuit design and global exponential stabilization of memristive neural networks with time-varying delays and general activation functions (Wen et al. 2015). By using the parallel-memristors connection corresponding to the capacitors and memristors synaptic connection in usual recurrent neural networks, general delayed memristive recurrent neural networks are proposed in Zhang et al. (2013). The investigation of synchronization for memristor-based neural networks with time-varying delay via an adaptive and feedback controller is studied in Li and Cao (2015). Mathiyalagan and his co-workers formulated and investigated the impulsive synchronization of memristor based bidirectional associative memory neural networks with time varying delays (Mathiyalagan et al. 2015).

In Mathiyalagan et al. (2016) the mixed H∞ and passivity based synchronization criteria for memristor-based recurrent neural networks with time-varying delays has been investigated. The impulsive synchronization of stochastic memristor-based recurrent neural networks with time delay is studied in Chandrasekar and Rakkiyappan (2016). Li and Cao presented the lag synchronization problem of memristor-based coupled neural networks with or without parameter mismatch using two different algorithms (Li and Cao 2016). A memristor-based complex Lorenz system and its modified projective synchronization have been introduced in Wang et al. (2016). Wen and his co-workers presented the sliding-mode control scheme of uncertain memristive Chua's circuits via the aforementioned method (Wen et al. 2016). Finally, a new memristor-based hyperchaotic complex Lü system and its adaptive complex generalized synchronization are presented in Wang et al. (2016).

3 Model of the Memristor-Based Shinriki's System

In this section, the memristor-based chaotic oscillator obtained by replacing the nonlinear positive conductance of the Shinriki's et al. (1981) circuit with a first order memristive diode bridge is considered, as it proposed by Kengne et al. (2015). The original Shinriki's oscillator, which is a modified van der Pol oscillator, has been introduced by Shinriki and co-workers in 1981 (Fig. 1). It consists of a resonant circuit and two nonlinear conductances, one negative, which is approximated by

$$i_{\mathrm{a}}(v_1) = -a_1 v_1 + a_3 v_1^3,\ a_1 > 0,\ a_3 > 0 \tag{1}$$

and another positive, which is approximated by

$$i_{\mathrm{d}}(v_2 - v_1) = b_1(v_2 - v_1) + b_3(v_2 - v_1)^3,\ b_1 > 0,\ b_3 > 0 \tag{2}$$

These approximations are quite reasonable from the qualitative viewpoint. The state equation of the Shinriki's circuit is written as:

$$\begin{cases} C_0 \dfrac{dv_1}{dt} = -G_1 v_1 + a_1 v_1 - a_3 v_1^3 + b_1(v_2 - v_1) + b_3(v_2 - v_1)^3 \\ C \dfrac{dv_2}{dt} = -i_L - G_2 v_2 - b_1(v_2 - v_1) - b_3(v_2 - v_1)^3 \\ L \dfrac{di_L}{dt} = v_2 \end{cases} \tag{3}$$

with $(v_1, v_2, i_L) \in \mathrm{R}^3$.

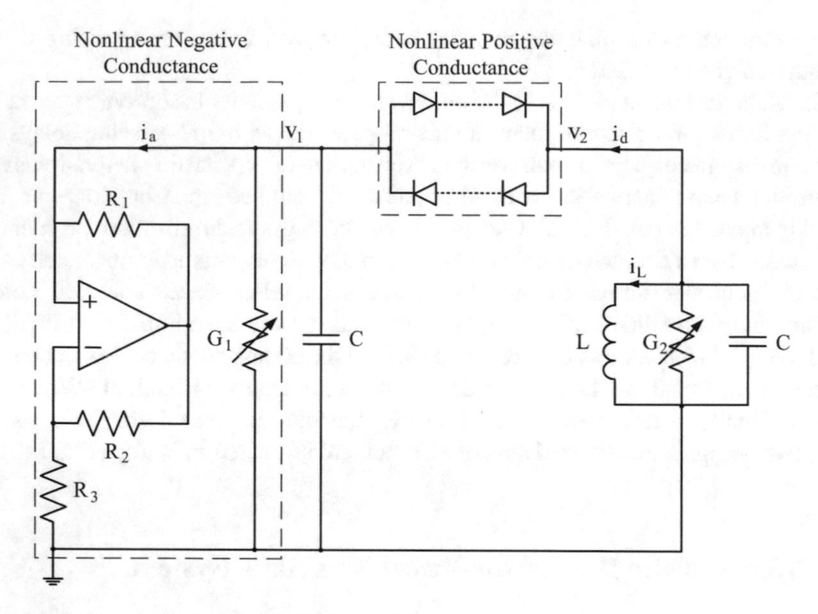

Fig. 1 The schematic of the original Shinriki's circuit

Shinriki and his co-workers showed that the circuit of Fig. 1 can generate oscillations with a random waveform or a periodic waveform depending on the chosen parameters.

In 1984, the dynamical behavior of the circuit of Fig. 1 has been further investigated in the work (Freire et al. 1984). The circuit is shown to develop a great variety of dynamical behaviors (equilibrium points, periodic oscillations, chaotic motions etc.) and the analysis proceeded to catalog all of them through a bifurcation study (pitchfork and Hopf bifurcations, flip bifurcations etc.). This study pointed out the interest devoted to the Shinriki's system (3).

Furthermore, in 2015 a novel memristor-based oscillator, obtained from Shinriki's circuit of Fig. 1, by substituting the nonlinear positive conductance with a first order memristive diode bridge, with a first order parallel *RC* filter, has been introduced (Kengne et al. 2015). The schematic diagram of the memristor-based Shinriki's circuit, which is an autonomous nonlinear circuit belonging to the memristive Chua's circuit family, is depicted in Fig. 2.

The mathematical model of the proposed memristor is given by the following equations:

$$i_m = g(v_{Cm}, v)v = 2I_S\exp(-kv_{Cm})\sinh(kv_m) \tag{4}$$

$$\frac{dv_{Cm}}{dt} = f(v_{Cm}, v) = \frac{2I_S\exp(-kv_{Cm})\cosh(kv_m)}{C_m} - \frac{v_{Cm}}{R_mC_m} - \frac{2I_S}{C_m} \tag{5}$$

where $k = 1/2nV_T$, while I_S, n, V_T denote the reverse current, the emission coefficient and the thermal voltage of the diode, respectively. Also, v_m, i_m represent the

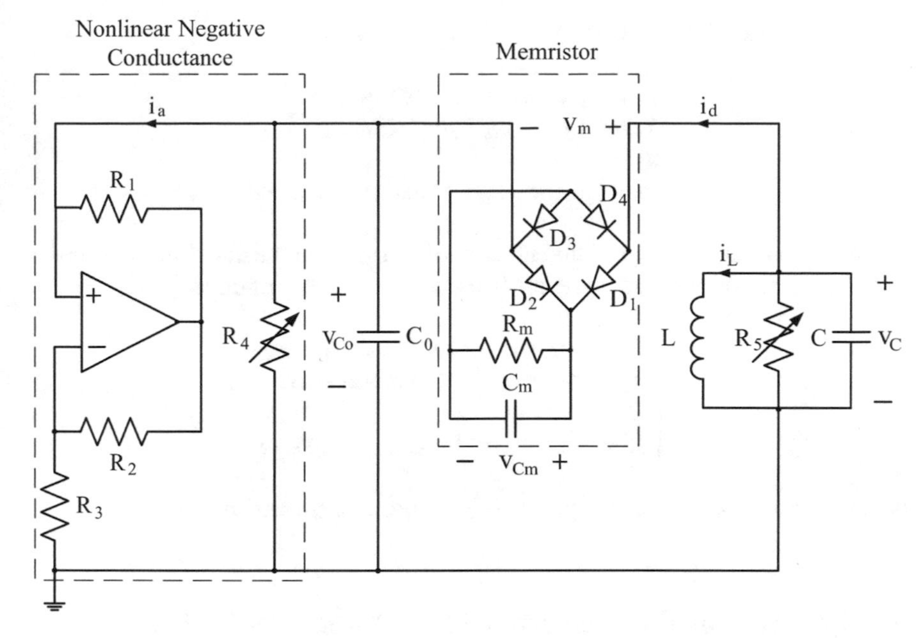

Fig. 2 The schematic of the memristor-based Shinriki's circuit

input voltage and current of the memristor, and v_{Cm} is the voltage of the capacitor C_m. The proposed memristor has been proved in Bao et al. (2014) that exhibits the three characteristic fingerprints for identifying a memristor (Adhikari et al. 2013).

By using the aforementioned memristor model, the Shinriki's system (3) has became a fourth order dynamical system described by the following set of differential equations:

$$\begin{cases} C\dfrac{dv_C}{dt} = -i_L - \dfrac{v_C}{R_5} - 2I_S e^{-kv_{Cm}}\sinh(kv_C - kv_{Co}) \\ C_0\dfrac{dv_{Co}}{dt} = \left(\dfrac{1}{R_3} - \dfrac{1}{R_4}\right)v_{Co} + 2I_S e^{-kv_{Cm}}\sinh(kv_C - kv_{Co}) \\ L\dfrac{di_L}{dt} = v_C \\ C_m\dfrac{dv_{Cm}}{dt} = -\dfrac{v_{Cm}}{R_m} + 2I_S e^{-kv_{Cm}}\cosh(kv_C - kv_{Co}) - 2I_S \end{cases} \quad (6)$$

Normalizing the system (6), by using the following change of variable and parameters:

$$\begin{gathered} x_1 = \frac{v_C}{V_{ref}}, x_2 = \frac{v_{Co}}{V_{ref}}, x_3 = \frac{\rho i_L}{V_{ref}}, x_4 = \frac{v_{Cm}}{V_{ref}}, \tau = \frac{t}{\sqrt{LC}}, \\ V_{ref} = 2\eta V_T, \rho = \sqrt{L/C}, \eta_1 = \frac{C}{C_o}, \eta_2 = \frac{C}{C_m}, \\ \alpha = \frac{\rho}{R_4}, \beta = \frac{\rho}{R_3}, \gamma = \frac{2\rho i_L}{V_{ref}}, \delta = \frac{\rho}{R_5}, \sigma = \frac{\rho}{R_m} \end{gathered} \quad (7)$$

The dimensionless circuit's system is defined as:

$$\begin{cases} \dot{x}_1 = -x_3 - \delta x_1 - \gamma e^{-x_4} \sinh(x_1 - x_2) \\ \dot{x}_2 = \eta_1 [(\beta - \alpha) x_2 + \gamma e^{-x_4} \sinh(x_1 - x_2)] \\ \dot{x}_3 = x_1 \\ \dot{x}_4 = \eta_2 [-\sigma x_4 + \gamma e^{-x_4} \cosh(x_1 - x_2) - \gamma] \end{cases} \tag{8}$$

where the over dots denotes differentiation with respect to the dimensionless time τ. Finally, for simplifying the system (8) further, it can be written as:

$$\begin{cases} \dot{x}_1 = -x_3 - d x_1 - c e^{-x_4} \sinh(x_1 - x_2) \\ \dot{x}_2 = (b - a) x_2 + p e^{-x_4} \sinh(x_1 - x_2) \\ \dot{x}_3 = x_1 \\ \dot{x}_4 = -l x_4 + m e^{-x_4} \cosh(x_1 - x_2) - m \end{cases} \tag{9}$$

where, $a = \eta_1 \alpha$, $b = \eta_1 \beta$, $c = \gamma$, $d = \delta$, $p = \eta_1 \gamma$, $l = \eta_2 \sigma$ and $m = \eta_2 \gamma$.

4 Dynamics of the Memristor-based Shinriki's System

The detailed analysis of the memristor-based Shinriki's system (8), regarding its fixed point's analysis, system's symmetry and numerical study, can be found in (Kengne et al. 2015). However, in this section the system's chaotic behavior will be explored, in order to study, in the next sections, its chaos control and synchronization schemes.

For this reason the system (9) is solved numerically using the classical fourth-order Runge-Kutta integration algorithm with time step $\Delta t = 0.005$ and the calculations are performed using variables and constants parameters. Also, the system is integrated for a sufficiently long time and the transient is cancelled. Two indicators are substantially exploited to define the type of scenario giving rise to chaos. The bifurcation diagram stands as the first indicator, while the second indicator is the graph of the Lyapunov exponents.

Furthermore, the numerical analysis is performed with the following values of circuit components: $L = 225$ mH, $C_o = 10$ nF, $C = 100$ nF, $C_m = 940$ nF, $R_1 = 5.6$ kΩ, $R_2 = 5.6$ kΩ, $R_3 = 10$ kΩ, $R_4 = 50$ kΩ, $R_5 = 20$ kΩ, R_m—tuneable, 1N4148 diodes, with $\eta = 1.9$, $V_T = 26$ mV and $I_S = 2.682$ nA. With this set of components values, the system's dimensionless parameters values are fixed to: $a = 0.3$, $b = 1.5$, $c = 8.143724 \times 10^{-5}$, $d = 0.075$, $p = 8.143724 \times 10^{-4}$, and $m = 8.6635298 \times 10^{-4}$, while l is the control parameter. The choice of l as a control parameter is done, because it is related with the memristor (via R_m). So, it is preferable as a control parameter, in order to see how a memristor's parameter affects system's behavior.

As it is known, the bifurcation diagram provides a useful tool in nonlinear science because it shows the change of system's dynamical behavior. So, the

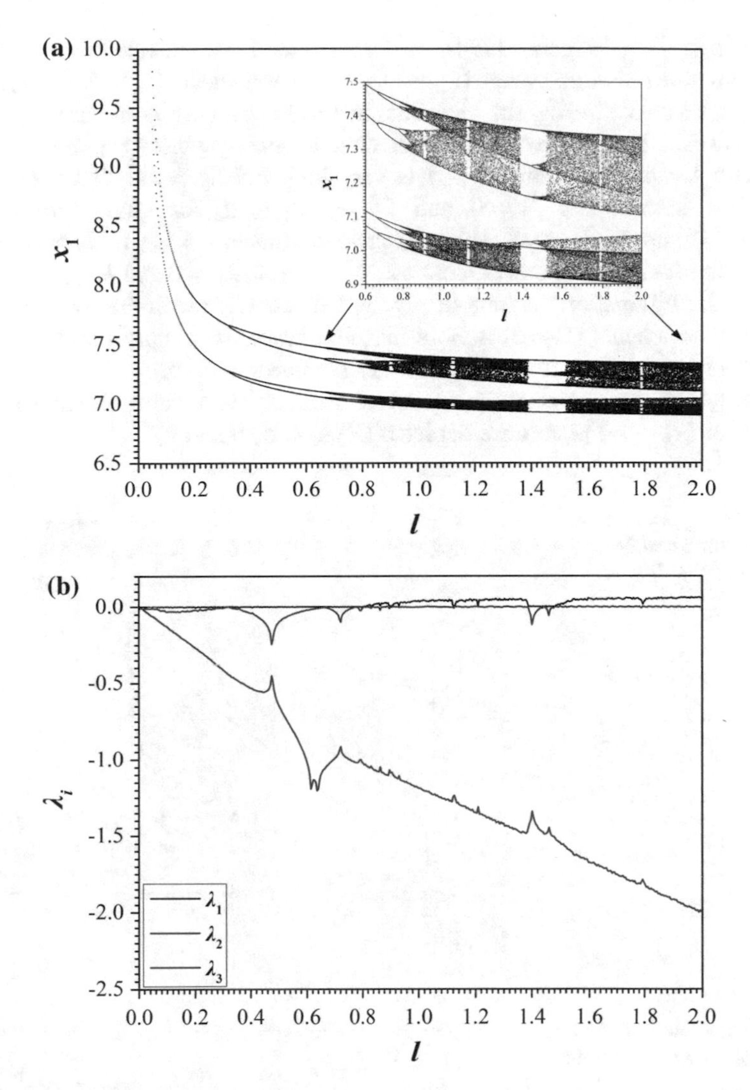

Fig. 3 **a** Bifurcation diagram of x_1 versus the parameter l and **b** the respective diagram of Lyapunov exponents versus the parameter l

bifurcation diagram of Fig. 3a has been obtained by plotting the value of variable x_1, when the trajectory intersects the section plane $x_3 = 0$, with $\dot{x}_3 > 0$, in terms of the bifurcation parameter l that is increased with step $\Delta l = 0.002$ in the range of $0 \leq l \leq 2$. From the observation of this diagram, the reader can see a period doubling route to chaos as the bifurcation parameter l is increased. The extended chaotic region is interrupted by tiny windows of periodic behavior sandwiched in the chaotic bands.

Also, it is well known, that Lyapunov exponents measure the exponential rates of the divergence and convergence of nearby trajectories in the phase space of the chaotic system (Strogatz 1994). In order to have detailed view of the memristor-based Shinriki system (9), the Lyapunov exponents (λ_i, with $i = 1, 2, 3, 4$) have been calculated using the algorithm in Wolf et al. (1985) and are depicted in Fig. 3b. In fact, Fig. 3b presents only the three largest Lyapunov exponents because the fourth Lyapunov exponent (λ_4) is very low. Briefly recall that for periodic orbits, the system has $\lambda_1 = 0$ and λ_2, λ_3, $\lambda_4 \leq 0$, for quasiperiodic orbits $\lambda_1 = \lambda_2 = 0$ and λ_3, $\lambda_4 \leq 0$, while for chaotic attractors $\lambda_1 \geq 0$, $\lambda_2 = 0$, and λ_3, $\lambda_4 \leq 0$, and for hyperchaotic attractors $\lambda_1 \geq \lambda_2 \geq 0$, $\lambda_3 = 0$ and $\lambda_4 \leq 0$. It can be seen that the bifurcation diagram of Fig. 3a well coincide with the spectrum of the Lyapunov exponents (Fig. 3b). Note that, the system is simply chaotic (and not hyperchaotic), although it is a fourth order nonlinear system.

Finally, for the aforementioned set of parameters, various numerical phase portraits in ($x_2 - x_1$) planes are depicted (Figs. 4, 5, 6 and 7).

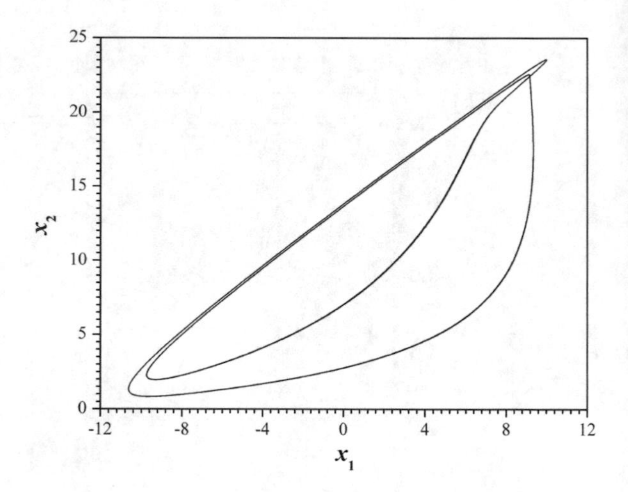

Fig. 4 Phase portrait of x_2 versus x_1, for $l = 0.2$ (period-2 state)

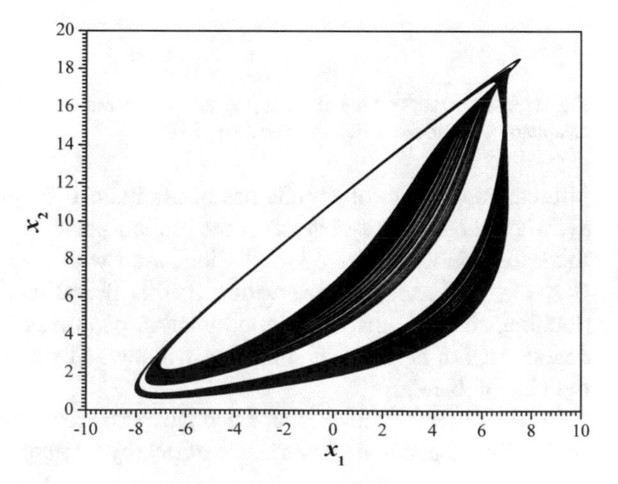

Fig. 5 Phase portrait of x_2 versus x_1, for $l = 1$ (chaotic state)

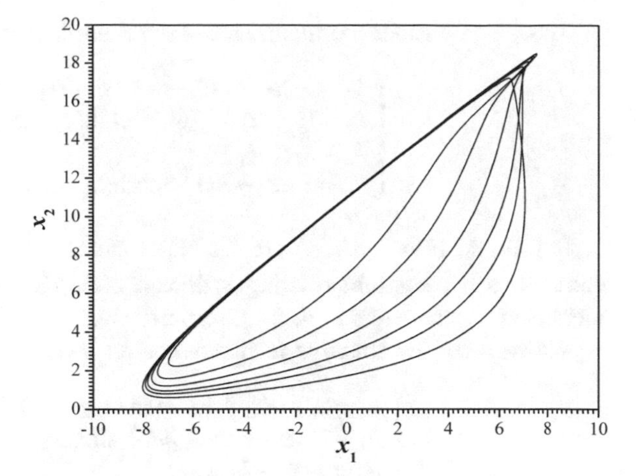

Fig. 6 Phase portrait of x_2 versus x_1 for $l = 1.4$ (period-6 state)

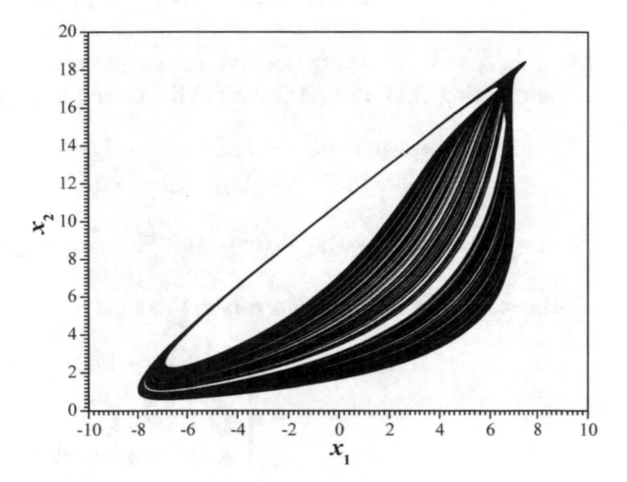

Fig. 7 Phase portrait of x_2 versus x_1 for $l = 2$ (chaotic state)

5 Adaptive Control of the Memristor-Based Shinriki's System

From the results of the aforementioned simulation process, it is obvious that the nature of the memristor add an extra complexity to system's dynamical behavior. So, it is useful to see if the proposed memristor-based Shinriki's system can be controlled by using the adaptive control method, in order to derive an adaptive feedback control law for globally stabilizing the system with memristor's unknown parameters.

Thus, we consider the memristor-based Shinriki's system given by

$$\begin{cases} \dot{x}_1 = -x_3 - dx_1 - ce^{-x_4}\sinh(x_1 - x_2) + u_1 \\ \dot{x}_2 = (b-a)x_2 + pe^{-x_4}\sinh(x_1 - x_2) + u_2 \\ \dot{x}_3 = x_1 + u_3 \\ \dot{x}_4 = -lx_4 + me^{-x_4}\cosh(x_1 - x_2) - m + u_4 \end{cases} \tag{10}$$

In (10), x_i, $(i = 1, \ldots, 4)$ are the states and u_i, $(i = 1, \ldots, 4)$ are the adaptive controls to be determined using estimates $\hat{c}(t)$, $\hat{p}(t)$, $\hat{l}(t)$, $\hat{m}(t)$ for the unknown memristor's parameters c, p, l, m, respectively.

We consider thc adaptive feedback control laws

$$\begin{cases} \dot{u}_1 = x_3 + dx_1 + \hat{c}e^{-x_4}\sinh(x_1 - x_2) - k_1x_1 \\ \dot{u}_2 = -(b-a)x_2 - \hat{p}e^{-x_4}\sinh(x_1 - x_2) - k_2x_2 \\ \dot{u}_3 = -x_1 - k_3x_3 \\ \dot{u}_4 = \hat{l}x_4 - \hat{m}e^{-x_4}\cosh(x_1 - x_2) + \hat{m} - k_4x_4 \end{cases} \tag{11}$$

where k_i, $(_i = 1, \ldots, 4)$ are positive gain constants.

Substituting (11) into (10), we get the closed-loop plant dynamics as:

$$\begin{cases} \dot{x}_1 = -(c-\hat{c})e^{-x_4}\sinh(x_1 - x_2) - k_1x_1 \\ \dot{x}_2 = (p-\hat{p})e^{-x_4}\sinh(x_1 - x_2) - k_2x_2 \\ \dot{x}_3 = -k_3x_3 \\ \dot{x}_4 = -(l-\hat{l})x_4 + (m-\hat{m})e^{-x_4}\cosh(x_1 - x_2) - (m-\hat{m}) - k_4x_4 \end{cases} \tag{12}$$

The parameter estimation errors are defined as:

$$\begin{cases} e_c(t) = c - \hat{c}(t) \\ e_p(t) = p - \hat{p}(t) \\ e_l(t) = l - \hat{l}(t) \\ e_m(t) = m - \hat{m}(t) \end{cases} \tag{13}$$

In view of (13), we can simplify the plant dynamics (12) as:

$$\begin{cases} \dot{x}_1 = -e_ce^{-x_4}\sinh(x_1 - x_2) - k_1x_1 \\ \dot{x}_2 = e_pe^{-x_4}\sinh(x_1 - x_2) - k_2x_2 \\ \dot{x}_3 = -k_3x_3 \\ \dot{x}_4 = -e_lx_4 + \mathrm{e}_me^{-x_4}\cosh(x_1 - x_2) - e_m - k_4x_4 \end{cases} \tag{14}$$

Differentiating (13) with respect to t, we obtain

$$\begin{cases} \dot{e}_c(t) = -\dot{\hat{c}}(t) \\ \dot{e}_p(t) = -\dot{\hat{p}}(t) \\ \dot{e}_l(t) = -\dot{\hat{l}}(t) \\ \dot{e}_m(t) = -\dot{\hat{m}}(t) \end{cases} \tag{15}$$

We use adaptive control theory in order to find an update law for the parameter estimates. We consider the quadratic candidate Lyapunov function defined by

$$V(x, e_c, e_p, e_l, e_m) = \frac{1}{2}(x_1^2 + x_2^2 + x_3^2 + x_4^2) + \frac{1}{2}(e_c^2 + e_p^2 + e_l^2 + e_m^2) \tag{16}$$

Differentiating V along the trajectories of (14) and (15), we obtain

$$\begin{aligned} \dot{V} &= x_1\dot{x}_1 + x_2\dot{x}_2 + x_3\dot{x}_3 + x_4\dot{x}_4 + e_c\dot{e}_c + e_p\dot{e}_p + e_l\dot{e}_l + e_m\dot{e}_m \\ &= -k_1x_1^2 - e_cx_1e^{-x_4}\sinh(x_1 - x_2) - k_2x_2^2 + e_px_2e^{-x_4}\sinh(x_1 - x_2) \\ &\quad - k_3x_3^2 - k_4x_4^2 - e_lx_4^2 + e_mx_4e^{-x_4}\cosh(x_1 - x_2) - e_mx_4 \\ &\quad - e_c\dot{\hat{c}} - e_p\dot{\hat{p}} - e_l\dot{\hat{l}} - e_m\dot{\hat{m}} \end{aligned} \tag{17}$$

or

$$\begin{aligned} \dot{V} = &- k_1x_1^2 - k_2x_2^2 - k_3x_3^2 - k_4x_4^2 + e_c[-x_1e^{-x_4}\sinh(x_1 - x_2) - \dot{\hat{c}}] \\ &+ e_p[x_2e^{-x_4}\sinh(x_1 - x_2) - \dot{\hat{p}}] + e_l\left[-x_4^2 - \dot{\hat{l}}\right] \\ &- e_m[x_4e^{-x_4}\cosh(x_1 - x_2) - x_4 - \dot{\hat{m}}] \end{aligned} \tag{18}$$

In view of (18), we take the parameter update law as

$$\begin{cases} \dot{\hat{c}} = -x_1e^{-x_4}\sinh(x_1 - x_2) \\ \dot{\hat{p}} = x_2e^{-x_4}\sinh(x_1 - x_2) \\ \dot{\hat{l}} = -x_4^2 \\ \dot{\hat{m}} = x_4e^{-x_4}\cosh(x_1 - x_2) - x_4 \end{cases} \tag{19}$$

Next, we state and prove the main result of this section.

Theorem 1 *The states* x_i, *(i = 1, …, 4) of the memristor-based Shinriki's system (5) with unknown system parameters are globally and exponentially stabilized for all initial conditions to the desired constant values c, p, l, m, respectively, by the adaptive control law* (11) *and the parameter update law* (19)*, where* k_1, k_2, k_3, k_4 *are positive gain constants.*

Proof This result will be proved by applying Lyapunov stability theory (Khalil 2001).

The quadratic Lyapunov function defined by (16), which is clearly a positive definite function on R^8 is considered.

By substituting the parameter update law (19) into (18), the time derivative of V is obtained as:

$$\dot{V} = -k_1 x_1^2 - k_2 x_2^2 - k_3 x_3^2 - k_4 x_4^2 \quad (20)$$

From (20), it is clear that dV/dt is a negative semi-definite function on R^8. Thus, the state vector $\boldsymbol{x}(t)$ and the parameter estimation error can be concluded that are globally bounded, i.e. $\left[x_1\, x_2\, x_3\, x_4\, e_c(t) e_p(t) e_l(t) e_m(t)\right]^T \in \mathbf{L}_\infty$.

If $k = \min\{k_1, k_2, k_3, k_4\}$, then it follows from (20) that

$$\dot{V} \le -k\|x(t)\|^2 \quad (21)$$

Thus

$$k\|x(t)\|^2 \le -\dot{V} \quad (22)$$

Integrating the inequality (22) from 0 to t, as:

$$k\int_0^t \|x(\tau)\|^2 d\tau \le V(0) - V(t) \quad (23)$$

From (23), it follows that $x \in \mathbf{L}_2$. Using (14), $\dot{x} \in \mathbf{L}_\infty$ can be concluded.

Also, by using Barbalat's lemma (Khalil 2001), the $\boldsymbol{x}(t) \to 0$ exponentially as $t \to \infty$ for all initial conditions $x(0) \in \mathbf{R}^4$ can be concluded. Hence, it follows that the states x_i, ($i = 1, \ldots, 4$) of the memristor-based Shinriki's system (5) with unknown memristor's parameters c, p, l, m are globally and exponentially stabilized for all initial conditions, by the adaptive control laws (11) and the parameter update law (19).

This completes the proof. ■

For the numerical simulations, the classical fourth-order Runge-Kutta method with step size $h = 10^{-8}$ is used to solve the systems (10) and (19), when the adaptive control laws (11) are applied. The parameter values of the memristor-based Shinriki's system (9) are taken as in the chaotic case, viz. $a = 0.3$, $b = 1.5$, $c = 8.143724 \times 10^{-5}$, $d = 0.075$, $p = 8.143724 \times 10^{-4}$, $m = 8.6635298 \times 10^{-4}$ and $l = 2$. Also, we take the positive gain constants as $k_i = 5$ for $i = 1, \ldots, 4$. Furthermore, as initial conditions of the memristor-based Shinriki's system (5), we take $x_1(0) = -0.2$, $x_2(0) = 0.3$, $x_3(0) = -0.3$ and $x_4(0) = 1$. Furthermore, as initial conditions of the parameter estimates $\hat{c}(t), \hat{p}(t), \hat{l}(t), \hat{m}(t)$, we take $\hat{c}(t) = 10^{-5}$, $\hat{p}(t) = 10^{-4}$, $\hat{l}(t) = 0.1$, $\hat{m}(t) = 8.143724 \cdot 10^{-5}$. In Fig. 8, the exponential convergence of the controlled states of the memristor-based Shinriki's system (10) is depicted.

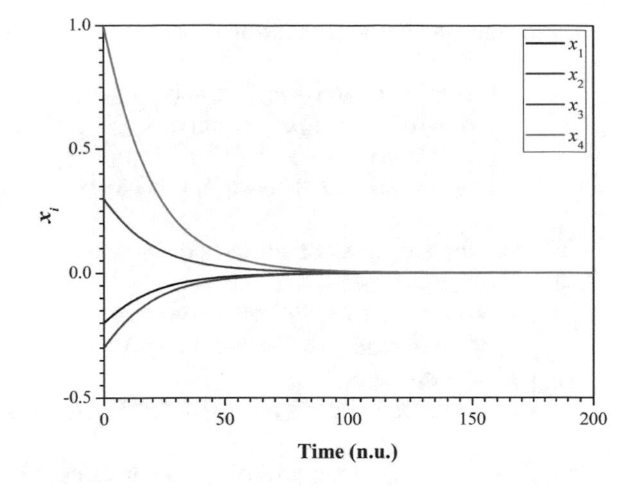

Fig. 8 Time-series of the states x_i, $(i = 1,\ldots, 4)$

6 Adaptive Synchronization of Identical Coupled Memristor-Based Shinriki's Systems

In this section, we derive an adaptive control law for globally and exponentially synchronizing the identical chaotic systems with unknown memristor's parameters. Thus, the master system is given by the chaotic memristor-based Shinriki's system (9), while the slave system is given by the following system dynamics.

$$\begin{cases} \dot{y}_1 = -y_3 - dy_1 - ce^{-y_4}\sinh(y_1 - y_2) + u_1 \\ \dot{y}_2 = (b-a)y_2 + pe^{-y_4}\sinh(y_1 - y_2) + u_2 \\ \dot{y}_3 = y_1 + u_3 \\ \dot{y}_4 = -ly_4 + me^{-y_4}\cosh(y_1 - y_2) - m + u_4 \end{cases} \tag{24}$$

where y_i, $(i = 1, \ldots, 4)$ are the states and u_i, $(i = 1, \ldots, 4)$ are the adaptive controls to be determined. In (9) and (24), the memristor's parameters c, p, l, m, are unknown and the design goal is to find adaptive feedback controls u_i that uses estimates $\hat{c}(t)$, $\hat{p}(t)$, $\hat{l}(t)$, $\hat{m}(t)$ for the parameters c, p, l, m respectively, so as to render the states of the systems (9) and (24) fully synchronized asymptotically.

The synchronization error between the chaotic systems (9) and (24) is defined as:

$$\begin{cases} e_1 = y_1 - x_1 \\ e_2 = y_2 - x_2 \\ e_3 = y_3 - x_3 \\ e_4 = y_4 - x_4 \end{cases} \tag{25}$$

Thus, the synchronization error dynamics is obtained as:

$$\begin{cases} \dot{e}_1 = -e_3 - de_1 - ce^{-y_4}\sinh(y_1 - y_2) + ce^{-x_4}\sinh(x_1 - x_2) + u_1 \\ \dot{e}_2 = (b-a)e_2 + pe^{-y_4}\sinh(y_1 - y_2) - pe^{-x_4}\sinh(x_1 - x_2) + u_2 \\ \dot{e}_3 = e_1 + u_3 \\ \dot{e}_4 = -le_4 + me^{-y_4}\cosh(y_1 - y_2) - me^{-x_4}\cosh(x_1 - x_2) + u_4 \end{cases} \tag{26}$$

We take the adaptive control laws defined by

$$\begin{cases} u_1 = e_3 + de_1 + \hat{c}e^{-y_4}\sinh(y_1 - y_2) - \hat{c}e^{-x_4}\sinh(x_1 - x_2) - k_1 e_1 \\ u_2 = -(b-a)e_2 + \hat{p}e^{-y_4}\sinh(y_1 - y_2) + \hat{p}e^{-x_4}\sinh(x_1 - x_2) - k_2 e_2 \\ u_3 = -e_1 - k_3 e_3 \\ u_4 = \hat{l}e_4 - \hat{m}e^{-y_4}\cosh(y_1 - y_2) + \hat{m}e^{-x_4}\cosh(x_1 - x_2) - k_4 e_4 \end{cases} \tag{27}$$

where k_i, ($i = 1, \ldots, 4$) are positive gain constants.

Substituting (27) into (26), we obtain the closed-loop error dynamics as:

$$\begin{cases} \dot{e}_1 = -(c-\hat{c})e^{-y_4}\sinh(y_1 - y_2) + (c-\hat{c})e^{-x_4}\sinh(x_1 - x_2) - k_1 e_1 \\ \dot{e}_2 = (p-\hat{p})e^{-y_4}\sinh(y_1 - y_2) - (p-\hat{p})e^{-x_4}\sinh(x_1 - x_2) - k_2 e_2 \\ \dot{e}_3 = -k_3 e_3 \\ \dot{e}_4 = -(l-\hat{l})e_4 + (m-\hat{m})e^{-y_4}\cosh(y_1 - y_2) - (m-\hat{m})e^{-x_4}\cosh(x_1 - x_2) - k_4 e_4 \end{cases} \tag{28}$$

The parameter estimation errors are defined as:

$$\begin{cases} e_c(t) = c - \hat{c}(t) \\ e_p(t) = p - \hat{p}(t) \\ e_l(t) = l - \hat{l}(t) \\ e_m(t) = m - \hat{m}(t) \end{cases} \tag{29}$$

Differentiating (29) with respect to t, we obtain

$$\begin{cases} \dot{e}_c(t) = -\dot{\hat{c}}(t) \\ \dot{e}_p(t) = -\dot{\hat{p}}(t) \\ \dot{e}_l(t) = -\dot{\hat{l}}(t) \\ \dot{e}_m(t) = -\dot{\hat{m}}(t) \end{cases} \tag{30}$$

By using (29), we rewrite the closed-loop system (28) as:

$$\begin{cases} \dot{e}_1 = -e_c e^{-y_4}\sinh(y_1 - y_2) + e_c e^{-x_4}\sinh(x_1 - x_2) - k_1 e_1 \\ \dot{e}_2 = e_p e^{-y_4}\sinh(y_1 - y_2) - e_p e^{-x_4}\sinh(x_1 - x_2) - k_2 e_2 \\ \dot{e}_3 = -k_3 e_3 \\ \dot{e}_4 = -e_l e_4 + e_m e^{-y_4}\cosh(y_1 - y_2) - e_m e^{-x_4}\cosh(x_1 - x_2) - k_4 e_4 \end{cases} \tag{31}$$

We consider the quadratic Lyapunov function given by

$$V(e, e_c, e_p, e_l, e_m) = \frac{1}{2}(e_1^2 + e_2^2 + e_3^2 + e_4^2) + \frac{1}{2}(e_c^2 + e_p^2 + e_l^2 + e_m^2) \tag{32}$$

Differentiating V along the trajectories of the systems (31) and (30), we obtain the following.

$$\begin{aligned}\dot{V} = & -k_1 e_1^2 - k_2 e_2^2 - k_3 e_3^2 - k_4 e_4^2 + e_c\left[-e_1 e^{-y_4}\sinh(y_1 - y_2) + e_1 e^{-x_4}\sinh(x_1 - x_2) - \dot{\hat{c}}\right] \\ & + e_p\left[e_2 e^{-y_4}\sinh(y_1 - y_2) - e_2 e^{-x_4}\sinh(x_1 - x_2) - \dot{\hat{p}}\right] + e_l\left[-e_4^2 - \dot{\hat{l}}\right] \\ & + e_m\left[e_4 e^{-y_4}\cosh(y_1 - y_2) - e_4 e^{-x_4}\cosh(x_1 - x_2) - \dot{\hat{m}}\right]\end{aligned} \tag{33}$$

In view of (33), we take the parameter update law as follows.

$$\begin{cases} \dot{\hat{c}} = -e_1 e^{-y_4}\sinh(y_1 - y_2) + e_1 e^{-x_4}\sinh(x_1 - x_2) \\ \dot{\hat{p}} = e_2 e^{-y_4}\sinh(y_1 - y_2) - e_2 e^{-x_4}\sinh(x_1 - x_2) \\ \dot{\hat{l}} = -e_4^2 \\ \dot{\hat{m}} = e_4 e^{-y_4}\cosh(y_1 - y_2) - e_4 e^{-x_4}\cosh(x_1 - x_2) \end{cases} \tag{34}$$

Next, we establish the main result of this section.

Theorem 2 *The memristor-based Shinriki's systems (9) and (24) with unknown parameters are globally and exponentially synchronized for all initial conditions by the adaptive feedback control law (27) and the parameter update law (34), were k_i, (i = 1, …, 4) are positive constants.*

Proof We prove this result via Lyapunov stability theory. We consider the quadratic Lyapunov function V defined by (32), which is positive definite on R^8. Next, by substituting the parameter update law (34) into (33), we obtain the time derivative of V as:

$$\dot{V} = -k_1 e_1^2 - k_2 e_2^2 - k_3 e_3^2 - k_4 e_4^2 \tag{35}$$

Thus, it is clear that $\dot{V}$ is a negative semi-definite function on R^8.

From (35), it follows that the synchronization error vector $e(t) = (e_1(t), e_2(t), e_3(t), e_4(t))$ and the parameter estimation error $(e_c(t), e_p(t), e_l(t), e_m(t))$ are globally bounded. We define $k = \min(k_1, k_2, k_3, k_4)$.

Then it follows from (35) that

$$\dot{V} \le -k\|e(t)\|^2 \tag{36}$$

Thus

$$k\|e(t)\|^2 \leq -\dot{V} \tag{37}$$

Integrating the inequality (37) from 0 to t, as:

$$k\int_0^t \|e(\tau)\|^2 d\tau \leq V(0) - V(t) \tag{36}$$

From (36), it follows that $e \in \mathbf{L}_2$, while from (28), it can be deduced that $\dot{e} \in \mathbf{L}_\infty$. Thus, using Barbalat's lemma (Khalil 2001), we can conclude that $\boldsymbol{e}(t) \to 0$ exponentially as $t \to \infty$ for all initial conditions.

This completes the proof. ∎

For numerical simulations, the classical fourth-order Runge-Kutta method with step size h = 10^{-8} is used to solve the system of differential Eqs. (9), (24) and (34), when the adaptive control laws (27) are applied.

The parameter values of the memristor-based Shinriki's systems (9) and (24) are taken as in the chaotic case of the previous section. The gain constants are taken as $k_i = 10$, for $i = 1, 2, 3, 4$.

Furthermore, as initial conditions of the master system (9), we take $x_1(t) = -0.2$, $x_2(t) = 0.3$, $x_3(t) = -0.3$ and $x_4(t) = 1$, while the initial conditions of the slave system (24), are $y_1(0) = 0.5$, $y_2(0) = -0.2$, $y_3(0) = -0.1$ and $y_4(0) = 0.7$.

Also, as initial conditions of the parameter estimates, we take $\hat{c}(t) = 10^{-5}$, $\hat{p}(t) = 10^{-4}$, $\hat{l}(t) = 0.1$, $\hat{m}(t) = 8.143724 \cdot 10^{-5}$. In Figs. 9, 10, 11 and 12, the synchronization of the states of the master system (9) and slave system (24) are depicted, when the adaptive control law (27) and parameter update law (34) are implemented. In Fig. 13, the time-history of the synchronization errors $e_1(t)$, $e_2(t)$, $e_3(t)$ and $e_4(t)$ is depicted.

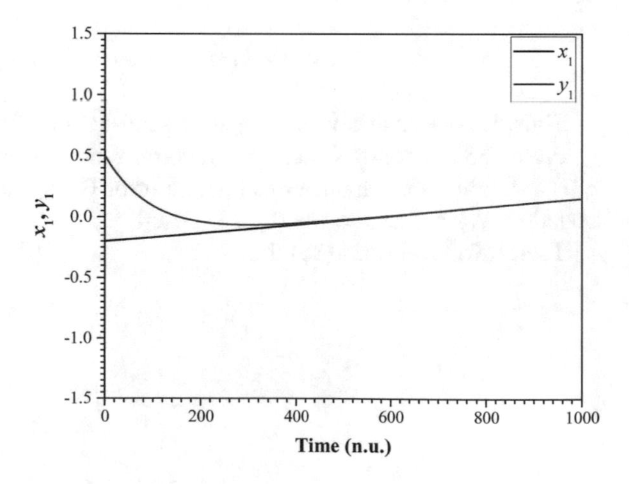

Fig. 9 Synchronization of the states $x_1(t)$ and $y_1(t)$

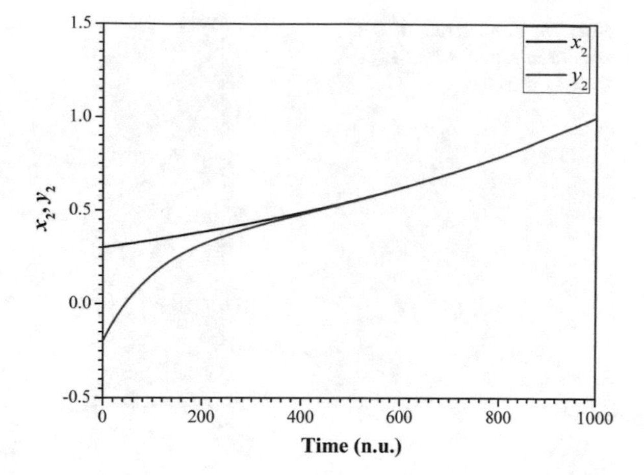

Fig. 10 Synchronization of the states $x_2(t)$ and $y_2(t)$

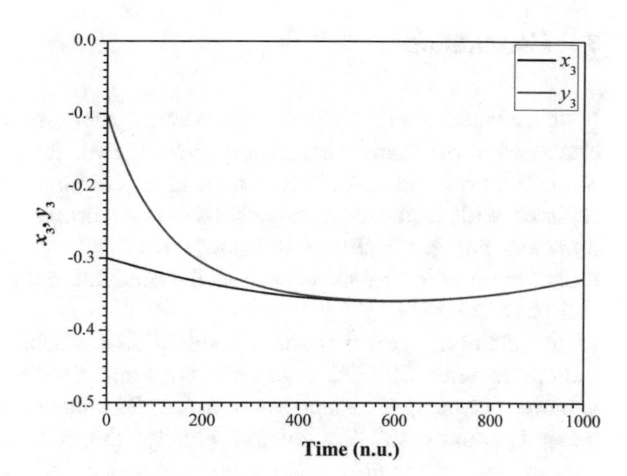

Fig. 11 Synchronization of the states $x_3(t)$ and $y_3(t)$

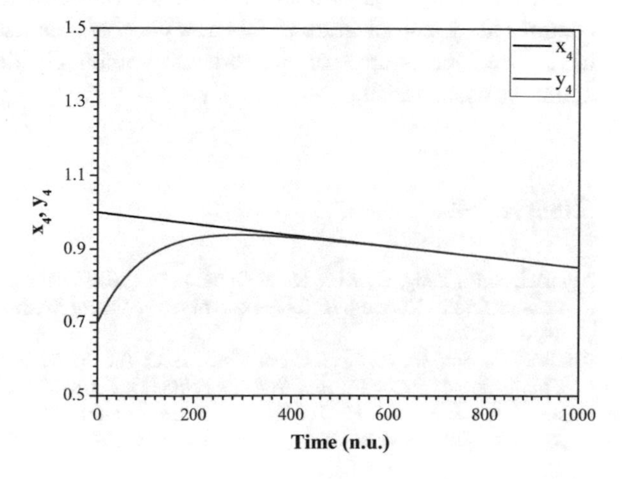

Fig. 12 Synchronization of the states $x_4(t)$ and $y_4(t)$

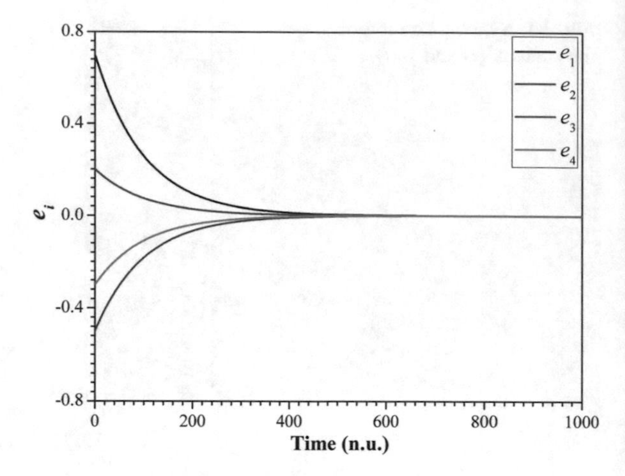

Fig. 13 Time-series of the synchronization errors $e_i(t)$, ($i = 1, 2, 3, 4$)

7 Conclusion

In this chapter a memristor-based chaotic system as well as its control and synchronization problems were mainly investigated. As a chaotic system, a modified Shinriki's nonlinear circuit, in which its nonlinear positive conductance has been replaced with a first order memristive diode bridge, was used. The study of its dynamics and especially of its chaotic behavior, was done by using well-known tools from nonlinear theory, such as the bifurcation diagram, Lyapunov exponents and phase portraits.

In addition, global control and global chaos synchronization of such memristor-based Shinriki's systems, with unknown memristor's parameters were achieved by using an adaptive controller. The main adaptive results were proved using Lyapunov stability theory. Finally, the simulation results confirmed the effectiveness of the proposed control and synchronization schemes.

So, this work is a step forward on the direction of studying the methods of control and synchronization of this new class of memristive chaotic systems, which have raised the interest of the research community due to memristor's intrinsic nonlinear characteristic.

References

Adhikari, S. P., Yang, C., Kim, H., & Chua, L. O. (2012). Memristor bridge synapse-based neural network and its learning. *IEEE Transactions on Neural Networks and Learning Systems, 23*(9), 1426–1435.

Adhikari, P., Sah, M. P., Kim, H., & Chua, L. O. (2013). Three fingerprints of memristor. *IEEE Transactions on Circuits and Systems I, 60*(11), 3008–3021.

Agiza, H. N., & Yassen, M. T. (2001). Synchronization of Rössler and Chen chaotic dynamical systems using active control. *Physics Letters A, 278,* 191–197.

Arena, P., Bucolo, M., Fazzino, S., Fortuna, L., & Frasca, M. (2005). The CNN paradigm: Shapes and complexity. *International Journal of Bifurcation and Chaos, 7,* 2063–2090.

Ascoli, A., & Corinto, F. (2013). Memristor models in a chaotic neural circuit. *International Journal of Bifurcation and Chaos, 23*(3), 1350052.

Ascoli, A., Corinto, F., Senger, V., & Tetzlaff, R. (2013). Memristor model comparison. *IEEE Circuits and Systems Magazine, 13*(2), 89–105.

Astakhov, V., Shabunin, A., & Anishchenko, V. (2000). Antiphase synchronization in symmetrically coupled self-oscillators. *International Journal of Bifurcation and Chaos, 10* (04), 849–857.

Bao, B., Yu, J., Hu, F., & Liu, Z. (2014). Generalized memristor consisting of diode bridge with first order parallel RC filter. *International Journal of Bifurcation and Chaos*, *24*(11), 1450143.

Barajas-Ramírez, J. G., Chen, G., & Shieh, L. S. (2003). Hybrid chaos synchronization. *International Journal of Bifurcation and Chaos, 13*(05), 1197–1216.

Bartoszewicz, A., & Patton, R. J. (2007). Sliding mode control. *International Journal of Adaptive Control and Signal Processing*, *21*(8–9), 635–637.

Bo-Cheng, B., Jian-Ping, X., Guo-Hua, Z., Zheng-Hua, M., & Ling, Z. (2011). Chaotic memristive circuit: Equivalent circuit realization and dynamical analysis. *Chinese Physics B, 20*(12), 120502.

Buscarino, A., Fortuna, L., Frasca, M., Gambuzza, L. V., & Sciuto, G. (2012a). Memristive chaotic circuits based on cellular nonlinear networks. *International Journal of Bifurcation and Chaos, 22*(03), 1250070.

Buscarino, A., Fortuna, L., Frasca, M., & Gambuzza, L. V. (2012b). A chaotic circuit based on Hewlett-Packard memristor. *Chaos: An Interdisciplinary Journal of Nonlinear Science, 22*(2), 023136.

Cao, L. Y., & Lai, Y. C. (1998). Antiphase synchronism in chaotic systems. *Physical Review E, 58* (1), 382–386.

Chandrasekar, A., & Rakkiyappan, R. (2016). Impulsive controller design for exponential synchronization of delayed stochastic memristor-based recurrent neural networks. *Neurocomputing, 173,* 1348–1355.

Chen, G. (1999). *Controlling chaos and bifurcations in engineering systems*. US: CRC Press.

Chen, G. (2011). A simple adaptive feedback control method for chaos and hyper-chaos control. *Applied Mathematics and Computation, 217*(17), 7258–7264.

Chen, S., & Lü, J. (2002a). Parameters identification and synchronization of chaotic systems based upon adaptive control. *Physics Letters A, 299*(4), 353–358.

Chen, S., & Lü, J. (2002b). Synchronization of an uncertain unified chaotic system via adaptive control. *Chaos, Solitons & Fractals, 14*(4), 643–647.

Chua, L. O. (1971). Memristor—the missing circuit element. *IEEE Transactions on Circuit Theory, 18*(5), 507–519.

Chua, L. O., & Kang, S. M. (1976). Memristive devices and systems. *Proceedings of the IEEE, 64,* 209–223.

Chua, L. O., & Yang, L. (1988a). Cellular neural networks: Theory. *IEEE Transactions on Circuits and Systems, 35,* 1257–1272.

Chua, L. O., & Yang, L. (1988b). Cellular neural networks: Applications. *IEEE Transactions on Circuits and Systems, 35,* 273–1290.

Chua, L. O. (1994). Chua's circuit: An overview ten years later. *Journal of Circuits Systems and Computers, 4,* 117–159.

Chua, L. O. (1998). *CNN: A paradigm for complexity*. Singapore: World Scientific.

Corinto, F., & Ascoli, A. (2012). Memristor based elements for chaotic circuits. *IEICE Nonlinear Theory and Its Applications, 3*(3), 336–356.

Driscoll, T., Quinn, J., & Klein, S. (2010). Memristive adaptive filters. *Applied Physics Letters, 97* (9), 093502.

Driscoll, T., Pershin, Y. V., Basov, D. N., & Di Ventra, M. (2011). Chaotic memristor. *Applied Physics A, 102*(4), 885–889.

Edwards, C., & Spurgeon, S. (1998). *Sliding mode control: Theory and applications*. US: CRC Press.

Freire, E., Franquelo, L. G., & Aracil, J. (1984). Periodicity and chaos in an autonomous electrical system. *IEEE Transactions on Circuits and Systems, 31*(3), 237–247.

Fujisaka, H., & Yamada, T. (1983). Stability theory of synchronized motion in coupled-oscillator systems. *Progress of Theoretical Physics, 69*(1), 32–47.

Huang, L., Wang, M., & Feng, R. (2005). Synchronization of generalized Henon map via backstepping design. *Chaos, Solitons & Fractals, 23*(2), 617–620.

Huang, J., Li, C., & He, X. (2013). Stabilization of a memristor-based chaotic system by intermittent control and fuzzy processing. *International Journal of Control, Automation and Systems, 11*(3), 643–647.

Idowu, B. A., Vincent, U. E., & Njah, A. N. (2009). Synchronization of chaos in nonidentical parametrically excited systems. *Chaos, Solitons & Fractals, 39,* 2322–2331.

Itoh, M., & Chua, L. O. (2008). Memristor oscillators. *International Journal of Bifurcation and Chaos, 18*(11), 3183–3206.

Joglekar, Y. N., & Wolf, S. J. (2009). The elusive memristor: Properties of basic electrical circuits. *European Journal of Physics, 30,* 661–675.

Karthikeyan, R., & Sundarapandian, V. (2014). Hybrid Chaos Synchronization of Four-Scroll Systems via Active Control. *Journal of Electrical Engineering, 65*(2), 97–103.

Kengne, J., Njitacke Tabekoueng, Z., Kamdum Tamba, V., & Nguomkam Negou, A. (2015). Periodicity, chaos, and multiple attractors in a memristor-based Shinriki's circuit. *Chaos, 25,* 103126.

Khalil, H. K. (2001). *Nonlinear systems*. New Jersey, USA: Prentice Hall.

Kim, C. M., Rim, S., Kye, W. H., Ryu, J. W., & Park, Y. J. (2003). Anti-synchronization of chaotic oscillators. *Physics Letters A, 320*(1), 39–46.

Kocarev, L., & Parlitz, U. (1996). Generalized synchronization, predictability, and equivalence of unidirectionally coupled dynamical systems. *Physical Review Letters, 76*(11), 1816–1819.

Landsman, A. S., & Schwartz, I. B. (2007). Complete chaotic synchronization in mutually coupled time-delay systems. *Physical Review E, 75*(2), 026201.

Laoye, J., Vincent, U., & Kareem, S. (2009). Chaos control of 4-D chaotic systems using recursive backstepping nonlinear controller. *Chaos, Solitons & Fractals, 39,* 356–362.

Li, Z., & Xu, D. (2004). A secure communication scheme using projective chaos synchronization. *Chaos, Solitons & Fractals, 22*(2), 477–481.

Li, C., Liao, X., & Wong, K. W. (2005). Lag synchronization of hyperchaos with application to secure communications. *Chaos, Solitons & Fractals, 23*(1), 183–193.

Li, G. H. (2007). Generalized projective synchronization between Lorenz system and Chen's system. *Chaos, Solitons & Fractals, 32*(4), 1454–1458.

Li, G. H., & Zhou, S. P. (2007). Anti-synchronization in different chaotic systems. *Chaos, Solitons & Fractals, 32*(2), 516–520.

Li, N., & Cao, J. (2015). New synchronization criteria for memristor-based networks: Adaptive control and feedback control schemes. *Neural Networks, 61,* 1–9.

Li, N., & Cao, J. (2016). Lag synchronization of memristor-based coupled neural networks via-measure. *IEEE Transactions on Neural Networks and Learning Systems, 27*(3), 686–697.

Lin, W., & He, Y. (2005). Complete synchronization of the noise-perturbed Chua's circuits. *Chaos: An Interdisciplinary Journal of Nonlinear Science, 15*(2), 023705.

Lin, W. (2008). Adaptive chaos control and synchronization in only locally Lipschitz systems. *Physics Letters A, 372*(18), 3195–3200.

Lin, D., Wang, X., Nian, F., & Zhang, Y. (2010). Dynamic fuzzy neural networks modeling and adaptive backstepping tracking control of uncertain chaotic systems. *Neurocomputing, 73*(16), 2873–2881.

Liu, Y., Takiguchi, Y., Davis, P., Aida, T., Saito, S., & Liu, J. M. (2002). Experimental observation of complete chaos synchronization in semiconductor lasers. *Applied Physics Letters, 80*(23), 4306–4308.

Liu, W., Xiao, J., Qian, X., & Yang, J. (2006). Antiphase synchronization in coupled chaotic oscillators. *Physical Review E, 73*(5), 057203.

Luo, X. S., Zhang, B., & Qin, Y. H. (2010). Controlling chaos in space-clamped FitzHugh–Nagumo neuron by adaptive passive method. *Nonlinear Analysis: Real World Applications, 11* (3), 1752–1759.

Mainieri, R., & Rehacek, J. (1999). Projective synchronization in three-dimensional chaotic systems. *Physical Review Letters, 82*(15), 3042–3045.

Mahmoud, G. M., Bountis, T., & Mahmoud, E. E. (2007). Active control and global synchronization of the complex Chen and Lü systems. *International Journal of Bifurcation and Chaos, 17*(12), 4295–4308.

Mahmoud, G. M., & Mahmoud, E. E. (2010). Complete synchronization of chaotic complex nonlinear systems with uncertain parameters. *Nonlinear Dynamics, 62*(4), 875–882.

Mathiyalagan, K., Park, J. H., & Sakthivel, R. (2015). Synchronization for delayed memristive BAM neural networks using impulsive control with random nonlinearities. *Applied Mathematics and Computation, 259,* 967–979.

Mathiyalagan, K., Anbuvithya, R., Sakthivel, R., Park, J. H., & Prakash, P. (2016). Non-fragile H∞ synchronization of memristor-based neural networks using passivity theory. *Neural Networks, 74,* 85–100.

Matsumoto, T. (1984). A Chaotic attractor from Chua's circuit. *IEEE Transactions on Circuits and Systems, 31,* 1055–1058.

Muthuswamy, B. (2010). Implementing memristor based chaotic circuits. *International Journal of Bifurcation and Chaos, 20*(05), 1335–1350.

Nbendjo, B. N., Tchoukuegno, R., & Woafo, P. (2003). Active control with delay of vibration and chaos in a double-well Duffing oscillator. *Chaos, Solitons & Fractals, 18*(2), 345–353.

Nbendjo, B. N., & Woafo, P. (2007). Active control with delay of horseshoes chaos using piezoelectric absorber on a buckled beam under parametric excitation. *Chaos, Solitons & Fractals, 32*(1), 73–79.

Pecora, L. M., & Carroll, T. L. (1990). Synchronization in chaotic systems. *Physical Review Letters, 64*(8), 821–825.

Pham, V. T., Volos, C. K., Vaidyanathan, S., Le, T. P., & Vu, V. Y. (2015). A memristor-based hyperchaotic system with hidden attractors: dynamics, synchronization and circuital emulating. *Journal of Engineering Science and Technology Review, 8*(2), 205–214.

Pikovsky, A. S., Rosenblum, M. G., Osipov, G. V., & Kurths, J. (1997). Phase synchronization of chaotic oscillators by external driving. *Physica D: Nonlinear Phenomena, 104*(3), 219–238.

Rosenblum, M. G., Pikovsky, A. S., & Kurths, J. (1996). Phase synchronization of chaotic oscillators. *Physical Review Letters, 76*(11), 1804–1807.

Rosenblum, M. G., Pikovsky, A. S., & Kurths, J. (1997). From phase to lag synchronization in coupled chaotic oscillators. *Physical Review Letters, 78*(22), 4193–4196.

Rosenblum, M., Pikovsky, A., Kurths, J., Schäfer, C., & Tass, P. A. (2001). Phase synchronization: From theory to data analysis. *Handbook of Biological Physics, 4,* 279–321.

Rulkov, N. F., Sushchik, M. M., Tsimring, L. S., & Abarbanel, H. D. (1995). Generalized synchronization of chaos in directionally coupled chaotic systems. *Physical Review E, 51*(2), 980–994.

Sarasu, P., & Sundarapandian, V. (2011). The generalized projective synchronization of hyperchaotic Lorenz and hyperchaotic Qi systems via active control. *International Journal of Software Computing, 6*(5), 216–223.

Shahverdiev, E. M., Sivaprakasam, S., & Shore, K. A. (2002). Lag synchronization in time-delayed systems. *Physics Letters A, 292*(6), 320–324.

Shin, S., Kim, K., & Kang, S. M. (2011). Memristor applications for programmable analog ICs. *IEEE Transactions on Nanotechnology, 10*(2), 266–274.

Shinriki, M., Yamato, M., & Mori, S. (1981). Multimode oscillations in a modified van der Pol oscillator containing a positive nonlinear conductance. *Proceedings of the IEEE, 69,* 394–395.

Strogatz, S. H. (1994). *Nonlinear dynamics and chaos: with applications to physics, biology, chemistry, and engineering*. Massachusetts, USA: Perseus Books.

Strukov, D., Snider, G., Stewart, G., & Williams, R. (2008). The missing memristor found. *Nature, 453,* 80–83.

Sun, J., Shen, Y., Yin, Q., & Xu, C. (2013). Compound synchronization of four memristor chaotic oscillator systems and secure communication. *Chaos: An Interdisciplinary Journal of Nonlinear Science, 23*(1), 013140.

Taherion, S., & Lai, Y. C. (1999). Observability of lag synchronization of coupled chaotic oscillators. *Physical Review E, 59*(6), R6247.

Tan, X., Zhang, J., & Yang, Y. (2003). Synchronizing chaotic systems using backstepping design. *Chaos, Solitons & Fractals, 16*(1), 37–45.

Tavazoei, M. S., & Haeri, M. (2008). Synchronization of chaotic fractional-order systems via active sliding mode controller. *Physica A: Statistical Mechanics and its Applications, 387*(1), 57–70.

Tetzlaff, R. (2014). *Memristors and memristive systems*. New York, USA: Springer.

Utkin, V. I. (1993). Sliding mode control design principles and applications to electric drives. *IEEE Transactions on Industrial Electronics, 40*(1), 23–36.

Utkin, V. I. (2004). Sliding mode control. In *Variable structure systems: from principles to implementation, IET control engineering series* (Vol. 66, pp. 3–17).

Utkin, V., Guldner, J., & Shi, J. (2009). *Sliding mode control in electro-mechanical systems*. US: CRC Press.

Vaidyanathan, S., & Rajagopal, K. (2011). Hybrid synchronization of hyperchaotic Wang-Chen and hyperchaotic Lorenz systems by active non-linear control. *International Journal of Signal System Control and Engineering Application, 4*, 55–61.

Vaidyanathan, S., & Rajagopal, K. (2012). Global chaos synchronization of hyperchaotic Pang and hyperchaotic Wang systems via adaptive control. *International Journal of Software and Computing, 7,* 28–37.

Vaidyanathan, S., & Volos, C. K. (2016a). *Advances and applications in nonlinear control systems*. Berlin, Germany: Springer.

Vaidyanathan, S., & Volos, C. K. (2016b). *Advances and applications in chaotic systems*. Berlin, Germany: Springer.

Vincent, U. E. (2008). Synchronization of identical and non-identical 4-D chaotic systems using active control. *Chaos, Solitons & Fractals, 37,* 1065–1075.

Wang, Y. W., & Guan, Z. H. (2006). Generalized synchronization of continuous chaotic system. *Chaos, Solitons & Fractals, 27*(1), 97–101.

Wang, Z. (2009). Anti-synchronization in two non-identical hyperchaotic systems with known or unknown parameters. *Communications in Nonlinear Science and Numerical Simulation, 14*(5), 2366–2372.

Wang, W., Li, L., Peng, H., Xiao, J., & Yang, Y. (2014). Synchronization control of memristor-based recurrent neural networks with perturbations. *Neural Networks, 53,* 8–14.

Wang, S., Wang, X., & Zhou, Y. (2015). A memristor-based complex Lorenz system and its modified projective synchronization. *Entropy, 17*(11), 7628–7644.

Wang, S., Wang, X., Zhou, Y., & Han, B. (2016). A memristor-based hyperchaotic complex Lü system and its adaptive complex generalized synchronization. *Entropy, 18*(2), 58.

Wedekind, I., & Parlitz, U. (2002). Synchronization and antisynchronization of chaotic power drop-outs and jump-ups of coupled semiconductor lasers. *Physical Review E, 66*(2), 026218.

Wen, S., Zeng, Z., & Huang, T. (2014a). Observer-based synchronization of memristive systems with multiple networked input and output delays. *Nonlinear Dynamics, 78*(1), 541–554.

Wen, S., Zeng, Z., Huang, T., & Zhang, Y. (2014b). Exponential adaptive lag synchronization of memristive neural networks via fuzzy method and applications in pseudorandom number generators. *IEEE Transactions on Fuzzy Systems, 22*(6), 1704–1713.

Wen, S., Huang, T., Zeng, Z., Chen, Y., & Li, P. (2015). Circuit design and exponential stabilization of memristive neural networks. *Neural Networks, 63,* 48–56.

Wen, S., Huang, T., Yu, X., Chen, M. Z., & Zeng, Z. (2016). Sliding-mode control of memristive Chua's systems via the event-based method. *IEEE Transactions on Circuits and Systems II: Express Briefs*, 1–5

Wolf, A., Swift, J. B., Swinney, H. L., & Wastano, J. A. (1985). Determining Lyapunov exponents from time series. *Physica D: Nonlinear Phenomena, 16,* 285–317.

Wu, A., Wen, S., & Zeng, Z. (2012). Synchronization control of a class of memristor-based recurrent neural networks. *Information Sciences, 183*(1), 106–116.

Wu, A., & Zeng, Z. (2013). Anti-synchronization control of a class of memristive recurrent neural networks. *Communications in Nonlinear Science and Numerical Simulation, 18*(2), 373–385.

Xie, Q., Chen, G., & Bollt, E. M. (2002). Hybrid chaos synchronization and its application in information processing. *Mathematical and Computer Modelling, 35*(1), 145–163.

Yan, J., & Li, C. (2005). Generalized projective synchronization of a unified chaotic system. *Chaos, Solitons & Fractals, 26*(4), 1119–1124.

Yang, S., & Duan, C. (1998). Generalized synchronization in chaotic systems. *Chaos, Solitons & Fractals, 9*(10), 1703–1707.

Yang, X., Cao, J., & Yu, W. (2014). Exponential synchronization of memristive Cohen-Grossberg neural networks with mixed delays. *Cognitive Neurodynamics, 8*(3), 239–249.

Yang, S., Guo, Z., & Wang, J. (2015). Robust synchronization of multiple memristive neural networks with uncertain parameters via nonlinear coupling. *IEEE Transactions on Systems, Man, and Cybernetics: Systems, 45*(7), 1077–1086.

Yau, H. T. (2004). Design of adaptive sliding mode controller for chaos synchronization with uncertainties. *Chaos, Solitons & Fractals, 22*(2), 341–347.

Yassen, M. T. (2006). Chaos control of chaotic dynamical systems using backstepping design. *Chaos, Solitons & Fractals, 27*(2), 537–548.

Yassen, M. T. (2007). Controlling, synchronization and tracking chaotic Liu system using active backstepping design. *Physics Letters A, 360*(4), 582–587.

Young, K. D., Utkin, V. I., & Ozguner, U. (1999). A control engineer's guide to sliding mode control. *IEEE Transactions on Control Systems Technology, 7*(3), 328–342.

Zhang, Y., & Sun, J. (2004). Chaotic synchronization and anti-synchronization based on suitable separation. *Physics Letters A, 330*(6), 442–447.

Zhang, D., & Xu, J. (2010). Projective synchronization of different chaotic time-delayed neural networks based on integral sliding mode controller. *Applied Mathematics and Computation, 217*(1), 164–174.

Zhang, G., Shen, Y., & Wang, L. (2013). Global anti-synchronization of a class of chaotic memristive neural networks with time-varying delays. *Neural Networks, 46*, 1–8.

Zhang, G., & Shen, Y. (2014). Exponential synchronization of delayed memristor-based chaotic neural networks via periodically intermittent control. *Neural Networks, 55,* 1–10.

Zheng, J. (2011). A simple universal adaptive feedback controller for chaos and hyperchaos control. *Computers & Mathematics with Applications, 61*(8), 2000–2004.

Canonic Memristor: Bipolar Electrical Switching in Metal-Metal Contacts

Gaurav Gandhi and Varun Aggarwal

Abstract In the work, by uncovering hitherto unknown electrical properties of a set of coherer and autocoherer, we find that extremely simple devices show memristive properties. Coherer and the auto-coherer are electrically-controllable state-dependent resistors, the state variable being the maximum current flown through the device. Bipolar switching in these devices, wherein the device can be programmed (electrically) to an older higher resistance state has also been observed. This shows that simple setting such as metallic contacts show memristive properties and constitute the canonic implementation of a memristor.

Keywords Memristor · Memristive system · Coherer · Resistive RAM

1 Introduction

Leon Chua defines a memristor as any two-terminal electronic device that is devoid of an internal power-source and is capable of switching between two resistance states upon application of an appropriate voltage or current signal that can be sensed by applying a relatively much smaller sensing signal (Chua 2011). A pinched hysteresis loop in the voltage versus current characteristics of the device serves as the fingerprint for memristors. Despite the simplicity of symmetry argument that predicts the existence of memristor (Chua 2011; Strukov et al. 2008), no simple physical device behaving as a memristor has been observed so far and thus considered to be non-existent (Pershin and Ventra 2011). Current memristor implementations use specialized materials such as transition metal oxides, chalcogenides, perovskites, oxides with valence defects, or a combination of an inert and an electrochemically active electrode.

G. Gandhi (✉) · V. Aggarwal
mLabs, New Delhi, India
e-mail: gaurav@mlabs.in

V. Aggarwal
e-mail: varun@mlabs.in

© Springer International Publishing AG 2017

S. Vaidyanathan and C. Volos (eds.), *Advances in Memristors, Memristive Devices and Systems*, Studies in Computational Intelligence 701,
DOI 10.1007/978-3-319-51724-7_11

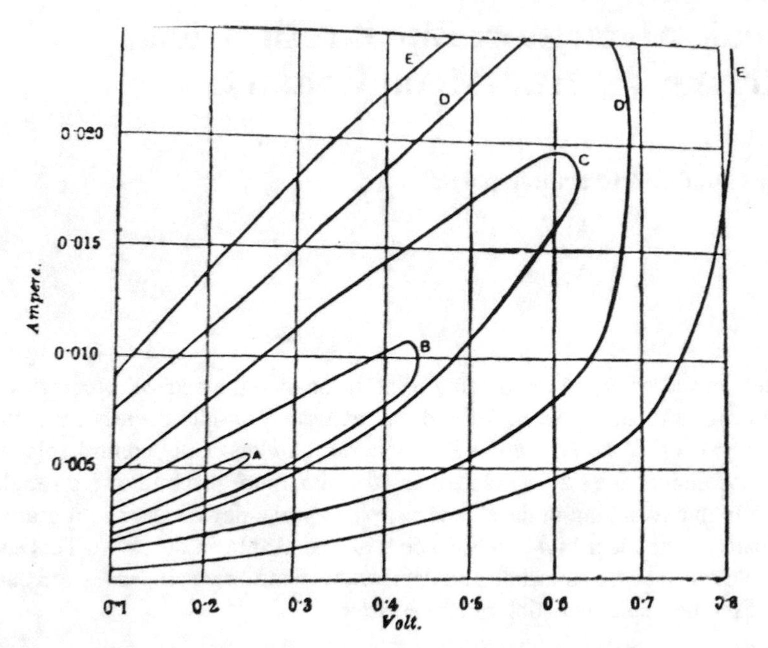

Fig. 55. **Cyclic Curves showing Conductivity *Hysteresis*. In each curve the right half is due to increasing, the left half to decreasing E. M. F.**

Fig. 1 Bose's observation of pinched hysteresis in current (*vertical axis*) and voltage (*horizontal axis*) for iron filing coherer. Interestingly, this reference has been missed by almost all the papers on memristors

On the other hand, coherer, invented by Edouard Branly (Dilhac 2009; Lodge 1897; Falcon and Castaing 2010) in the 19th century, in its many embodiments such as ball bearings, metallic filings (also referred to as granular media) in a tube or a point-contact exhibits an initial high-resistance state and coheres to a low-resistance state on the arrival of radio waves. The device attains its original resistance state on being tapped mechanically. The first electrically reset-able coherer, comprising a metal-mercury interface and named as an auto-coherer, (Bondyopadhyay 1998; Bose 1899a, 1901, 1904; Gandhi et al. 2014) did not require tapping and resets in the absence of radio waves. Although Bose observed that coherers demonstrate a pinched hysteresis I-V curve in the first quadrant (arguably the first such observation; Refer Fig. 1) (Bose 1901) and exhibit multiple stable resistance-states, he could not establish a systematic way to electrically reverse the diminution of resistance (Bose 1899b).[1]

Among several competing theories for explaining the coherer behavior, such as joule heating, molecular rearrangement, Seebeck and Peltier Effect (Béquin and Tournat 2010; Bose 1901; Eccles 1901, 1912; Lodge 1901), the most popular theory

[1]Cat's whisker was the first metal-semiconductor point contact device patented by JC Bose and was actively used in early radio research.

was that of current-induced heating resulting in the welding of metal-metal contacts that led to diminution of resistance. For the auto-coherer, Eccles (1909) postulated that current leads to the heating of oxide at the interface contacts and the change in resistance is a function of the temperature of the oxide. His thermistor equation for the said behavior is, in principle, the same as that prescribed by Chua and Kang for a thermistor (Chua and Kang 1976), and satisfies the conditions of Chua's memristive one-port (Chua 1971). The equation proposed by Eccles is being reproduced here:

$$\frac{d\theta}{dt} = k\rho c^2 - m\theta \tag{1}$$

$$L\frac{dc}{dt} + (r + \rho)c = \epsilon \tag{2}$$

$$\rho = \rho_0(1 + \alpha\theta) \tag{3}$$

where c represents the current, ρ represents the resistance of oxide, ϵ represents the voltage, θ is the temperature of the oxide and other variables pertain to the setup mentioned in Eccles (1909). L and r refer to the resistance and inductance in series with the coherer.

Thus the existence of electrically-controllable multiple resistance-states, and the possibility of a memristive constitutive relationship was known over a century ago. Though no one (including Eccles and Bose) observed a pinched hysteresis in both the quadrant. In works related to coherer in the last decade, its bi-stability has been reported and its multi-stable behavior has been confirmed (Béquin and Tournat 2010; Falcon and Castaing 2005). A thermal mechanism, similar to numerous others proposed a century back, has been postulated to explain the resistance change. All these studies affirm the unidirectionality of the resistance value (which fatigues with time) and propose no method to electrically recover the older, higher resistance states of the device. On the other hand, autocoherer has been shown to exhibit diode-like rectifying properties (Groenhaug 2001; Philips 1998). In the present work we have established, by uncovering hitherto unknown electrical properties of a set of coherer and autocoherer, that extremely simple devices show memristive properties (Fig. 2). We have found that the coherer and the auto-coherer are electrically-controllable state-dependent resistors, the state variable being the maximum current flown through the device. We have, for the first time observed bipolar switching in these devices, wherein the device can indeed be programmed (electrically) to an older higher resistance state. The state-map of the resistance of the device versus current is different for the two directions of the current (Refer Fig. 3d), which allows to write and erase it as a memory. The programmed resistance of the device can be read by another characteristic signal of small amplitude. Analogous to the phenomena of a wire showing resistive properties, a coil being inductive, and a set of conducting plates separated by a dielectric exhibiting capacitance, we show that two convex metallic surfaces in contact are memristive in nature and work as a fully-functional resistive RAM (Refer Fig. 4) (Gandhi et al. 2013).

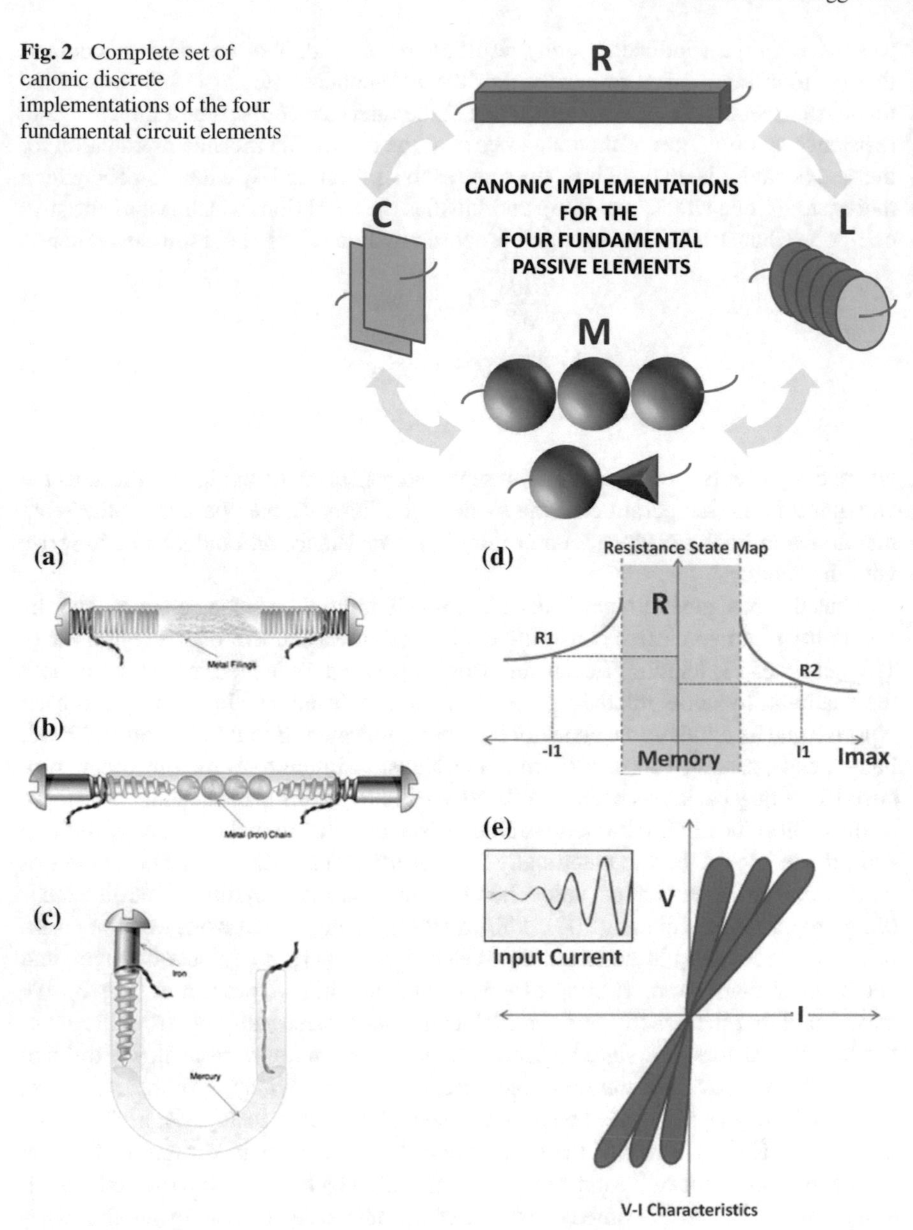

Fig. 2 Complete set of canonic discrete implementations of the four fundamental circuit elements

Fig. 3 *Left* Various embodiments of coherer used for experimentation. **a** Iron Filing Coherer (IFC), **b** Iron Chain Coherer (ICC), **c** Iron Mercury Coherer (IMC). *Right* **d** Resistance State Map for the device. Here *horizontal axis* refers to maximum current that has flowed through the device while *vertical axis* is the resistance of the device. **e** Current-voltage characteristics of the device for a current-mode sine wave signal of increasing amplitude. The device shows the famous pinched hysteresis loops and various possible current-voltage values for the same current

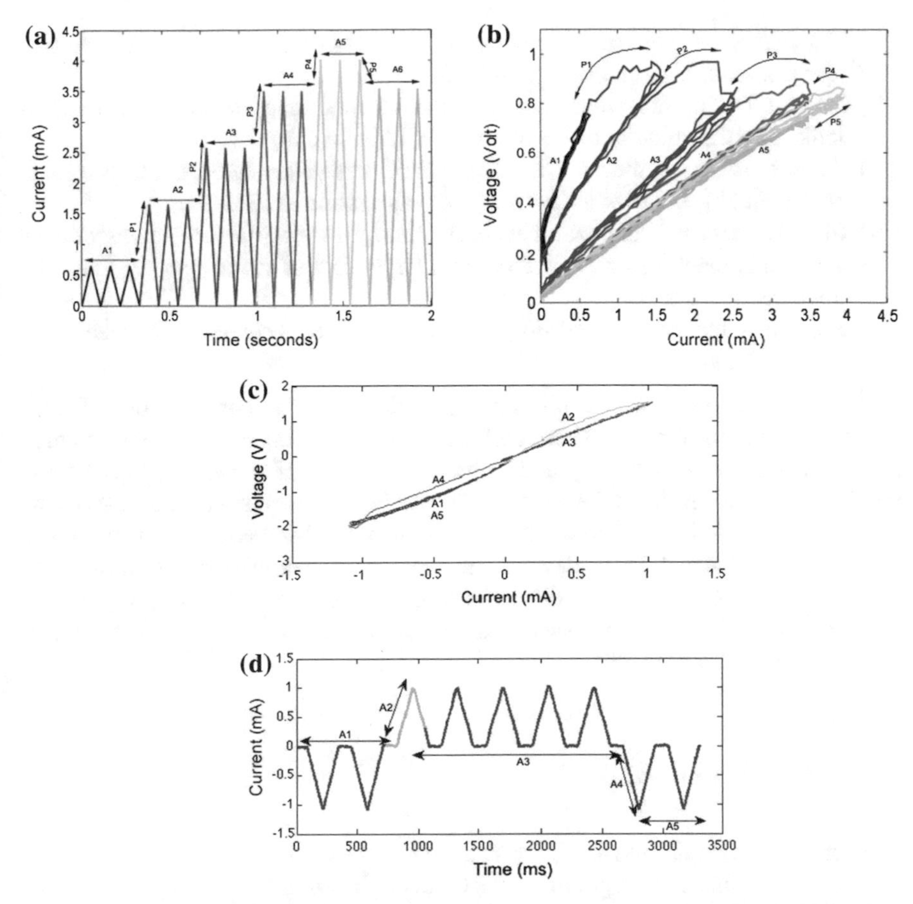

Fig. 4 Device behavior as a state-dependent resistance. **a** Input current versus time and **b** current-voltage plot. After configuring the device in the non-linear high-resistance mode, an input current pulse with varying amplitude is applied across it. It is observed that the maximum voltage across the device does not cross a threshold voltage, V_{th}. Bistable RRAM behavior, **c** input current versus time, **d** voltage across device versus input current. One clearly observes pinched hysteresis loop for Iron Filing Coherer

It is worth discussing what causes the resistance switching. The cause could be existence of certain impurities or oxide at the interface or merely by the geometry of the interface. We note, however, that no Metal-Insulator-Metal system containing just oxide and iron, or those built in a macro-dimension, has been reported to show memristive properties. We analytically deduce that the switching in our devices cannot be due to the different mechanisms observed in oxide-based bipolar memristors. Based on our material setup and experimental observations, our best understanding is that the resistance switching is caused by electric-field induced polarization at the interface of the metals (which may or may not contain impurities). This is discussed in detail in a later section. The paper makes the following contributions:

(i) Completes the entire set of canonic implementations of all the four known passive elements of circuit theory (Fig. 2),
(ii) Reports for the first time bipolar switching in simple metallic constructions indicating the ubiquity of the memristive phenomena,
(iii) Argue that thermal mechanism of resistance change in metallic contacts is inadequate and hypothesize a electric-field polarization as its cause,
(iv) Shows that memristor phenomena is not limited to specific materials assembled at small geometries, but is present in a large class of metals put together as a point contact and
(v) Provide an inexpensive and simple memristor for widespread experimentation, hitherto impossible.

This paper is organized as follows: Sect. 2 describes the constructions of three embodiments of the devices. These include those with a point contact between metals, a granular media assembly and a third comprising of a metal in liquid form. Section 3 describes in detail the electrical properties of these devices and their behavior under different electrical stimulations. Based on the observed behavior, we postulate an electrical model for the devices and identify the state-variable controlling the resistance change. In Sect. 4, we discuss the implication of our observations, hypothesize the physical mechanism governing the behavior of the devices and compare it with other memristor devices.

2 Materials and Methods

The current section discusses the construction of the devices which can be accomplished in any simple undergraduate electrical engineering lab. We replicated three embodiments of the coherer and autocoherer: an Iron Filing Coherer (IFC), an Iron Chain Coherer (ICC), and an Iron Mercury Coherer (IMC) (see Fig. 3a–c).[2]

The first embodiment, namely, Iron Filing Coherer (IFC), consists of a tube containing closely-packed iron filings with electrodes in contact with the metal filings at the two ends of the tube. In the second embodiment, called Iron Chain Coherer (ICC), iron filings are replaced by a chain (linear assembly) of iron beads and the third embodiment is an embodiment of the self-recovering coherer consisting of a U-tube filled with mercury forming contact with an iron screw on one side. In the third embodiment, henceforth referred as Iron Mercury Coherer (IMC), one electrode is connected to an iron screw, whereas the other dips into mercury on the other side of the U-tube. Depending on the packing density (IFC), pressure applied (ICC) and contact area (IMC), the devices show a continuum of states between a nonlinear high-resistance state and a more linear low-resistance state. The next section discusses the electrical behavior exhibited by the three devices.

[2]The experiment was repeated with several metals, including aluminum and magnesium flakes and nickel and zinc-coated ball bearings. Qualitatively similar results as reported herein were observed in all these experiments.

3 Experimental Results

These devices were activated by different current-mode input signals in their non-linear mode, and their transient behavior was recorded. We found that the three devices show similar qualitative behavior and that the mercury-iron system does not function as a diode, as previously reported (Groenhaug 2001; Philips 1998), but exhibits state-dependent resistance. All the observed behavior is common to the three devices. We have found that the devices exhibit three distinct behaviors: Cohering action, multi-stable memristive behavior, and bistable resistive RAM behavior.

3.1 Cohering Action

For any input current leading to a voltage below a specific threshold voltage, V_{th}, the devices exhibit a high non-linear resistance and may be used for rectification. Whereas IMC readily shows a moderate non-linear resistance that can be used for demodulation, IFC and ICC require considerable adjustment to do so. Due to this, only the IMC has historically been used for demodulation. In this region, the device remembers the resistance it had earlier, and continues to exhibit the same. We call this region as the memory state.

At a current higher than I_{th}, corresponding to a voltage V_{th}, the resistance of the device falls sharply (Refer P1 transition in Fig. 4), and the device exhibits lower conductance. Once the device takes this new state, it maintains the said resistance on excitation by current values below I_{th} as well. This is the well-known cohering behavior used for detecting electromagnetic waves. The device cannot be reset electrically to a resistance less than that shown at A2 (Refer Fig. 4). Contrary to earlier observations, this behavior is also exhibited by IMC Philips (1998).

3.2 Multistable Memristive Behavior

Once cohered, the device exhibits a state-dependent resistance, the state variable being the maximum current (I_{max}), i.e. $R_t = f([I_{max}]_{0-t})$. As the device is exposed to pulses of subsequently larger peak current (Refer Fig. 4[3]), it sets itself to new resistance values. The resistance remains non-linear, nonetheless. The maximum voltage across the device remains practically constant at V_{th}. This behavior is akin to that

---

[3]Note that the resistance changes appreciably only when the maximum current through the device has changed. This can be seen through color correspondence, where each color shows a new stable resistance-state and the resistance transitions are marked by the first pulse of higher amplitude: P1, P2, P3, P4 and P5. In case the maximum current passed through the device does not change, the resistance feebly oscillates around the same value, as seen in the region of A1, A2, A3, A4 and A5. Furthermore, one may observe that the resistance remains fixed even when the amplitude of the pulse is decreased (A6), since the maximum current has not changed.

of a diode, but unlike a diode the device remembers its changed resistance when taken to lower voltage levels. For input current pulses of same or lower amplitude than the maximum current experienced, the device shows hysteresis loops around the already-achieved resistance value, with small oscillations. In Béquin and Tournat (2010), some of these behaviors have been observed for ICC.

3.3 Bistable Resistive RAM

We have found that the resistance of the device is a function of the magnitude of I_{max} for either directions of current, but with a quantitatively different state-map, making it behave as a resistive RAM. This can be mathematically stated as:

Let

$$R_{p1} = f(magnitude([I_{max+}]_{0-t})) = I_1), \tag{4}$$

$$R_{n1} = f(magnitude([I_{max-}]_{0-t})) = I_1), \tag{5}$$

$$\implies R_{p1} \neq R_{n1} \tag{6}$$

where R_{p1} is the resistance of the device when activated by a maximum current of I_1 in positive direction, and R_{n1} is the resistance when activated by a maximum current of I_1 in the negative direction. $f(magnitude([I_{max+}]_{0-t}))$ implies the maximum current the device has experienced between time = 0 and time = t (Refer Fig. 3).

When activated by any two-sided current input, the device gets programmed into one state in the positive cycle, and a different state in the negative cycle. It keeps oscillating between these two stable states, forming the famous eight-shaped pinched hysteresis loop in its V-I characteristics (Refer Fig. 4[4]). It has been established that If it is pinched, it is memristive. Pinched hysteresis loop is the fingerprint of a memristor (Kim et al. 2012).

By using various stimuli with different maximum amplitudes on either sides, the device can be programmed to function in multiple stable resistance-states and move between them. When used as a resistive RAM, the memory can be read in the "memory" state by providing an excitation of a small amplitude. This fulfill the conditions of Chua's definition of memristor, and qualifies the century-old coherer as a canonical implementation of a memristor.

---

[4]It is worth noting that the transition in resistance value happens only when the polarity of the current is changed. For other pulses, the resistance remains constant. The device oscillates between two stable resistance-states for the same maximum current in opposite directions. It is evident by looking at regions depicted by A1 to A5 that the change in resistance happens at the first pulse of the transition. One may also note that these observations show recovery of resistance to a higher resistance state: A5 resistance is higher than A4 resistance. These results can be reproduced by careful experimentation for all the three devices.

4 Discussion

We have shown, through these results, that the century old coherer and auto-coherer function as a multi-state resistance RAM and is thus the canonic implementation of the elusive memristor. It intrigued the science of that era as much as memristor is exciting the scientists of the present day (Prodromakis et al. 2012). The present work shows that one does not require specific material or precise construction to implement memristors. It is a natural property of metallic point contacts. Note that there is another component called memistor which is an entirely different component and must not either be confused with coherer or memristor. It is rather an ill posed 3 terminal device (Kim and Adhikari 2012; Vaidyanathan and Volos 2016; Xia et al. 2011).

There are certain differences between the behavior of coherers and other present day memristors. Unlike Williams et al. memristor (Strukov et al. 2008), our devices do not behave as a charge-flux based memristor. Irrespective of the increase or decrease of flux, their resistance does not change till the maximum current or current polarity changes. Our device have similarities in behavior (Jo and Lu 2008; Kim et al. 2010) and construction (Kim et al. 2009) to that of some other memristors recently fabricated at nano-scale. However, none of these recent memristors have reported observation of multiple resistance states or dependence on I_{max}.

The question worth discussing is whether the resistive switching mechanism is due to existence of oxide at the interface of the metals. Our preliminary experiments with polished gold balls showed the said behavior, which indicates (but does not rule out) that the observations are not due to presence of oxide. From an analytical standpoint, the construction and behavior of our device doesn't fit those observed in oxide based memristors. The construction and mechanism of operation of oxide based memristors is discussed in detail in the review by Waser et al. (2009). One class of oxide based memristors switch unipolarly due to formation and melting of filaments thermally. This is similar to the explanation provided in coherer literature (Béquin and Tournat 2010; Falcon and Castaing 2010). Our memristor has bipolar switching and cannot be explained by a thermal process which is independent of current direction. Only the initial cohering action, akin to electroforming step reported in literature, may reasonably be explained by a thermal heating process.

Among bipolar oxide-based memristors, one class (Valence Change Mechanism) uses specific transition metal oxides or those with defects, whereas the other class has dissimilar electrodes (one active and one counter electrode) on the two sides of the oxides (or an electrolyte). In the latter case, the difference in the properties of the two electrodes leads to dependence on current direction. Our memristor has no explicitly introduced vacancy defects at the interface, doesn't require transition metal oxides and works perfectly well with the same metal across all contacts. Thus, its construction and behavior, put together, do not resemble any oxide based memristor configuration and behavior.

On the other hand, ferroelectric RAM containing a perovskite layer and nanoparticle assemblies (Kim et al. 2009) are symmetric, and yet show bipolar resistance-switching caused by electric field induced polarization. We believe that the behavior of our device is similar and a result of polarization at the contacts formed between the metals. It is still open to investigate whether this happens due to the geometry at the contact or due to impurities. The same requires to be investigated through material analysis and microscopic studies.

Our results show that bipolar switching can be observed in a large class of metals by a simple construction in form of a point-contact or granular media. It does not require complex construction, particular materials or small geometries. The signature of all our devices is an imperfect metal-metal contact and the physical mechanism for the observed behavior needs to be further studied. That the electrical behavior of these simple, naturally-occurring physical constructs can be modeled by a memristor, but not the other three passive elements, is an indication of its fundamental nature. By providing the canonic physical implementation for memristor, the present work not only fills an important gap in the study of switching devices, but also brings them into the realm of immediate practical use and implementation.

Acknowledgements Authors would like to thank Prof. Leon Chua for his continuous mentoring.

References

Béquin, P., & Tournat, V. (2010). Electrical conduction and joule effect in one-dimensional chains of metallic beads: Hysteresis under cycling dc currents and influence of electromagnetic pulses. *Granular Matter*, *12*(4), 375–385.

Bondyopadhyay, P. (1998). Sir J C Bose diode detector received Marconi's first transatlantic wireless signal of december 1901 (the Italian navy coherer scandal revisited). *Proceedings of the IEEE*, *86*(1), 259–285.

Bose, J. (1899a). On a self-recovering coherer and the study of the cohering action of different metals. *Proceedings of the Royal Society of London*, *65*(413–422), 166.

Bose, J. (1899b). On electric touch and the molecular changes produced in matter by electric waves. *Proceedings of the Royal Society of London*, *66*(424–433), 452.

Bose, J. (1901). On the change of conductivity of metallic particles under cyclic electromotive variation. *Originally presented to the British Association at Glasgow, September*.

Bose, J. (1904). Patent USA 755,840.

Chua, L. (1971). Memristor-the missing circuit element. *IEEE Transactions on Circuit Theory*, *18*(5), 507–519.

Chua, L. (2011). Resistance switching memories are memristors. *Applied Physics A: Materials Science and Processing*, *102*(4), 765–783.

Chua, L., & Kang, S. (1976). Memristive devices and systems. *Proceedings of the IEEE*, *64*(2), 209–223.

Dilhac, J. (2009). Edouard Branly, the coherer, and the Branly effect. *IEEE Communications Magazine*, *47*(9), 20–22.

Eccles, W. (1901). *On Filing Coherer*. PhD thesis, University of London.

Eccles, W. (1909). On coherers. *Proceedings of the Physical Society of London*, *22*, 289.

Eccles, W. (1912). Electrothermal phenomena at the contact of two conductors, with a theory of a class of radiotelegraph detectors. *Proceedings of the Physical Society of London*, *25*, 273.

Falcon, E., & Castaing, B. (2005). Electrical conductivity in granular media and branly's coherer: A simple experiment. *American Journal of Physics*, *73*, 302.
Falcon, E., & Castaing, B. (2010). El Efecto Branly. *Investigación y ciencia*, *404*, 80–86.
Gandhi, G., Aggarwal, V., & Chua, L. (2014). Coherer is the elusive memristor. In *2014 IEEE International Symposium on Circuits and Systems (ISCAS)* (pp. 2245–2248). IEEE.
Gandhi, G., Aggarwal, V., & Chua, L. O. (2013). The first radios were made using memristors!. *IEEE Circuits and Systems Magazine*, *13*(2), 8–16.
Groenhaug, K. (2001). Experiments with a replica of the bose detector. *IEEE GLOBECOM, 1*. http://www.home.online.no/~kgroenha/Marconi.pdf.
Jo, S., & Lu, W. (2008). CMOS compatible nanoscale nonvolatile resistance switching memory. *Nano Letters*, *8*(2), 392–397.
Kim, H., & Adhikari, S. (2012). Memistor is not memristor [express letters]. *IEEE Circuits and Systems Magazine*, *12*(1), 75–78.
Kim, H., Sah, M., & Adhikari, S. (2012). Pinched hysteresis loops is the fingerprint of memristive devices. arXiv preprint arXiv:1202.2437.
Kim, K., Jo, S., Gaba, S., & Lu, W. (2010). Nanoscale resistive memory with intrinsic diode characteristics and long endurance. *Applied Physics Letters, 96*.
Kim, T., Jang, E., Lee, N., Choi, D., Lee, K., Jang, J., et al. (2009). Nanoparticle assemblies as memristors. *Nano Letters*, *9*(6), 2229–2233.
Lodge, O. (1897). The history of the coherer principle. *The Electrician*, *40*, 86–91.
Lodge, O. (1901). Patent USA, 674,846.
Pershin, Y., & Ventra, M. (2011). Teaching memory circuit elements via experiment-based learning. *IEEE Circuits and Systems Magazine*, *12*(1), 64–74.
Philips, V. (1998). The italian navy coherer affair: A turn of the century scandal. *Reproduced in Proceedings of IEEE*, *86*(1).
Prodromakis, T., Toumazou, C., & Chua, L. (2012). Two centuries of memristors. *Nature Materials*, *11*(6), 478–481.
Strukov, D., Snider, G., Stewart, D., & Williams, R. (2008). The missing memristor found. *Nature*, *453*(7191), 80–83.
Vaidyanathan, S., & Volos, C. (2016). *Advances and applications in nonlinear control systems* (Vol. 635). Springer.
Waser, R., Dittmann, R., Staikov, G., & Szot, K. (2009). Redox-based resistive switching memories-nanoionic mechanisms, prospects, and challenges. *Advanced Materials*, *21*(25–26), 2632–2663.
Xia, Q., Pickett, M., Yang, J., Li, X., Wu, W., Medeiros-Ribeiro, G., et al. (2011). Two-and three-terminal resistive switches: Nanometer-scale memristors and memistors. *Advanced Functional Materials*, *21*(14), 2660–2665.

Distributed In-Memory Computing on Binary Memristor-Crossbar for Machine Learning

Hao Yu, Leibin Ni and Hantao Huang

Abstract The recent emerging memristor can provide non-volatile memory storage but also intrinsic computing for matrix-vector multiplication, which is ideal for low-power and high-throughput data analytics accelerator performed in memory. However, the existing memristor-crossbar based computing is mainly assumed as a multi-level analog computing, whose result is sensitive to process non-uniformity as well as additional overhead from AD-conversion and I/O. In this chapter, we explore the matrix-vector multiplication accelerator on a binary memristor-crossbar with adaptive 1-bit-comparator based parallel conversion. Moreover, a distributed in-memory computing architecture is also developed with according control protocol. Both memory array and logic accelerator are implemented on the binary memristor-crossbar, where logic-memory pair can be distributed with protocol of control bus. Experiment results have shown that compared to the analog memristor-crossbar, the proposed binary memristor-crossbar can achieve significant area-saving with better calculation accuracy. Moreover, significant speedup can be achieved for matrix-vector multiplication in the neuron-network based machine learning such that the overall training and testing time can be both reduced respectively. In addition, large energy saving can be also achieved when compared to the traditional CMOS-based out-of-memory computing architecture.

H. Yu (✉) · L. Ni · H. Huang
School of Electrical and Electronic Engineering, Nanyang Technological University, Singapore, Singapore
e-mail: haoyu@ntu.edu.sg

L. Ni
e-mail: nile0001@e.ntu.edu.sg

H. Huang
e-mail: hhuang013@e.ntu.edu.sg

© Springer International Publishing AG 2017

S. Vaidyanathan and C. Volos (eds.), *Advances in Memristors, Memristive Devices and Systems*, Studies in Computational Intelligence 701,
DOI 10.1007/978-3-319-51724-7_12

1 Introduction

Future cyber-physical system requires efficient real-time data analytics (Kouzes et al. 2009; Wolpert 1996; Hinton et al. 2006; Müller et al. 2008; Glorot and Bengio 2010) with applications in robotics, brain-computer interface as well as autonomous vehicles. The recent works in Huang et al. (2006), Coates et al. (2011) have shown a great potential for machine learning with significant reduced training time for real-time data analytics.

Hardware-based accelerator is currently practiced to assist machine learning. In traditional hardware accelerator, there is intensive data migration between memory and logic (Kumar et al. 2014; Park et al. 2013) caused both bandwidth and power walls. Therefore, for data-oriented computation, it is beneficial to place logic accelerators as close as possible to the memory to alleviate the I/O communication overhead (Wang et al. 2015). The cell-level in-memory computing is proposed in Matsunaga et al. (2009), where simple logic circuits are embedded among memory arrays. Nevertheless, the according in-memory logic that is equipped in memory cannot be made for complex logic function, and also the utilization efficiency is low as logic cannot be shared among memory cells. In addition, there is significant memory leakage power in CMOS based technology.

Emerging memristor (Akinaga and Shima 2010; Kim et al. 2011; Chua 1971; Williams 2008; Strukov et al. 2008; Shang et al. 2012; Fei et al. 2012) has shown great potential to be the solution for data-intensive applications. Besides the minimized leakage power due to non-volatility, memristor in crossbar structure has been exploited as computational elements (Kim et al. 2011; Liu et al. 2015). As such, both memory and logic components can be realized in a power- and area- efficient manner. More importantly, it can provide a true in-memory logic-memory integration architecture without using I/Os. Nevertheless, the previous memristor-crossbar based computation is mainly based on an analog fashion with multi-level values (Kim et al. 2012) or Spike Timing Dependent Plasticity (STDP) (Lu et al. 2011). Though it improves computation capacity, the serious non-uniformity of memristor-crossbar at nano-scale limits its wide applications for accurate and repeated data analytics. Moreover, there is significant power consumption from additional AD-conversion and I/Os mentioned in Lu et al. (2011).

In this chapter, we propose a distributed in-memory accelerator. Both computational energy efficiency and robustness are greatly improved by a binary memristor-crossbar for memory and logic units. The memory arrays are paired with the in-memory logic accelerators in a distributed fashion, operated with a protocol of control bus for each memory-logic pair. Moreover, different from the multi-leveled analog memristor-crossbar, a three-step digitalized memristor-crossbar is proposed in this chapter to perform a digital matrix-vector multiplication. In addition, a 3D CMOS-memristor accelerator is also proposed for machine learning. The area overhead can be reduced due to the 3D architecture. CMOS based operations can be implemented after the memristor-crossbar process in such architecture.

2 Background of Machine Learning

The current data analytics is mainly based on machine learning algorithm and computational intelligence to build a model to correlate input data with targeted output (Vaidyanathan and Volos 2016a, b). Features extraction are also performed to extract the key information for data analytics in Neural Network. Neural network is the common model to build (Haykin et al. 2009), and usually has two computational phases: training and testing. In the training phase, the weight coefficients of the neural network model are determined by minimizing the error between the trial and the targeted using the training input data. In the testing phase, the neural network with determined coefficients is utilized for the classification of the new testing data.

However, the input data may be in high dimension with redundant information. To facilitate the training, feature extraction is usually needed performed to represent the characteristic data with redundancy or dimension reduction.

To speed-up the training process, we tackle this challenge from two perspectives. Firstly, we propose a general incremental machine learning architecture with minimal tuning of parameters as shown in Fig. 1, which is mainly based on incremental least-squares solution. Secondly, we analyze the key complexity of each learning step and propose a hardware friendly algorithm to explore the parallelism with minimized hardware operational complexity.

2.1 *Feature Extraction*

In general, the feature of original data $\mathbf{X}$ can be extracted by projection,

$$\mathbf{X}' = \mathbf{R} \cdot \mathbf{X} \tag{1}$$

where $\mathbf{X}'$ is the extracted feature. The projection matrix $\mathbf{R}$ can be found with the use of principal/singular components, random embedding or convolution (Wold et al.

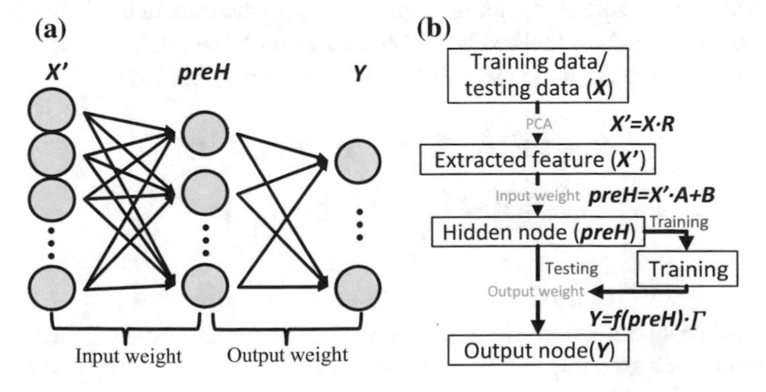

Fig. 1 **a** Single-layer neural network. **b** General flow of neural network training and testing

1987). Matrix **R** is computed off-line and used for dimension reductions. We can treat **R** as a new basis to represent columns of **X**; and remove those small values to minimize the total squared reconstruction error by

$$||\mathbf{X}' - \mathbf{R} \cdot \mathbf{X}||_2 \tag{2}$$

One can observe significant matrix-vector multiplications during the feature extraction as shown in (1).

2.2 Neural Network Based Learning

After feature extraction, one can perform various machine learning algorithms (Suykens and Vandewalle 1999; LeCun et al. 2012; Huang et al. 2006) for data analytics. As shown in Fig. 1 for a typical neural network model, one needs to determine the network weights from training and then practice testing. We use n to represent the number of features with training input $X_f \in R^{N \times n}$. n is the training data size. The extracted feature will be input to the neural network with following relationship for the first layer output **preH**:

$$\mathbf{preH} = \mathbf{X}_f \mathbf{A} + \mathbf{B},\ \mathbf{H} = g(\mathbf{preH}) = \frac{1}{1 + e^{-\mathbf{preH}}} \tag{3}$$

where $\mathbf{A} \in \mathbb{R}^{n \times L}$ and $\mathbf{B} \in \mathbb{R}^{N \times L}$ are randomly generated input weight and bias formed by a_{ij} and b_{ij} between $[-1, 1]$; **H** is the hidden-layer output matrix generated from the Sigmoid function $g(\cdot)$ for activation.

The training of neural network is to minimize error with an objective function below

$$\min_{\Gamma} ||\mathbf{H}\boldsymbol{\Gamma} - \mathbf{T}||_2^2 + \eta|||\Gamma||_2^2 \tag{4}$$

where η is the regularized parameter and **T** is the label of training data.

One can solve (4) either by iterative backward propagation method (Werbos 1990) or direct L2-norm solver method for least-squares problem (Huang et al. 2006). The output weight can be obtained as $||\tilde{\mathbf{H}}\Gamma - \tilde{\mathbf{T}}||$, and can be solved as

$$\begin{gathered} \boldsymbol{\Gamma} = (\tilde{\mathbf{H}}^T\tilde{\mathbf{H}})^{-1}\tilde{\mathbf{H}}^T\tilde{\mathbf{T}} = (\mathbf{H}^T\mathbf{H} + \eta\mathbf{I})^{-1}\mathbf{H}^T\mathbf{T} \\ \textit{where}\ \tilde{\mathbf{H}} = \begin{pmatrix} \mathbf{H} \\ \sqrt{\eta}\mathbf{I} \end{pmatrix},\ \tilde{\mathbf{T}} = \begin{pmatrix} \mathbf{T} \\ \mathbf{0} \end{pmatrix} \end{gathered} \tag{5}$$

Here $\tilde{\mathbf{H}} \in \mathbb{R}^{(N+L) \times L}$ is formed based on **H** and **I**. For matrix **Γ**, it is the solution of a least-square problem, where we adopt Cholesky decomposition to solve it (Higham 2009). We have also analyzed the major computations of Cholesky decomposition for least-square problem, which will be discussed in Sect. 5.

As a result, in the testing phase, output node $\mathbf{Y}$ is calculated by already determined hidden node value and output weight value as

$$\mathbf{Y} = \mathbf{H} \cdot \boldsymbol{\Gamma} \tag{6}$$

The index of the maximum value in $\mathbf{Y}$ represents the class that the test data belongs to.

Based on the computation analysis on feature extraction and neural network, we can observe that matrix-vector multiplication is the dominant operation as shown in (1), (5) and (6). As such, a hardware accelerator to facilitate the matrix-vector multiplication is indeed the critical requirement for the efficient machine-learning based data analytics.

2.3 *Incremental Least-Square Solver Based Learning*

The objective function (4) is a least-squares problem and can be solved using backwards propagations (BP) or direct solution based on matrix operations. Since our target is to have incremental learning with latest training samples, iterative gradient based backwards propagation is slow comparing to pseudo-inverse solutions (Huang et al. 2006), therefore, BP will not be elaborated in details. In fact, as discussed in the next sections, our proposed 3D multi-layer CMOS-memristor architecture can accelerate the matrix-vector multiplications, which will also benefit BP based neural network training method.

Equation 5 shows how to obtain the output weight $\boldsymbol{\Gamma}$. The symmetric positive definite matrix $\tilde{\mathbf{H}}^T\tilde{\mathbf{H}}$ is decomposed into $\mathbf{QPQ^T}$. $\mathbf{Q}$ is a lower triangular matrix with diagonal elements $q_{ii} = 1$ and $\mathbf{P}$ is a positive diagonal matrix. Such method can maintain the same memory space as Cholesky factorization but need not perform square root extraction, as the square root of $\mathbf{Q}$ is resolved by diagonal matrix $\mathbf{P}$ (Krishnamoorthy and Menon 2011). Here, we use $\mathbf{H}_l$ to represent the matrix decomposition at l iteration where $l \leq L$ as below

$$\begin{aligned} \tilde{\mathbf{H}}_l^T\tilde{\mathbf{H}}_l &= \left[\tilde{\mathbf{H}}_{l-1}\ h_l\right]^T \left[\tilde{\mathbf{H}}_{l-1}\ h_l\right] \\ &= \begin{pmatrix} \tilde{\mathbf{H}}_{l-1}^T\tilde{\mathbf{H}}_{l-1} & \mathbf{v}_l \\ \mathbf{v}_l^T & g \end{pmatrix} \end{aligned} \tag{7}$$

where $(\mathbf{v}_l, g)$ is a new column generated from new hidden node output $h_l^T h_l$, compared to $\tilde{\mathbf{H}}_{l-1}^T\tilde{\mathbf{H}}_{l-1}$. Therefore, we can find

$$\begin{aligned} &\mathbf{Q}_l\mathbf{P}_l\mathbf{Q}_l^T \\ &= \begin{pmatrix} \mathbf{Q}_{l-1} & 0 \\ \mathbf{z}_l^T & 1 \end{pmatrix} \begin{pmatrix} \mathbf{P}_{l-1} & 0 \\ 0 & p \end{pmatrix} \begin{pmatrix} \mathbf{Q}_{l-1}^T & \mathbf{z}_l \\ 0 & 1 \end{pmatrix} \end{aligned} \tag{8}$$

As a result, we can easily calculate the $\mathbf{z}_l$ and scalar p for Cholesky factorization as

$$\mathbf{Q}_{l-1}\mathbf{P}_{l-1}\mathbf{z}_l = \mathbf{v}_l,\ p = g - \mathbf{z}_l^T\mathbf{P}_{l-1}\mathbf{z}_l \tag{9}$$

where $\mathbf{Q}_l$ and $\mathbf{v}_l$ is known from (7), which means we can continue use previous factorization result and update only according part. Please note that Q_1 is 1 and P_1 is $\tilde{\mathbf{H}}_1^T\tilde{\mathbf{H}}_1$.

As a conclusion, we have elaborated the basic learning on neural network and optimize Cholesky decomposition to solve the incremental least-squares problem. We have found the major computations are matrix-vector multiplications such as layer output in Eqs. 3 and 6 and also Cholesky decomposition (Eqs. 7, 8 and 9). Therefore, the proposed 3D multi-layer CMOS-memristor architecture is designed to accelerate matrix-vector multiplication, which can be also extended to BP based training method, where matrix-vector operation is the major computation (Cong and Xiao 2014).

3 Memristor-Crossbar Based Accelerator

3.1 Distributed In-Memory Computing Architecture

Conventionally, processor and memory are separate components that are connected through I/Os. With limited width and considerable RC-delay, the I/Os are considered the bottleneck of system overall throughput. As memory is typically organized in H-tree structure, where all leaves of the tree are data arrays, it is promising to impose in-memory computation with parallelism at this level. In this work, we propose a distributed memristor-crossbar in-memory architecture (XIMA). Because both data and logic units have uniform structure when implemented on memristor-crossbar, half of the leaves are exploited as logic elements and are paired with data arrays. The proposed architecture is illustrated in Fig. 2. The distributed local data-logic pairs can form one local data path such that the data can be processed locally in parallel, without the need of being readout to the external processor.

Coordinated by the additional controlling unit called *in-pair control bus* the in-memory computing is performed in following steps. (1) logic configuration: processor issues the command to configure logic by programming logic memristor-crossbar into specific pattern according to the functionality required; (2) load operand: processor sends the data address and corresponding address of logic accelerator input; (3) execution: logic accelerator can perform computation based on the configured logic and obtain results after several cycles; (4) write-back: computed results are written back to data array directly but not to the external processor.

With emphasis on different functionality, the memristor crossbars for data storage and logic unit have distinctive interfaces. The data memristor-crossbar will have only one row activated at one time during read and write operations, and logic memristor-crossbar; however, we can have all rows activated spontaneously as rows are used to

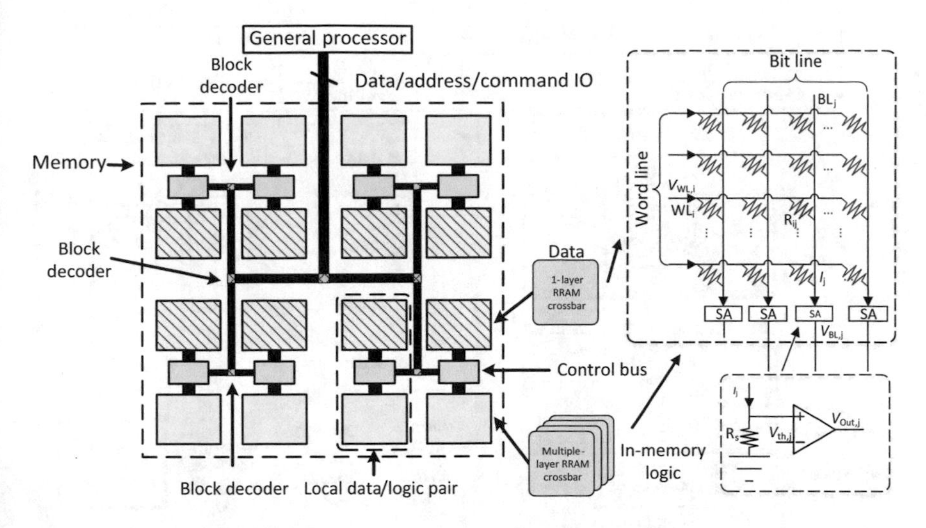

Fig. 2 Overview of distributed in-memory computing architecture on memristor-crossbar

take inputs. As such, the input and output interface of logic crossbar requires AD/DA conversions, which could outweigh the benefits gained. Therefore, in this paper, we propose a conversion-free digital-interfaced logic memristor crossbar design, which uses three layers of memristor crossbars to decompose a complex function into several simple operations that digital crossbar can tackle.

The conventional communication protocol between external processor and memory is composed of *store* and *load* action identifier, address that routes to different locations of data arrays, and data to be operated. With additional in-memory computation capacity, the proposed distributed in-memory computing architecture requires modifications on the current communication protocol. The new communication instructions are proposed in Table 1, which is called in-pair control.

In-pair control bus needs to execute instructions in Table 1. SW (store word) instruction is to write data into memristors in data array or in-memory logic. If target

Table 1 Protocols between external processor and control bus

Inst.	Op. 1	Op. 2	Action	Function
SW	Addr 1	Addr 2	Addr 1 data to Addr 2	Store data, configure logic, in-memory results write-back
	Data	Addr	Store data to Addr	
LW	Addr	–	Read data from Addr	Standard read
ST	Block Idx	–	Switch logic block on	Start in-memory computing
WT	–	–	Wait for logic block response	Halt while performing in-memory computing

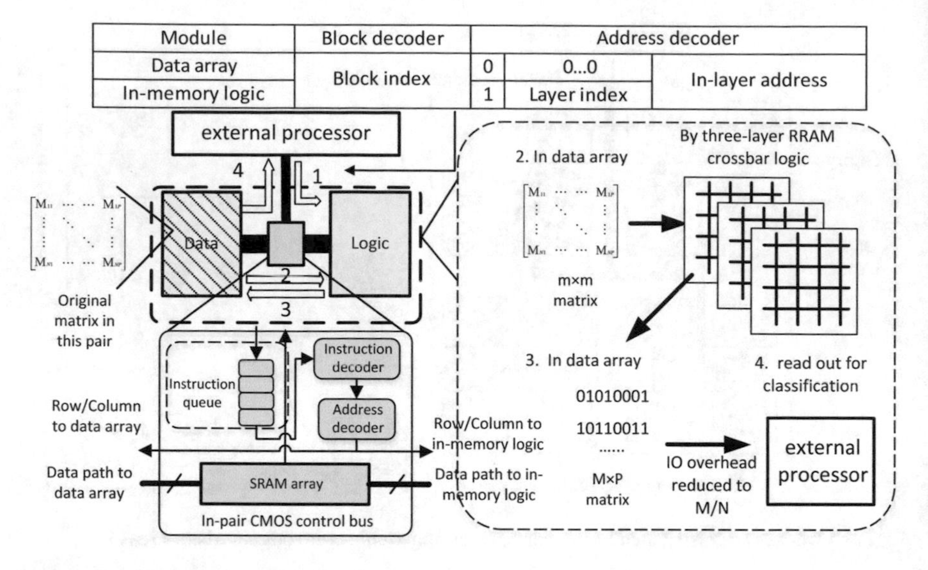

Fig. 3 Detailed structure of control bus and communication protocol

address is in data array, it will be a conventional write or result write-back; otherwise it will be logic configuration. LW (load word) instruction performs as conventional read operation. ST (start) instruction means to switch on the logic block for computing after the computation setup has been done. WT (wait) operation is to stop reading from instruction queue during computing.

Besides communication instructions, memory address format is also different from that in the conventional architecture. To specify a byte in the proposed architecture, address includes the following identifier segments. Firstly, the data-logic pair index segment is required, which is taken by block decoders to locate the target data-logic pair. Secondly, one-bit flag is needed to clarify that whether the target address is in data array or in-memory logic crossbar. Thirdly, if logic accelerator is the target, additional segment has to specify the layer index. Lastly, rest of address segment are row and column indexes in each memristor-crossbar. An address example for data array and in-memory logic is shown in Fig. 3.

To perform logic operation, the following instructions are required to performed. Firstly, we store the required input data and memristor values with SW operation. Secondly, an ST instruction will be issued to enable all the columns and rows to perform the logic computing. The WT instruction is also performed to wait for the completion of logic computing. At last, LW instruction is performed to load the data from the output of memristor-crossbar.

Given the new communication protocol between general processor and memory is introduced, one can design the according control bus as shown in Fig. 3. The control bus is composed of an instruction queue, an instruction decoder, an address decoder and a SRAM array. As the operation frequency of memristor-crossbar is slower than that of external processor, instructions issued by the external processor will be stored in the instruction queue first. They are then analyzed by instruction decoder on a first-come-first-serve (FCFS) basis. The address decoder obtains the row and column index from the instruction; and SRAM array is used to store temporary data such as computation results, which are later written back to data array.

3.2 3D CMOS-Memristor Architecture

Recent work (Topaloglu 2015) has shown that the 3D integration supports heterogeneous stacking because different types of components can be fabricated separately, and layers can be stacked and implemented with different technologies. Therefore, stacking non-volatile memories on top of microprocessors enables cost-effective heterogeneous integration. Furthermore, works in Chen et al. (2012), Liauw et al. (2012) have also shown the feasibility to stack memristor on CMOS to achieve smaller area and lower energy consumption.

The proposed 3D multi-layer CMOS-memristor accelerator with three layers is shown in Fig. 4a. This accelerator is composed of a two-layer memristor-crossbar and a one-layer CMOS circuit. As Fig. 4a shows, layer 1 of memristor-crossbar is implemented as a buffer to temporarily store input data to be processed. Layer 2 of memristor-crossbar performs logic operations such as matrix-vector multiplication and also vector addition. The details of implementation will be introduced in Sect. 4. Note that buffers are designed to separate resistive networks between layer 1 and layer 2. The last layer of CMOS contains read-out circuits for memristor-crossbar and

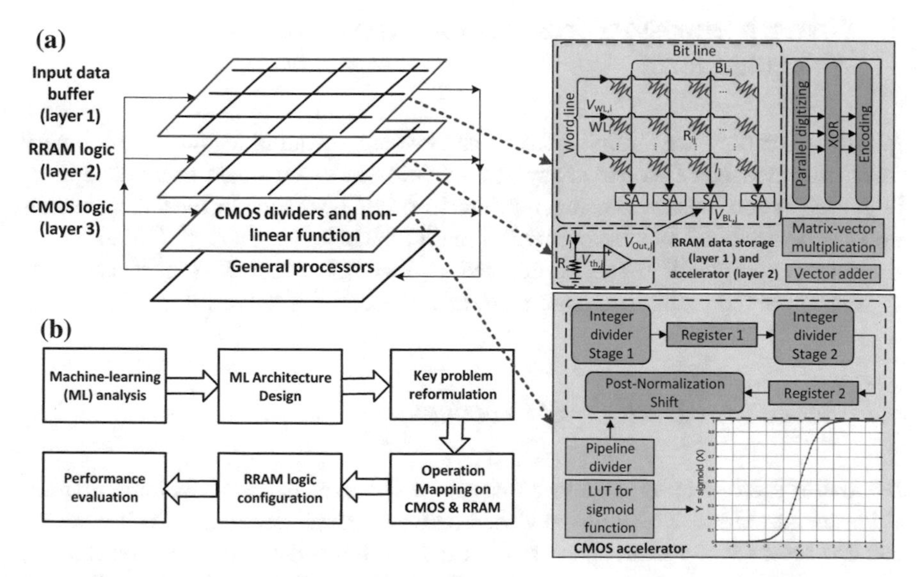

Fig. 4 **a** 3D multi-layer CMOS-memristor accelerator architecture; **b** Incremental machine learning algorithm mapping flow on proposed accelerator

performs as logic accelerators designed for other operations besides matrix-vector multiplication, including pipelined divider, look-up table (LUT) designed for division operation and activation function in machine learning.

Moreover, Fig. 4b shows the work flow for incremental machine learning based on the proposed architecture. Firstly, detailed architecture of machine learning (ML) (e.g. number of layers and activation function) is determined based on the accuracy requirements and data characteristics. Secondly, operations of this machine learning algorithm are analyzed and reformulated so that all the operations can be accelerated in 3D multi-layer CMOS-memristor architecture as illustrated in Fig. 4a. Furthermore, the bit-width operating on memristor-crossbar is also determined by balancing the accuracy loss and energy saving. Finally, logic operations on memristor-crossbar and CMOS are configured based on the reformulated operations, energy saving and speed-up.

Such a 3D multi-layer CMOS-memristor architecture has advantages in three manifold. Firstly, by utilizing memristor-crossbar for input data storage, leakage power of memory is largely removed. In a 3D architecture with TSV interconnection, the bandwidth from this layer to next layer is sufficiently large to perform parallel computation. Secondly, memristor-crossbar can be configured as computational units for the matrix-vector multiplication with high parallelism and low power. Lastly, with an additional layer of CMOS-ASIC, more complicated tasks such as division and non-linear mapping can be performed. As a result, the whole training process of machine learning can be fully mapped to the proposed 3D multi-layer CMOS-memristor accelerator architecture towards real-time training and testing.

4 Binary Memristor-Crossbar for Matrix-Vector Multiplication

In this work, we implement matrix-vector multiplication on binary memristor-crossbar. It is one always-on operation in various data-analytic applications such as compressive sensing, machine learning. For example, the feature extraction can be achieved by multiplying Bernoulli matrix in Wright et al. (2009).

Matrix multiplication can be denoted as $Y = \Phi X$, where $X \in \mathbb{Z}^{N \times P}$ and $\Phi \in \{0, 1\}^{M \times N}$ are the multiplicand matrices, and $Y \in \mathbb{Z}^{M \times P}$ is the result matrix.

4.1 *Memristor Device and Crossbar*

Memristor is a two-terminal device that can be observed in sub-stoichiometric transition metal oxides (TMOs) sandwiched between metal electrodes. Such a device can be used as non-volatile memory with state of ion resistance, which results in 2 non-volatile states: high resistance state HRS and low resistance state LRS. One can change the state from HRS to LRS or vice versa by applying a SET voltage (V_w) or a RESET voltage ($-V_w$).

The two states RHS and LRS represent 0 and 1, respectively. To read a memristor cell, one can apply a read voltage V_r to the memristor. The V_r and V_w follow

$$V_w > V_{th} > V_w/2 > V_r, \tag{10}$$

where V_{th} is the threshold voltage of the memristor.

Because of the high density of memristor device, one can build a crossbar structure as the array of memristor (Kim et al. 2011; Kang et al. 2014; Fan et al. 2014; Gu et al. 2015; Srimani et al. 2015; Wang et al. 2014). Such crossbar structure can be utilized as memory for high-density data storage. The memory array can be read or written by controlling the voltage of wordlines (WLs) and bitlines (BLs). For example, we can apply $V_w/2$ on the ith WL and $-V_w/2$ on the jth BL to write data into the memristor cell on ith row, jth column.

4.2 *Traditional Analog Memristor Crossbar*

The fabric of crossbar intrinsically supports matrix-vector multiplication where vector is represented by row input voltage levels and matrix is denoted by mesh of memristor resistances. As shown in Fig. 5, by configuring $\boldsymbol{\Phi}$ into the memristor crossbar, analog computation $y = \boldsymbol{\Phi}x$ by memristor crossbar can be achieved.

However, such analog memristor-crossbar has two major drawbacks. Firstly, the programming of continuous-valued memristor resistance is practically challenging due to large memristor process variation. Specifically, the memristor resistance is determined by the integral of current flowing through, which leads to a switching curve as shown in Fig. 6a. With the process variation, the curve may shift and leave intermediate values very unreliable to program, as shown in Fig. 6b. Secondly, the A/D and D/A converters are both timing-consuming and power-consuming. In our simulation, the A/D and D/A conversion may consume up to 85.5% of total operation energy in 65 nm as shown in Fig. 7.

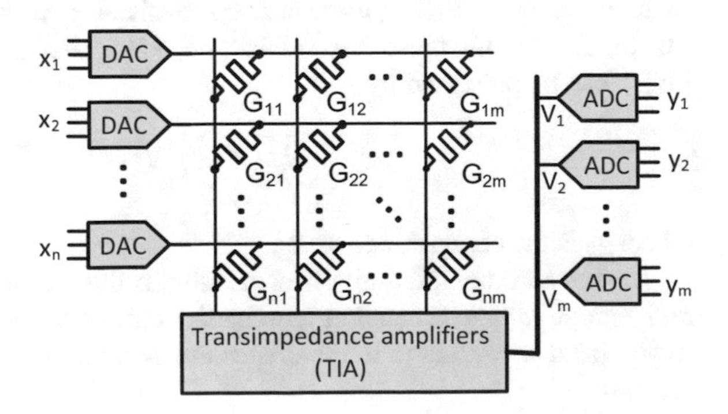

Fig. 5 Traditional analog-fashion memristor crossbar with ADC and DAC

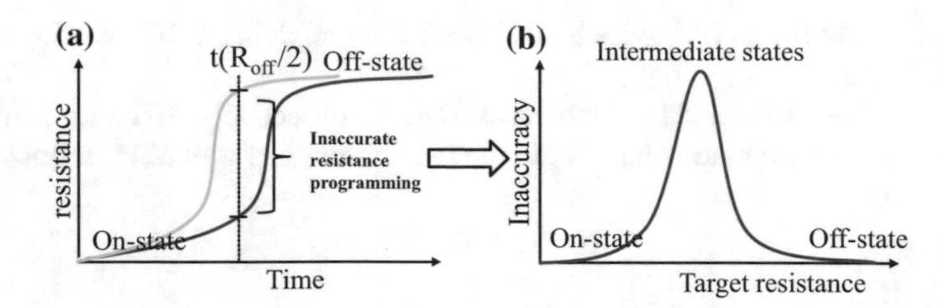

Fig. 6 **a** Switching curve of memristor under device variations. **b** Programing inaccuracy for different memristor target resistances

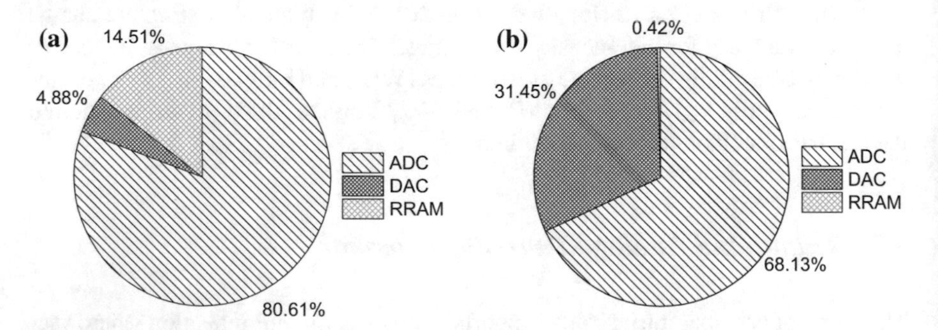

Fig. 7 **a** Power consumption of analog-fashion memristor crossbar. **b** Area consumption of analog-fashion memristor crossbar

4.3 Proposed Digitalized Memristor Crossbar

To overcome the aforementioned issues, we propose a full-digitalized memristor-crossbar for matrix-vector multiplication. Firstly, as ON-state and OFF-state are much more reliable than intermediate values shown in Fig. 6, only binary values of memristor are allowed to reduce the inaccuracy of memristor programming. Secondly, we deploy a pure digital interface without A/D conversion.

In memristor crossbar, we use V^i_{wl} and V^j_{bl} to denote voltage on ith wordline (WL) and jth bitline (BL). R_{off} and R_{on} denote the resistance of off-state and on-state. In each sense amplifier (SA), there is a sense resistor R_s with fixed and small resistance. The relation among these three resistance is $R_{off} \gg R_{on} \gg R_s$. Thus, the voltage on jth BL can be presented by

$$V^j_{bl} = \sum_{i=1}^{m} g_{ij} V^i_{wl} R_s \tag{11}$$

where g_{ij} is the conductance of R_{ij}.

The key idea behind digitalized crossbar is the use of comparators. As each column output voltage for analog crossbar is continuous-valued, comparators are used to digitize it according to the reference threshold applied to SA in Fig. 2,

$$O_j = \begin{cases} 1, & if\ V_{bl}^j \geq V_{th}^j \\ 0, & if\ V_{bl}^j < V_{th}^j \end{cases} \tag{12}$$

However, the issue that rises due to the digitalization of analog voltage value is the loss of information. To overcome this, three techniques are applied. Firstly, multi-thresholds are used to increase the quantization level so that more information can be preserved. Secondly, the multiplication operation is decomposed into three sub-operations that binary crossbar can well tackle. Thirdly, the thresholds are delicately selected at the region that most information can be preserved after the digitalization.

4.4 Implementation of Digital Matrix Multiplication

In this section, hardware mapping of matrix multiplication on the proposed architecture is introduced. The logic required is a matrix-vector multiplier by the memristor-crossbar. Here, a three-step memristor-crossbar based binary matrix-vector multiplier is proposed, in which both the input and output of the memristor-crossbar are binary data without the need of ADC. The three memristor-crossbar step: parallel digitizing, XOR and encoding are presented in details as follows. As the output of a memristor-crossbar array can be connected to the input of another memristor-crossbar array, we can use multiple memristor arrays in the logic block for the mappings. Here we use symbol s to denote the result of binary matrix-vector multiplication. Therefore, s follows

$$0 \leq s \leq N, \tag{13}$$

where N is the maximum result. To illustrate the three-step procedure more clearly, we will use the following matrix-vector multiplication as an example:

$$[00101011] \times [10111110]^{\mathrm{T}} = 3 \tag{14}$$

The output after the three-step procedure will be shown when $s = 3$ and $N = 8$.

4.4.1 Parallel Digitizing

The first step is called parallel digitizing, which requires $N \times N$ memristor crossbars. The idea is to split the matrix-vector multiplication to multiple inner-product operations of two vectors. Each inner-product is produced by one memristor crossbar. For each crossbar, as shown in Fig. 8, all columns are configured with same elements that correspond to one column in random Boolean matrix $\boldsymbol{\Phi}$, and the input voltages on word-lines (WLs) are determined by x. As $g_{on} \gg g_{off}$, current on memristors with high impedance are insignificant, so that the voltages on BLs approximately equal to $kV_r g_{on} R_s$ according to Eq. (11) where k is the number of memristor with in low-resistance state (g_{on}).

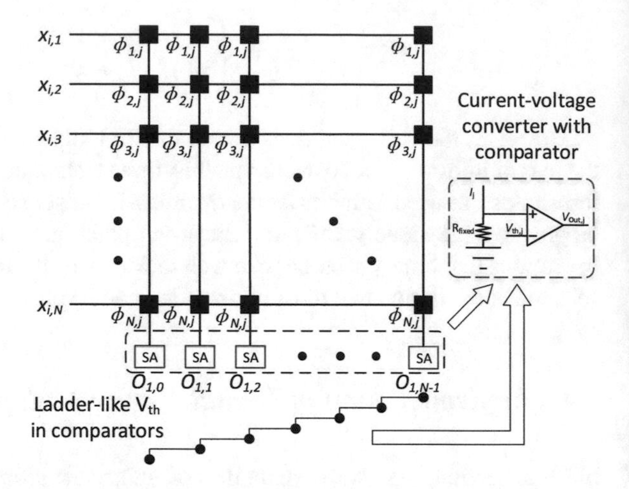

Fig. 8 Parallel digitizing step of memristor crossbar in matrix multiplication

It is obvious that voltages on bit-lines (BLs) are all identical. Therefore, the key to obtain the inner-product is to set ladder-type sensing threshold voltages for each column,

$$V_{th,j} = \frac{(2j+1)V_r g_{on} R_s}{2}, \tag{15}$$

where $V_{th,j}$ is the threshold voltage for the j_{th} column. The $O_{i,j}$ is used to denote the output of column j in memristor crossbar step i after sensing. Therefore, for the output we have

$$O_{1,j} = \begin{cases} 1, & j \leq s \\ 0, & j > s, \end{cases} \tag{16}$$

where s is the inner-product result. In other words, the first $(N - s)$ output bits are 0 and the rest s bits are 1 ($s <= N$). In our example, $x_i = [00101011]$ and $\phi_i = [10111110]$, and the corresponding output $O_1 = [11100000]$.

4.4.2 XOR

The inner-product output of parallel digitizing step is determined by the position where $O_{1,j}$ changes from 0 to 1. The XOR takes the output of the first step, and performs XOR operation for every two adjacent bits in $O_{1,j}$, which gives the result index. For the same example of $s = 3$, we need to convert the first-step output $O_1 = [11100000]$ to $O_2 = [001000000]$. The XOR operation based on memristor crossbar is shown in Fig. 9. According to parallel digitizing step, $O_{1,j}$ must be 1 if $O_{1,j+1}$ is 1. Therefore, XOR operation is equivalent to the AND operation $O_{1,j} \oplus O_{1,j+1} = O_{1,j}\overline{O_{1,j+1}}$, and therefore we have

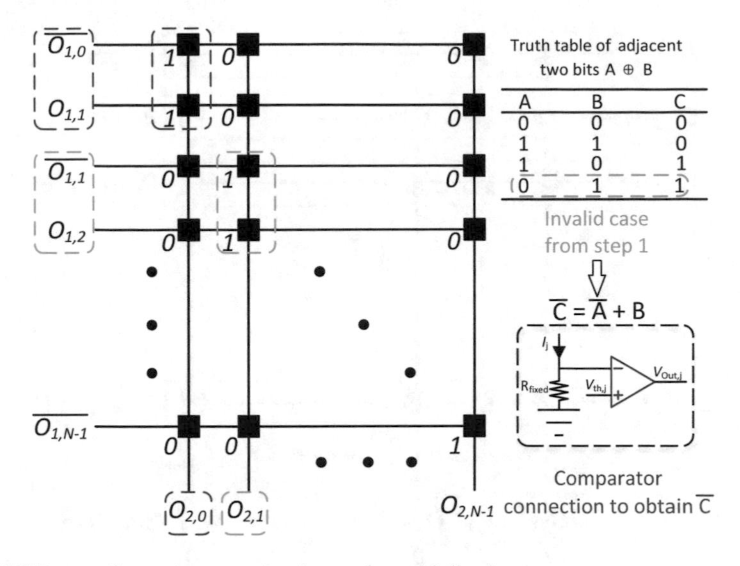

Fig. 9 XOR step of memristor crossbar in matrix multiplication

$$\overline{O_{2,j}} = \begin{cases} \overline{O}_{1,j} + O_{1,j+1}, & j < N - 1 \\ \overline{O}_{1,j}, & j = N - 1. \end{cases} \tag{17}$$

In addition, the threshold voltages for the columns have to follow

$$V_{th,j} = \frac{V_r g_{on} R_s}{2} \tag{18}$$

Eqs. (17) and (18) show that only output of s_{th} column is 1 on the second step, where s is the inner product result. Each crossbar in XOR step has the size of $N \times (2N - 1)$.

4.4.3 Encoding

The third step takes the output of XOR step and produces s in binary format as an encoder. Therefore, O_3 should be in the binary format of s. In our example, $O_3 =$ [00000011] when $s = 3$. In the output of XOR step, as only one input will be 1 and others are 0, according binary information is stored in corresponding row, as shown in Fig. 10. Encoding step needs $N \times n$ memristors, where $n = \lceil \log_2 N \rceil$ is the number of bits in order to represent N in binary format. The thresholds for the encoding step are set following Eq. 18 as well.

For activation function in (3), the exponentiation and division operations can be implemented by look-up table (LUT). The output of XOR layer is the index of **preH**. Therefore, the encoding layer performs as a LUT mapping process from **preH** to **H**. For example, if we want to map **preH** = 3 to **H** = 0.953, we can make the memristors

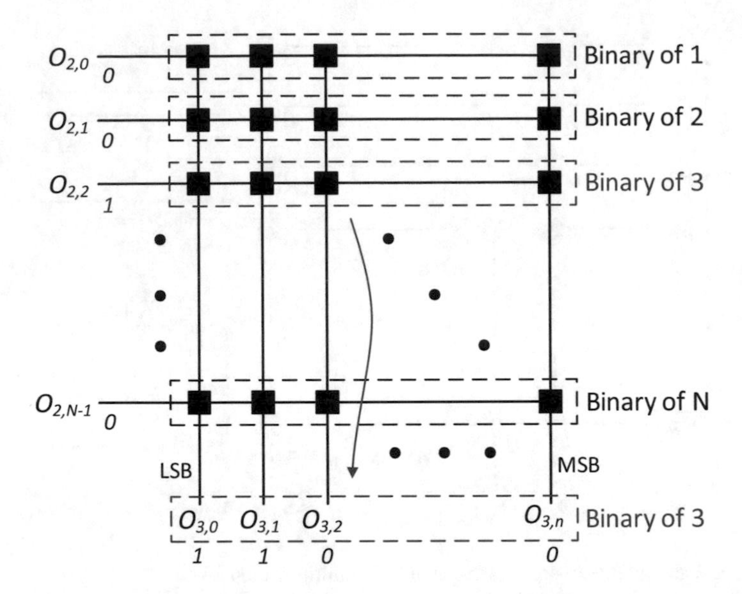

Fig. 10 Encoding step of memristor crossbar in matrix multiplication

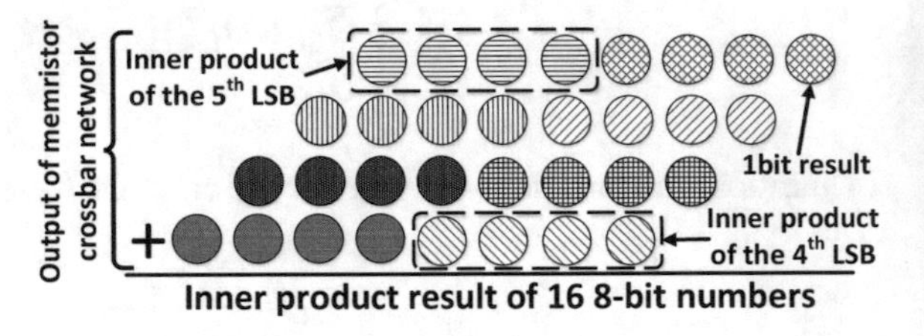

Fig. 11 Memristor-based inner-product operation

in column 3 as the binary format of 0.953 ([11110100], no-signed, 8-bit fixed point with 256 scaling factor). As a result, the mapping process can be included in the encoding step.

4.4.4 Adding and Shifting for Inner-Product Result

The output of encoding step is in binary format, but some processes are needed to obtain the final inner-product result. Adder and shifter are designed to complete this process as shown in Fig. 11. We suppose the original data is 8-bit and data dimension is 512, the workload of adder is 512 without any acceleration. With three-step memristor-crossbar accelerator as pre-processing, the workload of adder can be significantly reduced to 9 ($\log_2 512$). Detailed comparison results will be shown in Sect. 5.

5 Performance Evaluation

5.1 Experiment Settings

The hardware evaluation platform is implemented on a computer server with 4.0 GHz core and 16.0 GB memory. Feature extraction is implemented by general processor, CMOS-based ASIC, non-distributed and distributed in-memory computing based on digitalized memristor crossbar respectively. For the memristor-crossbar design evaluation, the resistance of memristor is set as 1 kΩ and 1 MΩ as on-state and off-state resistance respectively according to Lee et al. (2008).

The general processor solution is performed by Matlab on a 4.0 GHz desktop. For CMOS-ASIC implementation, we implement it by Verilog and synthesize with CMOS 65 nm low power PDK. For memristor-crossbar based solution, we verify the function in circuit level with SPICE tool NVMSPICE (Yu and Wang 2014). By analyzing the machine learning algorithm, we obtain the basic operations and the number of memristor-crossbar logic units required.

The working frequency of general processor implementation is 4.0 GHz while the CMOS ASIC feature extraction design frequency is 1.0 GHz. For in-memory computing based on the proposed memristor crossbar, write voltage V_w is set as 0.8 V and read voltage V_r is set as 0.1 V as well as duration time of 5 ns. In addition, the analog computation on memristor-crossbar is performed for comparison based on design in Singh et al. (2007).

In the followings, we will show the performance of matrix-vector multiplication on memristor-crossbar first. A scalability study is introduced to show the area, energy and computation delay with different matrix sizes. Afterwards, the evaluation of face recognition on in-memory architecture is presented. Finally, we will illustrate the object classification on 3D CMOS-memristor architecture. Performance of different bid-width configurations will also be shown. In addition, the 3D CMOS-memristor solution will be compared with CMOS-ASIC as well as GPU implementation.

5.2 Performance Comparison of Multiplication

To evaluate the performance of binary memristor-crossbar for matrix-vector multiplication, we use the proposed architecture to accelerate dimension reduction of fingerprint images. 1,000 fingerprint images selected from Tan and Sun (2010) are stored in memory with 328×356 resolution, with 8 bits in each pixel. To agree with patch size, random Bernoulli $N \times M$ matrix is with fixed N and M of 356 and 64, respectively. The original images can be seen as $X \in \mathbb{Z}^{N \times P}$ in matrix-vector multiplication. The detailed comparison is shown in Table 2 with numerical results including energy consumption and delay obtained for one image on average of 1,000 images. For the digitized XIMA implementation, we need to compute the area of memristor cell, adding and shifting as well as the control bus. For analog XIMA, the majority of area is consumed by ADC/DACs and area of memristor cell can be neglected.

Table 2 Matrix-vector multiplication performance comparison under among software and hardware implementation

Implementation	General purpose processor (MatLab)	CMOS ASIC	Non-distributed digitalized XIMA	Distributed digitalized XIMA	Distributed analog XIMA
Area	177 mm^2	5 mm^2	3.28 mm^2 (800 MBit memristors) + 0.088 mm^2 + 128 μ m^2	0.05 mm^2 (12 MBit memristors) + 0.088 mm^2 + 8192 μ m^2	8.32 mm^2
Frequency	4 GHz	1 GHz	200 MHz	200 MHz	200 MHz
Cycles	–	69,632	Computing: 984	Computing: 984	Computing: 328
			Pre-computing: 262,144	Pre-computing: 4,096	Pre-computing: 4,096
Time	1.78 ms	69.632 μs	Computing: 4,920 ns	Computing: 4,920 ns	Computing: 1,640 ns
			Pre-computing: 1.311 ms	Pre-computing: 20.48 μs	Pre-computing: 20.48 μs
Dynamic power	84 W	34.938 W	Memristor: 4.71 W	Memristor: 4.71 W	Memristor: 1.28 W
			Control-bus: 100 μW	Control-bus: 6.4 mW	Control-bus: 6.4 mW
Energy	0.1424 J	2.4457 mJ	Memristor: 23.17 μJ	Memristor: 23.17 μJ	Memristor: 2.1 μJ
			Control-bus: 0.131 μJ	Control-bus: 0.131 μJ	Memristor: 0.131 μJ

Among hardware implementations, in-memory computing based on the proposed XIMA achieves better energy-efficiency than CMOS-based ASIC. Non-distributed XIMA (only one data and logic block inside memory) needs fewer CMOS control bus but large data communication overhead on a single-layer crossbar compared to distributed memristor crossbar. Although distributed analog memristor crossbar can achieve the best in energy perspective but has larger area compared to the digitalized one. Shown in Table 2, memristor crossbar in analog fashion only consumes 2.1 μJ for one vector multiplication while the proposed architecture requires 23.17 μJ because most of power consumption comes from memristor in computing instead of ADCs. However, ADCs need more area so that memristor crossbar with analog fashion is 8.32 mm^2 while the proposed one is only 0.15 mm^2 because of the high density of memristor crossbar.

Calculation error of analog and digitalized memristor crossbar are compared in Fig. 12, where M and N are both set as 256. Calculation error is very low when memristor error rate is smaller than 0.004 for both analog and digitalized fashion memristor. However, when memristor error rate reaches 0.01, calculation error rate of analog memristor crossbar goes to 0.25, much higher than the other one with only 0.07. As such, computational error can be reduced in the proposed architecture compared to analog fashion memristor crossbar.

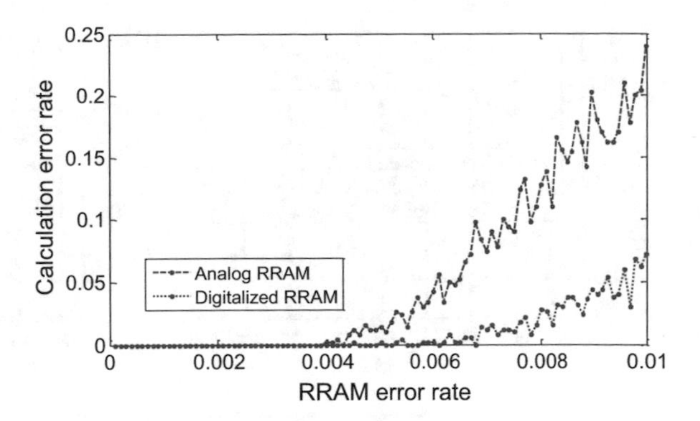

Fig. 12 Calculation error comparison between multi-leveled and binary memristor

5.3 *Scalability Study*

Hardware performance comparison among CMOS-based ASIC, non-distributed and distributed XIMA with varying M is shown in Fig. 13. From area consumption perspective shown in Fig. 13a, distributed memristor-crossbar is much better than the other implementations. With increasing M from 64 to 208, its total area is from 0.057 to 0.185 mm^2, approximately 100x smaller than the other two approaches. Non-distributed memristor crossbar becomes the worst one when $M > 96$. From delay perspective shown in Fig. 13b, non-distributed memristor crossbar is the worst because it has only one control bus and takes too much time on preparing of computing. Delay of non-distributed memristor crossbar grows rapidly while distributed memristor crossbar and CMOS-based ASIC implementation maintains on approximately 21 μs and 70μs respectively as the parallel design. For energy-efficiency side shown in Fig. 13c, both non-distributed and distributed memristor crossbar do better as logic accelerator is off at most of time. The proposed architecture also performs the best in energy-delay product (EDP) shown in Fig. 13d. Distributed XIMA performs the best among all implementation under different specifications. The EDP is from 0.3 to 2×10^{-9}sJ, which is 60× better than non-distributed memristor crossbar and 100× better than CMOS-based ASIC.

What is more, hardware performance comparison with varying N is shown in Fig. 14. Area and energy consumption trend is similar to Fig. 13. But for computational delay, the proposed architecture cannot maintain constantly as Fig. 13b because it needs much time to configure the input, but still the best among the three. Distributed XIMA still achieves better performance than the other two (Fig. 15).

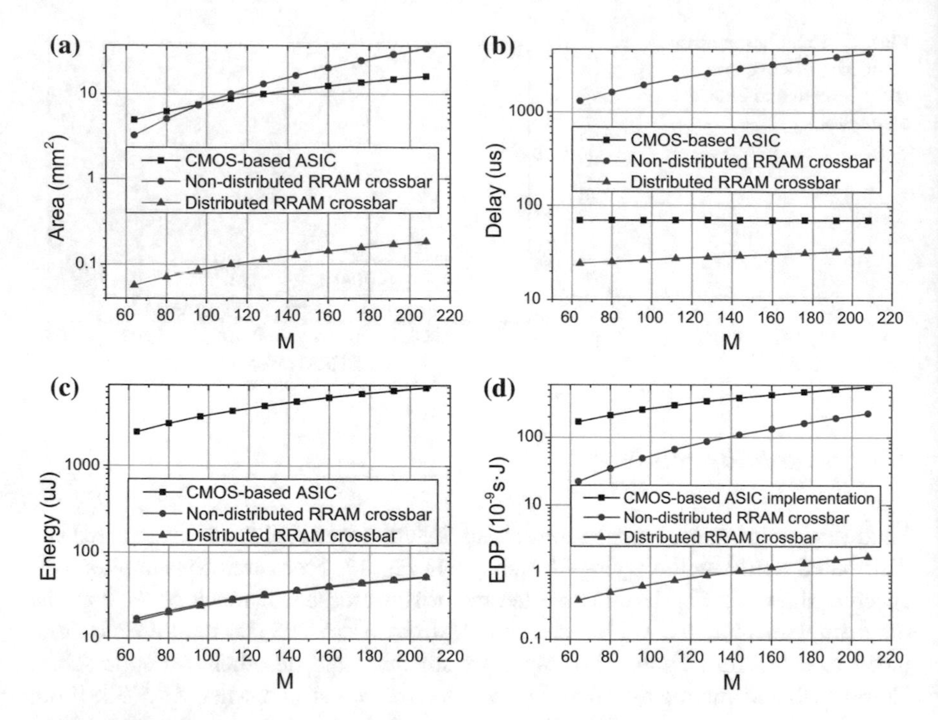

Fig. 13 Hardware performance scalability under different reduced dimension for **a** area; **b** delay; **c** energy; **d** EDP

5.4 *Performance of In-Memory Architecture*

In this work, we implement the face recognition application on the in-memory architecture. We will analyze the computation complexity of face recognition first, and then evaluate the performance.

In the experiment, 200 face images of 13 people are selected from Huang et al. (2007), with scaled image size 262 of each image **X**. In PCA, feature size of image is further reduced to 128 by multiplying the matrix **R**. The number of hidden node L and classes m are 160 and 13, respectively. Based on the experimental settings, computation complexity is analyzed with results shown in Fig. 16. 82% of computations are multiplication in output weight calculation, which is the most time-consuming procedure in neural network. Time-consumption of each process in neural network is introduced in Fig. 16b. Since processes except activation function involve matrix-vector multiplication, we extracted this operation in the whole algorithm and found that 64.62% of time is consumed in matrix-vector multiplication, shown in Fig. 16c.

We implement the face recognition in the distributed in-memory architecture. In Table 3, general performance comparisons among MatLab, CMOS-ASIC and memristor-crossbar accelerator are introduced, and the acceleration of each procedure as the formula described in Sect. 2 is also addressed. Among three

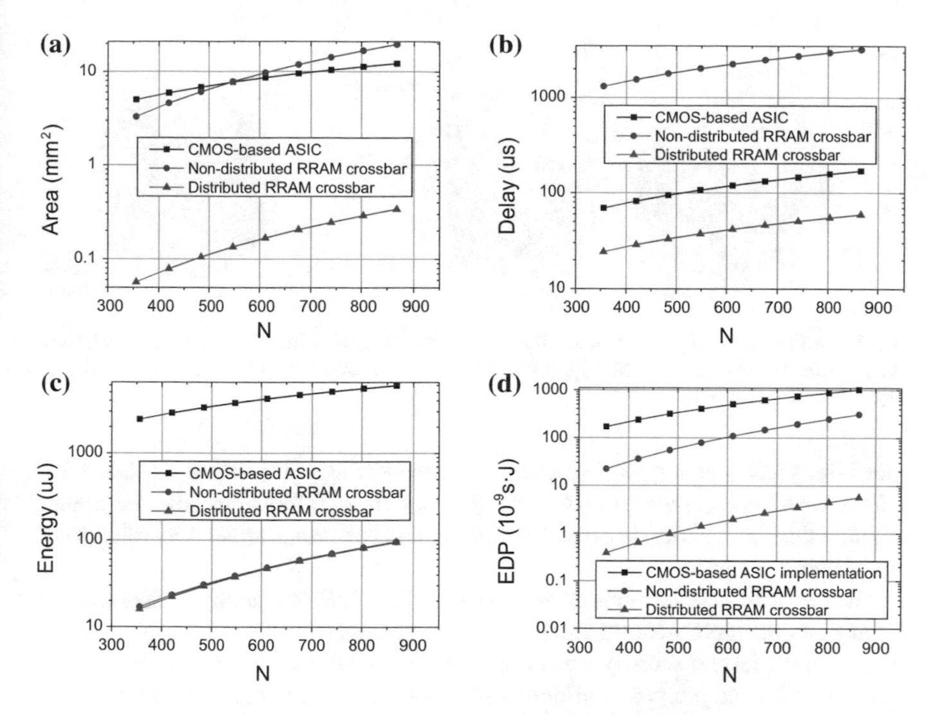

Fig. 14 Hardware performance scalability under different original dimension for **a** area; **b** delay; **c** energy; **d** EDP

Class 1
Class 2
Class 3
Class 4
Class 5

Test cases/ classes	Test case 1 (Y)	Test case 2 (Y)	Test case 3 (Y)
Class 1	1.518062	-0.79108	-0.58029
Class 2	-0.29803	-0.87155	-0.24397
Class 3	-0.95114	0.793256	-0.35867
Class 4	-0.65597	-0.44714	-0.70879
Class 5	-0.65955	0.262689	0.872497
⋮	⋮	⋮	⋮

Fig. 15 Training samples and prediction value **Y** (6) for face recognitions

implementations, memristor-crossbar architecture performs the best in area, energy and speed. Compared to MatLab implementation, it achieves 32.84× speed-up, 210.69× energy-saving and almost four-magnitude area-saving. We also design a CMOS-ASIC implementation with similar structure as memristor-crossbar with better performance compared to MatLab. memristor-crossbar architecture is 4.34×

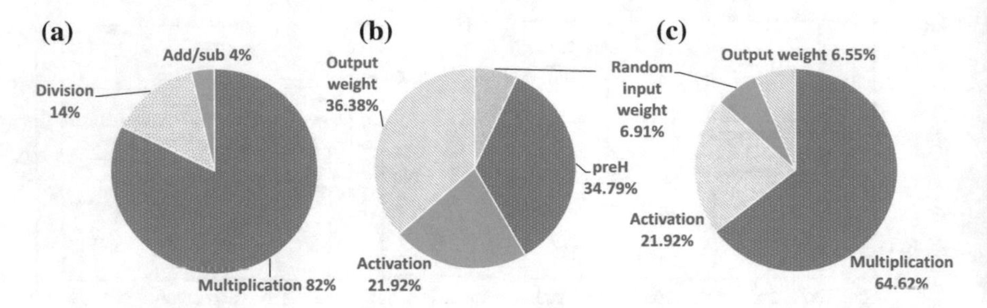

Fig. 16 **a** Time consumption breakdown for output weight calculation. **b** Neural network training computation effort analysis. **c** Multiplication analysis for neural network training ($N = 200$, $n = 128$, $L = 160$ and $m = 13$)

speed-up, 13.08× energy-saving and 51.3× area-saving compared to CMOS-ASIC. The performance comparison is quite different from Table 2, because we applied different designs (memristor-crossbar size) is this two experiments according to the dimension of matrices.

The result of face recognition is shown in Fig. 15. Five training classes are provided as an example with three test cases. Each test face will be recognized as the class with the largest score (prediction result as the index of $Max(\mathbf{Y})$, marked in red color). In this example, case 1 is identified as class 1, while case 2 and 3 are classified into class 3 and 5, respectively.

5.5 Performance of 3D CMOS-memristor Architecture

We implement the object classification in the 3D CMOS-memristor Architecture. Table 4 shows the testing accuracy under different datasets (Lichman 2013; Krizhevsky and Hinton 2009) and configurations for machine learning of support vector machine (SVM) and single layer feed-forward neuron network (SLFN). It shows that accuracy of classification is not very sensitive to the memristor configuration bits. For example, the accuracy of Iris dataset is working with negligible accuracy at 5 memristor bit-width. When the memristor bit-width increased to 6, it performs the same as 32 bit-width configurations. Similar observation is found in Chen et al. (2015) by truncating algorithms with limited precision for better energy efficiency. Please note that training data and weight related parameters are quantized to perform matrix-vector multiplication on memristor crossbar accelerator.

Figure 17 shows the energy comparisons under different bit-width configurations for CMOS and memristor under the same accuracy requirements. An average of 4.5× energy saving can be achieved for the same number of bit-width configurations. The energy consumption is normalized by the CMOS 4 bit-width configuration. Furthermore, we can observe that not always smaller number of bits achieves better energy saving. Fewer number of bit-width may require much larger neuron network to perform required classification accuracy. As a result, its energy consumption increases.

Table 3 Face recognition performance comparison under among software and hardware implementation

Implementation	General purpose processor (MatLab)				CMOS-ASIC				Distributed in-memory memristor-crossbar architecture			
Frequency	4.0 GHz				1.0 GHz				200 MHz			
Area (mm^2)	177				5.64				0.11			
Power (W)	844				39.4144				13.1			
Computation	PCA	Input layer	L2-norm	Output layer	PCA	Input layer	L2-norm	Output layer	PCA	Input layer	L2-norm	Output layer
Cycles	–	–	–	–	195,000	121,900	12,256,400	12,440	7,680	4,800	566,400	486
Time	1.56 ms	0.98 ms	92.5 ms	0.1 ms	195 μs	121.9 μs	12.26 ms	12.4 μs	38.4 μs	24 μs	2, 832 μs	2.4 μs
Energy (mJ)	131	82	7,770	8.4	7.68	4.80	483.2	0.49	0.5	0.3	37.1	0.03
Speed-up	–				7.56 × (95140 μs : 12589.3 μs)				32.84 × (95140 μs : 2896.8 μs)			
Energy-saving	–				16.11 × (7991.4 mJ : 496.17 mJ)				210.69 × (7991.4 mJ : 37.93 mJ)			

Table 4 Testing accuracy of ML techniques under different dataset and configurations (normalized to all 32 bits)

Datasets	Size	Feat.	Cl.	4 bit Acc.(%)		5 bit Acc.(%)		6 bit Acc.(%)	
				SVM‡	&SLFN	SVM‡	&SLFN	SVM‡	&SLFN
Glass	214	9	6	100.07	100.00	93.88	99.30	99.82	99.30
Iris	150	4	3	98.44	94.12	100.00	94.18	100.00	100.00
Seeds	210	7	3	97.98	82.59	99.00	91.05	99.00	98.51
Arrhythmia	179	13	3	96.77	97.67	99.18	98.83	99.24	100.00
Letter	20,000	16	7	97.26	53.28	98.29	89.73	99.55	96.13
CIFAR-10	60,000	1600†	10	98.75	95.71	99.31	97.06	99.31	99.33

†1600 features extracted from 60,000 32 × 32 color images with 10 classes. ‡Least-square SVM is used for comparison

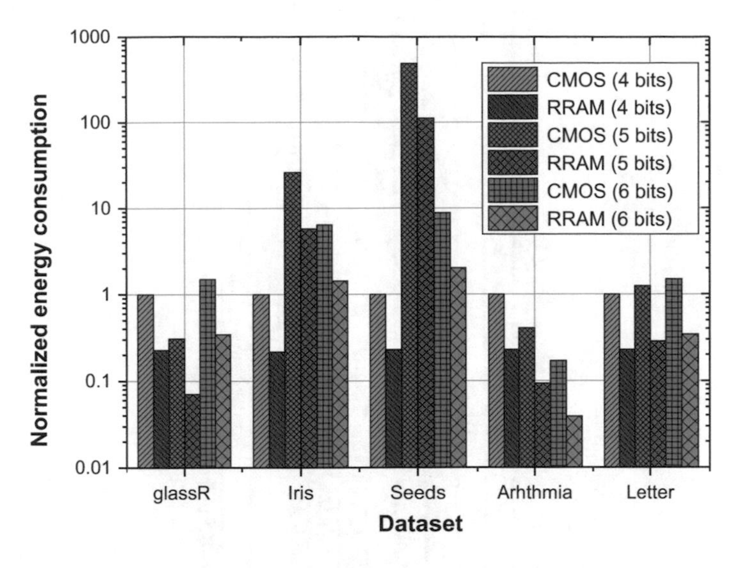

Fig. 17 Energy-saving comparison under different bit-width of CMOS and memristor with the same accuracy requirement

Test cases/ Classes	Test case 1	Test case 2	Test case 3
Class 1	-3.3363	-0.0037	2.2211
Class 2	-3.1661	0.6737	1.2081
Class 3	-3.5008	0.0613	1.4670
Class 4	-3.4521	-0.0527	1.5498
Class 5	-3.0085	-0.1861	1.0764
⋮	⋮	⋮	⋮

Class 1 ship
Class 2 dog
Class 3 airplane
Class 4 bird
Class 5 cat

Fig. 18 On-line machine-learning for image recognition on the proposed 3D multi-layer CMOS-memristor accelerator using benchmark CIFAR-10

5.6 Performance of Machine Learning Based Face Recognition

Figure 18 shows the classification values in (6) on image data (Krizhevsky and Hinton 2009) with an example of 5 classes. As mentioned in the Sect. 3, the index with maximum values (highlighted in red) is selected to indicate the class of test case. A few sample images are selected. Please note that 50,000 and 10,000 images are used for training and testing with 10 classes.

Table 5 Performance comparison under different software and hardware implementations

Implementation	General purpose processor (Matlab)						3D-CMOS-ASIC †						3D CMOS-memristor architecture †					
Frequency	3.46 GHz						1.0 GHz						200 MHz					
Area	240 mm² (Intel Xeon X5690)						1.86 mm² (65 nm global foundry)						1.46 mm² (65 nm memristor and CMOS)					
Power (W)	130						39.41						13.1					
Computation	Feature extraction		$IL^{\ddagger}$	ℓ_2 norm		OL^{*}	Feature extraction		$IL^{\ddagger}$	ℓ_2 norm		OL^{*}	Feature extraction		$IL^{\ddagger}$	ℓ_2 norm		OL^{*}
Operation	Sort	Mul.	–	Div.	Mul.	–	Sort	Mul.	–	Div.	Mul.	–	Sort	Mul.	–	Div.	Mul.	–
Time (s)	1473.2	736.6	729.79	294.34	1667.95	750.3	216.65	97.43	96.53	43.29	220.62	99.25	216.65	22.43	22.22	43.29	50.79	22.85
Energy (KJ)	191.52	95.76	94.87	38.26	216.83	97.54	0.078	3.84	3.80	0.015	8.69	3.91	0.078	0.293	0.291	0.015	0.67	0.30
Speed-up	–						7.30 × (5651.88 s : 773.77 s)						14.94 × (5651.88 s : 378.22 s)					
Energy-saving	–						36.12 × (734.78 KJ : 20.34 KJ)						447.17× (734.78 KJ : 1.643 KJ)					

†6 bit-width configuration is implemented for both CMOS and memristor. $IL^{\ddagger}$ is for input layer and OL^{*} is for output layer

In Table 5, performance comparisons among Matlab, 3D-CMOS-ASIC and 3D multi-layer CMOS-memristor accelerator are presented, and the acceleration of each procedure based on the formula described in Sect. 2 is also addressed. Among the three implementations, 3D multi-layer CMOS-memristor accelerator performs the best in area, energy and speed. Compared to Matlab implementation, it achieves 14.94× speed-up, 447.17× energy-saving and 164.38× area-saving. We also design a 3D-CMOS-ASIC implementation with similar structure as 3D multi-layer CMOS-memristor accelerator with better performance compared to Matlab. The proposed 3D multi-layer CMOS-memristor 3D accelerator is 2.05× speed-up, 12.38× energy-saving and 1.28× area-saving compared to 3D-CMOS-ASIC. To compare the performance with GPU, we also implemented the same code using Matlab GPU parallel toolbox. It takes 1163.42 s for training benchmark CIFAR-10, which is 4.858× faster than CPU. Comparing to our proposed 3D multi-layer CMOS-memristor architecture, our work is 3.07× speed-up and 162.86× energy saving (267.59 KJ :1.643 KJ). Detailed comparisons of each step is not shown due to the limited space of table.

6 Conclusion

In this chapter, we have presented the distributed in-memory matrix-vector multiplication accelerator using binary RRAM-crossbar for machine learning. The design of three-step digital matrix multiplier on the binary RRAM-crossbar is presented. In addition, the distributed in-memory computing architecture is introduced with the according control protocol of the digital memory-logic pair. The performance of the mapped machine learning can be boosted by the proposed accelerator with significant improvement in speed and energy efficiency.

Experiment results have shown that as for the matrix-vector multiplication, 72% smaller error can be observed when compared to the analog RRAM-crossbar. Moreover, 2.86× speedup and 105.6× power saving can be achieved when compared to the CMOS-ASIC. What is more, as for the machine learning based face recognition, 4.34× speedup and 13.08× power saving can be also achieved when compared to the CMOS-ASIC.

References

Akinaga, H., & Shima, H. (2010). Resistive random access memory (reram) based on metal oxides. *Proceedings of the IEEE*, *98*(12), 2237–2251.

Chen, P. Y., et al. (2015). Technology-design co-optimization of resistive cross-point array for accelerating learning algorithms on chip. In *IEEE date*.

Chen, Y.-C., Wang, W., Li H., & Zhang, W. (2012). Non-volatile 3d stacking rram-based fpga. In *22nd International conference on field programmable logic and applications (FPL)* (pp. 367–372). IEEE.

Chua, L. O. (1971). Memristor-the missing circuit element. *IEEE Transactions on Circuit Theory, 18*(5), 507–519.

Coates, A., Ng, A. Y., & Lee, H. (2011). An analysis of single-layer networks in unsupervised feature learning. In *International conference on artificial intelligence and statistics* (pp. 215–223).

Cong, J., & Xiao, B. (2014). Minimizing computation in convolutional neural networks. In *International conference on artificial neural networks* (pp. 281–290). Springer.

Fan, D., Sharad, M., & Roy, K., (2014). Design and synthesis of ultralow energy spin-memristor threshold logic. *IEEE Transactions on Nanotechnology, 13*(3), 574–583.

Fei, W., Yu, H., Zhang, W., & Yeo, K. S. (2012). Design exploration of hybrid cmos and memristor circuit by new modified nodal analysis. *IEEE Transactions on Very Large Scale Integration (VLSI) Systems, 20*(6), 1012–1025.

Glorot, X., & Bengio, Y. (2010). Understanding the difficulty of training deep feedforward neural networks. In *International conference on artificial intelligence and statistics* (pp. 249–256).

Gu, P., Li, B., Tang, T., Yu, S., Cao, Y., Wang, Y., & Yang, H. (2015). Technological exploration of rram crossbar array for matrix-vector multiplication. In *2015 20th Asia and South Pacific design automation conference (ASP-DAC)* (pp. 106–111). IEEE.

Haykin, S. S., Haykin, S. S., & Haykin, S. S. (2009). *Neural networks and learning machines* (Vol. 3). Pearson Education Upper Saddle River.

Higham, N. J. (2009). Cholesky factorization. *Wiley Interdisciplinary Reviews: Computational Statistics, 1*(2), 251–254. doi:10.1002/wics.18.

Hinton, G. E., Osindero, S., & Teh, Y. -W. (2006). A fast learning algorithm for deep belief nets. *Neural Computation, 18*(7), 1527–1554.

Huang, G.-B., Zhu, Q.-Y., & Siew, C.-K. (2006). Extreme learning machine: Theory and applications. *Neurocomputing, 70*(1), 489–501.

Huang, G. B., Ramesh, M., Berg, T., Learned-Miller, E. (2007). *Labeled faces in the wild: A database for studying face recognition in unconstrained environments*. Technical Report 07-49, University of Massachusetts, Amherst.

Kang, J., Gao, B., Chen, B., Huang, P.-Y., Zhang, F., & Deng, Y. et al. (2014). 3d rram: Design and optimization. In *2014 12th IEEE international conference on solid-state and integrated circuit technology (ICSICT)* (pp. 1–4). IEEE.

Kim, K. -H., Gaba, S., Wheeler, D., Cruz-Albrecht, J. M., Hussain, T., & Srinivasa, N., et al. (2011). A functional hybrid memristor crossbar-array/cmos system for data storage and neuromorphic applications. *Nano Letters, 12*(1), 389–395.

Kim, Y., Zhang, Y., & Li, P. (2012). A digital neuromorphic vlsi architecture with memristor crossbar synaptic array for machine learning. In *2012 IEEE international SOC conference (SOCC)* (pp. 328–333). IEEE.

Kouzes, R. T., Anderson, G. A., Elbert, S. T., Gorton, I., & Gracio, D. K. (2009). The changing paradigm of data-intensive computing. *Computer, 1*, 26–34.

Krishnamoorthy, A., & Menon, D. (2011). Matrix inversion using cholesky decomposition. arXiv preprint arXiv:11114144.

Krizhevsky, A., & Hinton, G. (2009). Learning multiple layers of features from tiny images.

Kumar, V., Sharma, R., Uzunlar, E., Zheng, L., Bashirullah, R., & Kohl, P., et al. (2014). Airgap interconnects: Modeling, optimization, and benchmarking for backplane, pcb, and interposer applications. *IEEE Transactions on Components, Packaging and Manufacturing Technology, 4*(8), 1335–1346.

LeCun, Y. A., Bottou, L., Orr, G. B., & Müller, K. -R. (2012). Efficient backprop. In *Neural networks: Tricks of the Trade* (pp. 9–48). Springer.

Lee, H., Che, P., Wu, T., Che, Y., Wan, C., & Tzen, P., et al. (2008). Low power and high speed bipolar switching with a thin reactive ti buffer layer in robust hfo2 based rram. In *IEEE international electron devices meeting, IEDM 2008* (pp. 1–4). IEEE.

Liauw, Y. Y., Zhang, Z., Kim, W., El Gamal, A., Wong, S. S. (2012). Nonvolatile 3d-fpga with monolithically stacked rram-based configuration memory. In *2012 IEEE international solid-state circuits conference* (pp. 406–408). IEEE.

Lichman, M. (2013). UCI machine learning repository. http://archive.ics.uci.edu/ml.

Liu, X., Mao, M., Liu, B., Li, H., Chen, Y., & Li, B., et al. (2015). Reno: A high-efficient reconfigurable neuromorphic computing accelerator design. In *2015 52nd ACM/EDAC/IEEE design automation conference (DAC)* (pp. 1–6). IEEE.

Lu, W., Kim, K. -H., Chang, T., & Gaba, S. (2011). Two-terminal resistive switches (memristors) for memory and logic applications. In *Design automation conference (ASP-DAC)*.

Matsunaga, S., Hayakawa, J., Ikeda, S., Miura, K., Endoh, T., & Ohno, H., et al. (2009). Mtj-based nonvolatile logic-in-memory circuit, future prospects and issues. In *Proceedings of the Conference on Design European Design and Automation Association: Automation and Test in Europe* (pp. 433–435).

Müller, K.-R., Tangermann, M., Dornhege, G., Krauledat, M., Curio, G., & Blankertz, B. (2008). Machine learning for real-time single-trial eeg-analysis: From brain-computer interfacing to mental state monitoring. *Journal of neuroscience methods*, *167*(1), 82–90.

Park, S., Qazi, M., Peh, L. -S., & Chandrakasan, A. P. (2013). 40.4 fj/bit/mm low-swing on-chip signaling with self-resetting logic repeaters embedded within a mesh noc in 45nm soi cmos. In *Proceedings of the Conference on Design, Automation and Test in Europe, EDA Consortium* (pp. 1637–1642).

Shang, Y., Fei, W., & Yu, H., (2012). Analysis and modeling of internal state variables for dynamic effects of nonvolatile memory devices. *IEEE Transactions on Circuits and Systems I: Regular Papers*, *59*(9), 1906–1918.

Singh, P. N., Kumar, A., Debnath, C., Malik, R. (2007). 20mw, 125 msps, 10 bit pipelined adc in 65nm standard digital cmos process. In *Custom integrated circuits conference, CICC'07* (pp. 189–192). IEEE.

Srimani, T., Manna, B., Mukhopadhyay, A. K., Roy, K., Sharad, M. (2015). Energy efficient and high performance current-mode neural network circuit using memristors and digitally assisted analog cmos neurons. arXiv preprint arXiv:151109085.

Strukov, D. B., Snider, G. S., Stewart, D. R., & Williams, R. S. (2008). The missing memristor found. *Nature*, *453*(7191), 80–83.

Suykens, J. A., & Vandewalle, J. (1999). Least squares support vector machine classifiers. *Neural processing letters*, *9*(3), 293–300.

Tan, T., & Sun, Z. (2010). CASIA-FingerprintV5. http://biometrics.idealtest.org/.

Topaloglu, R. O. (2015). *More than moore technologies for next generation computer design*. Springer.

Vaidyanathan, S., & Volos, C. (2016a). *Advances and applications in chaotic systems* (Vol. 636). Springer.

Vaidyanathan, S., Volos, C. (2016b). *Advances and applications in nonlinear control systems* (Vol. 635). Springer.

Wang, Y., Yu, H., & Zhang, W. (2014). Nonvolatile cbram-crossbar-based 3-d-integrated hybrid memory for data retention. *IEEE Transactions on Very Large Scale Integration (VLSI) Systems*, *22*(5), 957–970.

Wang, Y., Yu, H., Ni, L., Huang, G. -B., Yan, M., & Weng, C., et al.(2015). An energy-efficient nonvolatile in-memory computing architecture for extreme learning machine by domain-wall nanowire devices. *IEEE Transactions on Nanotechnology*, *14*(6), 998–1012.

Werbos, P. J. (1990). Backpropagation through time: What it does and how to do it. *Proceedings of the IEEE*, *78*(10), 1550–1560.

Williams, S. R. (2008). How we found the missing memristor. *Spectrum, IEEE*, *45*(12), 28–35.

Wold, S., Esbensen, K., & Geladi, P. (1987). Principal component analysis. *Chemometrics and Intelligent Laboratory Systems*, *2*(1–3), 37–52.

Wolpert, D. H. (1996). The lack of a priori distinctions between learning algorithms. *Neural Computation*, *8*(7), 1341–1390.

Wright, J., Yang, A. Y., Ganesh, A., Sastry, S. S., & Ma, Y., (2009). Robust face recognition via sparse representation. *IEEE Transactions on Pattern Analysis and Machine Intelligence*, *31*(2), 210–227.

Yu, H., & Wang, Y. (2014). *Design exploration of emerging nano-scale non-volatile memory*. Springer.

Memristive-Based Neuromorphic Applications and Associative Memories

C. Dias, J. Ventura and P. Aguiar

Abstract The recent realization of memristors opened the possibility to fabricate novel neuromorphic computational systems, including highly scalable and low power artificial neural networks. In fact, it has been shown that memristors can be used as an artificial synapse or to build the spiking core of an artificial neuron. The high resemblance between memristor and synaptic dynamics offers exciting possibilities in two major research fields: on one hand, memristors can be used to advance our understanding of the human brain, by supporting very-large-scale integration (VLSI) models where experiments can be performed and hypothesis tested in an in silico testbed. On the other hand, memristors have the potential to support novel advances in computing by providing the building blocks to bio-inspired computing paradigms, alternative to the von Neumann architecture, where storage and processing are supported by the same substrate. This chapter reviews the neuromorphic properties of memristors, comparing them with the key players of neuronal computations, synapses and neurons. The presentation is extended to more complex systems, where multiple computing units are combined in networks to achieve more elaborated dynamics. Emphasis is given to memristive-based associative memories, a bio-inspired content addressable memory system which relevant properties such as distributed storage and noise correction.

C. Dias (✉) · J. Ventura
Department of Physics and Astronomy, Faculty of Sciences, University of Porto,
IFIMUP and IN—Institute of Nanotechnology, Rua do Campo Alegre, 678,
4169-007 Porto, Portugal
e-mail: c.dias@fc.up.pt

J. Ventura
e-mail: joventur@fc.up.pt

P. Aguiar
i3S – Instituto de Investigação e Inovação em Saúde, Universidade do Porto,
Rua Alfredo Allen, 208, 4200-135 Porto, Portugal
e-mail: pauloaguiar@ineb.up.pt

P. Aguiar
INEB – Instituto de Engenharia Biomédica, Universidade do Porto,
Rua Alfredo Allen, 208, 4200-135 Porto, Portugal

© Springer International Publishing AG 2017

S. Vaidyanathan and C. Volos (eds.), *Advances in Memristors, Memristive Devices and Systems*, Studies in Computational Intelligence 701,
DOI 10.1007/978-3-319-51724-7_13

Keywords Memristor • Neuromorphic properties • Plasticity • Associative memories

1 Introduction

Building computers capable of learning and adapting to new environments has for long been an aspiration of the scientific community and is an essential step towards the realization of artificial intelligence. Modern-day computers, based on the deterministic von Neumann architecture, where memory and processing are physically separated, cannot accomplish this goal in an efficient and practical way. A novel approach is thus necessary and there is no better model than the human brain itself. The brain relies on a non-deterministic approach with massive parallelism of simple processing units—neurons—which take the role of essential building blocks in learning and decision-making. This distributed computing results in a significant power efficiency, adaptation and resilience to unit failure, all of which might be the keys to the creation of intelligent machines.

The communication between neurons is established by synapses and it is well known that synaptic strength is used to store information in brains. Learning is accomplished by modifying (either increasing or decreasing) this strength through a mechanism called synaptic plasticity. Emulating the biological synapse in an electronic circuit is the biggest challenge to the fabrication of a neuromorphic (brain-like) system. Fortunately this is now closer to becoming a reality, with the recent realization of memristors (Chua 1971; Strukov et al. 2008), a device with properties resembling those of biological synapses.

The memristor is a non-volatile two-terminal device, a metal-insulator-metal structure, characterized by a nonlinear relationship between the histories of current and voltage. Its response to a periodic voltage (or current) input is a "pinched hysteretic loop". The dynamic resistance and nanosized of memristors make them exciting candidates for electronic synapse applications and have inspired the neuromorphic community to explore their potential for building low-power intelligent machines. Noteworthy, different experimentally verified synaptic learning rules, such as spike timing dependent plasticity (STDP), have been faithfully reproduced in memristive devices.

Similar to what happens in biology with synaptically connected neural networks, synthetic memristive-based neural networks can be developed for purposes of machine learning and cognitive science. The presence of memristors allow these synthetic networks to be trained and to learn how to perform complex computations such as pattern recognition. One particular type of network architectures with computational interest is associative memory networks. These networks have the ability to store (learn) and recall (remember) associations between unrelated data items. Importantly, they work as content-addressable memories, where pairs of input/output patterns are stored in such a way that, if the input is presented, the output is readily given. Moreover, associative memory networks have robustness to noise: the system is able to recall the correct output pattern even if the presented input is a partial/noisy version of the original pattern (learnt during storage phase). The memristor fast speed, adaptive properties and small size meet the necessary specifications for these type of applications. Thereupon, memristor technology has the potential to revolutionize computing and scientific research in the coming decades.

The rest of the chapter is organized as follows. The next Sect. 2 provides a brief introduction to the key neurobiological mechanisms supporting information processing and information storage. Section 3 discusses the neuromorphic properties of memristors and their ability to mimic fundamental computational properties from single neurons and synapses. Then, Sect. 4 extends the discussion to the network level, where more complex computational capabilities can be achieved by combining a large number of memristive-based units. Section 5 is dedicated to a particularly relevant network architecture, the content-addressable associative memories. Finally, Sect. 6 outlines the conclusions of this chapter and discusses some of the open challenges for memristive-based neuromorphic computations.

2 Neural Computation

2.1 Brain Architecture and Operation

Two fundamental units of the human brain, the neuron and the synapse (Fig. 1), play essential roles in learning and memory formation. Neurons are electrically excitable cells which are able to respond to stimuli, to conduct impulses, and to communicate with each other. Synapses are specialized junctions between neurons that allow the rapid transmission of electrical and chemical signals so that neurons can communicate with each other (Shi et al. 2011). When an action potential generated by a neuron reaches a pre-synaptic terminal, a cascade of events leads to the release of neurotransmitters that give rise to a flow of ionic currents into or out of the post-synaptic neuron. The magnitude of these currents are subject to modulation—driven by synaptic plasticity mechanisms—allowing learning in the neuronal circuits. The amplitude or strength of a synapse is usually described as *synaptic weight*.

Figure 1 illustrates the basic structure of neurons and of their connection by synapses. The pre-synaptic neuron sends an action potential through one of its axons to the synaptic junction producing a response in the post-synaptic neuron. Action potentials, commonly named *spikes*, are fast (stereotyped) depolarizations in the neuron's membrane electrical potential which propagate, without attenuation, through axons. At rest, the membrane potential is close to −70 mV (and the membrane is said to be polarized; the reference is the extracellular space), and during a spike the membrane potential may be close to +30 mV (giving an amplitude of roughly 100 mV).

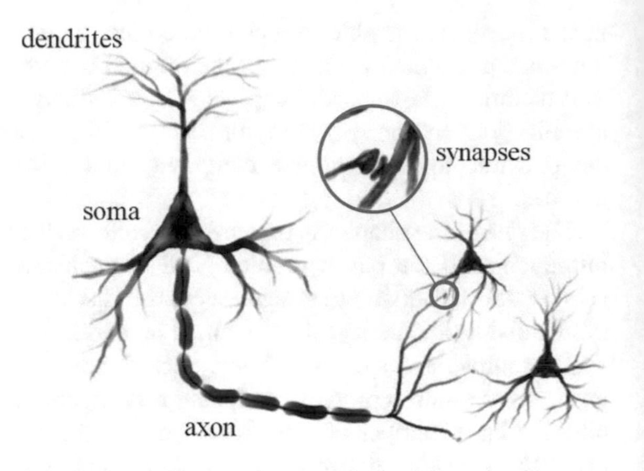

Fig. 1 Structure of a neuron showing the cell body, dendrites and synapses

2.2 *Learning and Memory*

Learning and memory are the capabilities to gain new information and store it in a recallable way. Although alternative mechanisms to store information have been studied (Conde-Sousa and Aguiar 2013), the current dogma in neuroscience is that information in the human brain is mostly stored in the synaptic strength (weight), with learning being accomplished by modifying (either increasing or decreasing) this strength. Synaptic plasticity makes it then possible to store information and to react to inputs based on past knowledge (Kandel et al. 2003).

2.2.1 Hebbian Learning

Conditions which are consistently experienced together tend to become associated. As a consequence, future exposure to one of the conditions automatically activates the response to the second condition (Kandel et al. 2003). One of the most famous experiments related to associative memory is Pavlov's experiment on classical conditioning. In this experiment, salivation of a dog is first set by the sight of food. Then, if the sight of food is accompanied by a sound (bell) over a certain period of time, the dog learns to associate the sound to the food, and salivation can start to be triggered by the sound alone (Kozma et al. 2012). In 1949, D. Hebb addressed this learning postulate at the neural level: "neurons that fire together, wire together" (Shi et al. 2011). Thus, when two connected neurons are active at the same time, the weight of the connecting synapse increases to reinforce that correlation. This learning rule, however, is incomplete since it provides a rule for increasing synaptic weight but not for decreasing it; also it does not specify the effective time window between pre- and post-synaptic activity that will result in potentiation.

2.2.2 Spike Timing Dependent Plasticity

Spike timing dependent plasticity (STDP) is an experimentally verified biological phenomenon in which the precise timing of spikes affects the sign and magnitude of changes in synaptic strength. STDP contains two distinct plasticity windows defining long-term potentiation (LTP) and long-term depression (LTD). In the former, synapses increase their efficiency as a pre-neuron is activated shortly (in the order of milliseconds) before a post-neuron; in the latter synapses decrease their efficiency as a post-neuron is activated shortly before a pre-neuron (Seo et al. 2011). While the association rule produced by Hebbian learning can be seen as supporting a correlation measure, STDP goes a step further in complexity by providing a (temporal) causality measure. As depicted in Fig. 2, the interspike interval between action potentials in the pre- and post-synaptic cells modulates STDP. The smaller the timing difference between pre- and post-synaptic spikes, the larger will be the induced plasticity change in either LTP or LTD. On the other hand, longer intervals (above 50 ms) produce little or no change in synaptic strength (Choi et al. 2011; Karmarkar and Buonomano 2002). This shows the existence of a causality window within the events and that, outside this causality temporal window, no changes are produced. The relative synaptic conductance change $\Delta G = (G_{after} - G_{before})/G_{before}$ [where G_{before} (G_{after}) is the conductance before (after) the pre- and post-spike pair], has a range of [0, +1[for potentiation and [1, 0] for depression (Choi et al. 2011; Yu et al. 2011). Another reason for the importance of STDP relies on the fact that it addresses both questions left open by Hebb: it establishes a critical time window in which pre- and

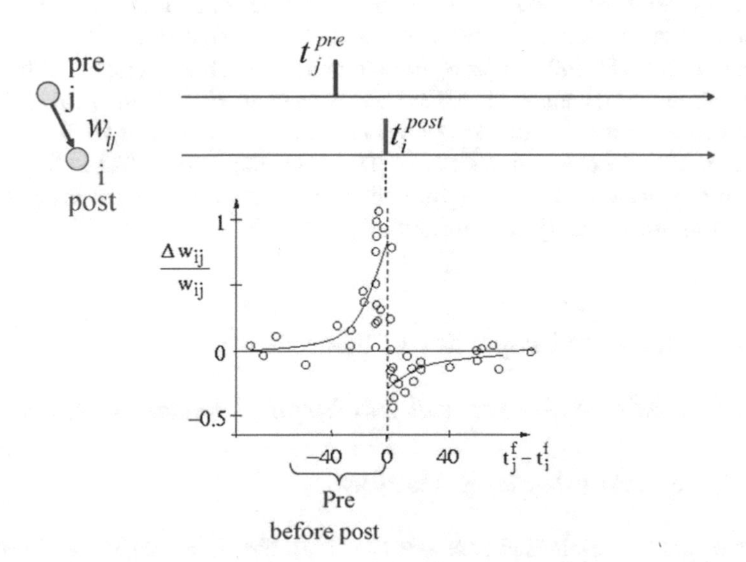

Fig. 2 STDP diagram showing the relative change in synaptic connection as a function of the time between pre- (t_j) and post-synaptic spikes (t_i) (Courtesy of *Scholarpedia Copyright* Owner © Wulfram Gerstner)

post-synaptic activity must occur to produce long-term changes in synaptic strength, and it provides a simple learning rule that decreases synaptic strength.

2.2.3 Short- and Long-Term Plasticity

The synaptic plasticity mechanisms in the human brain are diverse and occur at different time scales. Important computing properties can be achieved with synaptic modifications which are lost not long after the onset conditions. Therefore, in addition to long-term potentiation/depression (LTP/D) mechanisms, neurons also exhibit short-term potentiation/depression (STP/D). Short-term plasticity is mostly a modulation of the release probability of the pre-synaptic vesicles. As opposed to most synapses in the peripheral nervous system, which are deterministic, many types of synapses in the central nervous system are probabilistic: upon the arrival of an action potential at the synapse there is a probability p that a signal will be carried to the post-synaptic neuron. This stochasticity has deep implications in learning and optimization, especially because this release probability is modulated by short-term plasticity rules: a progressive, but temporary increase in p (increase in vesicle availability) is associated with STP, whereas a progressive decrease in p (depletion of the vesicles' pool) is associated with STD. In either cases, the changes in the release probability p are temporary, decaying to its initial state in scales of seconds to minutes. However, in situations where there is a strong consistency in the stimulation pattern, a different plasticity mechanism can be triggered (involving modifications in both in pre- and post-synaptic sites) and long-term changes can be produced, lasting from hours to days, and even beyond (lasting a lifetime). For example, repeated stimulation can cause a permanent change in the synaptic connection to reach LTP and shorter repetition intervals enable efficient LTP formation from fewer stimuli (Wang et al. 2012; Hasegawa et al. 2012; Ohno et al. 2011). On the other hand, consistent low frequency stimulation can lead to LTD. Note however that it is erroneous to associate S/LTP to learning and S/LTD to forgetting—information storage and memory formation in the nervous system involves plasticity in both directions (just as forgetting).

3 Neuromorphic Properties of Memristors

3.1 Memristive Synapses and Bio-inspired Learning Rules

3.1.1 Spike Timing Dependent Plasticity

Since the first proposals that resistive switching structures could mimic relevant properties of biological synapses, including spike timing dependent plasticity, short- and long-term potentiation/depression (Hu et al. 2013a, b; Snider 2008;

Kim et al. 2012a, b) and their integration in hybrid memristor-CMOS neuromorphic circuits (Indiveri et al. 2013), there has been a tremendous effort in experimentally demonstrating learning rules in these novel devices. To relate memristance to traditional biological spike timing dependent plasticity, one requires a voltage/flux controlled bipolar memristor with voltage threshold (below which no variation of the resistance is observed) and an exponential behavior beyond threshold (above which a continuous increment/decrement of the resistance takes place). It was also found that the strength of STDP learning in memristors (i.e. the induced conductance change) can be modulated by the amplitude or shape of the electric spikes. This means that the conductivity can be tuned depending on the precise timing between the pre- and post-synaptic spikes and the learning window by changing the shape of the pulses (Zamarreño-Ramos et al. 2011). Different STDP learning rules can also be obtained depending on the physical origin of resistance switching (Serrano-Gotarredona et al. 2013). For filamentary switching, where the memristor's conductance varies due to the formation and rupture of metallic filaments within the oxide, one expects a quadratic STDP learning rule, in which the synaptic strength update is proportional to the square of the synaptic strength. On the other hand, in interfacial or *domain wall* switching, related with uniform variations in the conductance of the oxide, the synaptic strength update is independent of the memristor's conductance (additive STDP update rule). Recently, a SPICE model able to account for STDP and synaptic dynamics in memristors was also implemented (Li et al. 2015).

Jo et al. were the first to demonstrate STDP in nanoscale memristors (Jo et al. 2010). Their crossbar memristive structure consisted in bottom tungsten and top chrome/platinum nanowire electrodes and a co-sputtered Ag and Si active layer in which the Ag/Si ratio varied along the depth (Fig. 3a). The authors then showed that the sample's conductance continuously increased (decreased) during the positive (negative) voltage sweeps, and that the current-voltage (I-V) slope of each subsequent sweep picks up where the last sweep left of (Fig. 3b). Instead of the usual abrupt, two level switching, in the co-sputtered memristor, the applied bias led to analog switching in which there is a continuous motion of the conduction front (Ag ions moving from the Ag-rich region to the Ag-poor region or vice-verse). This allowed the first demonstration of STDP in a memresistive structure. Figure 3c shows the change in the conductance (synaptic weight) of the memristor synapse as a function of the timing difference (Δt) between spikes arising from pre-synaptic and post-synaptic CMOS integrate-and-fire neurons. When the pre-synaptic neuron spikes before the post-synaptic neuron, the memristor conductance increases (potentiation behavior). The opposite (depressing behavior) is observed when the post-synaptic neuron spikes before the pre-synaptic neuron. Furthermore, the smaller the timing difference between the pulses, the larger is the change in the memristor conductance. Both these characteristics follow extremely well the STDP function of biological synapses, as can be seen in Fig. 2.

Since then, synaptic behavior was observed in several memristive structures, including in TiO_{2-x}/TiO_y bilayer systems showing multilevel conductance due to the movement of oxygen between the TiO_{2-x} and TiO_y layers (Seo et al. 2011),

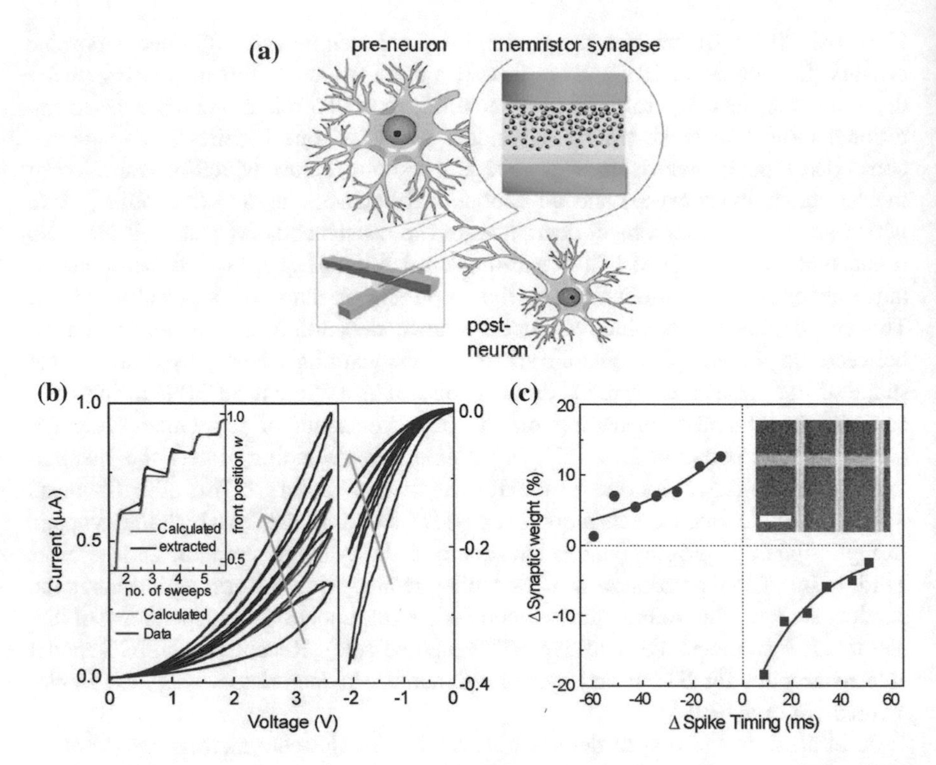

Fig. 3 **a** Schematics of the structure of the fabricated memristor with a gradient of Ag concentration. **b** Measured (*blue*) and calculated (*orange*) I-V characteristics and **c** corresponding memristor synaptic weight as a function of the relative timing of the neuron spikes. *Inset* SEM image of the crossbar array (scale: 300 nm; Reprinted with permission from (Jo et al. 2010). Copyright 2010 American Chemical Society)

$PrCaMnO_3$-based memristors, in which the fabrication of a 1 kbit synaptic array allowed to confirm the possibility to build a neuromorphic system for pattern recognition (Park et al. 2012), asymmetric memristors showing single-sided hysteresis (Williamson et al. 2013), Ag/conducting polymer/Ta memristive systems (Li et al. 2013), volatile and nonvolatile rectification in WO_{3-x}-based nanoionic devices (Yang et al. 2012), Ni-doped graphene oxide (Pinto et al. 2012), nanoparticle organic memory field-effect transistors (Alibart et al. 2012) or exploring multilevel switching in metal oxide memories (Yu et al. 2011).

Choi et al. fabricated $Pt/Cu_2O/W$ metal-insulator-metal (MIM) structures and experimentally demonstrated the successful storing of biological synaptic weight variations (Choi et al. 2011). They also showed the reliability of plasticity by varying the amplitude and pulse-width of the input voltage signal, matching their results with biological plasticity. A practical issue before the industrial implementation of memristors in actual large-scale neural networks is the dependence of the change of the memristive synaptic weight on its initial conductance. Nevertheless, experiments carried out with $Pt/Al_2O_3/TiO_{2-x}/Ti/Pt$ memristors in a

12 × 12 crossbar array, were able to demonstrate STDP behavior with self-adaptation of the average memristor conductance, making plasticity insensitive to the initial conductance state of the devices (Prezioso et al. 2016).

The implementation of STDP was also achieved in a second-order memristor (Kim et al. 2015). Kim et al. showed that the dynamics of $Pd/Ta_2O_{5-x}/TaO_y/Pd$ structures could be well explained considering two sets of state-variables. The first-order state-variable is related with the size of the conduction filament region (*w*) due to the diffusion of oxygen vacancies that sets the memristor conductance, while the second-order state-variable is the local temperature of the device that changes the dynamics of the diffusion process (Fig. 4a). The characteristic short times of temperature dynamics give rise to an internal timing mechanism and the possibility to affect *w* by providing close enough pulsed stimuli that device

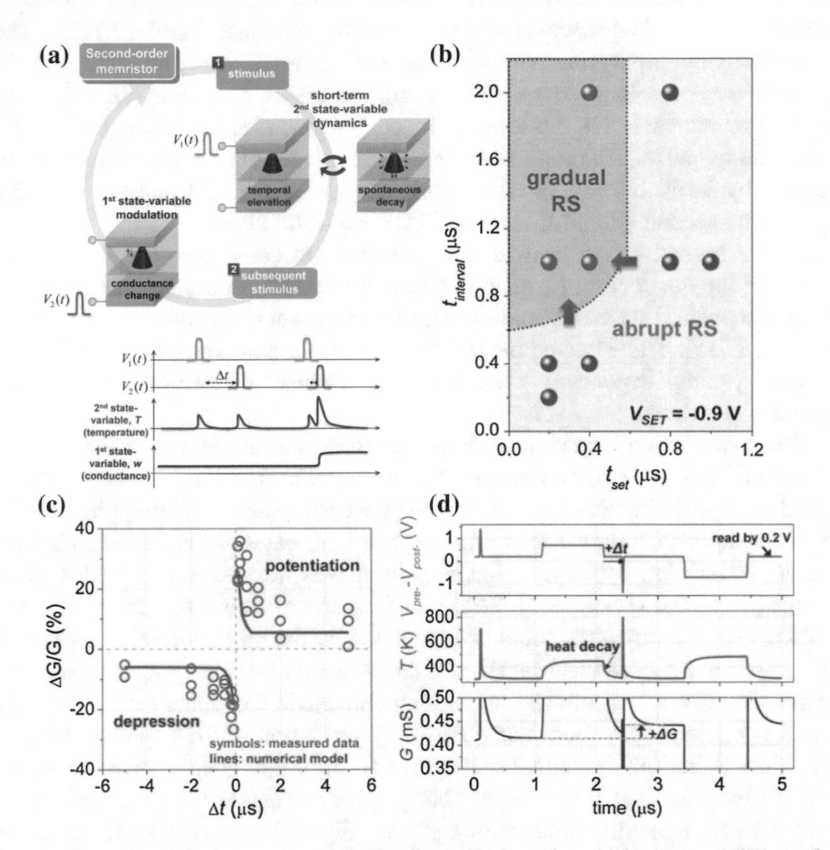

Fig. 4 **a** Operation principles of a second-order memristor in which the modulation of the 2nd-order state-variable (temperature) can trigger changes in the 1st-order state-variable (electrical conductance). **b** The obtained t_{Set} versus $t_{interval}$ operating window for gradual/abrupt resistive switching. **c** Observation of spike-timing dependent plasticity in a second-order memristor (measurements—*symbols* and simulation—*solid lines*). **d** Simulations on the evolution of the internal temperature of a second-order memristor when stimulated by two consecutive spikes (Reprinted with permission from (Kim et al. 2015). Copyright 2015 American Chemical Society)

temperature is still above its steady state value when the second pulse is triggered. This is closer to the case of biological than the first-order memristor, as in biological synapses the weight is also not directly modulated by the spikes, but rather by secondary state-variable(s) such as the postsynaptic calcium ion (Ca^{2+}) concentration. By providing a set of pulses to the fabricated structures, the authors determined the conditions (Set pulse duration t_{Set} and time interval between pulses, $t_{interval}$) for which the conductance showed a gradual variation which enables plasticity (Fig. 4b). Long pulses (>800 ns) always led to abrupt switching because of the large amount of generated heat. The same was observed for sets of short Set pulses at low $t_{interval}$ due to the accumulation of heat that fastens filament growth and thus leads to abrupt switching. However, for sufficiently large $t_{interval}$ and small t_{Set}, the heat generated by each pulse is dissipated before the next one arrives and intermediate states can be obtained (gradual switching occurs). This allowed the observation of frequency-dependent weight changes and STDP using non-overlapping input spikes with a biorealistic implementation (Fig. 4c). The input consisted in a long, low voltage heating pulse which resulted in a significant temperature increase, but no change in conductance and a short, high voltage programming pulse. Although each pulse could not change the memristor conductance by itself, the rise in temperature caused by the first pulse enhanced the effect of the second (Fig. 4d), so that STDP could be obtained.

Another second-order memristor implementation considered as state variables the area of the conducting filament between the two electrodes and the mobility of oxygen vacancies that was found to increase when a stimulation pulse was applied (Du et al. 2015). This allowed the observation of different synaptic behaviors such as spike timing dependent plasticity, paired-pulse facilitation or experience dependent plasticity.

STDP was also replicated in a purely electronic Ti/ZnO/Pt memristor (Pan et al. 2016). In this case, regulation of the current compliance during Set and maximum applied voltage during Reset allowed tuning the carrier trapping/detrapping level and thus the sample's conductance. Resistive switching based on homogeneous barrier modulation induced by oxygen ion migration was also shown to lead to neuromorphic properties (Wang et al. 2015). In fact, it was found that the resistance of $Ta/TaO_x/TiO_2/Ti$ structures could be modulated by the migration of O_2^- towards the TaO_x layer under negative bias (Reset) and towards the Ta/TaO_x interface under positive bias (Set). This allowed measuring and modelling spike timing dependent plasticity related with non-filamentary O_2^- evolution during potentiation and depression. Paired-pulse facilitation (PPF), a form of biological short-term synaptic plasticity in which synaptic weight changes are correlated with the time interval between two consecutive potentiating pulses (Fig. 5a) was also observed in these $Ta/TaO_x/TiO_2/Ti$ structures. The paired-pulse facilitation ratio was calculated using:

$$PPF = \frac{(G_2 - G_1)}{G_1} = C_1 e^{-t/\tau_1} + C_2 e^{-t/\tau_2}, \quad (1)$$

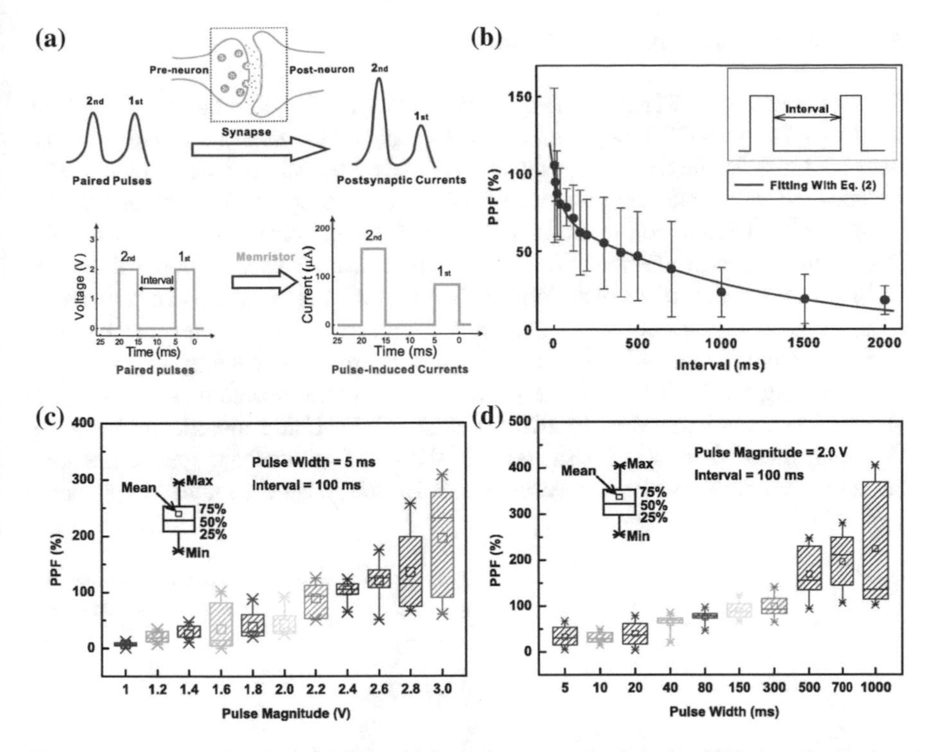

Fig. 5 Paired-pulse facilitation of **a** a biological synapse (*top*) and a NiO_x-based memristor (*bottom*). Dependence of the paired-pulse facilitation [Eq. (2)] of the memristor on **b** the pulse interval (for a pulse magnitude and width of 1.8 V and 5 ms), **c** pulse magnitude (for a pulse width and interval of 5 ms and 100 ms), and **d** pulse width (for a pulse magnitude and interval of 2.0 V and 100 ms; Reprinted from (Hu et al. 2013a, b), with the permission of AIP Publishing)

where G_1 and G_2 are the sample's conductance after the first and second pulses, respectively and C_1, C_2, τ_1 and τ_2 are fitting constants. The values obtained for the fast ($\tau_1 = 45$ ms) and slow ($\tau_2 = 800$ ms) time constants are in good agreement with those found in biological synapses. The emulation of biological PPF was also achieved in NiO_x-based memristors (Hu et al. 2013a, b). As in biological synapses, the change in the memristor's conductance induced by the second pulse is enhanced when compared with that of the first one (Fig. 5a). Again, the PPF magnitude decreases as the interval between the two pulses increases (Fig. 5b), going from 105.4% for 5 ms to 18.3% for 2000 ms. PPF also depended on the magnitude and width of the pulse pair (Fig. 5c, d). With both increasing pulse magnitude (from 1 to 3 V) and width (from 5 to 1000 ms) the average PPF value was also found to increase. These results were explained by the formation of conductive filaments of metallic Ni phases whose number or size increased with the application of the second pulse. However, the effect of the first pulse somehow fades away gradually, as the conductance change induced by the second pulse decreases with the pulse interval. Higher pulse voltages or longer pulses could also lead to the formation of more filaments or the growth of existing ones.

3.1.2 Short- and Long-Term Memory

Short-term memory (STM) and long-term memory (LTM), and the transition from STM to LTM through repetitions (rehearsal), have also been recreated in memristive devices. While STM can only be sustained by constantly rehearsing the same stimulus, LTM, despite the presence of natural forgetting, can be maintained for a longer period without follow-up stimuli (Fig. 6a). The terms short-term plasticity (STP) and long-term plasticity (LTP) are used in neuroscience, whereas STM and LTM are used to describe psychological phenomena (Ohno et al. 2011); plasticity is typically used in the context of localized changes (e.g. a synapse) and memory typically refers to system level changes (e.g. a neuronal population).

Depending on the input voltage pulses, different memorization behaviors were observed in memristive devices (Chang et al. 2011). Using memristors based on WO_x thin-films whose low resistance state showed a spontaneous loss of retention due to the random motion of oxygen vacancies following a stretched-exponential

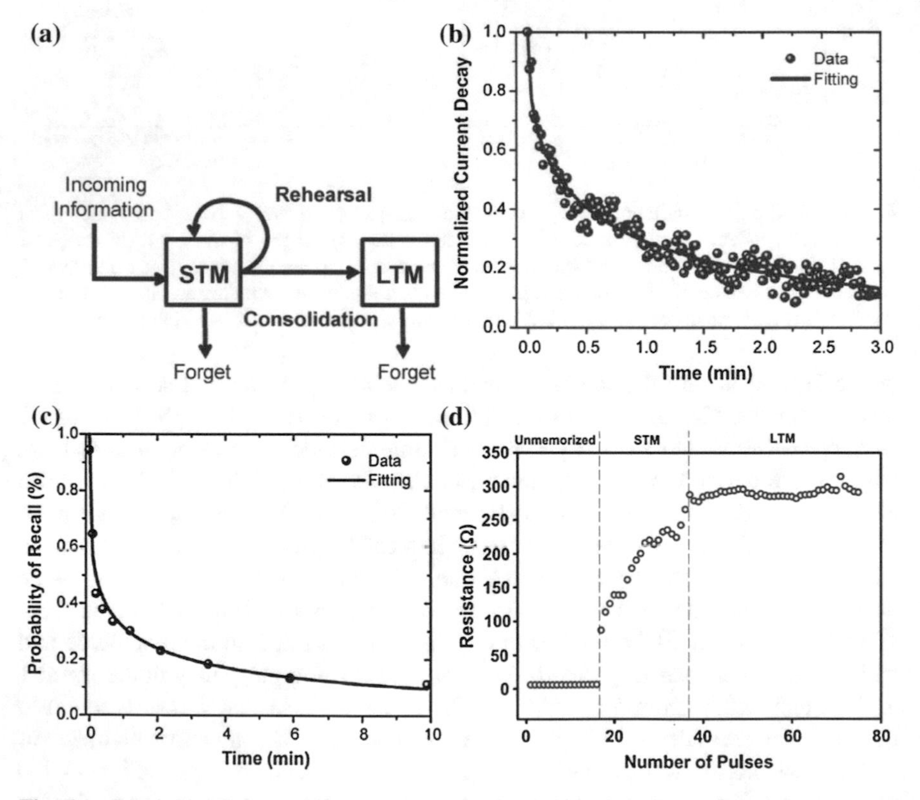

Fig. 6 **a** Schematic of the multi-store memory model, **b** WO_x-based memristor retention curve and **c** forgetting curve of human memory (Reprinted with permission from (Chang et al. 2011). Copyright 2011 American Chemical Society). **d** Three memory stages observed in NiO_x memristors upon application of consecutive voltage pulses (0.5 V in height, 0.1 s in width; Reprinted with permission from (Liu et al. 2011). Copyright 2011 American Chemical Society)

function law (Fig. 6b), Chang et al. demonstrated artificial properties close to those of memory loss in biological structures (Fig. 6c). However, if repeated stimulus are given, both the overall conductance and the retention time of the device are enhanced with increasing number of stimulus and stimulation rate, showing similarities with paired-pulse facilitation and post-tetanic potentiation. The transition between short-term and long-term memory was then suggested based on the existence of two memory regimes with short and long retention times.

As shown in Fig. 6d, three memory stages (unmemorized, STM and LTM) were also observed in a Ni-rich nickel oxide device by Liu et al. (2011). Memorization from STM to LTM is obtained by increasing the number of pulses. Behaviors such as STM, LTM, STDP and spike-rate dependent plasticity were also observed in Ta/PEDOT:PSS/Ag (Li et al. 2013). Tsuruoka et al. found LTM in an $Ag/Ta_2O_5/Pt$ cell under voltage bias for a high repetition rate of input pulses, which is analogous to the behavior of biological synapses (Tsuruoka et al. 2012) and Wan et al. mimicked STM and LTM in nanogranular phosphorus-doped SiO_2 films by tuning the pulse gate voltage amplitude (Wan et al. 2013). The compliance current during electroforming was also shown to have an impact on the retention and analog properties of FeO_x based memristors (Wang et al. 2016).

Wang et al. reported a spontaneous decay of the synaptic weight in an amorphous InGaZnO memristor (Wang et al. 2012). The decay is very fast in the initial stage and then gradually slows down, which is consistent with the human "forgetting curve". Their results also indicate that in synaptic devices with "learning-experience", re-learning is easily achieved (Fig. 7). Also, with increasing

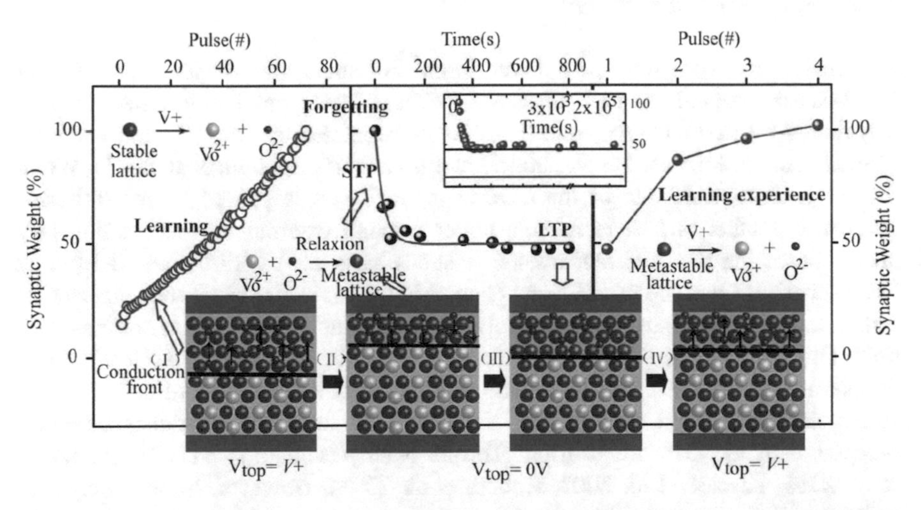

Fig. 7 **a** InGaZnO memristor conductance (synaptic weight) increase with the number of applied pulses. **b** Spontaneous relaxation of the conductivity (forgetting curve) upon removal of the stimulus and corresponding fit to an exponential function. **c** Re-stimulation upon voltage decay, showing now the need for only four pulses to reach the high conductance state. *Bottom inset* shows the proposed oxygen ion migration model (From (Wang et al. 2012), Advanced Functional Materials, Copyright © 2012 by [John Wiley & Sons, Inc.]. Reprinted by permission of [John Wiley & Sons, Inc.])

number of stimulation pulses, the relaxation time increases from several seconds to tens of seconds and tends to saturate beyond 100 stimulations, indicating a decreasing forgetting rate (Wang et al. 2012).

Using TiO_2 bipolar memristors, R. Berdan et al. demonstrated short-term plasticity with protocols similar to those used for biological synapses (Berdan et al. 2016). Figure 8a shows that the application of voltage pulses initially leads to volatile switching, in which the memristors conductance decays back to its low value. However, the successive application of voltage pulses eventually leads to non-volatile switching to the high conductance state. Two cases of short-term facilitation were emulated by providing three consecutive voltage pulses: in the first (Fig. 8b), the memristor's conductance is increased by each of the applied pulses (classical short-term facilitation; STP-F); however, in some particular situations, the conductance after the second and third pulses decreased with respect to that obtained after the first voltage spike (Fig. 8c); saturation short-term facilitation STP-S]. The latter effect was explained in terms of a mobility saturation of oxygen vacancies near a partially formed filament. Spatio-temporal computation was also achieved using these memristive synapses and an exponential integrate-and-fire neuron. The built circuit was able to differentiate two spatio-temporal patterns each consisting of three −4 V, 10 μs pulses separated by 250 ms applied first to a static resistance and then to the memristive synapse (AB) or vice-versa (BA). A discrimination success rate of 67.5% with 15% false positives was obtained.

3.1.3 Spintronic Memristors

Memristive behavior has also been found in spintronic devices, particularly MgO-based magnetic tunnel junctions (MTJs). These are two-terminal devices constituted by two ferromagnetic (FM) layers separated by an MgO thin insulator (Parkin et al. 2004; Yuasa et al. 2004; Ikeda et al. 2010; Teixeira et al. 2011). While the magnetization of one of the ferromagnetic layers is pinned by an antiferromagnet, the other is free to reverse under a small external magnetic field. This allows obtaining two different resistance states associated with the parallel (when the magnetization of the two FM layers point in the same direction) and anti-parallel (when they are opposite) configurations by simply applying an external magnetic field. Furthermore, the discovery of the Spin Transfer Torque (STT) effect opened the possibility to switch the magnetization of the MTJ by passing a sufficiently high spin polarized current through the stack or to induce persistent magnetization precession in the GHz range (Spin Torque Nano Oscillators; STNOs) (Locatelli et al. 2014; Kiselev et al. 2003; Kubota et al. 2008). However, besides magnetic switching, these structures also display resistance variations of non-magnetic origin arising from ionic migration within the ultra-thin insulating barrier: resistive switching (Krzysteczko et al. 2009a, b, 2012; Teixeira et al. 2009; Ventura et al. 2007; Yoshida et al. 2008).

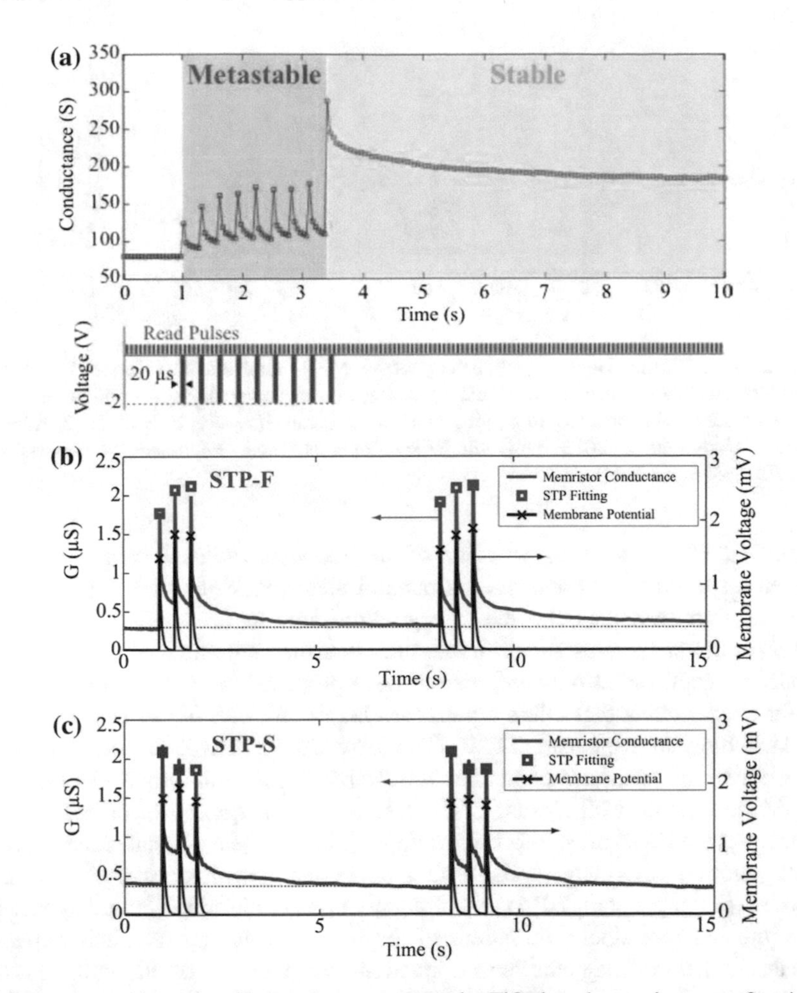

Fig. 8 **a** Metastable and stable conductance states in TiO_2-based memristors as function of voltage stimulus. Short-term plasticity and of a memristor acting as a **b** facilitating (STP-F) and **c** saturated (STP-S) synapse and corresponding fitting (Reprinted from Berdan et al. 2016; used in accordance with the Creative Commons Attribution (CC BY) license)

Krzysteczko et al. were the first to demonstrate that MTJs possess the characteristics of both of synapses (arising from non-magnetic resistive switching) and neurons (driven by STT effects; see below) (Krzysteczko et al. 2012). Figure 9a shows their observation of STDP in magnetic tunnel junctions given a proper choice of the amplitudes of the voltage function. When the time interval between the pulses (Δt) is high, the joint effect of the two spikes does not exceed the threshold voltage for switching, so that no variation of the conductance is seen. On the other hand, for low Δt, the applied voltage goes above the threshold and increasingly larger conductance variations are seen with decreasing pulse interval. The same work also reported that, by applying an electrical current large enough to

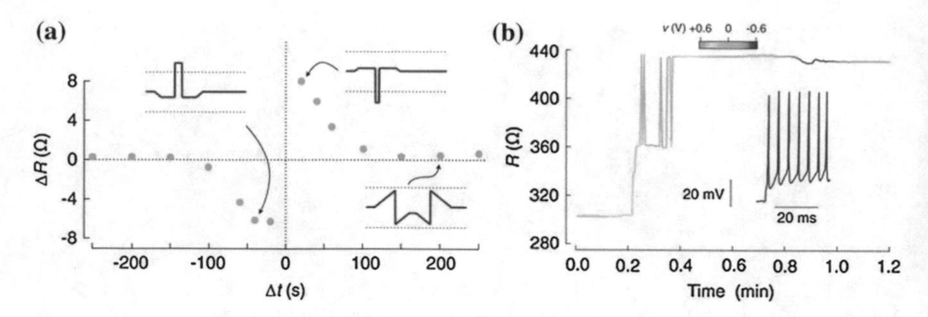

Fig. 9 **a** Spike Timing Dependent Plasticity of an MgO-based magnetic tunnel junction and **b** Spin Transfer Torque driven stochastic switching between two magnetic configurations. The *inset* shows the spiking behavior of a pyramidal neuron (From Krzysteczko et al. 2012, Advanced Materials, Copyright © 2012 by [John Wiley & Sons, Inc.]. Reprinted by permission of [John Wiley & Sons, Inc.])

induce STT effects, the magnetization of the free layer switched back-and-forward between an intermediate and the antiparallel state (back-hopping; Fig. 9b). Such stochastic current spikes showed a large similitude to the spiking of biological neurons and were proposed to emulate inter-neuronal communication.

Fully magnetic spintronic memristors were proposed in 2009 based on the spin transfer torque effect that offers a path for energy efficient, high-speed magnetization switching in nanoscale MTJs (Dimitrov 2009). These included magnetic tunnel junctions with perpendicular anisotropy under spin-torque excitations or STT-driven domain wall motion of the free layer of a spin valve or MTJ. A four terminal device based on spin-orbit torque by joining a heavy metal and a magnetic tunnel junction displaying spike-timing dependent plasticity was also recently proposed (Sengupta et al. 2015). An alternative proposal for spin based synapses is to use the inherent stochastic nature of the binary switching (parallel/antiparallel resistance states) of magnetic tunnel junctions that depends on the voltage amplitude and duration of the STT pulse (Vincent et al. 2015; Kavehei and Skafidas 2014). Following such proposal, it was possible to implement a simplified STDP rule, in which STDP occurs only when an output neuron spikes and the STT-switching of the MTJ conductance is stochastic rather than deterministic. System level simulations of a crossbar array of MTJs following the developed model demonstrated the possibility to train the network to learn to detect vehicle in a video. The impact of device-to-device variations of the parallel/antiparallel conductances and tunnel magnetoresistance on the detection rate of the vehicle counter was also studied and found to be robust for relative standard deviation (one-sigma) up to 17%.

3.1.4 Atomic Switches

In atomic switches, sometimes also called electrochemical metallization cells, or conductive-bridge random-access memories, resistive switching is related with the migration of metallic cations (M) from the active electrode (usually Cu or Ag) through an ionic conductor and the formation/annihilation of a conductive filament (Hasegawa et al. 2012; Jana et al. 2015; Goux and Valov 2016; Valov et al. 2011). Under the action of a positive bias (with the inert electrode grounded), ions from the active electrode (anode) are oxidized and diffuse towards the inert cathode where they are reduced. This ultimately results in the formation of a metallic filament and thus the high conductance state. The application of a reverse voltage results in the oxidation of the metallic ions from the filament and their reduction at the active electrode, resulting in the dissociation of the metallic filament and thus the low conductance state.

The first observation of learning abilities in atomic switches was achieved in silver rich Ag_2S-based structures (Hasegawa et al. 2010). In this case, an Ag protrusion grows between the Ag_2S and Pt electrodes across a vacuum gap by the action of an applied voltage, ultimately bridging the two, with learning/unlearning occurring with the widening/thinning of the Ag atomic bridge. With the same type of structure, Ohno et al. were able to emulate both STP and LTP depending on the stimulation rate (Fig. 10) (Ohno et al. 2011). When stimulated with a low repetition rate (at 20 s intervals), the conductance of the atomic switch increases to a high conductance state but spontaneously decreases to the low conductance state after each pulse is removed rate (Fig. 10a). This was related with the formation of an incomplete metallic bridge that dissolves when the applied bias voltage is removed and can be associated with short-term plasticity of biological synapses. On the other hand, the application of high repetition rates (2 s intervals) results in the formation of a robust atomic bridge between the electrodes (persistent high conductance state) and thus to a transition to LTP (Fig. 10b). The authors also showed the possibility to implement the so called multistore model: sensory information is initially stored as a sensory memory and then selected information is transferred from a temporary short-term state to a permanent long-term state through rehearsal, depending on the amplitude and width of the voltage stimulus (Ohno et al. 2011). For this, two images were stored in a 7×7 array of Ag_2S atomic switches (Fig. 10c). While the image of the number ‘2’ was stored using well separated voltages pulses (20 s; corresponding to the STP case), that of number ‘1’ was stored using closer pulses (2 s; which led to LTP). Initially both numerals were present, but ‘2’ started to disappear as soon as the voltage pulses were removed due to the prompt decay of the conductance associated with the STP state and only the ‘1’ image was transferred to the LTM mode and remained after stimulus stopped. The short-term dynamics of Ag_2S atomic switches was also used to encode input spike patterns (Ma et al. 2015).

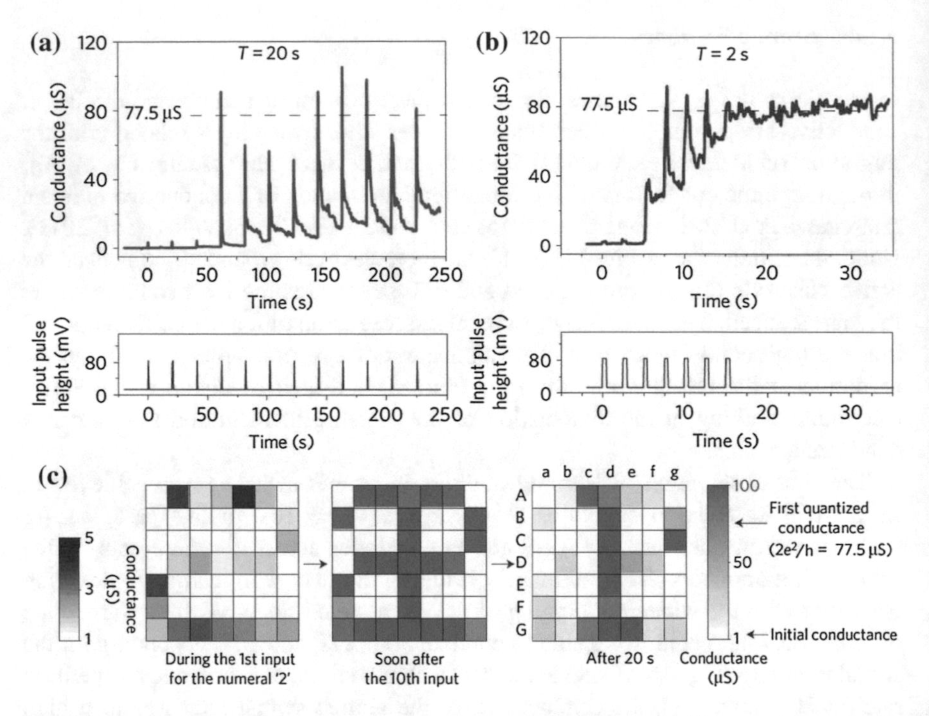

Fig. 10 Conductance variation of an Ag_2S atomic switch for **a** low and **b** high repetition rates (for 80 mV pulses) showing STP and LTP, respectively. **c** Storage of the numerals '1' (in the LTM mode) and '2' (in the STM) in a 7 × 7 array and the decay of the '2' memory 20 s after short-term memorization (Reprinted by permission from Macmillan Publishers Ltd: Nature Materials (Ohno et al. 2011), copyright 2011)

Furthermore, Barbera et al. showed that the STP/LTP transition could be tuned by both the used current compliance during SET (I_c; switching from the low to the high conductance states) and the number of excitatory pulses that can be independently changed (Fig. 11) (Barbera et al. 2015). It was shown that higher current compliances led to larger ON state conductance due to an increased density or width of the formed (dendritic) filaments. By measuring the relaxation time (τ) of the ON state conductance after the device was exposed to a train of potentiation pulses with varying number of pulses (from 15 to 150), it was possible to observe that samples with higher conductance before relaxation (G_{max}; obtained using either higher number of pulses or compliance current) had shorter relaxation time constants (Fig. 11a). As G_{max} increased, so did the stability of the filaments which then lead to higher τ-values and to a transition from STP to LTP. The same overall result can be seen in (Fig. 11b), where G_{100s} stands for the conductance value 100 s after the train of potentiation pulses was applied and $G_{100s} \approx G_{max}$ indicates that no relaxation of the conductance took place, i.e. long-term plasticity.

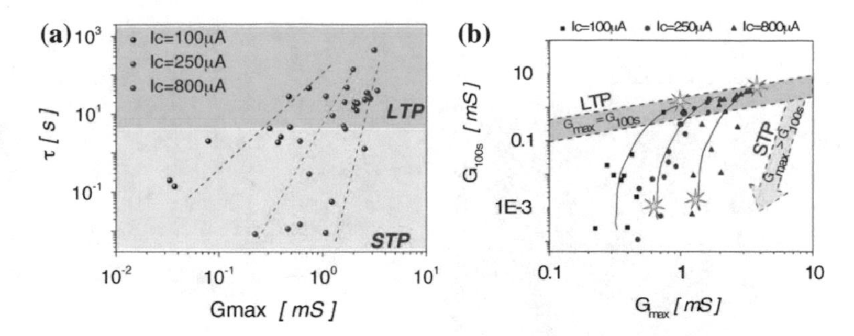

Fig. 11 **a** Relaxation time constant as a function of the conductance value after a train of potentiation pulses with varying number of pulses (from 15 to 150) is applied (G_{max}) for different compliance currents. **b** Conductance measured 100 s after the application of the train of potentiation pulses for different compliance currents (Adapted with permission from (Barbera et al. 2015). Copyright 2015 American Chemical Society)

An alternative proposal was recently reported based on stochastic learning rules using GeS_2-atomic switches (Suri et al. 2013). Taking into account the intrinsic stochastic switching behavior of atomic switches at low applied voltage, a stochastic STDP rule, similar to that presented for spintronic magnetic tunnel junctions (Sect. 3.1.3), was developed and used for auditory and visual pattern extraction.

3.2 Neuristors

A neuristor is a circuit capable of performing neural functions, since it successfully generates an action potential upon sufficient excitation, and emulates its propagation through an axon following the conditions of threshold action, refractory period and constant propagation. The term was carved in 1960 by H.D. Crane and, to implement it, one only requires three components: an energy source, an energy-storage element and a negative resistance (active) device (Crane 1960; Lu 2012; Pickett and Williams 2013).

The neuristor is a dynamical spiking device where the logic ‘1’ is the existence of a spike and ‘0’ its absence. It is a system able to compute all Boolean functions, so that it is said to be logically complete, allowing the duplication of any logic system (Crane 1960) and enabling massive parallel bio-inspired computing architectures (Pickett and Williams 2013). Following this idea, in 2012 Pickett et al. fabricated a neuristor based on two Mott memristors exhibiting neural functions, where an insulating-to-conducting phase transition (in NbO_2) takes place due to Joule heating and results in the creation of a conductive channel between the two electrodes (Fig. 12) (Pickett et al. 2012). The implemented circuit (Fig. 12a) is composed by two units: one equivalent to the sodium channels and the other to the potassium ones. Each capacitor serves to build up charge and the memristor in

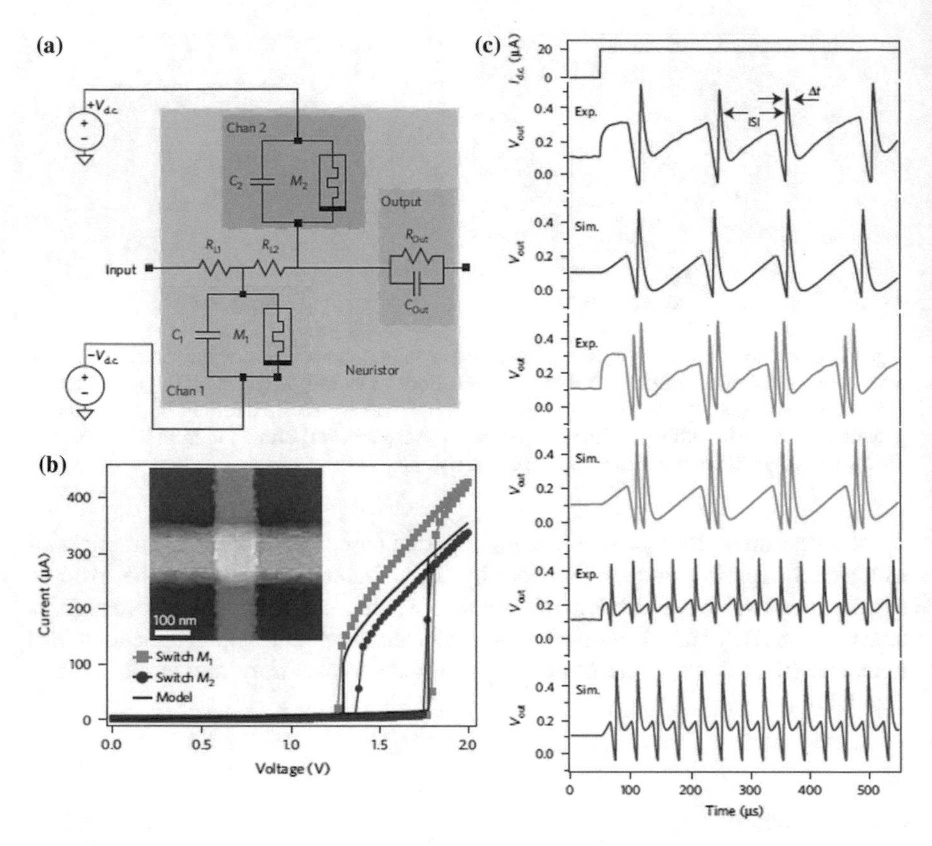

Fig. 12 **a** Neuristor diagram and **b** Mott memristors characteristics, **c** experimental and simulated spikes for different inter-spike interval (ISI) and spike width (Δt) (Reprinted by permission from Macmillan Publishers Ltd: Nature Materials (Pickett et al. 2012), copyright 2012)

parallel to release it suddenly. When a voltage threshold is exceeded, the Mott insulators change into the metallic phase (Fig. 12b), discharging the capacitors and spikes of activity that emulate an axon action potential are produced (Fig. 12c). Changing the components values, a variety of spiking behaviors were achieved, both experimentally and in simulations. Mott memristors turning ON and OFF can therefore be seen as electronic and inorganic analogues of the biological neuronal ion channels opening and closing. These neuristors can be downscaled and either be integrated with existing circuits or implemented in transistorless designs. With this scalable implementation, there is no need for a comparator and a logic element in the artificial neuron design (Chabi et al. 2014; Zhou and Ramanathan 2015). Furthermore, these schemes should allow a deep study of the nervous system (Pickett et al. 2012).

Mehonic et al. also implemented a simplified circuit model to emulate the neuron electrical activity using a SiO_x memristor (Mehonic and Kenyon 2016). They observed voltage spiking and a dynamic voltage output when applying a constant or pulsed current input. Gale et al. studied non-ideal memristive networks and concluded that the richness and dynamics complexity of spiking circuits increases for three memristors in anti-series and/or anti-parallel, when compared to systems with only two memristors (Gale et al. 2014). However, if the three memristors have the same polarity, the circuit is stable and does not spike. They further simulated eight circuit compositions of up to three memristors both in series and in parallel. Notice that, in all these cases, memristive devices are used to assume the functions of a neuron, a complement to the synapse behavior to be used in neuromorphic systems.

4 Synthetic Neural Networks

As in conventional electronics, significant processing power and complexity is achieved in the human brain by combining multiple simpler computational units (i.e. synapses and neurons) into large network systems with well-defined architectures. There has been, therefore, several attempts to mimic the biological learning rules in artificial synapses and to construct artificial neural networks (ANNs) capable of performing complex functions. A network is based on the transmission of events from one source node (neuron) to multiple nodes by edges (synapses; see examples in Fig. 13). In most ANN models, synapses are dynamical two-terminal entities that connect a pre- (source) to a post-synaptic neuron (sink). The source emits a signal that is modified by a synaptic transfer function and delivered to the sink. To facilitate the communication between neurons, the action potential is propagated as a digital pulse (Schemmel and Grubl 2006). The output of a neural network node is a function of the sum of all input signals (Ha and Ramanathan 2011). The sink has a state variable that partially depends upon the history of incoming signals received from synapses that drive it. This variable along with the source signal determine the evolution of the synaptic state variable.

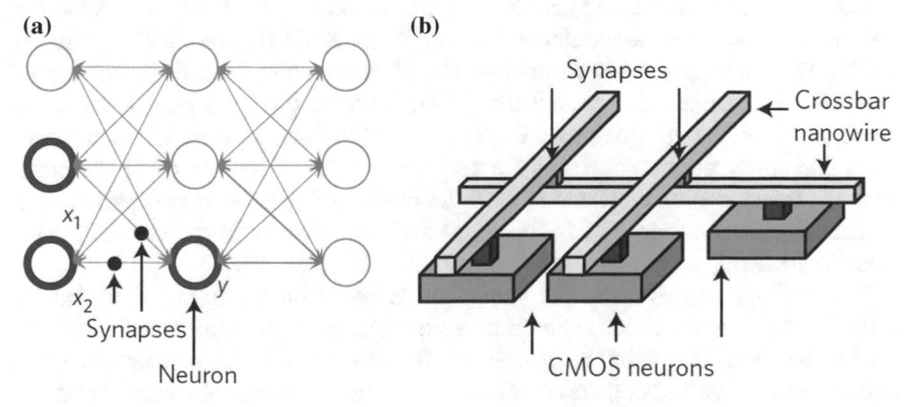

Fig. 13 **a** Graph and **b** crossbar network architectures (Reprinted by permission from Macmillan Publishers Ltd: Nature Nanotechnology (Yang et al. 2013), copyright 2013)

A radical approach in the construction of artificial neural networks is to use very large scale integration (VLSI) to implement directly in silicon the required computational model of a neural system. IBM researchers built a complex chip using 5.4 billion transistors to simulate 1 million neurons and 256 million synapses (Merolla et al. 2014).

In neuromorphic implementations, the key challenge is to design circuits with large time constants while keeping the neuronal structure simple, occupying small silicon area and using only one electronic device as an artificial synapse. However, the silicon area occupied by the synaptic circuit can vary significantly, as it depends on the choice of layout design solutions and more conservative solutions use large transistors. Implementing the large connectivity of the brain with transistors on a single chip is a major challenge, since a large number of transistors are needed (Shi et al. 2011; Seo et al. 2011). Therefore, the electronic conventional implementation is not practical and a simple and scalable device able to emulate synaptic functions is required (Seo et al. 2011). As we have seen, the memristor displays such properties, making it the most promising candidate to be used in scalable neural networks.

Examples of different network architectures are discussed here, presenting configurations, which support operations such as classification (perceptron) or in-formation storage/memories.

4.1 Perceptron

In 1957, F. Rosenblatt developed a simple NN for pattern classification problems of linearly separable patterns: the perceptron (Fig. 14) (Rosenblatt 1957). It is the simplest kind of NN capable of learning and parallel processing (Wang et al. 2013). It consists of a main neuron that accepts several inputs from sensory neurons, connected by adjustable synaptic weights, and sums all weighted inputs (Fig. 14a). Depending on the result this neuron will fire (or not) if the result is positive (or negative), based on error-correlation learning (Park 2006; Haykin 2009; Daumé III 2012). The learning process for pattern classification occupies a finite number of iterations. Rosenblatt also proved that, if the patterns (vectors) used to train the perceptron are drawn from two linearly separable classes, then the perceptron algorithm converges and positions the decision surface in the form of a hyperplane between the two classes (Haykin 2009), a division of space into two halves by a straight line, where one half is "positive" and the other is "negative" (Fig. 14b) (Daumé III 2012).

The learning algorithm of the perceptron is based on the information that, in brains, many neurons encode stimuli intensity in terms of their firing rate (which in turn defines how "activated" the neuron is). Therefore, based on how much the input neurons fire and how strong the neural connections are, the main neuron will respond accordingly. Note that learning is nothing more than neurons becoming connected with each other and adapting their connection strength over time (Daumé III 2012).

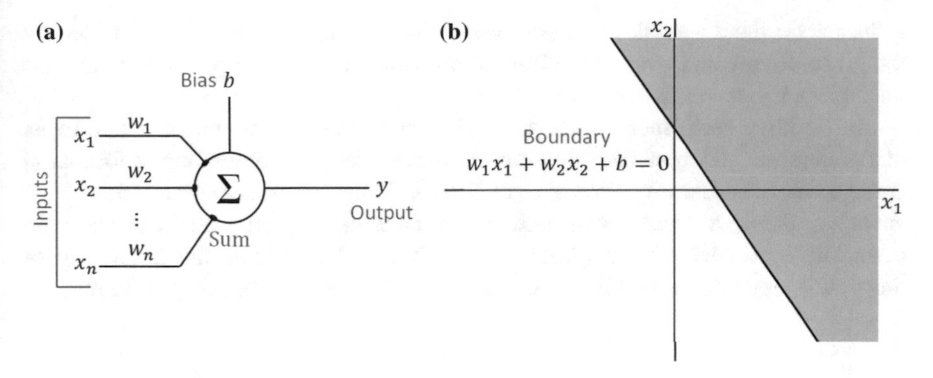

Fig. 14 **a** Single layer perceptron and **b** illustration of the hyperplane for a two-dimensional classification problem

Mathematically, the perceptron can be modulated as an input vector $x = x_1, x_2, \ldots, x_n$ arriving from n neurons, with n stored weights, $w_1, w_2, \ldots, w_n$, at the main neuron that computes a sum, a. Also, it is often convenient to have a non-zero threshold, which is achieved introducing a single scalar bias term into the neuron, so that activation is always increased by some fixed value b. The overall sum a, parameterized by n weights, and a bias value b is given by (Daumé III 2012):

$$a = \left[\sum_{i=1}^{n} w_i x_i\right] + b. \tag{2}$$

The weights are easy to interpret: if one input has a zero weight, then the activation is the same regardless of its value. Furthermore, positive (negative) weights are indicative of positive (negative) examples because they cause the activation to increase (decrease). This is the first learning algorithm in which the abilities of learning and pattern classification were achieved by Artificial Intelligence (Park 2006) and one of those algorithms that is incredibly simple and yet works amazingly well for some types of problems (Daumé III 2012).

As a classifier, perceptron applications include pattern recognition (fingerprint and iris) (Wang et al. 2013; Ashidi et al. 2011; Gatet et al. 2009), classification of medical images (Wang et al. 2013) (cancer classification for example Rosenblatt 1957) or gene array analysis (Bo et al. 2006), surface classification, object detection, distance measurement (Gatet et al. 2009), and forecast ozone and nitrogen dioxide levels measurement in real-time (Agirre-Basurko et al. 2006). The perceptron can thus be determinant in many fields such as obstacle detection for autonomous robots or vehicles, identification, surveillance, security systems, medical applications, industrial processes and navigation (Wang et al. 2013; Gatet et al. 2009). As an example, imagine for the medical case that you have databases from previous diagnosed patients for different disease indicators, each with a certain influence on the diagnosis. You can then train your perceptron to learn to identify the given disease and even make it more probable to give false positives than false negatives or vice-versa.

As we have seen, memristors are highly attractive for this purpose due to its reconfigurable and analogue resistance, nanoscale size, automatic information storage and non-volatility (Wang et al. 2013). Alibart et al. achieved pattern classification using a single-layer perceptron network implemented with a TiO_2 memrisitive crossbar circuit (Alibart et al. 2013). This shows the possibility of fabricating new, dense and high performance information processing systems.

4.2 Memristive Artificial Neural Network

The recent progress in the experimental realization of memristive devices has renewed the interest in artificial neural networks (Thomas 2013) and many different learning laws have been proposed for edges (Snider 2007). Adjustable edge weights are the defining characteristic of neural networks and are the origin of their broad adaptive functionality (Ha and Ramanathan 2011). An edge's conductance changes as a function of the voltage drop across the edge induced by forward spikes from the source node and back spikes from the sink node. Using memristive nanodevices to implement edges, conventional analog and digital electronics to implement nodes, and pairs of bipolar pulses (spikes), to implement communication, it is possible to develop fully electronic neural networks (Snider 2007). For a very large number of synapses, a practical implementation of an artificial network using memristors allows the weights to be updated in parallel due to the high interconnectivity. For this, the logic value of the input ('1' and '0') is multiplied by the memristance value (Ha and Ramanathan 2011; Strukov and Kohlstedt 2012; Rose et al. 2011). It should be noted that, for ANNs, the density of the memristive devices is the most important property. Also, ANNs are resilient to variations in synapses and neurons (Strukov and Kohlstedt 2012). For example, instead of a single pulse, the average effect of hundreds of parallel synapse inputs into one neuron determines whether the neuron will fire or not (Yu et al. 2011).

One of the possible applications of memristor-based ANNs is to carry out position detection. This was simulated in Ref. (Ebong and Mazumder 2012), using ANNs that combine winner-take-all and STDP learning rules. Random networks of polymer coated Ag and oxide passivated Ni nanowires (Fig. 15), where their placement is not important and differences in properties are averaged out, also presented I-V memristive-like behavior (Nirmalraj et al. 2012). Atomic switch networks of $Ag/Ag_2S/Ag$ with random topology nanowires similar to Turing's B-Type unorganized machine were also fabricated using SU-8 photoresist and Ag_2S nanowires where filaments formation takes place at the atomic level (Fig. 16a, b; see Sect. 3.1.4) (Stieg et al. 2012; Stieg and Avizienis 2014). After an electroforming step, the electrical characterization of the whole network of highly interconnected atomic switches using macroscopic electrodes reproducibility

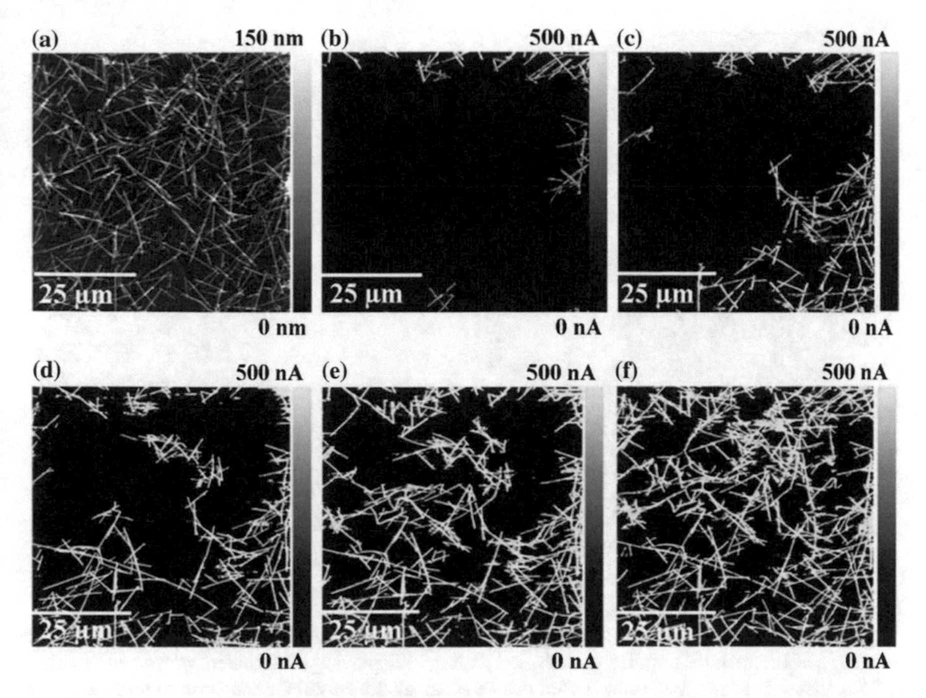

Fig. 15 **a** Topography of a random network of Ag nanowires. A metal coated atomic force microscope tip was used to locally activate sites in the network by applying a voltage pulse. The current maps shown in **b–f** are the result of applying the voltage pulses at selected regions (marked 1–5 on the topographic map; Reprinted with permission from (Nirmalraj et al. 2012). Copyright 2012 American Chemical Society)

showed pinched hysteresis loops similar to those of single devices (Fig. 16c). Furthermore, infra-red imaging confirmed the presence of distributed power dissipation throughout the network (Fig. 16d) and thus the formation of a functional network. Finally, emergent behavior, i.e. a behavior that is not found or associated with a single unit of the network, similar to that of neuron assemblies was also inferred from the observation of a large number of metastable conductance states resulting from discrete configurations of the network (Fig. 16e, f).

4.2.1 Crossbar Memory Arrays

A possible architecture for brain-based nanoelectronic computation is the crossbar array. Passive crossbar memory arrays are simple matrices consisting only of pre- and post-neuron connecting lines and a resistive switch at each junction acting as a synapse (Kügeler et al. 2011; Linn et al. 2012). In a crossbar structure, a two-terminal memristor synapse is formed at each crosspoint connecting pre- and post-synaptic neurons (Jo et al. 2010; Linn et al. 2012). Every neuron in the pre-neuron layer of the crossbar configuration is directly connected to every neuron

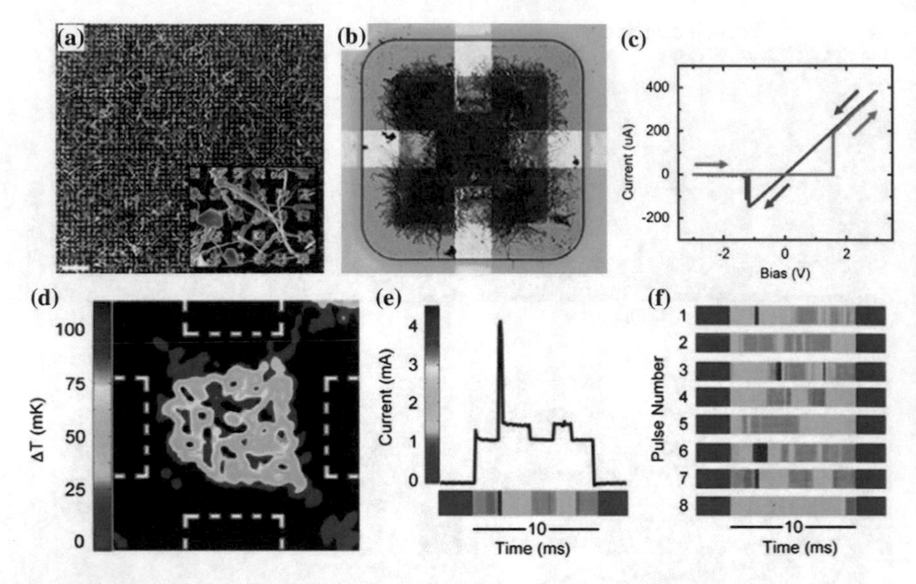

Fig. 16 **a** Zoomed and **b** enlarged view of an $Ag/Ag_2S/Ag$ network of random nanowires. **c** Pinched hysteresis loops associated with the electrical behavior of the whole network. **d** Infra-red imaging of the network under an applied voltage displaying distributed power dissipation. **e**, **f** Switching between metastable conductance states as the network's response to voltage pulses (From (Stieg et al. 2012). Advanced Materials, Copyright © 2012 by [John Wiley & Sons, Inc.]. Reprinted by permission of [John Wiley & Sons, Inc.])

in the post-neuron layer (Jo et al. 2010; Kügeler et al. 2011). This 2D ANN can be seen in a generic manner as mapping sets of input patterns into specific output patterns fully connected through adaptive synapses. The information can be stored changing the value at the edges through an applied voltage between the nodes that they link and can have the meaning of a memory representation (data storage) or of a function representation (computation). Furthermore, higher memory density can be achieved in crossbar architectures than in CMOS architectures and, with 3D stacking, the memory density can be further increased (Chen 2011).

As shown in Fig. 17a, Prezioso et al. used Al_2O_3/TiO_{2-x} to fabricate a highly dense 12 × 12 (200 nm × 200 nm) memristors crossbar (free of transistors) (Prezioso et al. 2015). They used it to implement a stable single layer perceptron 10 (inputs) × 3 (outputs) that successfully performed pattern classification of 3 × 3-pixel images corresponding to three classes of letters, by in situ learning of 23 iterations on average. The operation is based on the self-tuning of the memristor conductance that is assured by its STDP behavior (Prezioso et al. 2016). A critical step was the optimization of the current-voltage nonlinearity through the aluminum oxide thickness to achieve devices with very low variability (Fig. 17b).

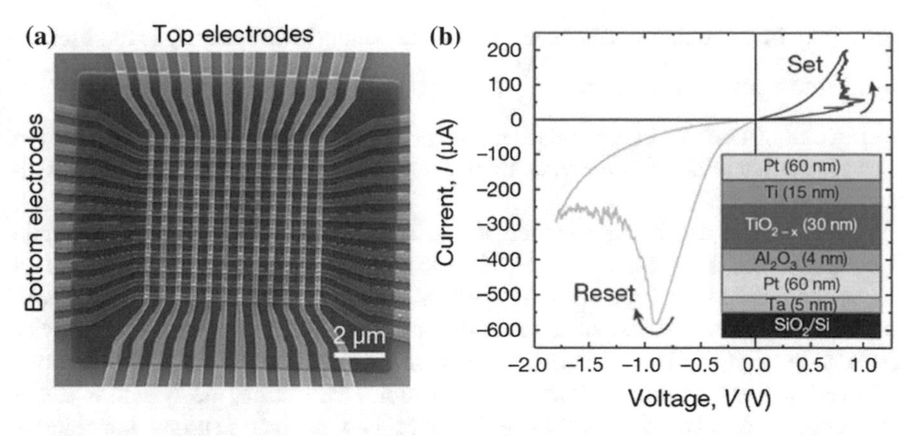

Fig. 17 **a** Memristors crossbar and **b** I-V curve of a single memristor (Reprinted by permission from Macmillan Publishers Ltd: Nature (Prezioso et al. 2015). copyright 2015)

Mostafa et al. also implemented a 8 × 8 (60 μm × 60 μm) hybrid CMOS-memristor perceptron based on an extended STDP rule connecting each two silicon neurons by a TiO_{2-x} memristor (Mostafa et al. 2015). Their rule has the advantage of not involving a post-synaptic spike timing, but instead a correlation between pre-synaptic spikes and signals from the post-synaptic neuron (membrane potential). This means that there is no need to generate temporally long waveforms on both synaptic sides. In both cases, each memristor was electrically prepared before being used as an artificial synapse by an individual electroforming process.

5 Associative Memories

In biological systems, neuronal circuits are deeply involved in the transmission, processing and storage of information. Importantly, and as previously mentioned, these three operations are supported by the same wetware substrate. While the mechanisms supporting efficient and reliable information transmission in neurons are known already since the early 50 s (Hodgkin and Huxley 1952) only more recently were the first steps given on understanding how information may be stored in neuronal populations (Willshaw et al. 1969; Hopfield 1982) The simplest models for information storage in neuronal populations assume that each unit (neuron) can be in one of two states: active or inactive. The information content of a neuronal population can then be seen as the spatiotemporal patterns of activity of its neurons: given a population of neurons, which neurons are active (spatial coding) at a specific time (temporal coding) encodes the information being handled in the network. This section focuses on memory systems relying of spatial coding only, in other words, static neuronal activation patterns in which a particular memory state is represented by a specific constellation of active neurons (no temporal dynamics included). In a population of N neurons, and assuming each memory is represented by M active neurons, a total of $\binom{N}{M}$ different patterns can be considered (number of M-combinations from a population with N elements). Assuming a 5% activation in a population with 1000 neurons, there are on the order of 10^{85} different (spatial) activation patterns.

With information being encoded in the spatial patterns of activity, a simple associative memory system consists of associating particular activity patterns in an input population of neurons to other particular activity patterns in an output population of neurons. A list of associations may be stored, linking input activity patterns to output activity patterns, and the information is stored in the connections between neurons. In this feedforward connection architecture, the system is called an hetero-associative memory as it involves two distinct neurons populations (Willshaw et al. 1969). The alternative is when a single population, with recurrent connectivity, acts both as input and output, in which case the system take the name of auto-associative memory (Hopfield 1982).

The storage of information and memory formation in the nervous system has multiple properties which are very interesting from the engineering point of view. In particular, and as opposed to the standard man-made storage systems, neuronal based memory systems are distributed, meaning that elements of information are encoded across groups of units. These memory systems are therefore capable of graceful degradation: the progressive loss of individual storage units (neurons) does not lead to an abrupt failure of the memory system. Another important feature is that neuronal memory systems are content-addressable memories—information retrieval is achieved not with the presentation of a reference/memory address, but instead it is triggered by the presentation of part of the stored information (cue). Standard man-made memory systems are reference addressable instead. In the case of a hard drive, this implies the existence of a file allocation table (FAT, or a modern form of it) associating file contents to disk cluster addresses—simply damaging/eliminating the allocation table impairs information retrieval (a strategy which was commonly used by computer viruses). Neuronal based memory systems are also intrinsically robust to noise and capable of auto-completion; that is, a modified version of the original input pattern can be used to trigger the recall of the originally associated output pattern. This important property links associative networks to the problem of pattern recognition. All together, these properties (summarized in Table 1) make biology inspired associative networks extremely attractive from the engineering point of view.

Memristors are in a unique favorable position to support the fabrication of efficient and reliable associative memories mimicking the key features of their biological counterparts. Memristors have been shown to be adequate for the implementation of both hetero-associative (Dias et al. 2015) and auto-associative (Hu et al. 2015a, b; Duan et al. 2016; Guo et al. 2015) memories.

Table 1 Comparison between neuronal and standard man-made memory systems

	Neuronal associative memory systems	Standard man-made memory systems
Access	Content-addressable	Reference-addressable
Storage	Distributed	Local
Intrinsic robustness to noise	Yes	No
Intrinsic auto-completion	Yes	No
Low power	Yes	Yes, in some implementations
Response to units loss	Graceful degradation	Potential memory corruption
Storage capacity	Low[a]	High

[a]Current experimental/theoretical evidence points to a storage capacity (as a function of the number of storage units or number of connections) which is lower than standard man-made memory systems (Treves and Rolls 1991) (but see also Alme et al. 2014)

5.1 Principles of Operation

In both types of associative memory architectures, full connectivity is typically assumed between the input and output populations: this means that each neuron in the output population receives one connection from each neuron in the input population (in the case of the auto-associative memory all neurons are connected with all neurons except with themselves). In the simplest associative networks, not only the neuronal state is binary (silent or active) but also the connection strength (weak or strong). Analogously with the read/write modes in man-made systems, current models for associative memories also assume two operation modes: storage and recall (or retrieval). A proper separation, in terms of mechanism, between these operations is necessary to avoid information corruption. Memory storage and retrieval is governed by the neuronal units' dynamics as well as by synaptic modification (learning) rule. The common (and the original) choice for the neuronal dynamics in the discrete associative memory is the McCullock-Pitts model (McCulloch and Pitts 1943) in which the output state of the neuron is 1 (active) only if the weighted sum of its inputs is equal or above a predefined threshold value T; otherwise the output state is 0 (silent) (Fig. 18). As for the learning rule applied in the storage phase, the typical choice is the Hebb rule. It should be noted, nonetheless, that in general this is not an optimal learning rule in terms of storage capacity – higher memory storage capacities can be achieved using slightly modified learning rules (see for example Refs. Dayan and Willshaw (1991); Storkey (1997)). The Hebb rule has however the advantage of being easily recreated with simple memristor dynamics.

In memristive-based implementations of associative memories, memristors are used to establish the network connections while controlled voltage sources (Dias et al. 2015) or operational amplifiers (Hu et al. 2015a, b) are used to represent the state of input neuron units. A simple memristive-based hetero-associative memory

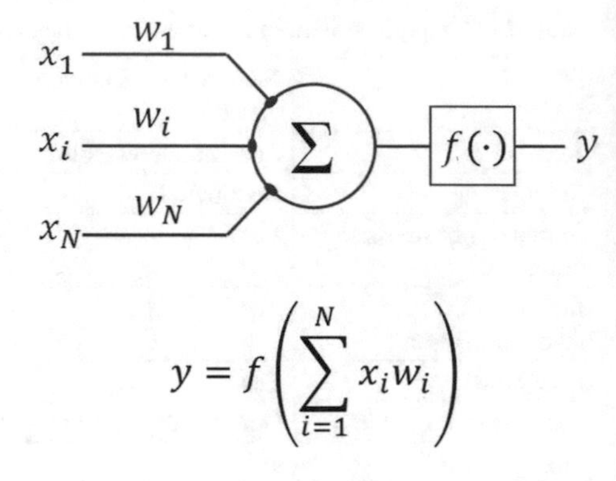

Fig. 18 Diagram of the McCulloch-Pitts neuron. Input signals x are weighted by the corresponding synaptic strengths *w* and a non-linear transfer function f (e.g. a threshold function) is used to set the output signal *y*

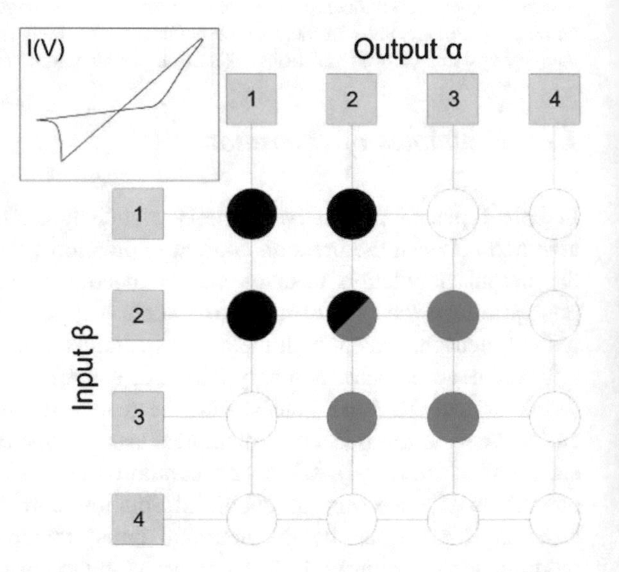

Fig. 19 Hetero-associative memory architecture. *Squares* represent neuron units and *circles* represent memristors (low conductance state, *empty circles*; and high conductance states, *shaded circles*). Two associations were stored in this network: $\{\beta_1, \beta_2\} \rightarrow \{\alpha_1, \alpha_2\}$ and $\{\beta_2, \beta_3\} \rightarrow \{\alpha_2, \alpha_3\}$ (Reprinted from (Dias et al. 2015). with the permission of AIP Publishing)

is show in Fig. 19. Squares represent neuron units from two populations (input β and output α) while circles represent memristors (empty circles represent low conductance state and shaded circles represent high conductance state). In this representation, the following associations are intended: $\{\beta_1, \beta_2\} \rightarrow \{\alpha_1, \alpha_2\}$ and $\{\beta_2, \beta_3\} \rightarrow \{\alpha_2, \alpha_3\}$. During storage, memristors conductances are modified according to the learning rule. In the case of Hebb learning and binary synapses, the memristor state is changed from low to high conductance state only when both input and output units are active. During retrieval, the activation of a stored input pattern leads (with the appropriate threshold T settings) to the recall of the associated output pattern. Naturally, with the increasing number of stored associations comes interference between patterns. With random, sparse activity patterns, the capacity of this memory system scales with $\frac{N_\alpha N_\beta}{M_\alpha M_\beta}$, where N is the total number of units in each population, and M is the number of active units in each patterns

(Willshaw et al. 1969). Intuitively, maximum capacity is achieved when half of the memristors where modified to the high conductance state. After this stage, storage of additional association leads to a fast memory performance degradation.

5.2 *Coping with Faulty Memristors*

Presently, memristor fabrication is still far from a 100% yield (Strukov and Likharev 2007) and some variability is expected between units. Implementations of associative memories based on memristor technology cannot rely on the assumption that all memristor units perform equally well. A source of inspiration and insights on how to cope with faulty memristors comes from the associative memories literature addressing the issues of partial connectivity and different threshold setting strategies (Buckingham and Willshaw 1993; Graham and Willshaw 1996).

In the context of memristors emulating connections with binary weights, a common defect resulting from lithography processes is the inability to switch between resistive states. This leaves the memristors device permanently ("stuck at") on the high, or in the low, resistive state. The situation of stuck at high resistance is comparable to the absence of connections. Given the intrinsic resilience of associative memories (distributed information, soft degradation, noise robustness) it has been shown (Dias et al. 2015) that under appropriate operation strategies, the presence of a significant fraction of faulty units (5%) does not necessarily imply catastrophic failure of the memory system. Moreover, knowing a priori the expected percentage of faulty units after the fabrication process, it is possible to devise threshold setting strategies (different thresholds for each output unit) (Buckingham and Willshaw 1993) which mitigate the reduction in memory capacity.

5.3 *Unlearning and Palimpsest Memories*

The information retrieval performance in conventional associative memory systems collapses after the storage capacity has been exceeded. Statistically speaking, if R is the memory system capacity, the number of correctly recalled patterns approaches R as storage/recall cycles progress, and then abruptly falls to very low values. If one wants to produce effective memristor-based associative memories, one needs to surpass this problem. In the field of neuronal networks, several approaches have been proposed to address this, introducing different forms of erasing, unlearning or "forgetting" old and no more used associations. In the case of a memristive device with filamentary switching, unlearning can be seen for example as the thinning of the filament by applying an opposite bias polarity (Hasegawa et al. 2010). In general, one interesting approach to address the memory collapse problem introduces the concept of palimpsest memories (Sandberg et al. 2000; Sterratt and Willshaw 2008). A palimpsest was a medieval manuscript that was repeatedly reused over time by inscribing text over the previous existing, being only the most recent writing visible. The analogy with new memories replacing older ones can be done (Henson 1993). There is not a catastrophic forgetting, but instead these forgetful learning rules select the older memories to forget in order to store the new ones (Storkey 2015).

There are three main strategies using palimpsest considering forgetting approaches in the Willshaw network. The first is random resetting, where random switches are turned off with a (small) random probability. The second is weight ageing, in which switches are turned off with a probability that depends on the age of the switch which in turn is the time since it was last triggered. The last one is generalized learning, the only that is not random but dependent on the patterns presented at a given time (Henson 1993).

While memristors can in principle support the construction of palimpsest associative memory systems, this subject has not yet been appropriately addressed.

6 Conclusions and Outlook

Memristors exhibit noteworthy properties which make them excellent building blocks to construct neuromorphic systems. They are low power devices, nanosized, have good scalability properties and above all they mimic core aspects of the dynamics of ion channels present at synapses and neuron membranes. As a result, they are able to reproduce, *in silico*, many of the mechanisms associated with neural computation, such as classification, decision-making, learning and memory. The human brain is the result of more than 4.5 billion years of evolution and optimization, and is therefore Nature's solution to fast and efficient information processing and information storage. By copying features of the human brain uncovered by recent advances in neuroscience, one hopes to produce innovations in computing technology and memory storage devices. Current computing paradigms face major scalability problems, one of them being electric power demand. It is forecasted that, by 2040, computing electric power demand will surpass the amount generated (ITRS 2015). The human brain, capable of pattern recognition, precise motor control, and other complex information processing tasks, has a remarkably low power consumption (Sengupta and Stemmler 2014), in the range of 25 watts (Kandel et al. 2003). Memristors have the potential to enable novel bio-inspired computing paradigms.

Since the memristor hypothesis by Leon Chua, the realization of a physical model by HP's researchers, and through the many recent exciting results regarding theory and fabrication, the field has made tremendous progress and the time is ripe to explore the possibilities opened by memristors. Simulating or even discriminating actual neural activity (Gupta et al. n.d.) of the human brain, and getting inspiration of it to produce novel computing systems is certainly one of the most promising applications. But there are still many open challenges in memristive-based neuromorphic systems: mathematical models of memristors' dynamics need to be consolidated and validated, the physical implementation of small memristive-based circuits needs to become more consistent and reliable, circuits to drive learning processes (as well and information storage/retrieval) in memristive systems need to be improved, efficient coding/decoding mechanisms for memristor memory systems need to be explored, just to name a few. As in all other fields in science, challenges are also opportunities, and memristive-based neuromorphic systems offers many exciting possibilities to the research groups willing to tackle them.

Acknowledgements This work was supported in part by projects PTDC/CTM-NAN/122868/2010, PTDC/CTM-NAN/3146/2014 and POCI-01-0145-FEDER-016623. This work was also partially supported by FEDER (Fundo Europeu de Desenvolvimento Regional) funds through the COMPETE 2020 Operational Programme for Competitiveness and Internationalisation (POCI), Portugal 2020 and by Portuguese funds through FCT (Fundação para a Ciência e a Tecnologia) and Ministério da Ciência, Tecnologia e Inovação in the framework of the project "Institute for Research and Innovation in Health Sciences" (POCI-010145 FEDER-007274) and through the Associated Laboratory—Institute of Nanoscience and Nanotechnology. J. V. acknowledges financial support through FSE/POPH. C. Dias is thankful to FCT for grant SFRH/BD/101661/2014.

References

Agirre-Basurko, E., Ibarra-Berastegi, G., & Madariaga, I. (2006). Regression and multilayer perceptron-based models to forecast hourly O_3 and NO_2 levels in the Bilbao area. *Environmental Modelling & Software, 21*(4), 430–446.

Alibart, F., et al. (2012). A memristive nanoparticle/organic hybrid synapstor for neuroinspired computing. *Adv. Funct. Mater., 22*(3), 609–616.

Alibart, F., Zamanidoost, E., & Strukov, D. B. (2013). Pattern classification by memristive crossbar circuits using ex situ and in situ training. *Nature Communications, 4* (2072).

Alme, C. B., et al. (2014). Place cells in the hippocampus: Eleven maps for eleven rooms. *Proceedings of the National Academy of Sciences of the USA, 111*(52), 18428–18435.

Ashidi, N., et al. (2011). Clustered-hybrid multilayer perceptron network for pattern recognition application. *Applied Soft Computing, 11*(1), 1457–1466.

Barbera, S. La, Vuillaume, D., & Alibart, F. (2015). Filamentary switching: Synaptic plasticity through device volatility. *ACS Nano, 9,* 941–949.

Berdan, R., et al. (2016). Emulating short-term synaptic dynamics with memristive devices. *Scientific Reports, 6,* 18639 (November 2015).

Bo, L., Wang, L. & Jiao, L. (2006). Multi-layer Perceptrons with embedded feature selection with application in Cancer Classification. *Chinese Journal of Electronics*.

Buckingham, J., & Willshaw, D. (1993). On setting unit thresholds in an incompletely connected associative net. *Network, 4,* 441–459.

Chabi, D., et al. (2014). Robust learning approach for neuro-inspired nanoscale crossbar architecture. *ACM Journal on Emerging Technologies in Computing Systems, 10*(1), 1–20.

Chang, T., Jo, S., & Lu, W. (2011). Short-term memory to long-term memory transition in a nanoscale memristor. *ACS Nano, 5*(9), 7669–7676.

Chen, A. (2011). Ionic memory technology. In *Solid state electrochemistry II: Electrodes, interfaces and ceramic membranes* (pp. 1–18).

Choi, S. J., et al. (2011). Synaptic behaviors of a single metal-oxide-metal resistive device. *Applied Physics A, 102*(4), 1019–1025.

Chua, L. (1971). Memristor-the missing circuit element. *IEEE Transactions on Circuit Theory, 18* (5), 507–519.

Conde-Sousa, E., & Aguiar, P. (2013). A working memory model for serial order that stores information in the intrinsic excitability properties of neurons. *Journal of Computational Neuroscience, 35*(2), 187–199.

Crane, H. D. (1960). The Neuristor. *IRE Transactions On Electronic Computers, 9,* 370–371.

Daumé III, H. (2012). The perceptron. In *A course in machine learning* (pp. 37–50).

Dayan, P., & Willshaw, D. J. (1991). Optimising synaptic learning rules in linear associative memories. *Biological cybernetics, 65*(4), 253–265.

Dias, C., et al. (2015). Memristor-based Willshaw network: Capacity and robustness to noise in the presence of defects. *Applied Physics Letters, 106*(22), 223505.

Dimitrov, D. (2009). Spintronic memristor through spin-torque-induced magnetization motion. *IEEE Electron Device Letters, 30*(3), 294–297.

Du, C., et al. (2015). Biorealistic implementation of synaptic functions with oxide memristors through internal ionic dynamics. *Advanced Functional Materials, 25*(27), 4290–4299.

Duan, S., et al. (2016). Small-world Hopfield neural networks with weight salience priority and memristor synapses for digit recognition. *Neural Computing and Applications, 27*(4), 837–844.

Ebong, I. E., & Mazumder, P. (2012). CMOS and memristor-based neural network design for position detection. *Proceedings of the IEEE, 100*(6), 2050–2060.

Gale, E., Costello, B. D. L., & Adamatzky, A. (2014). Emergent spiking in non-ideal memristor networks. *Microelectronics Journal, 45*(11), 1401–1415.

Gatet, L., Tap-Béteille, H., & Bony, F. (2009). Comparison between analog and digital neural network implementations for range-finding applications. *IEEE Transactions on Neural Networks, 20*(3), 460–470.

Goux, L., & Valov, I. (2016). Electrochemical processes and device improvement in conductive bridge RAM cells. *Physica Status Solidi (A) Applications and Materials Science*, *213*(2), pp. 274–288.

Graham, B., & Willshaw, D. (1996). Information efficiency of the associative net at arbitrary coding rates. In *Artificial neural networks—ICANN 96* (pp. 35–40).

Guo, X., et al. (2015). Modeling and experimental demonstration of a hopfield network analog-to-digital converter with hybrid CMOS/memristor circuits. *Frontiers in Neuroscience*, *9*, 1–8(Dec).

Gupta, I., et al. (n.d.) Memristive integrative sensors for neuronal activity. arXiv:1507.06832.

Ha, S. D., & Ramanathan, S. (2011). Adaptive oxide electronics: A review. *Journal of Applied Physics, 110*(7), 071101.

Hasegawa, T., et al. (2012). Atomic switch: Atom/ion movement controlled devices for beyond von-neumann computers. *Advanced materials (Deerfield Beach, Fla.)*, *24*(2), 252–67.

Hasegawa, T., Ohno, T., Terabe, K., Tsuruoka, T., Nakayama, T., Cimzewski, J. K., et al. (2010). Learning abilities achieved by a single solid-state atomic switch. *Advanced Materials, 22*(16), 1831–1834.

Haykin, S. (2009). Rosenblatt's perceptron. In *Neural networks and learning machines* (pp. 47–67). Pearson.

Henson, R. (1993). *Short-term associative memories*.

Hodgkin, A. L., & Huxley, A. F. (1952). A quantitative description of membrane current and its application to conduction and excitation in nerves. *Journal of Physiology, 117,* 500–544.

Hopfield, J. J. (1982). Neural networks and physical systems with emergent collective computational abilities. *Proceedings of the National Academy of Sciences of the USA, 79*(8), 2554–2558.

Hu, S. G., et al. (2013a). Design of an electronic synapse with spike time dependent plasticity based on resistive memory device. *Journal of Applied Physics*, *113*(11).

Hu, S. G., et al. (2013b). Emulating the paired-pulse facilitation of a biological synapse with a NiO_x-based memristor. *Applied Physics Letters*, *102*(18).

Hu, S. G., et al. (2015a). A memristive Hopfield network for associative memory. *Nature Communications, 6,* 7522.

Hu, S. G., et al. (2015b). Associative memory realized by a reconfigurable memristive Hopfield neural network. *Nature Communications, 6*(May), 7522.

Ikeda, S., et al. (2010). A perpendicular-anisotropy CoFeB–MgO magnetic tunnel junction. *Nature Materials, 9*(9), 721–724.

Indiveri, G., et al. (2013). Integration of nanoscale memristor synapses in neuromorphic computing architectures. *Nanotechnology, 24*(38), 384010.

ITRS. (2015). *International technology roadmap for semiconductors 2.0.*

Jana, D., et al. (2015). Conductive-bridging random access memory: Challenges and opportunity for 3D architecture. *Nanoscale Research Letters, 10*(1), 188.

Jo, S. H., et al. (2010). Nanoscale memristor device as synapse in neuromorphic systems. *Nano Letters, 10*(4), 1297–1301.

Kandel, E. R., Schwartz, J. H., & Jessell, T. M. (2003). *Principles of neural science*. Manole.

Karmarkar, U. R., & Buonomano, D. V. (2002). A model of spike-timing dependent plasticity: One or two coincidence detectors? *Journal of Neurophysiology, 88*(1), 507–513.

Kavehei, O., & Skafidas, E. (2014). Highly scalable neuromorphic hardware with 1-bit stochastic nano-synapses. In *Proceedings—IEEE International Symposium on Circuits and Systems*, pp. 1648–1651.

Kim, H., Sah, M. P., et al. (2012a). Memristor bridge synapses. *Proceedings of the IEEE, 100*(6), 2061–2070.

Kim, H., Sah, M. P., et al. (2012b). Neural synaptic weighting with a pulse-based memristor circuit. *IEEE Transactions on Circuits and Systems, 59*(1), 148–158.

Kim, S., et al. (2015). Experimental demonstration of a second-order memristor and its ability to biorealistically implement synaptic plasticity. *Nano Letters, 15*(3), 2203–2211.

Kiselev, S. I., et al. (2003). Microwave oscillations of a nanomagnet driven by a spin-polarized current. *Nature, 425*(6956), 380–383.

Kozma, R., Pino, R. E., & Pazienza, G. E. (2012). *Advances in neuromorphic memristor science and applications*, Springer Publishing Company, Incorporated.

Krzysteczko, P., Kou, X., et al. (2009a). Current induced resistance change of magnetic tunnel junctions with ultra-thin MgO tunnel barriers. *Journal of Magnetism and Magnetic Materials, 321*(3), 144–147.

Krzysteczko, P., Reiss, G., & Thomas, A. (2009b). Memristive switching of MgO based magnetic tunnel junctions. *Applied Physics Letters, 95*(11), 112508.

Krzysteczko, P., et al. (2012). The memristive magnetic tunnel junction as a nanoscopic synapse-neuron system. *Advanced Materials (Deerfield Beach, Fla.), 24*(6), 762–766.

Kubota, H., et al. (2008). Quantitative measurement of voltage dependence of spin-transfer torque in MgO-based magnetic tunnel junctions. *Nature Physics, 4*(1), 37–41.

Kügeler, C., et al. (2011). Materials, technologies, and circuit concepts for nanocrossbar-based bipolar RRAM. *Applied Physics A, 102*(4), 791–809.

Li, Q., et al. (2015). A memristor SPICE model accounting for synaptic activity dependence. *PLoS ONE, 10*(3), e0120506.

Li, S., et al. (2013). Synaptic plasticity and learning behaviours mimicked through Ag interface movement in an Ag/conducting polymer/Ta memristive system. *Journal of Materials Chemistry C, 1*(34), 5292.

Linn, E., et al. (2012). Beyond von Neumann–logic operations in passive crossbar arrays alongside memory operations. *Nanotechnology, 23*(30), 305205.

Liu, Y., et al. (2011). Self-learning ability realized with a resistive switching device based on a Ni-rich nickel oxide thin film. *Applied Physics A, 105*(4), 855–860.

Locatelli, N., Cros, V., & Grollier, J. (2014). Spin-torque building blocks. *Nature Materials, 13*(1), 11–20.

Lu, W. (2012). Memristors: Going active. *Nature Materials, 12*(2), 93–94.

Ma, W., et al. (2015). Temporal information encoding in dynamic memristive devices. *Applied Physics Letters, 107*(19), 193101.

McCulloch, W. S., & Pitts, W. (1943). A logical calculus of the ideas immanent in nervous activity. *The Bulletin of Mathematical Biophysics, 5*(4), 115–133.

Mehonic, A., & Kenyon, A. J. (2016). Emulating the electrical activity of the neuron using a silicon oxide RRAM cell. *Frontiers in Neuroscience, 10*, article 57.
Merolla, P. A., et al. (2014). A million spiking-neuron integrated circuit with a scalable communication network and interface. *Science, 345*(6197), 668–673.
Mostafa, H., et al. (2015). Implementation of a spike-based perceptron learning rule using TiO_{2-x} memristors. *Frontiers in Neuroscience, 9,* 357.
Nirmalraj, P. N., et al. (2012). Manipulating connectivity and electrical conductivity in metallic nanowire networks. *Nano Letters, 12*(11), 5966–5971.
Ohno, T., Hasegawa, T., Nayak, A., et al. (2011a). Sensory and short-term memory formations observed in a Ag_2S gap-type atomic switch. *Applied Physics Letters, 99*(20), 203108.
Ohno, T., Hasegawa, T., Tsuruoka, T., et al. (2011b). Short-term plasticity and long-term potentiation mimicked in single inorganic synapses. *Nature Materials, 10*(8), 591–595.
Pan, R., et al. (2016). Synaptic devices based on purely electronic memristors. *Applied Physics Letters, 108*(1), 013504.
Park, H. (2006). Multilayer perceptron and natural gradient learning. In *New generation computing* (pp. 79–95).
Park, S., et al. (2012). RRAM-based synapse for neuromorphic system with pattern recognition function. In *2012 international electron devices meeting* (pp. 10.2.1–10.2.4). IEEE.
Parkin, S. S. P., et al. (2004). Giant tunnelling magnetoresistance at room temperature with MgO (100) tunnel barriers. *Nature Materials, 3*(12), 862–867.
Pickett, M. D., Medeiros-Ribeiro, G., & Williams, R. S. (2012). A scalable neuristor built with Mott memristors. *Nature Materials, 12*(2), 114–117.
Pickett, M. D., & Williams, R. S. (2013). Phase transitions enable computational universality in neuristor-based cellular automata. *Nanotechnology, 24*(38), 384002.
Pinto, S., et al. (2012). Resistive switching and activity-dependent modifications in Ni-doped graphene oxide thin films. *Applied Physics Letters, 101*(6), 063104.
Prezioso, M., et al. (2016). Self-adaptive spike-time-dependent plasticity of metal-oxide memristors. *Scientific Reports, 6*(February), 21331.
Prezioso, M., et al. (2015). Training and operation of an integrated neuromorphic network based on metal-oxide memristors. *Nature, 521*(7550), 61–64.
Rose, G. S., Pino, R., & Wu, Q. (2011). A low-power memristive neuromorphic circuit utilizing a global/local training mechanism. In *The 2011 international joint conference on neural networks* (pp. 2080–2086). IEEE.
Rosenblatt, F. (1957). *The perceptron, a perceiving and recognizing automation.*
Sandberg, A., et al. (2000). A palimpsest memory based on an incremental Bayesian learning rule. *Neurocomputing, 32–33,* 987–994.
Schemmel, J., & Grubl, A. (2006). Implementing synaptic plasticity in a VLSI spiking neural network model. In *International joint conference on neural networks* (pp. 1–6).
Sengupta, A., et al. (2015). Spin-orbit torque induced spike-timing dependent plasticity. *Applied Physics Letters, 106*(9), 093704.
Sengupta, B., & Stemmler, M. B. (2014). Power consumption during neuronal computation. *Proceedings of the IEEE, 102*(5), 738–750.
Seo, K., et al. (2011). Analog memory and spike-timing-dependent plasticity characteristics of a nanoscale titanium oxide bilayer resistive switching device. *Nanotechnology, 22*(25), 254023.
Serrano-Gotarredona, T., et al. (2013). STDP and sTDP variations with memristors for spiking neuromorphic learning systems. *Frontiers in Neuroscience*, *7*(7 Feb), 1–15.
Shi, L. P., et al. (2011). Artificial cognitive memory-changing from density driven to functionality driven. *Applied Physics A, 102*(4), 865–875.
Snider, G. (2008). Spike-timing-dependent learning in memristive nanodevices. In *Nanoscale architectures, 2008. NANOARCH 2008* (pp. 85–92).
Snider, G. S. (2007). Self-organized computation with unreliable, memristive nanodevices. *Nanotechnology, 18*(36), 365202.
Sterratt, D. C., & Willshaw, D. (2008). Inhomogeneities in heteroassociative memories with linear learning rules. *Neural Computation, 20*(2), 311–344.

Stieg, A., & Avizienis, A. (2014). Self-organized atomic switch networks. *Japanese Journal of Applied Physics, 53*, 01AA02.

Stieg, A. Z., et al. (2012). Emergent criticality in complex turing B-Type atomic switch networks. *Advanced Materials, 24*(2), 286–293.

Storkey, A. (1997). Increasing the capacity of a Hopfield network without sacrificing functionality. *Artificial Neural Networks – ICANN'97* (pp. 451–456).

Storkey, A., 2015. Palimpsest memories: A new high-capacity forgetful learning rule for Hopfield networks. *Electrical Engineering* (July), 1–14.

Strukov, D. B., et al. (2008). The missing memristor found. *Nature, 453*(7191), 80–83.

Strukov, D. B., & Kohlstedt, H. (2012). Resistive switching phenomena in thin films: Materials, devices, and applications. *MRS Bulletin, 37*(02), 108–114.

Strukov, D. B., & Likharev, K. K. (2007). Defect-tolerant architectures for nanoelectronic crossbar memories. *Journal of Nanoscience and Nanotechnology, 7,* 151–167.

Suri, M., et al. (2013). Bio-inspired stochastic computing using binary CBRAM synapses. *IEEE Transactions on Electron Devices, 60*(7), 2402–2409.

Teixeira, J. M., et al. (2009). Electroforming, magnetic and resistive switching in MgO-based tunnel junctions. *Journal of Physics. D. Applied Physics, 42*(10), 105407.

Teixeira, J. M., et al. (2011). Resonant tunneling through electronic trapping states in thin MgO Magnetic junctions. *Physical Review Letters, 106*(19), 196601.

Thomas, A. (2013). Memristor-based neural networks. *Journal of Physics. D. Applied Physics, 46,* 093001.

Treves, A., & Rolls, E. (1991). What determines the capacity of autoassociative memories in the brain? *Network: Computation in Neural Systems, 2*(4), 371–397.

Tsuruoka, T., et al. (2012). Conductance quantization and synaptic behavior in a Ta_2O_5-based atomic switch. *Nanotechnology, 23*(43), 435705.

Valov, I., et al. (2011). Electrochemical metallization memories-fundamentals, applications, prospects. *Nanotechnology, 22*(28), 289502.

Ventura, J., et al. (2007). Three-state memory combining resistive and magnetic switching using tunnel junctions. *Journal of Physics D Applied Physics, 40*(19), 5819–5823.

Vincent, A. F., et al. (2015). Spin-transfer torque magnetic memory as a stochastic memristive synapse for neuromorphic systems. *IEEE Transactions on Biomedical Circuits and Systems, 9* (2), 166–174.

Wan, C. J., et al. (2013). Memory and learning behaviors mimicked in nanogranular SiO_2-based proton conductor gated oxide-based synaptic transistors. *Nanoscale, 5*(21), 10194–10199.

Wang, C., et al. (2016). Investigation and manipulation of different analog behaviors of memristor as electronic synapse for neuromorphic applications. *Scientific Reports*, 6(November 2015), p. 22970.

Wang, L., Duan, M., & Duan, S. (2013). Memristive perceptron for combinational logic classification. *Mathematical Problems in Engineering, 2013*(1), 1–7.

Wang, Y.-F., et al. (2015). Characterization and modeling of nonfilamentary $Ta/TaO_x/TiO_2/Ti$ analog synaptic device. *Scientific Reports, 5,* 10150.

Wang, Z. Q., et al. (2012). Synaptic learning and memory functions achieved using oxygen ion migration/diffusion in an amorphous InGaZnO memristor. *Advanced Functional Materials, 22* (13), 2759–2765.

Williamson, A., et al. (2013). Synaptic behavior and STDP of asymmetric nanoscale memristors in biohybrid systems. *Nanoscale, 5*(16), 7297–7303.

Willshaw, D. J., Buneman, O. P., & Longuet-Higgins, H. C. (1969). Non-holographic associative memory. *Nature, 222*(5197), 960–962.

Yang, J. J., Strukov, D. B., & Stewart, D. R. (2013). Memristive devices for computing. *Nature Nanotechnology, 8*(1), 13–24.

Yang, R., et al. (2012). On-demand nanodevice with electrical and neuromorphic multifunction realized by local ion migration (SI). *ACS Nano, 6*(11), 9515–9521.

Yoshida, C., Kurasawa, M., & Lee, Y. (2008). Unipolar resistive switching in CoFeB/MgO/CoFeB magnetic tunnel junction. *Applied Physics Letters, 92,* 113508.

Yu, S., Wu, Y., & Jeyasingh, R. (2011). An electronic synapse device based on metal oxide resistive switching memory for neuromorphic computation. *IEEE Transactions on Electron Devices, 58*(8), 2729–2737.

Yuasa, S., et al. (2004). Giant room-temperature magnetoresistance in single-crystal Fe/MgO/Fe magnetic tunnel junctions. *Nature Materials, 3*(12), 868–871.

Zamarreño-Ramos, C., et al. (2011). On spike-timing-dependent-plasticity, memristive devices, and building a self-learning visual cortex. *Frontiers in neuroscience, 5,* 26.

Zhou, Y., & Ramanathan, S. (2015). Mott memory and neuromorphic devices. *Proceedings of the IEEE, 103*(8), 1289–1310.

Experimental Analogue Implementation of Memristor Based Chaotic Oscillators

R. Jothimurugan, S. Sabarathinam, K. Suresh and K. Thamilmaran

Abstract The theory of memristor was postulated in the year of 1971 by Leon O. Chua. The intensive interest on memristive systems is given by the researchers since after the physical realization of the hysteresis behavior in a nanoscale TiO_2 memristor in 2008 by a group of researchers at HP Labs lead by Stanley Williams. Research on memristive systems has been carried out on various capacities such as understanding the mathematics of memristor, finding new materials which have memristive properties, studying the underlying dynamics of memristive systems and revisiting the existing concepts with memristor as a nonlinear element. As a result, memristors have potential applications in various domains. It ranges from neural networks, memory devices, artificial intelligence, high speed computing, nano batteries and human skin modeling, etc. In the recent times, much attention is given to explore the nonlinear dynamics of memristor based circuits. In this chapter, we consider a smooth continuous cubic memristor as nonlinear element. It is applied to (a) an autonomous and (b) a non-autonomous dynamical systems namely, the Chua's circuit and Duffing Oscillator, to study the associated dynamics of these systems. The numerical simulation of the circuit systems as well as its hardware experimental studies are

R. Jothimurugan
Department of Science and Humanities, Karpagam University, Coimbatore 641 021, Tamilnadu, India
e-mail: jothi.nld@gmail.com

S. Sabarathinam · K. Thamilmaran (✉)
Centre for Nonlinear Dynamics, School of Physics, Bharathidasan University, Tiruchirappalli 624 024, Tamilnadu, India
e-mail: maran.cnld@gmail.com

S. Sabarathinam
e-mail: sabacnld@gmail.com

K. Suresh
Department of Physics, Anjalai Ammal Mahalingam Engineering College, Kovilvenni 614 403, Tamil Nadu, India
e-mail: sureshscience@gmail.com

© Springer International Publishing AG 2017

S. Vaidyanathan and C. Volos (eds.), *Advances in Memristors, Memristive Devices and Systems*, Studies in Computational Intelligence 701,
DOI 10.1007/978-3-319-51724-7_14

performed in the laboratory. An inductor free realization and volume expanded period doubling scenario in a memristive Chua's circuit is studied. The complex behaviors, like, bifurcations and chaos, three-tori, transient chaos and intermittency in a memristive Duffing oscillator are described. In addition, "0–1 test" for the experimental time series data characterizing the regular and chaotic dynamics of the proposed circuits are also discussed.

Keywords Chua's circuit · Duffing Osillator · Chaos · Transient chaos · Intermittency · 0–1 test · Memristor · Electronic analogy · Analogue circuit · PSpice simulation

1 Introduction

Memristor was postulated by Leon O. Chua in the year of 1971 based on the symmetry arguments. It is hypothesized that the memristor would be the fourth fundamental passive circuit element. The resistor, capacitor and inductor are the other three fundamental circuit elements (Chua 1971). Theory of memristors is recognised by the scientific community since after the physical realization of nanoscale memristor by Stanley William's group from HP labs in 2008 (Strukov et al. 2008). By definition, memristor is a passive two-terminal electronic element described by a nonlinear constitutive relation which takes either one of the following two forms of the voltage-current relationship, (i) $v = M(q)i$ and (ii) $i = W(\phi)v$, where $M(q)$ and $W(\phi)$ are nonlinear functions regarded as *memristance* and *memductance* respectively. These relations are defined by $M(q) = d\,\phi(q)/dq$ and $W(\phi) = dq(\phi)/d\,\phi$ (Itoh and Chua 2008). The memristor has potential applications in almost all branches of science such as physics, mathematics, engineering, biology, and various other fields. Recent researches explicitly demonstrate the potentiality of the memristor such as mathematical modeling and analysis (Botta et al. 2011; Bao et al. 2010; Teng et al. 2014; Slipko et al. 2013; Corinto and Ascoli 2012; Secco et al. 2015), application of memristor theory to the nano-battery (Valov et al. 2013), synthesis of memristive devices (Strukov et al. 2008; Yang and Pickett 2008; Fouda and Radwan 2014; Kyriakides and Georgiou 2014; Zidan et al. 2014), memory effects on light emitting diodes (Zakhidov 2010), memristive behavior of electrical properties of the human skin (Martinsen et al. 2010), and synaptic connections among brain cells of the human (Thomas 2013). The analysis of nonlinear systems using experimental circuits is always an active area of research that provides a better understanding of the theoretical concepts. The direct experimental implementation of the memristor element has various challenges. Hence, the emulators and equivalent circuits of the memristor has been used to replace memristor element in experimental circuit implementations. The Chua's circuit was first investigated with analogue integrator based smooth memristor (Muthuswamy 2010). The occurrence of chaotic beats in a driven Chua's circuit and the nonsmooth bifurcations, transient hyperchaos and hyperchaotic beats in a Murali-Lakshmanan-Chua circuit (Ishaq Ahamed and

Lakshmanan 2013) are demonstrated using time varying resistor based memristor (Ishaq Ahamed et al. 2011). On the otherhand, analogue circuit simulation is another useful idea to accurately synthesize the memristor emulators (Valsa et al. 2011; Kim et al. 2012; Bao et al. 2013; Li et al. 2013, 2014; Sánchez-Lópeza et al. 2015).

The inductor is an another critical element in the experimental implementation since it is a low accuracy component in tuning and also has internal resistance, which compromises the circuit operation. Usually, the inductor is a component with large dimensions, which can be a problem to implement compact circuits on PCB/VLSI. Although, a core is added to increase the inductance with less number of coil windings, it causes distortion to the signal via hysteresis (Bharathwaj et al. 2009). Again, it is less likely preferable for a circuit which is highly sensitive to initial conditions. Even the commercially available discrete inductors (looks like resistors) are also offers high internal resistance (Jothimurugan et al. 2014). Many alternatives to replace a physical inductor can be proposed for the design of oscillator, such as the use of Wien-bridge (Morgül 1995), active band pass filter (Banerjee 2012), and inductance emulator (Tôrres and Aguirre 2000). An interesting alternative to overcome the difficulties on physical realization of an inductorless memristive circuit is the construction of analogue circuit of the original system. The advantages of using electronic analogue oscillator are (a) it do not have either physical or synthetic inductors, (b) frequency of operation of the circuit can be easily varied and (c) any given mathematical model described using differential equations can be accurately implemented. Further, it is cost effective and easy to analyse and reproduce. We consider two universally famous chaotic systems such as (1) the Chua's circuit and (2) the Duffing oscillator. The memristor is substituted in these circuits as a nonlinearity and the dynamics of these two systems is discussed in this chapter in detail. This chapter is broadly segmented into two major parts. The first part, we discusses with the memristor based Chua's circuit while the rest is dealing about the memristor based Duffing oscillator.

2 The Cubic Memristor Nonlinearity

The form of the memristor used here is characterized by smooth continuous cubic monotonic increasing nonlinearity (Itoh and Chua 2008; Muthuswamy 2010)

$$q(\phi) = a\,\phi + b\phi^3, \tag{1}$$

where a and b are constants which are assumed to be greater than zero. From Eq. (1), the memductance $W(\phi)$ is obtained as

$$W(\phi) = \frac{dq(\phi)}{d\,\phi} = a + 3b\phi^2. \tag{2}$$

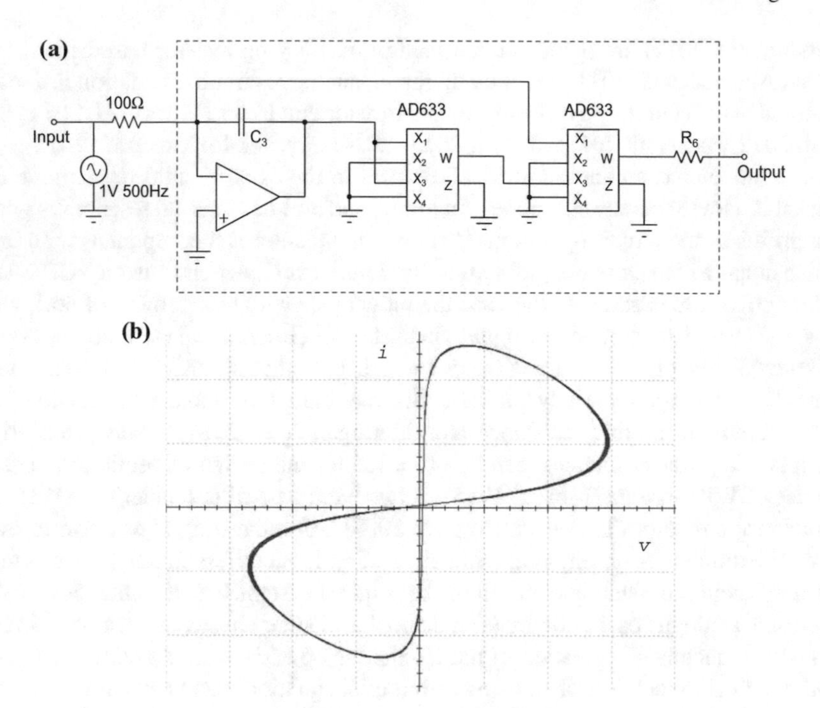

Fig. 1 Pspice analysis: **a** Schematic of the memristor nonlinearity with AC sweep analysis (1 V, 500 Hz) and **b** ($v - i$) characteristic curve of memristor

The schematic of the memristor emulator and its characteristic curve are shown in the Fig. 1. The pinched hysteresis loop of Fig. 1b is obtained from PSpice simulation.

3 The Memristor Based Chua's Oscillator

The memristor oscillator derived from the famous Chua's circuit is shown in Fig. 2. Here, N_R is an active nonlinear element which is the parallel combination of negative conductance ($-G$) and two terminal passive flux controlled memristor. The state equations for the memristor based Chua's circuit is written using Kirchhoff's laws as follows (Itoh and Chua 2008),

$$\frac{dv_1}{dt} = \frac{1}{RC_1}(v_2 - v_1 + GRv_1 - RW(\phi)v_1), \tag{3a}$$

$$\frac{dv_2}{dt} = \frac{1}{RC_2}(v_1 - v_2 + Ri_L), \tag{3b}$$

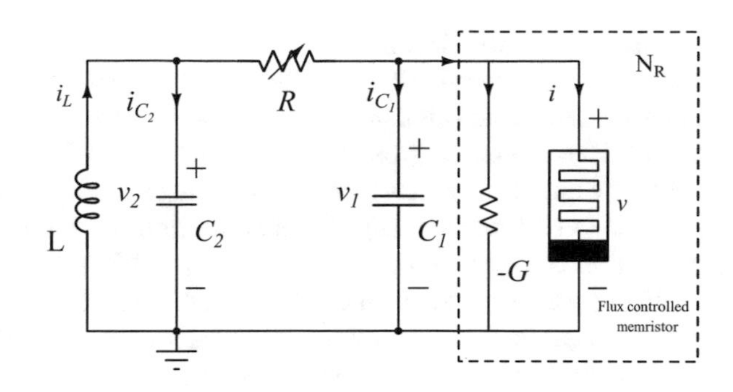

Fig. 2 The memristor based Chua's circuit

$$\frac{di_L}{dt} = -\frac{1}{L}v_2, \tag{3c}$$

$$\frac{d\phi}{dt} = v_1, \tag{3d}$$

where, v_1, v_2, i_L and ϕ are voltage across the capacitors C_1, C_2, current flowing through the inductor L and the flux developed in the memristor, respectively. Substituting the functional form $W(\phi)$ into Eq. (3a) yields,

$$\frac{dv_1}{dt} = \frac{1}{RC_1}(v_2 - v_1 + GRv_1 - R(a + 3b\phi^2)v_1). \tag{3e}$$

A linear transformation is applied to Eq. (3). The transformations are $x_1 = v_1, x_2 = v_2$, $x_3 = i_L$, $x_4 = \phi$, $\alpha = 1/C_1$, $\beta = 1/L$, $\epsilon = G$, $C_2 = 1$ and $R = 1$. Now the circuit Eq. (3) becomes,

$$\frac{dx_1}{dt} = \alpha(x_2 - x_1 + \epsilon x_1 - (a + 3bx_4^2)x_1), \tag{4a}$$

$$\frac{dx_2}{dt} = x_1 - x_2 + x_3, \tag{4b}$$

$$\frac{dx_3}{dt} = -\beta x_2, \tag{4c}$$

$$\frac{dx_4}{dt} = x_1. \tag{4d}$$

The dynamics of Eq. (4) is studied through numerical simulation. We used Runge-Kutta fourth order algorithm to numerically solving the equations. For the following specific choice of system parameters in the Eq. (4), $\alpha = 9.8$, $\beta = 100/7$, $\epsilon = 9/7$, $a = 1/7$, $b = 2/7$ and initial conditions $\{x_1, x_2, x_3, x_4\} = \{0, 0, 0.1, 0\}$, the double band chaotic attractor is obtained.

Scaling of System Variables
The maximum amplitude of oscillation of variables x_1, x_2, x_3 and x_4 of Eq. (4) are measured from the simulated chaotic time series after terminating a long initial transients. They are $|x_1| = 1.3965$ V, $|x_2| = 0.4772$ V, $|x_3| = 2.3797$ A and $|x_4| = 0.9441$ Wb. The direct implementation of memristor oscillator for the Eq. (4) is not possible due to two reasons. (a) The quantity flux can not directly implemented; at the same time (b) value of the current i_L or x_3 is 2.3 A, which is well above the current rating of the usual clcctronic circuit design. Hence analogue circuit simulation of Eq. (4) is preferred whereby all the variables are converted into voltages. Since the maximum amplitudes of the state variables x_2 and x_4 are less than 1, the signals of these variables are more prone to affected by the inherent noise while implementing the circuit. To avoid this limitation, we uniformly upscale the maximum value of all the variables to ± 5 V. This scaling of variables neither increase the complexity nor alter the original dynamics of the system, rather it produces the oscillation with higher magnitude than original system. It is useful in the circuit design. This range is adequate to avoid noise and saturation effects over the op-amps considering a power supply of ± 9 V. Further, it is convenient to visualize the time waveform in the oscilloscope traces as well as to capture the data through data acquisition systems. Applying the amplitude scaling to the Eq. (4), the new variables are defined as $x = 3.3x_1, y = 10x_2, z = 2x_3$ and $w = 5x_4$. Rewriting the Eq. (4) to the new variables yields the following upscaled system,

$$\frac{dx}{dt} = \alpha(0.33y - x + \epsilon x - ax + 0.12bw^2x), \tag{5a}$$
$$\frac{dy}{dt} = 3.03x - y + 5z, \tag{5b}$$
$$\frac{dz}{dt} = -0.2\,\beta\, y, \tag{5c}$$
$$\frac{dw}{dt} = 1.5152x. \tag{5d}$$

Considering a probability of any intermediate signal to surpass the saturation voltage limits of the op-amps, the entire equation system is divided by the largest parameter. In the present case, the largest parameter is $\beta = 100/7 = 14.28$, we divide the entire system by the factor 15 on both sides of the Eq. (5). Although this procedure alters the operational frequency of the circuit, it does not change the dynamics of the system. Hence, the rescaled final set of equations is given by,

$$\frac{dx}{dt} = 0.2156y + 0.09342x - 2.24109\frac{w^2}{100}x, \tag{6a}$$
$$\frac{dy}{dt} = 0.202x - 0.067y + 0.33z, \tag{6b}$$
$$\frac{dz}{dt} = -0.0133\,\beta\, y \tag{6c}$$

$$\frac{dw}{dt} = 0.10101x. \tag{6d}$$

During this rescaling, the parameter values used for numerical simulation of Eq. (4) is applied for all the parameters except β, which is used as the control parameter for the analysis.

3.1 Numerical Results

The Eq. (6) is numerically simulated. The phase portraits of double band chaotic oscillations in different projections for $\beta = 100/7$ are shown in Fig. 3. The phase portraits for different values of β during the period doubling transitions to chaos are shown in Fig. 4. Further, the local maxima of oscillations in the variable 'w' is calculated to plot the one parameter bifurcation plot in the ($\beta - w$) plane. The nature of the system is quantified by calculating the Lyapunov exponent spectrum. The one parameter bifurcation diagram and largest three Lyapunov exponents (λ_i for $i = 1, 2, 3$) are plotted in Fig. 5 for $12 < \beta < 20$. The presence of one positive exponent confirms the chaotic nature of the system. All these numerical investigations are carried out using a particular initial conditions ($\{x, y, z, w\} = \{0, 0.1, 0, 0\}$). The role of initial conditions on the dynamics of the system is also investigated, since present state of the memristor is depends on its previous state. Figure 6 shows the bifurcation plots for (a) ($x(0) - x$) and (b) ($w(0) - w$) planes for a wide range of initial conditions.

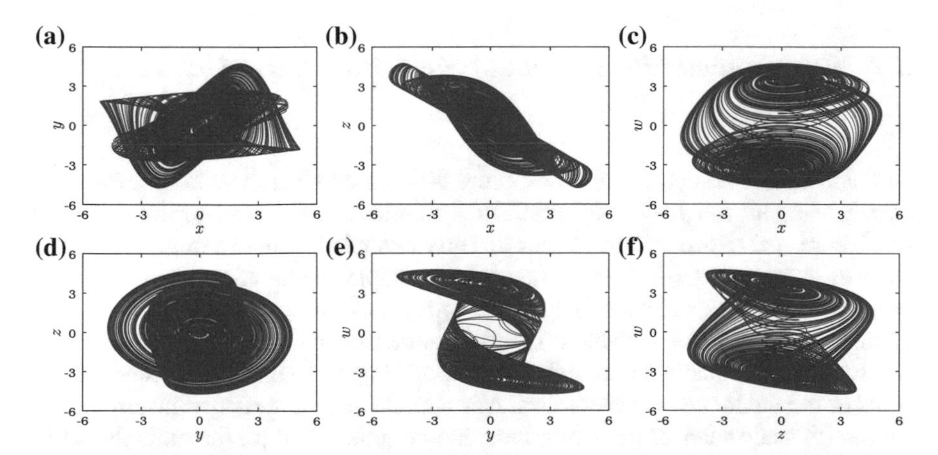

Fig. 3 Different projections of numerical phase portraits of Eq. (4) in the **a** ($x - y$), **b** ($x - z$), **c** ($x - w$), **d** ($y - z$), **e** ($y - w$) and **f** ($z - w$) planes. The parameters of the system are fixed as $\alpha = 9.8$, $\beta = 100/7$, $\gamma = 9/7$, $a = 1/7$ and $b = 2/7$. The initial conditions are $\{x, y, z, w\} = \{0, 0, 1.0, 0\}$

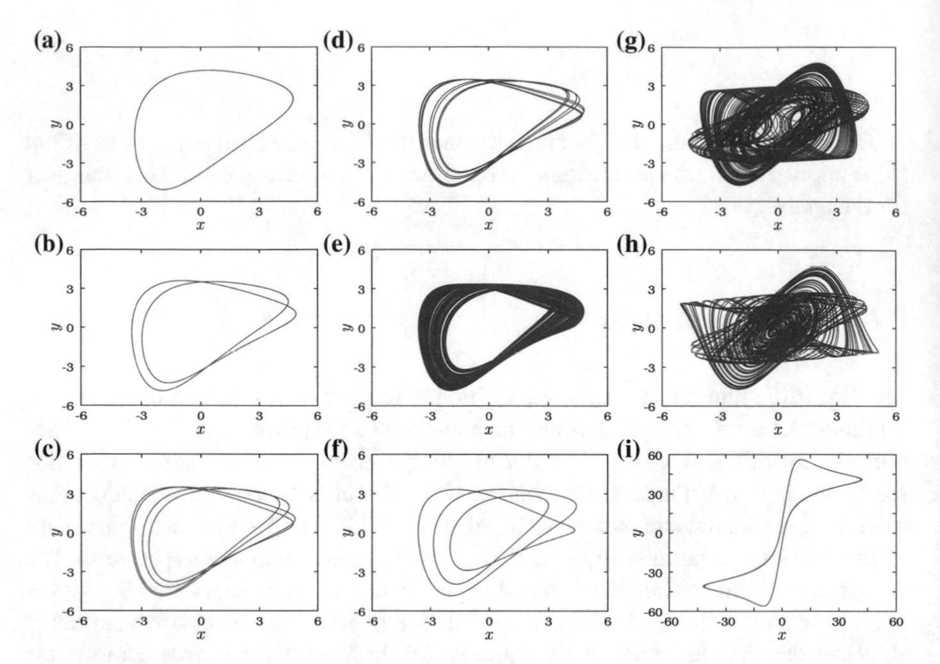

Fig. 4 The period doubling bifurcation sequence obtained using numerical simulation of phase portraits of Eq. (6) in the $(x - y)$ plane for various values of β. **a** period-1 limit cycle (β = 20), **b** period-2 limit cycle (β = 17.7), **c** period-4 limit cycle (β = 17.3), **d** period-8 limit cycle (β = 17.268), **e** one band chaos (β = 17), **f** period-3 window (β = 16.9), **g** double band chaos (β = 15.93), **h** double band chaos (β = 14.4) and **i** boundary limit cycle (β = 12). Rest of the parameters are same used for plotting Fig. 3

3.2 Experimental Realization of Memristor Based Chua's Oscillator

In the analogue computation, a first order differential equation can be solved primarily using an weighted integrator. The dynamics of the memristor based Chua's oscillator is described by a set of four first order coupled nonlinear differential equations as given in Eq. (6). Hence, the analogue implementation of memristor oscillator will have four weighted integrators. Unit gain inverting amplifiers are used to change the sign of the variables. To incorporate the nonlinear function present in the Eq. (6a), the analogue multipliers are used. Figure 7 shows three fundamental units of the analogue memristor oscillator namely, (a) the weighted inverting integrator, (b) scale changer (unit gain inverting amplifier) and (c) the multiplier chip. The transfer function of the weighted integrator is given by

$$v_o = -\frac{1}{R_{in}C}\int v_{in}\mathrm{d}t + v_c(0), \tag{7}$$

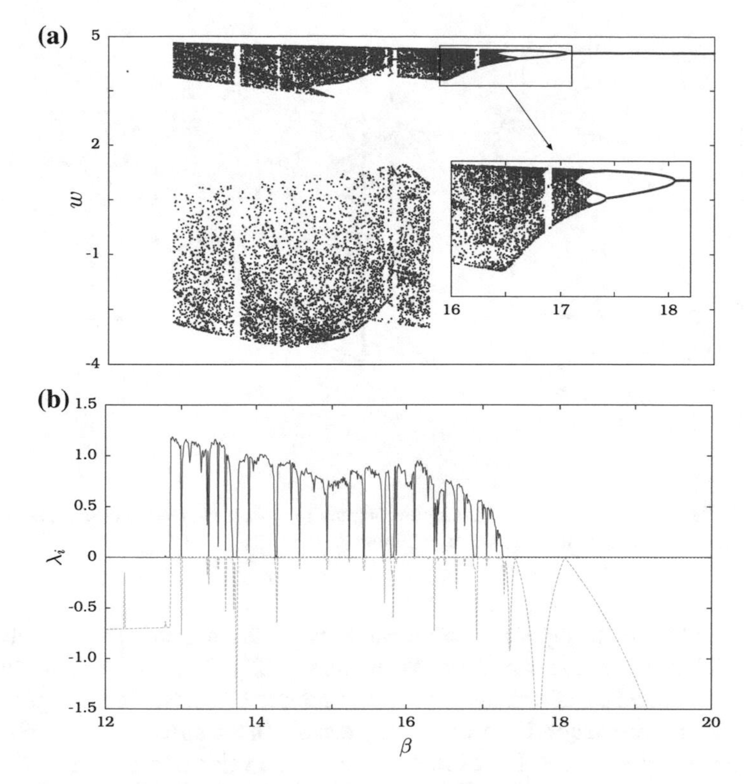

Fig. 5 **a** One parameter bifurcation diagram in the ($\beta - w$) plane, (A clear depiction of period doubling sequence is shown in inset) and **b** corresponding Lyapunov exponents spectrum for first three exponents (λ_i for $i = 1, 2$ and 3 with 'red', 'blue' and 'green' colours, respectively) are plotted in the ($\beta - \lambda_i$) plane. The fourth exponent is skipped to view the rest conveniently. The positive values in the λ_1 indicates the chaotic nature of the system obtained

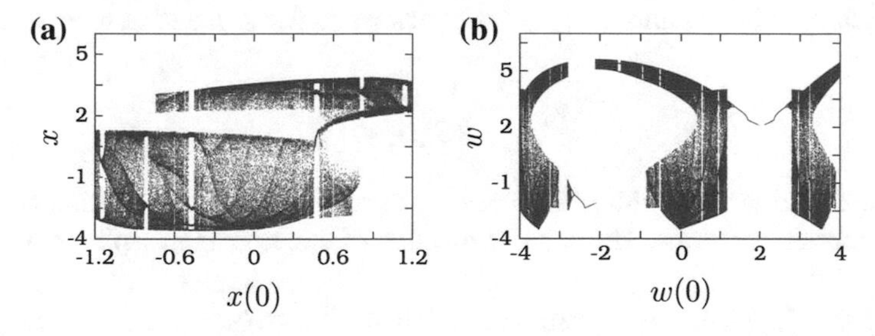

Fig. 6 The bifurcation plots for **a** ($x(0) - x$), $\{y, z, w\} = \{0.1, 0, 0\}$ and **b** ($w(0) - w$), $\{x, y, z\} = \{0, 0.1, 0\}$ planes of the Eq. (6). ($\{\cdots\} = \{\cdots\}$) represents the initial conditions of other three variables

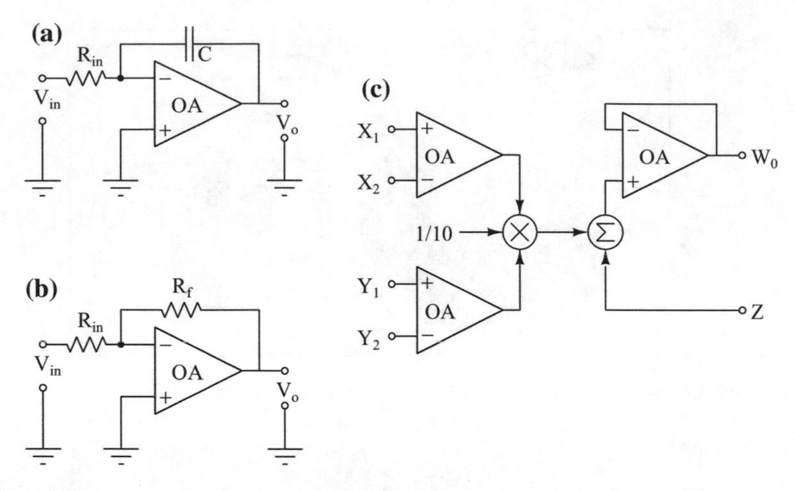

Fig. 7 The fundamental units of the analogue computation. Namely, **a** the integrator, **b** the inverting amplifier and **c** schematic of an analog multiplier

where v_{in} and v_0 are the input and output voltages of the integrator, respectively. $v_c(0)$ is the initial charge in the capacitor. We assume $v_c(0) = 0$ for the present study. R_{in} is divided in to two parts as R and R^*. The R^* refers a real positive number and R is denoted as base resistance.The value of any resistor in the circuit can be represented as the product of R^* and R. For example, assume the value of a resistor is 47 kΩ. It can be represented as 4.7×10 kΩ. This conversion is useful to determine the value of the resistor when an analogue circuit is designed for a given set of differential equations. The value of R^* is simply the reciprocal of corresponding coefficient in the equation. The values of the base resistance R and the capacitance C will determine the frequency of operation of the analogue circuit. It does not change the dynamics of the system rather it can slower or faster the oscillations. Thus we rewrite the Eq. 7 as,

$$v_o = -\frac{1}{RC}\int \frac{v_{in}}{R^*}\,\mathrm{d}t, \tag{8}$$

Inverting amplifier with unit gain is used to change the sign of any state variable in the differential equations. The transfer function of the inverting amplifier at unit gain is,

$$v_o = -v_{in}, \qquad \text{when} \quad R_{in} = R_f, \tag{9}$$

where R_{in} and R_f are the input and feedback resistors of an amplifier unit. These R_{in} and R_f are generally fixed as equal to the base resistance R in the analogue circuit design. The analog multiplier is useful component to multiply two analog signals. We use this component to get the nonlinear part present in the Eq. 6(a).

The product output W_0 of a typical multiplier can be written as,

$$W_0 = \frac{(X_1 - X_2) * (Y_1 - Y_2)}{10} + Z, \quad (10)$$

Here, X_1 and X_2 are *X-multiplicand* of non-inverting and inverting inputs, Y_1 and Y_2 are *Y-multiplicand* of non-inverting and inverting inputs and Z is the summing input. The dividing factor 10 is used to avoid the overflow of product output.

The analogue circuit to emulate the Eq. (6) is implemented using the above described three electronic entities is shown in Fig. 8. In the circuit diagram $U1$, $U2$, $U3$ and $U4$ are the weighted integrators, $U5$ and $U6$ are inverting amplifiers and $U7$ and $U8$ represents the multiplier chips. A feedback resistor with high value (typically $R = 2$ MΩ) is connected to each integrator in order to fix the gain at low frequencies and to reduce the effect of the offset voltages in the op-amps. This feedback resistor should be at least 10 times greater than the input resistance of its respective integrator. For similar reasons, resistors are placed at the Z terminals of the multipliers also. The coefficient value of the Eq. (6) are inversely proportional to R^* values in the circuit. To explain the calculation of the R^* value from the coefficient of

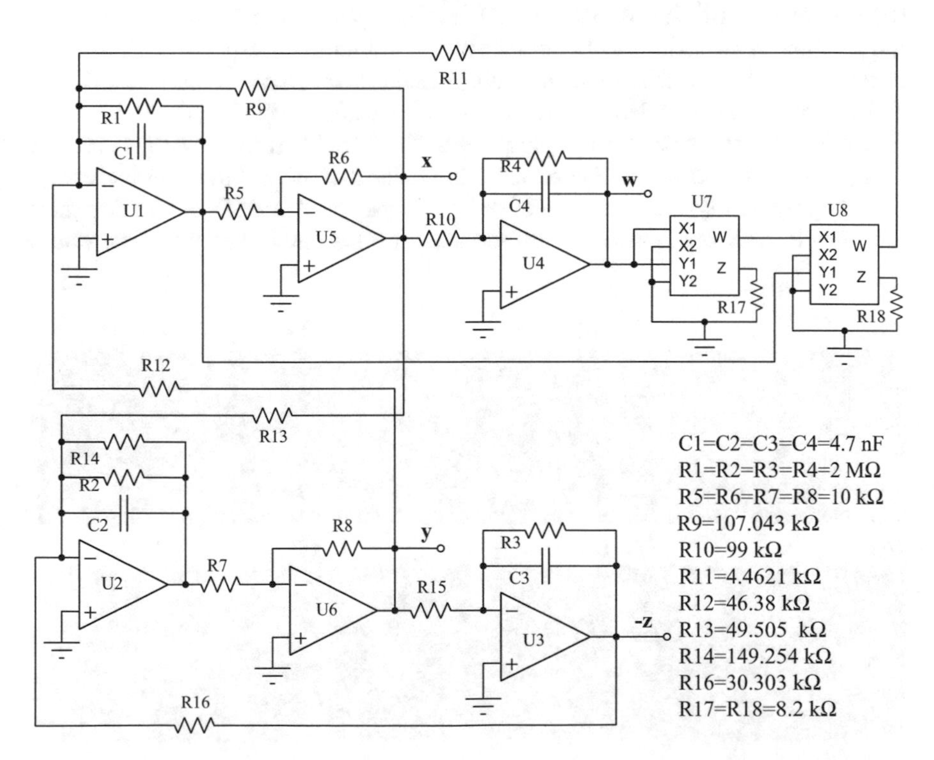

Fig. 8 The circuit diagram of the analogue oscillator based on memristor. The component values given in the diagram are used for PSpice simulation. For U1–U6, TL081 op-amps and for U7 and U8, AD633JN multiplier chips are used

Eq. (6), we consider Eq. 6(d). The variable x in the right hand side has the coefficient value 0.10101. The corresponding R^* value to the input resistance of the W cell is $1/0.10101 = 9.9$. The value of corresponding resistor in the circuit is then directly calculated as $(R^* \times R)$ 9.9×10 kΩ = 99 kΩ. In the similar way, values of the resistance corresponding to all other coefficients in Eq. (6) are determined.

3.3 Experimental Results

Initially the experimental implementation of the memristor oscillator is done using PSpice simulation. A double band chaotic attractor is obtained in PSpice simulation for $R15 = 53.5$ kΩ. In order to design the hardware circuit, the resistor values used for PSpice simulation are modified to its nearest "*off-the-shelf*" components values. The values of the components are chosen as $C1 = C2 = C3 = C4 = 4.7$ nF, $R1 = R2 = R3 = R4 = 2$ MΩ, $R5 = R6 = R7 = R8 = 10$ kΩ, $R9 = 106.8$ kΩ, ($R9A$ = 100 kΩ + $R9B$ = 6.8 kΩ), $R10 = 100$ kΩ, $R11 = 4.46$ kΩ, $R12 = 47$ kΩ, $R13 = 49.2$ kΩ, ($R13A = 47$ kΩ + $R13B = 2.2$ kΩ), $R14 = 150$ kΩ, $R16 = 30$ kΩ, ($R16A = 10$ kΩ + $R16B = 10$ kΩ + $R16C = 10$ kΩ) and $R17 = R18 = 6.8$ kΩ. The capacitors and resistors have maximum of tolerance about 10% and 1%, respectively. To get the accurate reproduction of numerical results on experiment, the resistors $R9$, $R13$, and $R16$ are split into parts whereby the sum of the resistance will yield the closest possible value towards the numerical parameters. The variable resistor "$R15$" is used as the control parameter which is corresponding to the parameter β in numerical analysis. The polyester type capacitors, TL081 op-amps and AD633JN multiplier chips biased with ±9 V dual power supply, are used for this implementation. The values

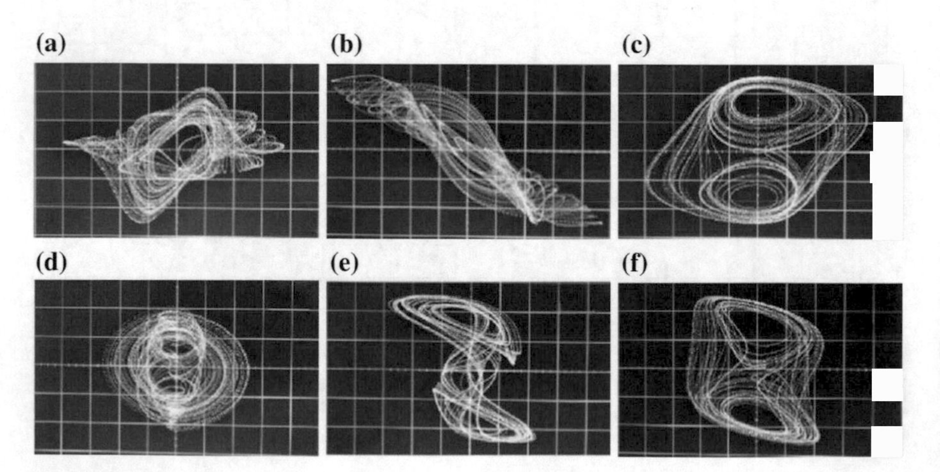

Fig. 9 Different projections of phase portraits for double band chaos obtained from experiment in the **a** $(x - y)$, **b** $(x - z)$, **c** $(x - w)$, **d** $(y - z)$, **e** $(y - w)$ and **f** $(z - w)$ planes. *Scale* Horizontal axis = 1.36 V/div, Vertical axis = 1.52 V/div

X, Y, Z and W are the four variables in the circuit. The unit of these variables are in 'volts'.

By varying the value of $R15$, different dynamical states of the circuit are captured on 'Agilent DSO-6014A' oscilloscope. For instance, when $R15 = 54.6\ \text{k}\Omega$, the circuit generates the double band chaotic attractor in the phase plane as shown is Fig. 9 on different projections which are in good agreement with numerical results presented in Fig. 3. The corresponding time waveforms are plotted in Fig. 10. It clearly shows that all the variables are oscillating in the range of ± 5 V as defined in the circuit design. The dynamics of the circuit for different values of the control parameter $R15$ in the range of 60–50 kΩ are given in Fig. 11. It is observed that this circuit exhibits the familiar period-doubling bifurcation route to chaos. Most of the plots shown in Fig. 11 are also captured on numerical simulations as shown in Fig. 4.

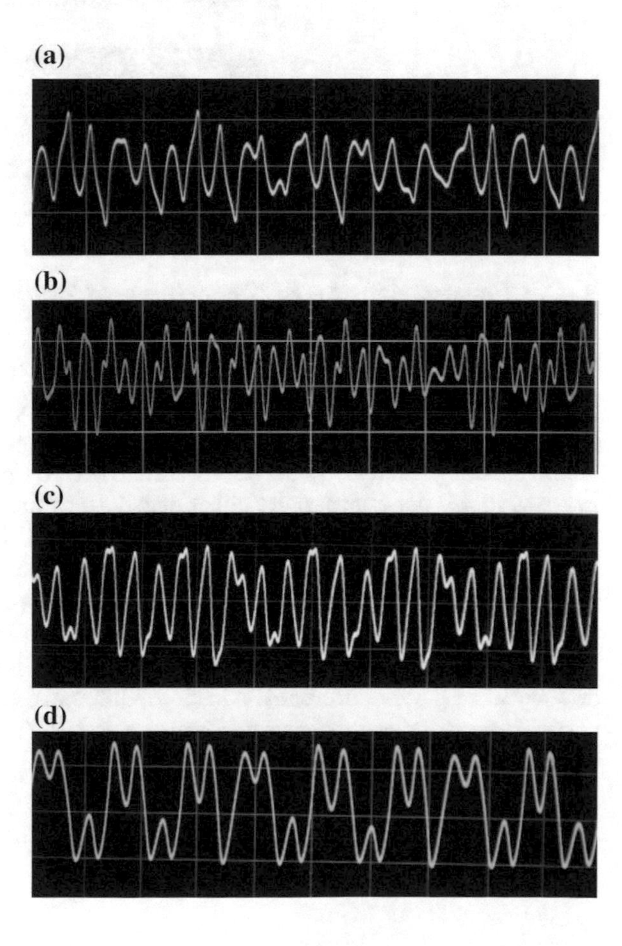

Fig. 10 Experimental time waveform of the variables $x(t), y(t), z(t)$ and $w(t)$ operating in the chaotic regime. *Scale* Horizontal axis = 5 ms/div, Vertical axis = **a** 2.5 V/div, **b** 3.0 V/div, **c** 4.0 V/div and **d** 2.5 V/div

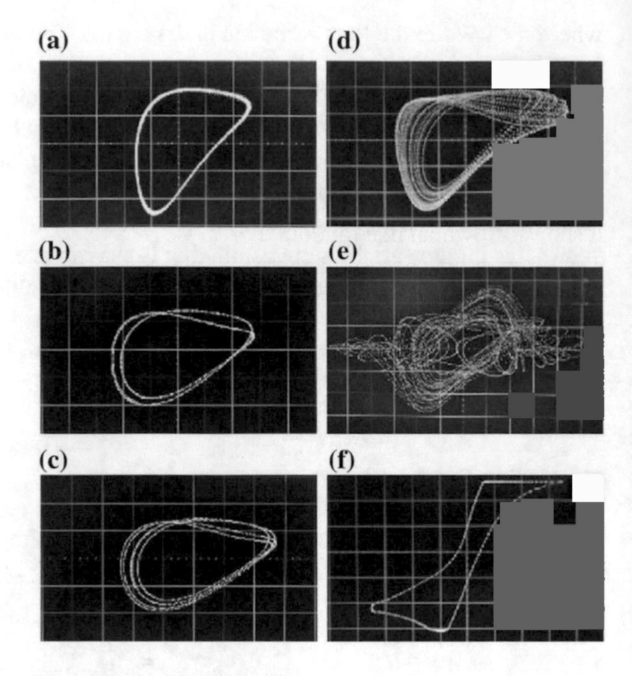

Fig. 11 Period doubling scenario of experimental phase portraits in the $(x - y)$ plane for the variation of $R15$. **a** period-1 limit cycle, ($R15 = 59.3$ kΩ); **b** period-2 limit cycle, ($R15 = 57.2$ kΩ); **c** period-4 limit cycle, ($R15 = 55.9$ kΩ); **d** Single band chaos, ($R15 = 55.1$ kΩ); **e** double band chaos, ($R15 = 54.6$ kΩ); and **f** saturated limit cycle, ($R15 = 53.1$ kΩ)

3.4 Characterization of Experimental Time Series Using '0–1' Test

Recently, G. A. Gottwald and I. Melbourne has developed a new kind of test which they named as "0–1 test" to distinguish the periodic and chaotic dynamics of any deterministic dynamical system (Gottwald and Melbourne 2004). This test is distinct in a way that it does not require either equations of the system or the large data. It also not requires the phase space reconstruction as well. The dimension of the system is also irrelevant. Its applicability is tested for periodic, quasi-periodic, chaotic and strange nonchaotic motions (Gopal et al. 2013). The '0–1 test' returns a single scalar value either '0' for periodic or '1' for chaotic motion for a set of time series data (n). The test applies to the data originated from any form of system governed by differential or partial differential equations, difference equations etc. Consider a set of discrete one dimensional data $\Phi(n)$ where $n = 1, 2, 3, \ldots, N$. Define the translation components $p(n)$ and $s(n)$ as

$$p(n) = \sum_{j=1}^{n} \Phi(j)cos(\theta(j)), \qquad n = 1, 2, 3, \ldots, N \tag{11}$$

$$s(n) = \sum_{j=1}^{n} \Phi(j)sin(\theta(j)), \qquad n = 1, 2, 3, \ldots, N \tag{12}$$

where

$$\theta(j) = jc + \sum_{i=1}^{j} \Phi(i), \qquad j = 1, 2, 3, \ldots, n. \tag{13}$$

where, the value c is constant and it can be chosen arbitrarily. The translation components (p, s) shows bounded dynamics in the $(p - s)$ plane for periodic motion, while it shows Brownian like random dynamics for chaotic dynamics of the given data set. The plots in the $(p - s)$ plane itself gives an visual confirmation to the type of the motion. Further, mean square displacement $M(n)$ is defined on the basis of $p(n)$, and $s(n)$, which is given below

$$M(n) = \lim_{N\to\infty} \frac{1}{N} \sum_{j=1}^{N} [p(j+n) - p(j)]^2 + [q(j+n) - q(j)]^2. \tag{14}$$

Here, it is noted that $n \ll N$. For the bounded dynamics of $p(n)$, the $M(n)$ is also bounded whereas for the Brownian like dynamics of $p(n)$, the $M(n)$ linearly grows with time. From this the asymptotic growth rate K is defined by

$$K = \lim_{N\to\infty} \frac{log\ M(n)}{log\ n}. \tag{15}$$

For a given data set, K yields '0' for periodic and '1' for chaos. To characterize the other dynamical states of a system one can follow the procedure elaborated in reference (Gopal et al. 2013).

In general, experimental time series data would contain a small amount of noise arises due to inherent thermal fluctuations present within the circuit. It is proven that "0–1 test" is suitable for the time series data with low noise (Kulp and Smith 2011). We perform the "0–1 test" for the experimental data collected for the periodic and chaotic motion of the circuit. The set of 'n'-data points of the time series 'w' after the initial transients vanished, is collected using 'Agilent-U2531A' data acquisition module with the sampling rate of 500 kSa/s. The value of the constant 'c' is randomly varied between $(0 - \pi)$. The computed translation components (p, s), the mean square displacement $M(n)$ and the asymptotic growth rate K for regular (period-1 limit cycle at $R15 = 59.3$ kΩ) and chaotic time series (double band chaotic attractor at $R15 = 54.6$ kΩ) are plotted in Fig. 12. These plots clearly differentiates the regular and chaotic motion of the circuit.

4 The Memristor Based Duffing Oscillator

One way of solving the differential equation is through their analogue simulation of electronic circuits is very important in nonlinear science. Analogue electronic simulation studies provide for a quick scan of the entire parameter space in real

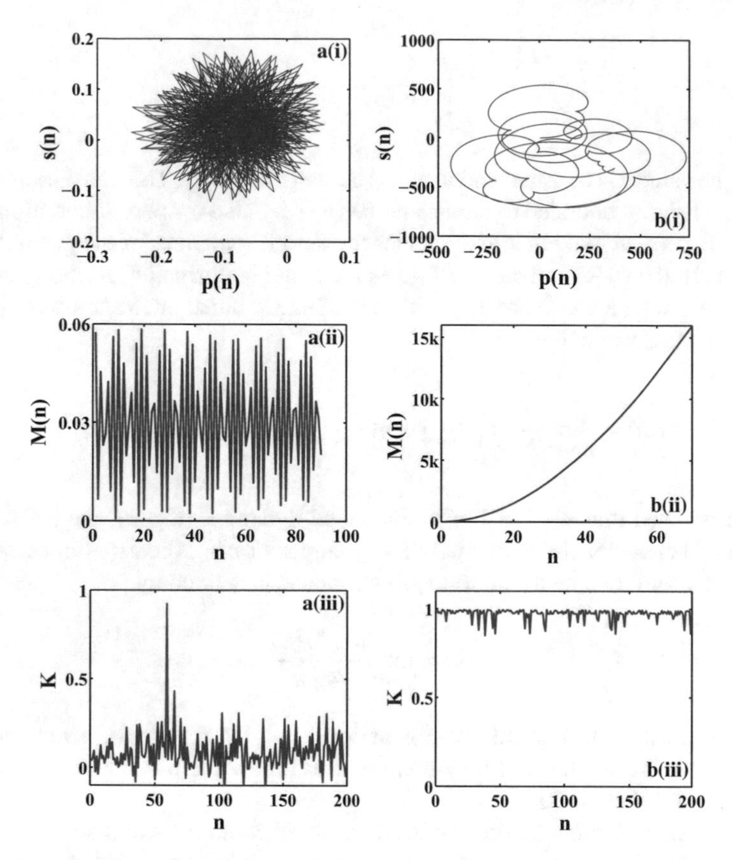

Fig. 12 Application of the 0–1 test to the experimental time series for **a** regular ($R15 = 59.3$ kΩ) and **b** chaotic motion ($R15 = 54.6$ kΩ): (*i*) translation dynamics of $(p(n), s(n))$, (*ii*) mean square displacement $M(n)$ and (*iii*) asymptotic growth rate K

time, using simple instruments that are available even in an undergraduate laboratory and off the shelf electronic circuit components. Further, as they are purely RC based circuits the drawbacks of LC networks are done away with. Their simplicity and ease of implementation have led do not only validation of past theoretical predictions but also discoveries of many new phenomena. In this section, we present the investigations on the memristor based Duffing oscillator. To design the memristor based non-autonomous chaotic circuit, we consider Duffing oscillator system (George 1918; Kovacic and Brennan 2011) by replacing the cubic nonlinearity with a flux-controlled memristor characteristic nonlinear equation (1). The state equation for the memristor based Duffing oscillator is

$$\ddot{x} + \alpha \dot{x} + W(\phi)x = fsin(\omega t), \tag{16}$$

For convenience, the Eq. (16) is rewritten as a system of coupled first order equations

$$\begin{aligned} \dot{x_1} &= x_2 \\ \dot{x_2} &= x_3 \\ \dot{x_3} &= -\alpha x_3 - W(\phi)x_2 + fsin(x_4) \\ \dot{x_4} &= \omega\,. \end{aligned} \tag{17}$$

where, $W(\phi) = \omega_0^2 + 3\beta x_1^2$ is the memristance function. Thus by replacing the cubic nonlinearity in the classical Duffing oscillator, a new memristor based oscillator is defined, upon which its dynamical behaviour is investigated in detail. The electronic implementation of analogue circuit model of the memristive Duffing oscillator is shown in Fig. 13. In this figure, the memristor part is boxed in '*red colour*'. The dynamical equations for the proposed memristor based Duffing oscillator are

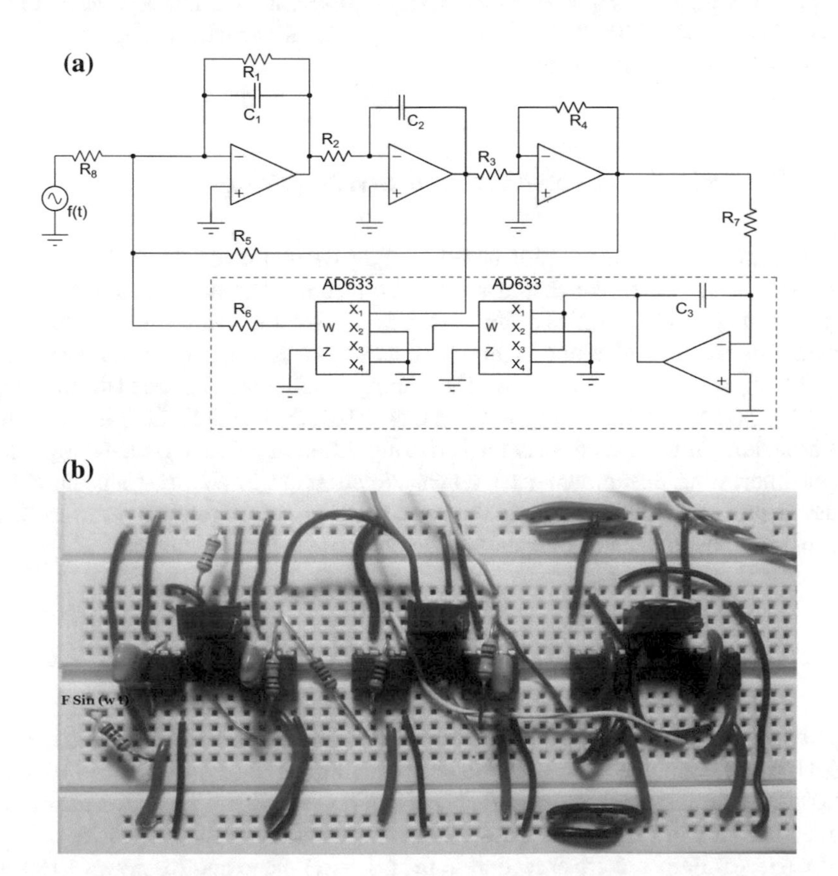

Fig. 13 **a** Schematic of the memristor based Duffing oscillator with external forcing $f(t) = f\sin(\omega t)$ (*red box* indicates the realization of the memristor nonlinearity) and **b** Photograph of the experimental circuit realization

obtained by using Kirchhoff's law at voltages (v_1, v_2, v_3) across the capacitors $(C_1, C_2, C$
respectively

$$\frac{d^3v_3}{dt^3} = -\frac{1}{R_1C_1}\frac{d^2v_3}{dt^2} + \frac{1}{R_2R_3C_1C_2}\frac{dv_3}{dt} - \frac{0.01v_3^2}{R_2R_7C_1C_2}\frac{dv_3}{dt} + \frac{1}{R_2R_3R_8C_1C_2C_3}F\sin(\Omega t). \tag{18}$$

The normalization parameters are defined as $d^2v_3/dt^2 = x_3$, $dv_3/dt = x_2$, $v_3(\tau)$ as x_1, $\tau = R_2C_1t$, $\Omega = \omega R_2C_1$, $\alpha = R_2/R_1$, $\omega_0^2 = R_2/R_5$ and $\beta = 0.01R_2/R_6$. The memristive Duffing circuit used for this investigation is shown in Fig. 13b. In order to study the dynamics experimentally, the values of the circuit components are fixed as, $R_1 = 1$ MΩ, $R_2 = R_3 = R_4 = R_7 = R_8 = 10$ kΩ, $R_5 = 28.57$ kΩ, $R_6 = 125$ Ω ($\alpha = 0.001$, $\omega_0^2 = -0.35$, $\beta = 0.8$), C_1, C_2, $C_3 = 10$ nF, provided $\pm 1\%$ of tolerance is permitted in all the elements.

4.1 Numerical and Experimental Investigations

The dynamics of the memristor based Duffing oscillator represented by normalized system Eq. (17) is sketched using the fourth-order Runge-Kutta algorithm. An approximate global dynamics of the system is constructed in a two parameter phase diagram in the $(f - \omega)$ plane in Eq. (17), as shown in Fig. 14 for the parameters $\alpha = 0.1$, $\omega_0^2 = -0.35$, and $\beta = 0.8$. The forcing amplitude f is varied in the range $(0.01 \geq f \geq 2.5)$ and frequency ω in the range $(0.01 \geq \omega \geq 2.5)$. Different dynamical behaviour of the system is described using different colours. Specifically, some of the interesting phenomena such as chaotic attractor, three tori, transient chaos, intermittency and various periodic states are sketched. In the following sub sections, few of such phenomena are subjected to appropriate statistical analysis.

4.2 Chaotic Attractor

By fixing the value of the parameters of system Eq. (17) as $\alpha = 0.0001$, $\omega_0^2 = -0.35$, $\beta = 0.8$ and $\omega = 1.0$, with the initial conditions $(x_1, x_2, x_3, x_4) = (0.01, 0.03, 0.003, 0.0)$, the system shows chaotic behaviour for red. Figure 15, shows the numerical observation of (a) phase portraits and its corresponding (b) time series of ($x_2(t)$ and $x_3(t)$), (c) Poincaré cross section in the $(x_2 - x_3)$ plane and (d) power spectrum of the chaotic dynamics of the system. Calculation of the Lyapunov exponent values, $\lambda_{1,2,3,4} = (0.1335, 0.0, -0.1327, 0.0)$ which confirms the chaotic behaviour quantitatively by means of one positive exponent.

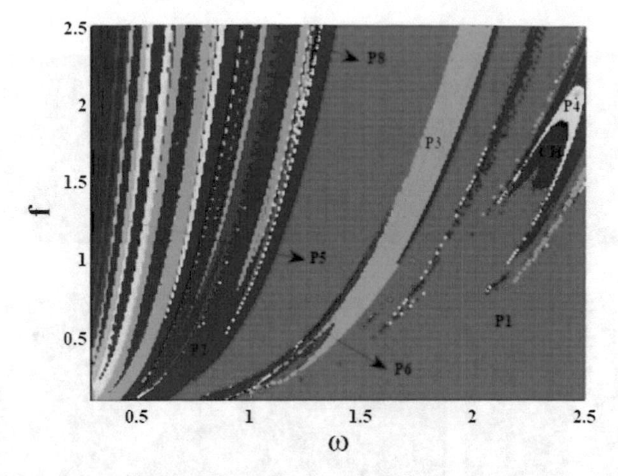

Fig. 14 Numerically computed two phase diagram in the $(f - \omega)$ plane with the initial conditions $x_1 = 0.0001$; $x_2 = 0.0006$; $x_3 = 0.01$; $x_4 = 0.0$ and fixed parameters of $\alpha = 0.1$, $\omega_0^2 = -0.35$, $\beta = 0.8$ of system Eq. (17). The different color indicates various dynamical behaviors of the system, such as Period-1 (*green*), Period-2 (*blue*), Period-3 (*light green*), Period-4 (*yellow*), Period-5 (*sky blue*), Period-6 (*Pink*), Period-8 (*black*) and chaotic attractor (*red*)

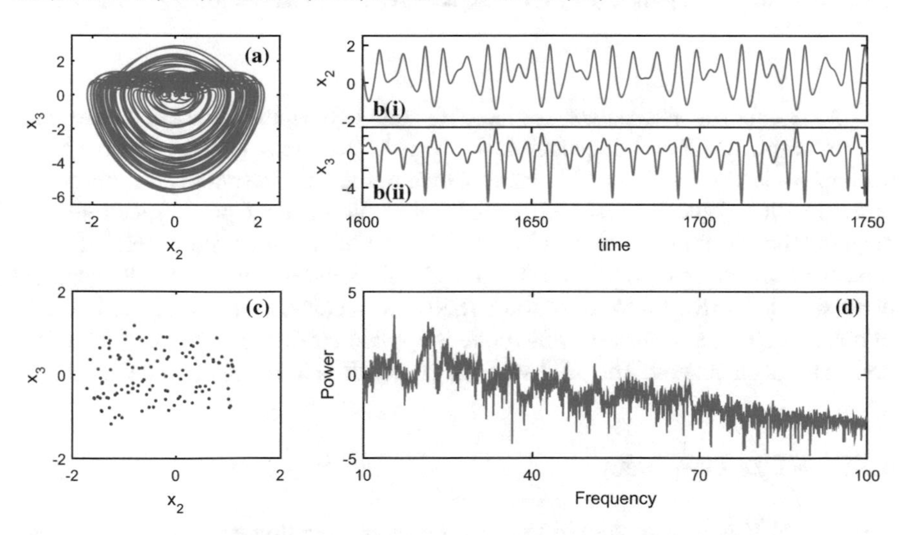

Fig. 15 Numerical observation of chaos: **a** phase portrait and its corresponding **c** Poincaré cross section in the $(x_2 - x_3)$ plane, **b** time evolution of (*i*) $x_2(t)$, (*ii*) $x_3(t)$, and **d** Power spectrum of the chaotic time series, obtained with the initial conditions $x_1 = -0.01, x_2 = 0.03, x_3 = 0.003, x_4 = 0.0$ and fixed parameters of $\alpha = 0.0001$, $\omega_0^2 = -0.35$, $\beta = 0.8$, $\omega = 1.0$ and $f = 0.8$ of Eq. (17)

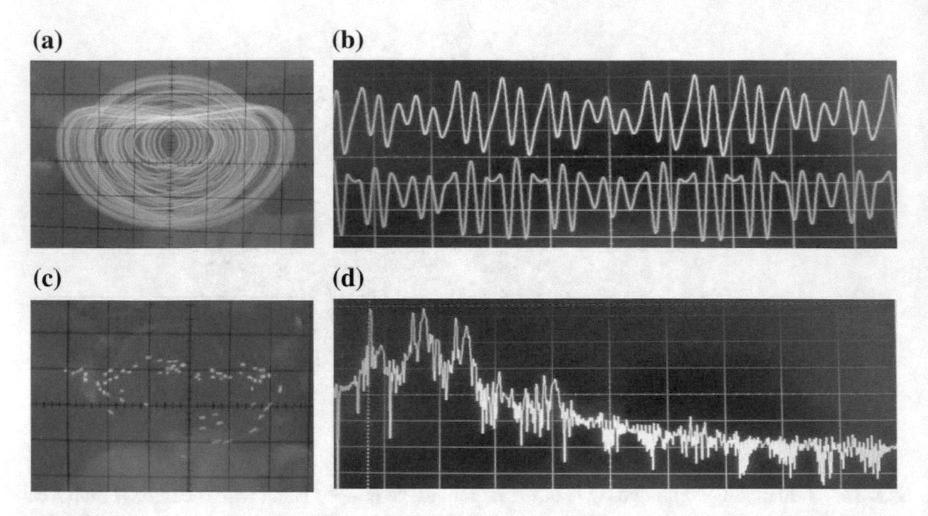

Fig. 16 Experimental observation of chaos: **a** phase portrait and its corresponding, **c** Poincaré cross section in the $(v_2 - v_3)$ plane, **b** time evolution of voltages of $(v_2(t))$ [*yellow*], $(v_3(t))$ [*green*], and **d** Power spectrum of the chaotic time series, the *dashed red line* crossing which indicating the external forcing frequency, (i, e) 780 Hz. The observation made by fixing $F = 1.120$ V and $\Omega = 784$ Hz

For the experimental demonstration of chaotic behaviour of the circuit the parameters, amplitude F and frequency Ω are varied. The circuit shows chaotic dynamics for $F = 1.120$ V and $\Omega = 784$ Hz as shown in Fig. 16. The phase portraits and its corresponding Poincare cross section, time series plots and the power spectrum indicates the chaotic motion which is similar to the numerical observations (Fig. 16). The robustness of the chaotic dynamics of the circuit is tested with externally supplied noise signal. The Signal to Noise Ratio (SNR) is calculated for the various intensity of noise (zero mean Gaussian white noise is applied for this test), the stability of the behaviour is calculated. The SNR value of 20.06 dB is measured for chaotic signal.

4.2.1 0–1 Test for Chaos

The '0–1 test' is also applied to the time series of both numerical and experimental data to detect the chaotic dynamics of the system. The dynamics of translational components $p(n)$, $q(n)$, mean square displacement $(m(n))$ and asymptotic growth rate (K) are plotted in Fig. 17. The value of $K = 0.9696$ for numerical data shown in Fig. 17a(iii), and $K = 0.9780$ for experimental data as shown in Fig. 17b(iii) indicates tested time series has the chaotic flow over the time.

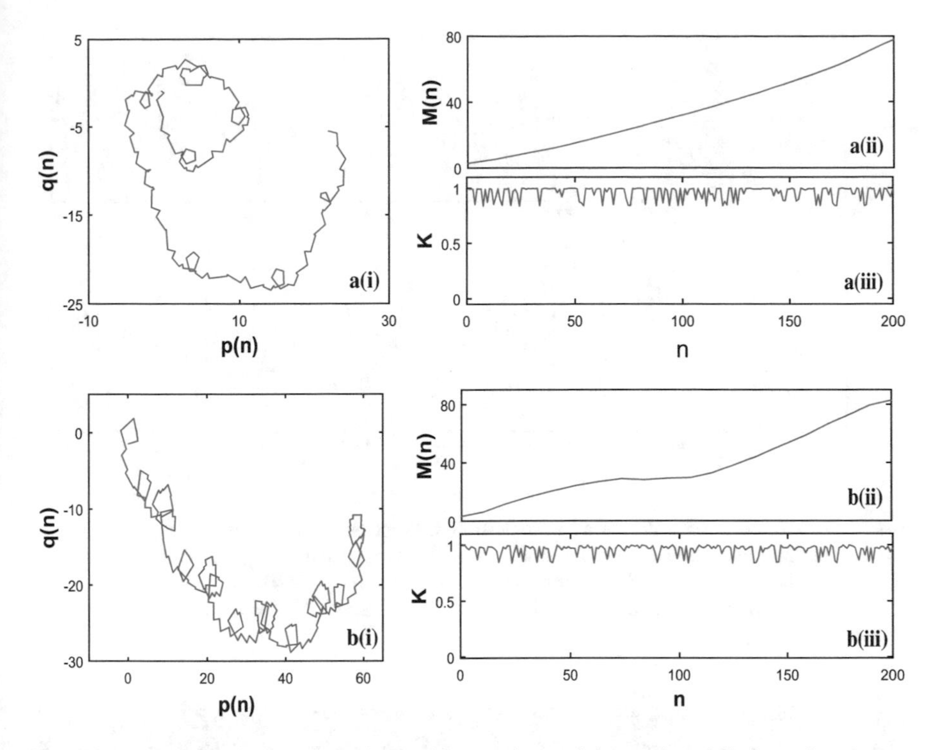

Fig. 17 0–1 test for chaos: **a** numerical and **b** experimental time series of (*i*) dynamics of translation components (p(n), q(n)), (*ii*) mean square displacement (M(n)) and (*iii*) asymptotic growth rate (K) of the chaotic time series data

4.3 Three Tori

Torus is a phenomenon which in general occurs in a system where there exists a two incommensurate frequencies. Three tori is a phenomena rarely encountered even in a coupled dynamical systems. We observe three tori in a single system of memristor based Duffing oscillator. Under the absence of external forcing, *i.e.* for $f = 0$ and $\omega = 0$ dynamics of the system asymptotically approaches a stable fixed point. An external periodic force applied to the circuit modulates the environment in which the intrinsic mode of oscillations lives. As a result of this modulation, the nonlinearity and dissipation will again tend to change the intrinsic mode of oscillation. In the interval of $0.01 \geq f \geq 0.2$, three tori is identified. The presence of 3-tori has characterized by Lyapunov exponents, by which three of them must be zero. In the present case, the calculated Lyapunov exponents values are $\lambda_1 = 0.0001$, $\lambda_2 = 0.0$, $\lambda_3 = -0.0005$ and $\lambda_4 = 0.0$ for $f = 0.06$ and $\omega = 0.2009$ with the initial conditions $x_1 = 0.0001; x_2 = 0.0006; x_3 = 0.01; x_4 = 0.0$. The other parameters for the numerical simulation are $\alpha = 0.000119$, $\omega_0^2 = 0.35$, and $\beta = 0.51$. Figure 18 shows, numerical observation of the phase portraits (Fig. 18a), the Poincaré cross section (Fig. 18b)

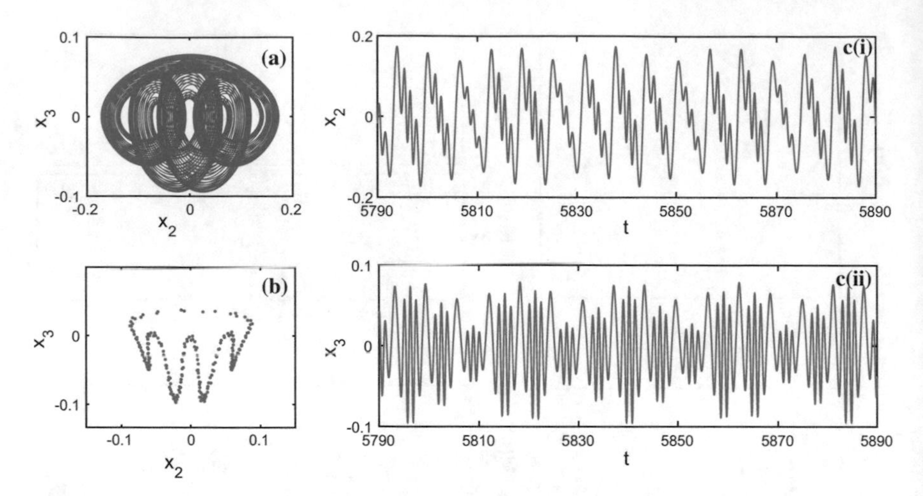

Fig. 18 Numerical observation of **a** Phase portrait, **b** Poincaré cross section in the $(x_2 - x_3)$ plane and its corresponding **c** time series plot of state variable *(i)* $x_2(t)$ and *(ii)* $x_3(t)$ explores the presence of three tori attractor observed with the initial conditions $x_1 = 0.0001$; $x_2 = 0.0006$; $x_3 = 0.01$; $x_4 = 0.0$ and fixed parameters of $\alpha = 0.000119$, $\omega_0^2 = 0.35$, $\beta = 0.51, f = 0.06$ and $\omega = 0.2009$ of system Eq. (17)

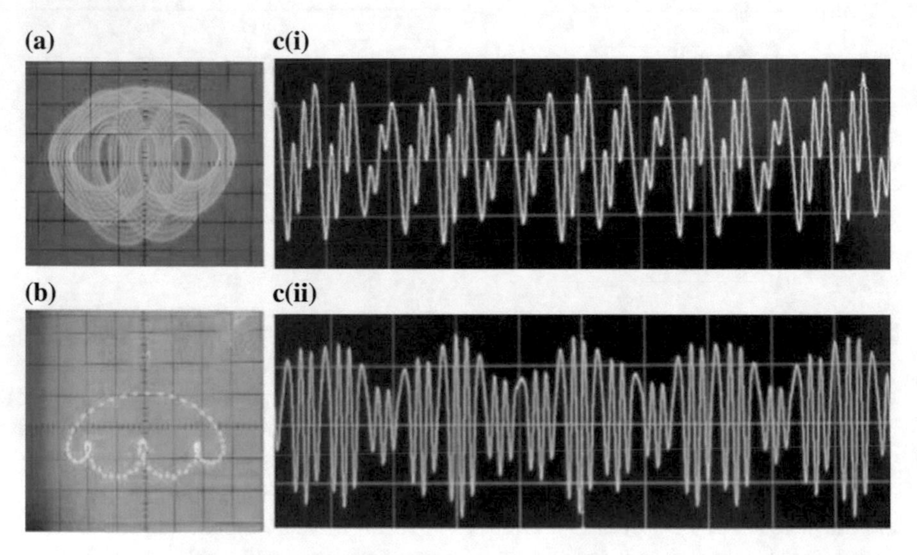

Fig. 19 Experimental observation of **a** Phase portrait, **b** Poincaré cross section in the $(v_2 - v_3)$ plane and its corresponding **c** time series plot of state variable *(i)* $v_2(t)$ and *(ii)* $v_3(t)$ explores the presence of three tori

in the $(x_2 - x_3)$ plane and (c) time series of $(x_2(t), x_3(t))$ of typical three tori attractor. The corresponding experimental plots obtained for $F = 1.907$ V and $\Omega = 540$ Hz are shown in Fig. 19.

4.4 Transient Chaos

The transient chaos termed as the trajectories starting from random initial conditions have chaotic behaviour for a finite time length, and then quite abruptly switches to an attractor which is usually dynamically nonchaotic (Lai and Tél 2011). This arises due to the presence of nonattracting chaotic saddles in phase space. It is known that chaotic saddles and transient chaos are responsible for important physical phenomena such as chaotic scattering (Bleher et al. 1990), and particle transport in open hydrodynamical flows (Jung et al. 1993). They are believed to be the culprit for catastrophic phenomena such as voltage collapse in electrical power systems (Dhamala and Lai 1999) and species extinction in ecology. An extensive study of chaotic transients in spatially extended systems was studied by Tél and Lai (2008). It has also been reported in a memristive Chua's oscillator by Bao et al. (2010). We have found both numerically and experimentally the behaviour of transient chaos in the memristive Duffing oscillator. To observe this, we fixed parameters of $\alpha = 0.01$, $\omega_0^2 = -0.35$, $\beta = 0.8$, $\omega = 0.6$ and $f = 1.0$, while the initial conditions of system Eq. (17) are chosen as $x_1 = 0.01$, $x_2 = 0.03$, $x_3 = 0.003$, $x_4 = 0.0$. The numerical time series of $(x_1(t))$ variable is shown in the Fig. 20a and the

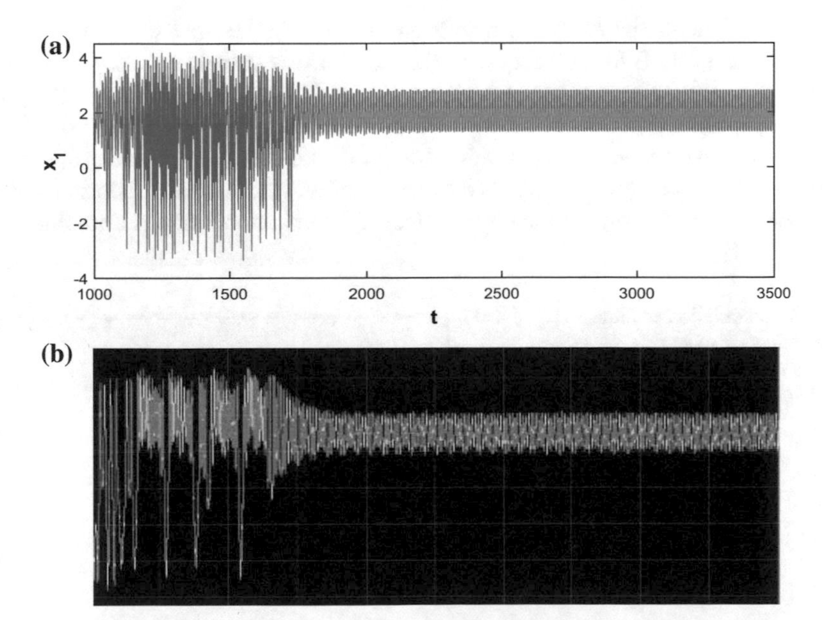

Fig. 20 Transient chaos regime: **a** Numerical observation of time series of $x_1(t)$ with initial conditions $x_1 = 0.01$; $x_2 = 0.03$; $x_3 = 0.003$; $x_4 = 0.0$ and fixed parameters of $\alpha = 0.01$, $\omega_0^2 = -0.35$, $\beta = 0.8$, $\omega = 0.6$ and $f = 1.0$ of system Eq. (17). **b** Experimentally observed time series of $v_1(t)$ with $F = 1.230$ V, and $\Omega = 600$ Hz

corresponding experimental time series for the variable ($v_1(t)$) is shown in Fig. 20b for $F = 1.230\ V$ and $\Omega = 600$ Hz. We do not have the control on initial conditions of the circuit, hence we have repeated the experiment several times to obtain transient chaos time series which is very similar to the numerical observation.

4.4.1 Finite Time Lyapunov Exponent

The existence of transient chaos can be characterized by using Finite Time Lyapunov Exponent (FTLE). The FTLE is a statistical measure of the amount of stretching (or folding) of a trajectory over a finite time interval (Sabarathinam and Thamilmaran 2015). The FTLE is defined as,

$$\lambda_j^m = log||e_j^m||, m = 1, 2, \ldots, k \tag{19}$$

The reorthonormalization vector (e_j^m) is denoted as,

$$e_j^m = JM(x_j, y_j, \Theta_j, \phi_j)\hat{e}_j^m. \tag{20}$$

Here, *JM*- denotes the Jacobean matrix and *j*-refers to the time step, and we have calculated the FTLE for system Eq. (17). The FTLE is calculated using $M = 1000$ data points, as functions of time are shown in Fig. 21. For the value of $t < 1000$, the largest Lyapunov exponent is positive while it becomes zero when $t > 1000$. Other three exponents becomes negative for $t > 1000$, indicates the periodic dynamics. This shows initially system exhibits a chaotic motion which then transforms to periodic behavior after a finite time period. Hence, the chaotic regime is transient.

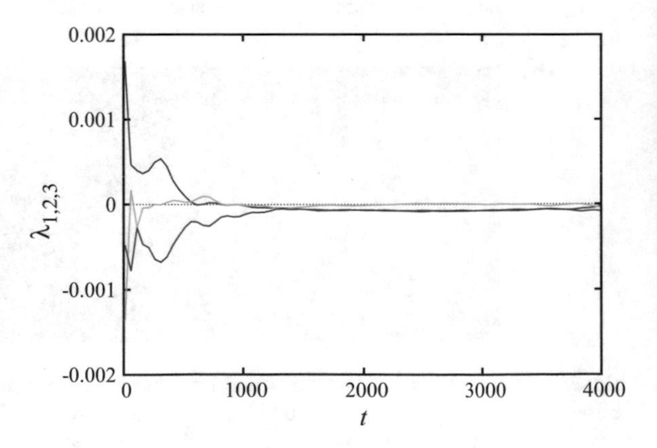

Fig. 21 Calculation of finite time Lyapunov exponent spectra with the same initial condition and parameters as in Figs. 18 and 19

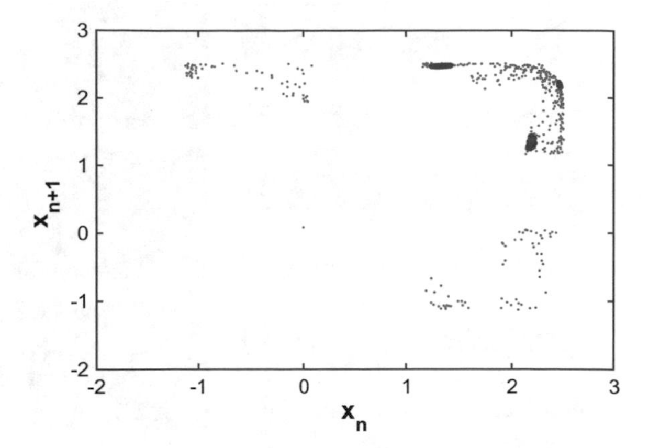

Fig. 22 Poincaré return map plotted of $x_1(t)$ variable in the $(x_n - x_{n+1})$ plane for transient chaos for 5000 data points with fixed step size of 0.01. The initial conditions are $x_1 = 0.01$, $x_2 = 0.003$, $x_3 = 0.1$, $x_4 = 0.0$ and the fixed parameters are $\alpha = 0.01$, $\omega_0^2 = -0.35$, $\beta = 0.8$, $\omega = 0.6$ and $f = 1.0$ of system Eq. (17)

The Poincaré return map constructed using the maxima of the trajectory x_1. Two successive maximas are denoted as x_n and $x(n+1)$, respectively, are plotted in the $(x_n - x_{n+1})$ plane as shown in Fig. 22. The initial chaotic motion is reflected by random scattering of poincare points (light blue dots), which are then confined to fixed dots as times evolves (dark blue dots).

4.5 *Intermittency*

Intermittency is the irregular alternation of laminar phases of apparently periodic and chaotic dynamics or different forms of chaotic dynamics. Pomeau and Manneville described three routes to intermittency where a nearly periodic system shows irregularly spaced bursts of chaos (Manneville and Pomeau 1979). This kind of behaviour is observed in the present memristive Duffing oscillator circuit by varying either amplitude f or frequency ω. Figure 23 shows, the numerical (Fig. 23a) as well as the experimental (Fig. 23b) time series of type-I intermittency. In numerical observation, we fix the initial condition at $x_1 = 0.01$, $x_2 = 0.003$, $x_3 = 0.1$ and the fixed system parameters are $\alpha = 0.1$, $\omega_0^2 = -0.35$, $\beta = 0.6$, and $\omega = 0.928$, while the external forcing amplitude f is the controlling parameter. We get intermittent behaviour in the range of parameter $f \in (0.140, 0.155)$. In the corresponding experimental observations, the circuit parameter of either amplitude F or frequency Ω of the external periodic force are varied, while the other circuit values are fixed. We obtained the entire intermittent regions at $\Omega = 240$ Hz, when the amplitude $F \in (400$ mV, 600 mV$)$.

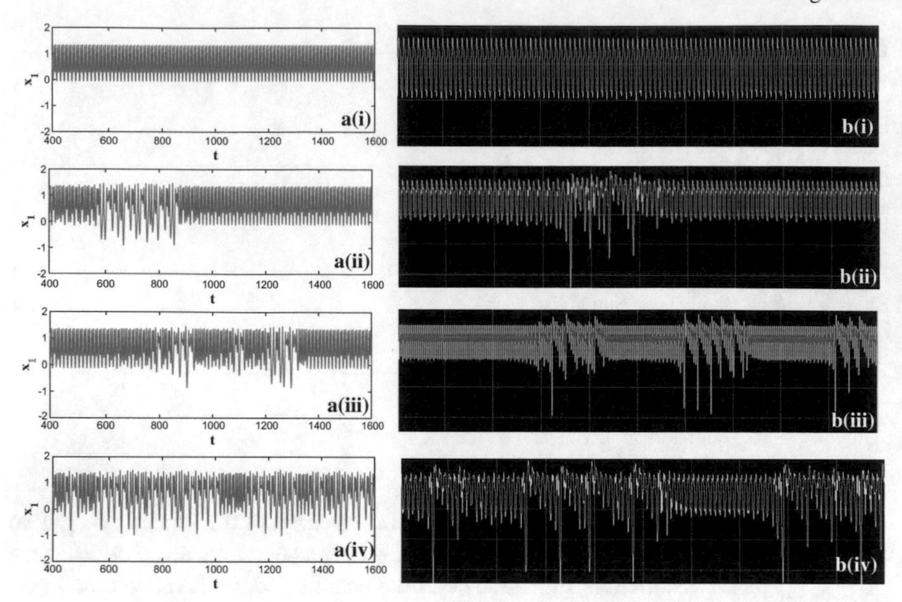

Fig. 23 Intermittency regime: **a** Numerical observation of time series in the $x_1(t)$ with the initial conditions $x_1 = 0.01$; $x_2 = 0.003$; $x_3 = 0.1$; $x_4 = 0.0$ and the fixed parameters of $\alpha = 0.1$, $\omega_0^2 = -0.35$, $\beta = 0.6$, and $\omega = 0.928$ of system Eq. (17). *a(i)* laminar ($f = 0.1400$), *a(ii)* burst-1 ($f = 0.1480$), *a(iii)* burst-3 ($f = 0.1495$), *a(iv)* chaos ($f = 0.1550$), **b** The corresponding experimental observation of the equivalent time series is produced with $F = 400$–600 mV and $\Omega = 240$ Hz

5 Summary

In summary, two chaotic oscillators such as an autonomous Chua's oscillator and a non-autonomous Duffing oscillator are investigated in detail using cubic memristor as nonlinear functional part. Both numerical and experimental investigations are performed. Period doubling scenario in Chua's circuit is demonstrated. Different types of dynamical behaviours such as chaos, transient chaos, three tori and intermittency are studied in Duffing oscillator. The applicability of '0–1 test' is examined in both the oscillators which clearly characterise the chaotic dynamics from periodic one. All the experimental results are corroborated with numerical simulation results.

Acknowledgements R.J and S.S knowledges the financial support of University Grants Commission, India through UGC (BSR)-RFSMS scheme. K.S is thankful to the financial support of the Department of Science and Technology (DST), India through PURSE scheme. K.T is grateful to DST, India for the financial support in form of major research project.

References

Banerjee, T. (2012). Single amplifier biquad based inductor-free Chua's circuit. *Nonlinear Dynamics*, *68*, 565–573.

Bao, B.-C., Xu, J.-P., & Liu, Z. (2010). Initial state dependent dynamical behviours in a memristor based chaotic circuit. *Chinese Physics Letters*, *27*(7), 070504.

Bao, B.-C., Fei, F., Wei, D., & Pan, S.-H. (2013). The voltage-current relationship and equivalent circuit implementation of parallel flux-controlled memristive circuits. *Chinese Physics B*, *22*(6), 068401.

Bharathwaj, M., Blain, T., & Sundqvist, K. (2009). *A synthetic inductor implementation of Chua's circuit*. EECS Department, University of California, Berkeley. UCB/EECS-2009-20.

Bleher, S., Grebogi, C., & Ott, E. (1990). Bifurcation to chaotic scattering. *Physica D: Nonlinear Phenomena*, *46*, 87–121.

Botta, V. A., Nespoli, C., & Messias, M. (2011). Mathematical analysis of a third order memristor based Chua's oscillator. *TEMA Tendencias em Matematica Aplicada e Computacional*, *12*(2), 91–99.

Chua, L. O. (1971). Memristor-the missing circuit element. *IEEE Transactions on Circuit Theory*, *18*, 507–519.

Corinto, F., & Ascoli, A. (2012). Memristive diode bridge with LCR filter. *Electronics Letters*, *48*(14), 824–825.

Dhamala, M., & Lai, Y.-C. (1999). Controlling transient chaos in deterministic flows with applications to electrical power systems and ecology. *Physical Review E*, *59*, 1646.

Fouda, M. E., & Radwan, M. G. (2014). Memristor-based voltage-controlled relaxation oscillators. *International Journal of Circuit Theory and Applications*, *42*, 1092–1102.

George, D. (1918). Erzwungene schwingung bei vernderlicher eigenfrequenz und ihre technische bedeutung. Vieweg.

Gopal, R., Venkatesan, A., & Lakshmanan, M. (2013). Applicability of 0–1 test for strange nonchaotic attractors. *Chaos*, *23*, 023123.

Gottwald, G. A., & Melbourne, I. (2004). A new test for chaos in deterministic systems. *Proceedings of the Royal Society of London A*, *460*, 603–611.

Ishaq Ahamed, I., & Lakshmanan, M. (2013). Nonsmooth bifurcations, transient hyperchaos and hyperchaotic beats in a memristive Murali-Lakshmanan-Chua's circuit. *International Journal of Bifurcation and Chaos*, *23*(6), 1350098.

Ishaq Ahamed, I., Srinivasan, K., Murali, K., & Lakshmanan, M. (2011). Observation of chaotic beats in a driven memristive Chua's circuit. *International Journal of Bifurcation and Chaos*, *21*(3), 737–757.

Itoh, M., & Chua, L. O. (2008). Memristor oscillators. *International Journal of Bifurcation and Chaos*, *18*(11), 3183–3206.

Jothimurugan, R., Suresh, K., Ezhilarasu, M., & Thamilmaran, K. (2014). Improved realization of canonical Chua's circuit with synthetic inductor using current feedback operational amplifiers. *AEU—International Journal of Electronics and Communications*, *68*(5), 413–421.

Jung, C., Tél, T., & Ziemniak, E. (1993). Application of scattering chaos to particle transport in a hydrodynamical flow. *Chaos: An Interdisciplinary Journal of Nonlinear Science*, *3*, 555–568.

Kim, H., Sah, M., Yang, C., Cho, S., & Chua, L. (2012). Memristor emulator for memristor circuit applications. *IEEE Transactions on CAS*, *I*(59), 2422–2431.

Kovacic, I., & Brennan, M. J. (2011). The Duffing equation: Nonlinear oscillators and their behaviour. Wiley.

Kulp, C. W., & Smith, S. (2011). Characterization of noisy symbolic time series. *Physical Review E*, *83*, 026201.

Kyriakides, E., & Georgiou, J. (2014). A compact, low-frequency, memristor-based oscillator. International Journal of Circuit Theory and Applications.

Lai, Y.-C., & Tél, T. (2011). Transient chaos: Complex dynamics on finite time scales (Vol. 173). Springer Science & Business Media.

Li, Y., Huang, X., & Guo, M. (2013). The generation, analysis, and circuit implementation of a new memristor based chaotic system. *Mathematical Problems in Engineering*, 398306.

Li, H., Wang, L., & Duan, S. (2014). A memristor-based scroll chaotic system—design, analysis and circuit implementation. *International Journal of Bifurcation and Chaos*, *24*, 1450099.

Manneville, P., & Pomeau, Y. (1979). Intermittency and the Lorenz model. *Physics Letters A*, *75*, 1–2.

Martinsen, O. G., Grimnes, S., Lütken, C. A., & Johnsen, G. K. (2010). Memristance in human skin. *Journal of Physics: Conference Series*, *224*, 012071.

Morgül, Ö. (1995). Inductorless realization of Chua's oscillator. *Electronics Letters*, *31*, 1424–1430.

Muthuswamy, B. (2010). Implementing memristor based Chaotic circuits. *International Journal of Bifurcation and Chaos*, *20*(5), 1335–1350.

Sabarathinam, S., & Thamilmaran, K. (2015). Transient chaos in a globally coupled system of nearly conservative Hamiltonian duffing oscillators. *Chaos, Solitons & Fractals*, *73*, 129–140.

Sánchez-Lópeza, C., Carrasco-Aguilara, M., & Muñiz-Montero, C. (2015). A 16 Hz-160 kHz memristor emulator circuit. *International Journal of Electronics and Communications (AEU)*, *69*, 1208–1219.

Secco, J., Biey, M., Corinto, F., Ascoli, A., & Tetzlaff, R. (2015). Complex behavior in memristor circuits based on static nonlinear two-ports and dynamic bipole. *IEEE—European Conference on Circuit Theory and Design (ECCTD)*, 1–4.

Slipko, V. A., Pershin, Y. V., & Di Ventra, M. (2013). Changing the state of a memristive system with white noise. *Physical Review E*, *87*, 042103.

Strukov, D. B., Snider, G. S., Stewart, D. R., & Stanley Williams, R. (2008). The missing memristor found. *Nature*, *453*, 80–83.

Tél, T., & Lai, Y.-C. (2008). Chaotic transients in spatially extended systems. *Physics Reports*, *460*, 245–275.

Teng, L., Iu, H. H. C., Wang, X., & Wang, X. (2014). Chaotic behavior in fractionalorder memristor-based simplest chaotic circuit using fourth degree polynomial. *Nonlinear Dynamics*, *77*, 231–241.

Thomas, A. (2013). Memristor-based neural networks. *Journal of Physics D: Applied Physics*, *46*, 093001.

Tôrres, L., & Aguirre, L. (2000). Inductorless Chua's circuit. *Electronics Letters*, *36*, 1915–1916.

Valov, I., Linn, E., Tappertzhofen, S., Schmelzer, S., van den Hurk, J., Lentz, F., et al. (2013). Nanobatteries in redox-based resistive switches require extension of memristor theory. *Nature Communications*, *4*, 1771.

Valsa, J., Biolek, D., & Biolek, Z. (2011). An analogue model of the memristor. *International Journal of Numerical Modelling*, *24*, 400–408.

Yang, J. J., Pickett, M. D., Li, X., Ohlberg, D. A. A., Stewart, D. R., & Stanley Williams, R. (2008). Memristive switching mechanism for metal/oxide/metal nanodevices. *Nature Nanotechnology*, *3*, 429–433.

Zakhidov, A. A., Jung, B., Slinker, J. D., Abruña, H. D., & Malliaras, G. G. (2010). A light-emitting memristor. *Organic Electronics*, *11*, 150–153.

Zidan, M. A., Omran, H., Smith, C., Syed, A., Radwan, A. G., & Salama, K. N. (2014). A family of memristorbased reactance-less oscillators. *International Journal of Circuit Theory and Applications*, *42*, 1103–1122.

Memristor and Inverse Memristor: Modeling, Implementation and Experiments

Mohammed E. Fouda, Ahmed G. Radwan and Ahmed Elwakil

Abstract Pinched hysteresis is considered to be a signature of the existence of memristive behavior. However, this is not completely accurate. In this chapter, we are discussing a general equation taking into consideration all possible cases to model all known elements including memristor. Based on this equation, it is found that an opposite behavior to the memristor can exist in a nonlinear inductor or a nonlinear capacitor (both with quadratic nonlinearity) or a derivative-controlled nonlinear resistor/transconductor which we refer to as the inverse memristor. We discuss the behavior of this new element and introduce an emulation circuit to mimic its behavior. Connecting the conventional elements with the memristor and/or with inverse memeristor either in series or parallel affects the pinched hysteresis lobes where the pinch point moves from the origin and lobes' area shrinks or widens. Different cases of connecting different elements are discussed clearly especially connecting the memristor and the inverse memristor together either in series or in parallel. New observations and conditions on the memristive behavior are introduced and discussed in detail with different illustrative examples based on numerical, and circuit simulations.

Keywords Circuit theory · Memristor · Inverse memristor · Pinched hysteresis

M.E. Fouda · A.G. Radwan (✉)
Faculty of Engineering, Engineering Mathematics and Physics Department,
Cairo University, Giza 12613, Egypt
e-mail: agradwan@ieee.org

M.E. Fouda
e-mail: m_elneanaei@ieee.org

A.G. Radwan · A. Elwakil
NISC Research Center, Nile University, Giza, Egypt
e-mail: elwakil@ieee.org

A. Elwakil
Department of Electrical and Computer Engineering, University of Sharjah,
Sharjah, United Arab Emirates

© Springer International Publishing AG 2017

S. Vaidyanathan and C. Volos (eds.), *Advances in Memristors, Memristive Devices and Systems*, Studies in Computational Intelligence 701,
DOI 10.1007/978-3-319-51724-7_15

1 Introduction

Since postulating the existence of the memristor in 1971 by Leon Chua, a huge number of publications have been published. These publications address modeling (Radwan et al. 2010a, b), and analysis. Chua (1971); Kozma et al. (2012); Radwan and Fouda (2015); Adamatzky and Chua (2013); Fouda and Radwan (2015a, b). In addition, memristors have been used in many applications such as sinusoidal oscillators (Talukdar et al. 2011a, b, 2012), relaxation oscillators (Fouda et al. 2013; Khatib et al. 2012; Fouda and Radwan 2015c; Zidan et al. 2011, 2014), nonlinear control systems (Vaidyanathan and Volos 2016b), chaotic systems (Vaidyanathan and Volos 2016a; Gambuzza et al. 2015; Radwan et al. 2011), digital and analog circuits (Radwan and Fouda 2015; ElSlehdar et al. 2015). Memristors have a unique behavior which distinguish them from voltage–current other nonlinear devices. They exhibit pinched hysteresis in the plane. The hysteresis lobe area of memristor decreases monotonically as the excitation frequency increases. Also, the pinched hysteresis loop should shrink to a single-value when the frequency tends to infinity. This means that the lobe area declines with increased frequency. These characteristics should exist in a device to be referred to as a memristor (Biolek et al. 2011; Adhikari et al. 2013).

There are two categories of memristor models, current- and voltage-controlled models. The current-controlled memristor has a state variable that is function of the current passing through the memristor such as the HP model (Joglekar and Wolf 2009), and Picket model (Pickett et al. 2009). On the other hand, the state variable of the voltage-controlled memristor is a function of the voltage across the memristor (Kozma et al. 2012; Radwan and Fouda 2015). Some of these models are simple and some of them are complex but all of them capture the main memristor characteristics which should exist regardless of the memristor type.

There is another way to distinguish between memristor types which is based on the ideality of the device. According to this, we have two main types of the memristor; ideal and nonideal memristors. The memristor is called ideal if the self cross point is the origin. Otherwise, it is called nonideal memristor (Biolek et al. 2015). The non-ideality may appear in the devices due to the existence of reactive elements as will be discussed in detail in this chapter.

The inverse memristor is a system that exhibits self crossing pinched hysteresis. Contrarily with memristors, the memristor hysteresis widens with increasing the applied frequency (Fouda et al. 2015). The inverse memristor can be modeled as a nonlinear inductor or a nonlinear capacitor (with quadratic nonlinearity) in series with a resistor. As a conclusion, pinched hysteresis is a necessary but not a sufficient condition to prove the memristivity. Other conditions should be satisfied as well.

Considering memristor and inverse memristor in circuit theory is essential. That's why, many publications have been published to discuss the memristor inside conventional circuits. The memory existence inside memristive devices gives new characteristics. By adding the memristor to well known circuits, new responses and behaviors are obtained due to the unique behavior of the memristor (Radwan and Fouda 2015).

This chapter is organized as follows: section I introduces a generalized mathematical model for all the possible cases for circuit elements. Memristor is the special case of this modeling equation as deduced in section II. In section III, based on the generalized equation, we find a new element which has the opposite characteristics of the memristor. This element is discussed in details with some circuit emulators to proof the concept. In section IV, different circuit configurations are discussed and its effect on the hysteresis. Finally, conclusions are given.

2 Generalized Equation Model

A general equation can be defined as follows:

$$y = ax + (b + ex)\frac{dx}{dt} + (d + cx)\int_0^t x(\tau)d\tau \tag{1}$$

where y is a normalized output, x is a normalized input signal, and (a, b, c, d, e) are scaling constants. Equation (1) describes the different cases of applying an input signal and/or effect of integrating and differentiating the input signal.

This equation contains the definitions of all known circuit elements as we will discuss later. But firstly, let's study the behavior of this modeling equation under a sinusoidal excitation assuming $x(t) = ksin(\omega t + \phi)$. Therefore,

$$\frac{dx}{dt} = k\omega cos(\omega t + \phi) = \pm\omega\sqrt{k^2 - x^2} \tag{2}$$

and

$$\int_0^t x(\tau)d\tau = \frac{1}{\omega}\left(kcos(\phi) \mp \sqrt{k^2 - x^2}\right) \tag{3}$$

Substituting into (1) and using trigonometric identities, one obtains

$$y = ax + k(d + cx)\frac{cos(\phi)}{\omega} \pm \left(\left(e\omega - \frac{c}{\omega}\right)x + \left(b\omega - \frac{d}{\omega}\right)\right)\sqrt{k^2 - x^2} \tag{4}$$

This equation has the following properties:

1. There exists a line of odd-symmetry given by the first order relation between y and x
$$y = ax + k(d + cx)\frac{cos(\phi)}{\omega} \tag{5}$$
2. A pinched-double loop hysteresis behavior is observed in the x-y plane. The double-loop intersects itself at a point known as the pinch-point (x_p, y_p) obtained

by equating $\sqrt{k^2 - x^2}$ to zero; yielding

$$x_p = \frac{b\omega^2 - d}{c - e\omega^2}, \; y_p = ax_p + k(d + cx_p)\frac{cos(\phi)}{\omega} \tag{6}$$

At high frequency, this pinched point reduces to $\left(\frac{-b}{e}, \frac{-ab}{e}\right)$ while at low frequency it reduces to $\left(\frac{-d}{c}, 0\right)$. It is obvious that some scaling coefficient are amplified with increasing the frequency; namely b and e while other coefficients vanish with increasing the applied frequency like c and d. Changing the frequency does not affect the coefficient a.

3. Generally, (4) will pass by the four boundary points:

$$\left(0, \frac{dkcos(\phi)}{\omega} \pm \left(b\omega - \frac{d}{\omega}\right) k\right) \; and \; \left(\pm k, k\left(\mp a + \frac{cos(\phi)(d \mp ck)}{\omega}\right)\right). \tag{7}$$

The basic circuit elements can be obtain easily from this equation as follows:

- Resistor: Resistance can be obtained by putting all the coefficients equal to zero except a. Either x or y represents the current or the voltage. Then we have a linear relation between current and voltage representing the resistor.
- Capacitor: Capacitance is the linear relation between voltage and charge. So, by putting $x(t) = i(t)$ and $y(t) = v(t)$, the capacitance is $1/d$ where the other coefficients are zero.
- Inductor: Inductance can be given by putting $x(t) = i(t)$ and $y(t) = v(t)$, then inductance is b when the other coefficients are zero.

Based on (1), we can generate other elements such as memristor with symmetric and asymmetric behavior. Also we can anticipate new behaviors such as inverse memristor.

3 Deduced Memristive Equation

The self-crossing (pinched) hysteresis loop was shown to be a necessary characteristic of all memristive devices. However, (Adhikari et al. 2013) added two more conditions on memristive devices which are (i) starting from some critical frequency, the hysteresis lobe area should decrease monotonically as the excitation frequency increases and (ii) the pinched hysteresis loop should shrink to a single-valued function when the frequency tends to infinity. This means that the lobe area declines with increased frequency. So, any memristor should have these characteristics.

As a special case of (1), a simple equation for the symmetrical and asymmetric double-loop hysteresis behavior can be developed which was introduced in (Elwakil et al. 2013; Radwan and Fouda 2015). This equation has the basic memristor characteristics and is given as follows:

$$y(t) = x(t)\left(a + c\int_0^t x(\tau)d\tau\right) + b\frac{dx(t)}{dt}, \tag{8}$$

where a represents the initial state of the memristor.

This equation represents a symmetric behavior when $b = 0$, so let's discuss the symmetric case first.

3.1 Symmetrical Memristive Model

One of the test bench marks of the memristor it that the pinched hystersis decreases with increasing the applied input frequency. Figure 1 shows the observed double-loop behavior for $a = c = 1$ when $x(t) = cos(\omega t)$ and $\omega = 1$. Note that two cases are plotted in Fig. 1; namely the positive/ negative a and c cases of (8) with $b = 0$ resulting in either a positively inclined loop or a negatively inclined loop, respectively. Two more cases; (a, c) are$(+, -)$ and $(-, +)$ are also possible and lead, respectively, to similar positively inclined and negatively inclined loops. It is clear that for $x(t) = cos(\omega t)$, $\frac{y(t)}{x(t)} = a + \frac{1}{T\omega}sin(\omega t) \in [a - \frac{1}{T\omega}, a + \frac{1}{T\omega}]$. Therefore, $y = ax$ is a symmetry line and the polarity of a determines the quadrant in which the hysteresis loop appears.

The implementation of the double-loop hysteresis could be done using current or voltage signals; when $x(t)$ is represented by a current and $y(t)$ is represented by a voltage the implementation is current-controlled. Alternatively when $y(t)$ is represented by a current and $x(t)$ is represented by a voltage; a voltage-controlled memristive device is obtained. It is worth noting that the authors of (Biolek et al. 2011) have

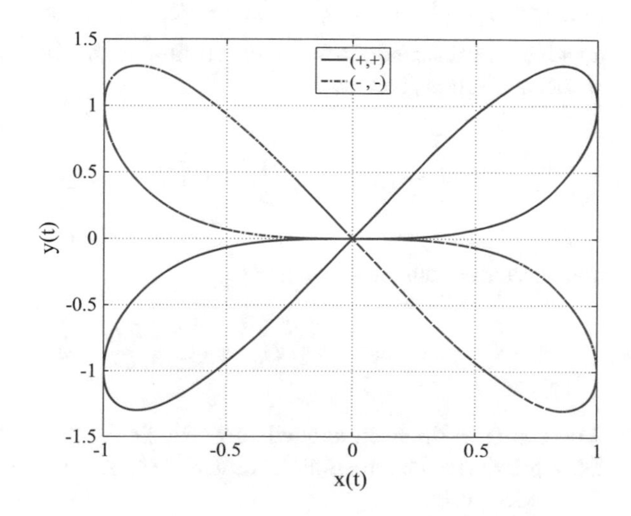

Fig. 1 Double-loop hysteresis for $a = c = 1$ and $x(t) = cos(t)$ in (8)

recently proposed conditions for symmetric pinched hysteresis. The model above satisfies these conditions and is simpler than the one in (Biolek et al. 2011).

3.1.1 Current-Controlled Memristor

Setting $x(t) = \frac{i(t)}{I_{ref}}$, $y(t) = \frac{v(t)}{I_{ref}R_s}$, where I_{ref} is an arbitrary reference current and R_s is an arbitrary resistance, and substituting into (8), the current-controlled memristor equation is given by

$$v(t) = \pm i(t)R_s \pm \frac{i(t)R_s}{TI_{ref}} \int_0^t i(\tau)d\tau = \pm i(t)R_s \pm \frac{i(t)R_s}{TI_{ref}} q(t), \tag{9}$$

and hence the memristance $R_m = v(t)/i(t)$ is given by

$$R_m = \pm R_s \pm \frac{R_s}{TI_{ref}} q(t). \tag{10}$$

It is seen here that R_m is a function of the accumulated current which is essentially the charge $q(t)$; similar to HP modeling equation (Elwakil et al. 2013). In terms of the four different possibilities for R_m, which (a, c) are $(+, +), (+, -), (-, +)$ and $(-, -)$ they respectively represent incremental/decremental R_m and incremental/decremental negative R_m; as demonstrated below.

3.1.2 Voltage-Controlled Memristor

Setting $x(t) = \frac{v(t)}{V_{ref}}$, $y(t) = \frac{i(t)}{V_{ref}G_s}$, where V_{ref} is an arbitrary reference voltage and G_s is an arbitrary transconductance, and substituting into (8), the voltage-controlled memristor equation is given by

$$i(t) = \pm v(t)G_s \pm G_s\frac{v(t)}{TV_{ref}} \int_0^t v(\tau)d\tau = \pm v(t)G_s \pm G_s\frac{v(t)}{TV_{ref}} \varphi(t), \tag{11}$$

and hence the trans-memristance G_m is

$$G_m = \pm G_s \pm \frac{G_s}{TV_{ref}} \varphi(t), \tag{12}$$

where $\varphi(t)$ is the accumulated flux. Similarly, there are four different possibilities representing incremental/decremental G_m and incremental/decremental negative G_m, respectively.

Fig. 2 I-V characteristics of an incremental R_m at different frequencies with $I_{ref} = 1\,\mu\text{A}$ and $R_s = 10\,\text{k}\Omega$

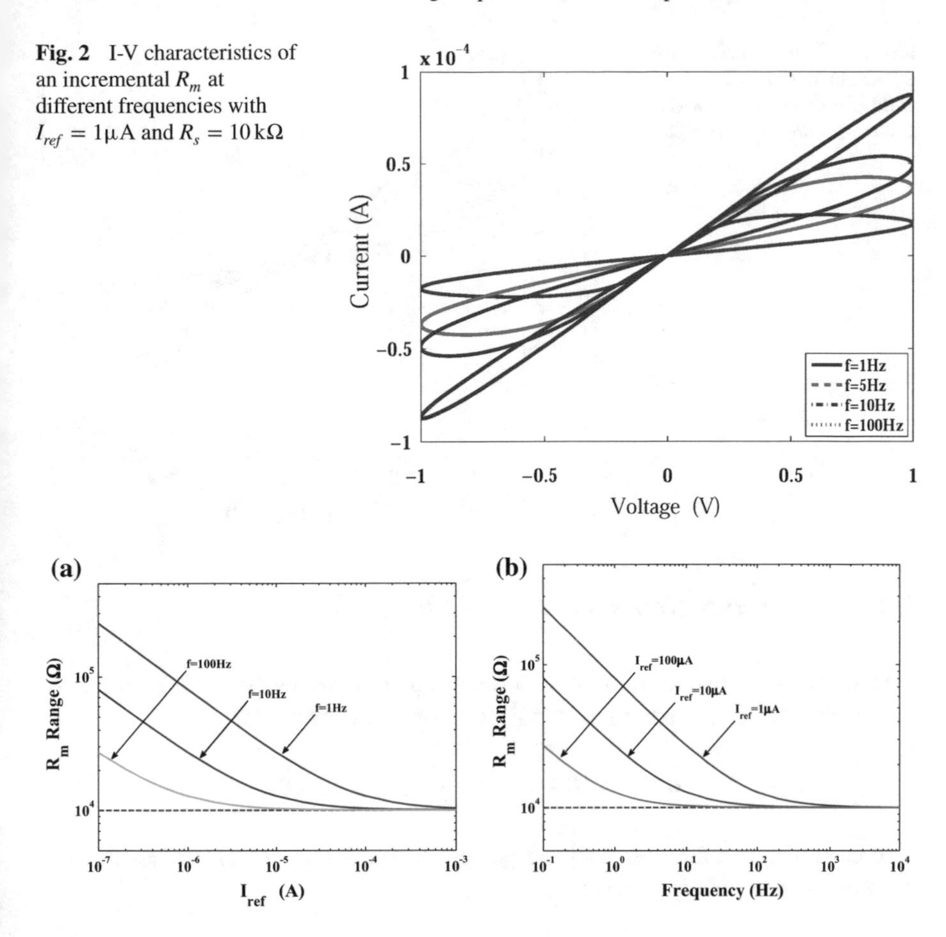

Fig. 3 Maximum and minimum incremental R_m when $R_s = 10\,\text{k}\Omega$

Figure 2 shows the I-V characteristics for an incremental R_m for four different frequencies of the sinusoidal input current $i(t)$ with $I_{ref} = 1\,\mu\text{A}$ and $R_s = 10\,\text{k}\Omega$. The maximum and minimum values of R_m are shown respectively in Fig. 3; plotted once for the range of I_{ref} spanning from $1\,\mu\text{A}$ to 0.1 mA and another for the frequency range 1–100 Hz of the input signal. In Fig. 4, the I-V characteristics for an incremental but negative R_m is also shown for four different frequencies of the sinusoidal input current $i(t)$ with $I_{ref} = 1\,\mu\text{A}$ and $R_s = 10\,\text{k}\Omega$.

As a proof of concept, different circuit emulators have been introduced based on this model and discussed in detail (Radwan and Fouda 2015).

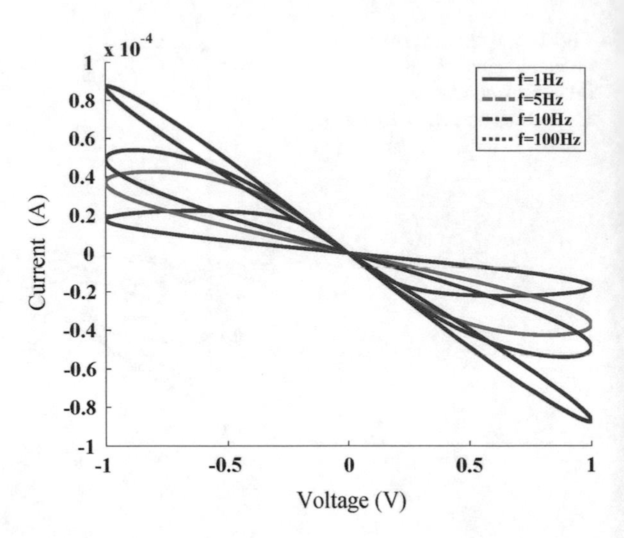

Fig. 4 I-V characteristics of an incremental negative R_m at different frequencies with $I_{ref} = 1\mu A$ and $R_s = 10\,k\Omega$

3.2 Continuous Non-symmetrical Model

Adding the derivative term to the symmetric equation makes the pinched hysteresis asymmetric. If $x(t) = cos(\omega t)$ then (8), with nonzero elements, yields

$$y(t) = ax(t) \mp (b\omega - \frac{c}{\omega}x(t))\sqrt{1 - x^2(t)}. \tag{13}$$

It can be shown that the pinch-off point, which corresponds to the vanishing of the second term of (13), is given by

$$[x_p, y_p] = \left[\frac{b}{c}\omega^2, (\frac{ab}{c}\omega^2)\right], \tag{14}$$

where $x_p \leq 1$. Moreover, the generated loop always passes by the three points: $(x, y) = (1, a), (-1, -a)$ and $(x, y) = (0, \pm b\omega)$. Figure 5a shows the observed non-symmetrical loops for different values of a when $b = c = 1$ at $f = 0.1$ Hz. A 3D view of these non-symmetrical loops for different values of c when $a = 0, b = 1$ is shown in Fig.5b while Fig. 5c shows the case when $a = b = 1$; both figures at $f = 0.5$ Hz. Note that if $x_p = \frac{b}{c}\omega^2 > 1$ then there is no pinched point and a single loop is observed.

4 Deduced Inverse Memristive Equation

A simple inverse memristive equation can be deduce from (1) by putting $b = c = 0$ as follows:

$$y = ax + (b + ex)\frac{dx}{dt} \tag{15}$$

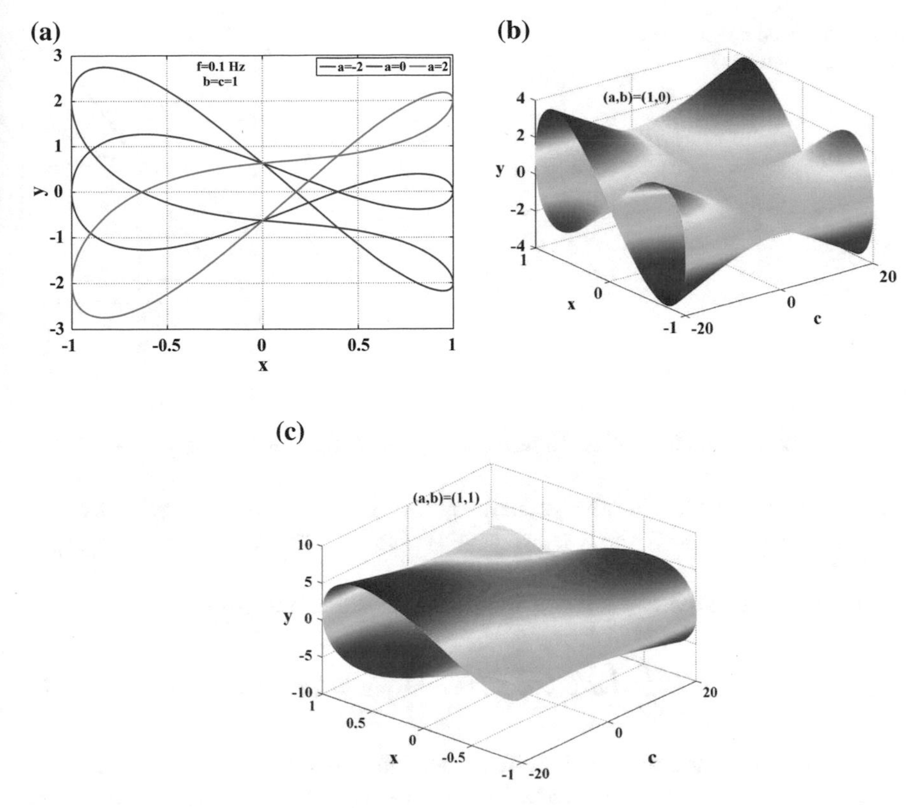

Fig. 5 Non-symmetrical loops when **a** $(b, c) = (1, 1)$, **b** $(a, b) = (0, 1)$, and **c** $(a, b) = (1, 1)$

Under a general sinusoidal excitation where $x(t) = k.sin(\omega t + \varphi)$, and by using trigonometric identities, we obtain

$$y = ax \pm (b + ex)\sqrt{k^2 - x^2} \tag{16}$$

4.1 *Inverse Memristor Properties*

This equation has the following properties:

1. There exists a line of symmetry given by the first order equation $y = ax$. Evidently, for $a = 0$, the y-axis is the line of symmetry.
2. A pinched double-loop hysteresis behavior is observed in the xy plane. The double-loop intersects itself at a point known as the pinch-point (x_p, y_p) given by

$$(x_p, y_p) = \frac{-b}{e}(1, a) = (0, 0)|_{b=0} \tag{17}$$

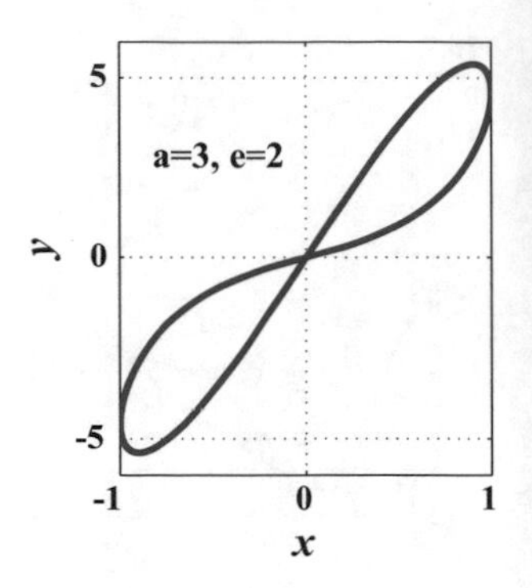

Fig. 6 Pinched hysteresis from (15) when $b = 0$

which is independent of ω. Figure 6 is a plot of the pinched loop for $(a, b, e) = (3, 0, 2)$.

3. The double-loop will always pass by the boundary points: $\omega(0, \pm bk)$ and $k(\pm 1, \mp a)$. For $b = 0$, the first two points coincide with the pinch point $(x_p, y_p) = (0, 0)$.
4. The area inside the two lobes of the pinched hysteresis is given by

$$A = 4 \int_0^k \left(\omega(b + ex)\sqrt{k^2 - x^2}\right) dx = 2k^2 \left(\pi b + \frac{2}{3}ek\right) \tag{18}$$

Hence, it is clear that A is directly proportional to ω ; i.e. maximizing the hysteresis loop area requires increasing ω. This represents inverse-memristor frequency characteristics since the condition in (Adhikari et al. 2013) implies that for a memristor the lobe area should decrease monotonically as the excitation frequency increases; shrinking to a single valued function when the frequency tends to infinity.

A non-symmetrical-loop may be obtained using (15) and also by adding an integral term in the form. For example, if $b = 0$, then

$$y = ax + ex\frac{dx}{dt} + d\int_0^t x(\tau)d\tau \tag{19}$$

for which the line of symmetry and pinch-point are, respectively, given by

$$y = ax + kd\frac{cos(\varphi)}{\omega} \quad and \quad (x_p, y_p) = \left(\frac{d}{e\omega^2}, \left(\frac{ad}{e\omega^2} + kd\frac{\varphi}{\omega}\right)\right) \tag{20}$$

both of which are frequency dependent. Since $|x_p| < k$, then $|d/e\omega^2| < k$ and therefore the frequency

$$\omega_c = \sqrt{\frac{|d/e|}{k}} \tag{21}$$

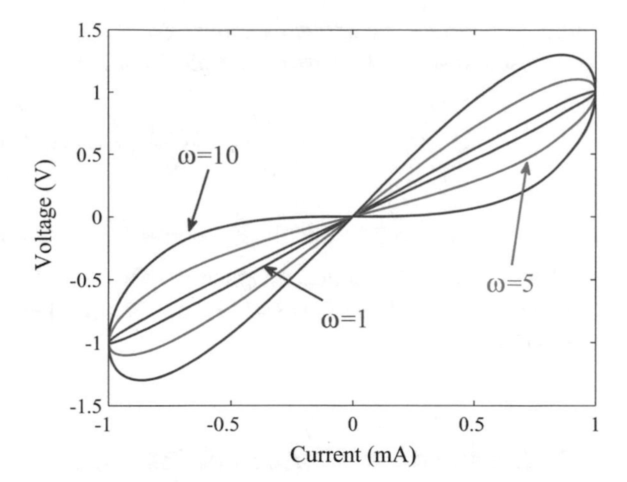

Fig. 7 Pinched hysteresis of an inverse memristor widens as frequency is increased

is the critical frequency at which the non-symmetrical double loop is born. The area of the this double-loop increases for $\omega > \omega_c$. This will be demonstrated further in the experimental results section.

From an electrical circuit point of view, Eq. (15) can represent different types of circuits based on the nature of x and y. Restricting ourselves to the $v(t) - i(t)$ plane, the possible choices of $x(t)$ and $y(t)$ are either a voltage $v(t)$ or a current $i(t)$. When $x(t) = i(t)$ and $y(t) = v(t)$ then (15) can be translated into series connected components. Alternatively, if $x(t) = v(t)$ and $y(t) = i(t)$ then (15) can be translated into parallel connected components as mentioned previously.

It is important to note that in case of $x(t) = i(t)$ and $y(t) = v(t)$, a nonlinear inductor with quadratic current dependence can be obtained where $\int v(i)dt = e/2i^2(t)$ where e has the units of Henry/Ampere and can be termed pseudo-inductance. Note that if $b = 0$, then (15) can be considered to collectively represent single derivative-controlled nonlinear resistor $R_d(i(t))$ where

$$v(t) = \left(a + e\frac{di(t)}{dt}\right).i(t) = R_d.i(t) \tag{22}$$

Figure 7 shows the observed pinched hysteresis loop in this special case for three different frequencies with $a = 1k\Omega, e = 100H/mA, b = 0$ and $k = 1mA$. The values of a, e and k were chosen to obtain a current in mA and a voltage in Volts. Note the widening of the loop as ω is increased since according to (18), $A_{b=0} = \frac{4}{3}ek^3\dot{\omega} \approx \omega/562.5$. In a conventional memristor, the loop area declines as ω is increased. Recall that $\left(\frac{4}{3}ek^3\right)$ represents the energy stored in the device and has the units of $(\mu H \times A^2)$. If we compare this to the expression of the energy stored in an effective inductor ($E_L = 0.5L_{eff}i^2$), we can calculate the effective inductance $L_{eff} = 267\mu\text{H}$.

In case of $x(t) = v(t)$ and $y(t) = i(t)$, a nonlinear capacitor with quadratic voltage dependence is obtained, where $\int i(t)dt = e/2v^2(t)$ and e has the units of Farad/Volt and can be termed pseudo-capacitance. Also, if $b = 0$, then (15) can be considered to represent a single derivative-controlled nonlinear transconductor $G_d(v(t))$ where

$$i(t) = \left(a + e\frac{dv(t)}{dt}\right).v(t) = G_d.v(t) \quad (23)$$

The energy stored in this device is also $\left(\frac{4}{3}ek^3\right)$ with the units of $(F \times V^2)$. As expected, this device mimics a capacitor with a stored energy of $(0.5C_{eff}v^2)$ resulting in $C_{eff} = (8e/3)F$. This device will be emulated and experimentally validated in the next subsection.

4.2 Inverse Memristor Circuit Emulator

Due to the lack of solid-state samples, researchers are developing emulation circuits to mimic the behavior of either current-controlled memristors or voltage-controllled memristor (Hussein and Fouda 2013; Radwan and Fouda 2014; Alharbi et al. 2015a, b, c). In (Fouda et al. 2015), a simple emulator circuit for inverse memristor is developed based on (23) where an applied voltage V is differentiated using a floating differentiator circuit and then used to control a voltage-controlled transconductance G_m through its control voltage V_c. Transconductance is implemented using an LM13700 chip where G_m is proportional to a bias voltage V_c given by

$$G_m = \left(0.64V_c + 8.6885\right)\frac{R_A}{R_B} \quad (m\Omega^{-1}) \quad (24)$$

and R_A, R_B are external biasing resistors. If the control voltage V_c is forced to be equal to the derivative of the applied voltage V then G_m in (24) can realize G_d in (23). This is achieved using the circuit shown in Fig. 8 with three op amps (TL084) controlling the bias voltage Vc of the LM13700. Consequently, (23) is realized with $a = 8.6885R_A/R_B(m\Omega^{-1})$ and $e = 064R_A/(R_BRC)(m\Omega^{-1}V^{-1}s)$.

This circuit was experimentally constructed as shown in Fig. 8 after selecting (C, R, R_B, R_A) equal to $(1mF, 10\,\text{k}\Omega, 100\,\text{k}\Omega, 10\,\text{k}\Omega)$. A $0.25V$ input voltage was applied at different frequencies. A current-to-voltage converter with equivalent resistance 56 kΩ was used to observe the current flowing into the two-terminal device. The observed loop is confirmed to widen as the frequency is increased in the sequence 300, 500, 700 and 900 Hz as shown in Fig. 9a, b. Further, we verified (19), which indicates that by adding a capacitor in series with G_d, non-symmetrical pinched loops can be obtained. This is shown in Fig. 10 using a 0.047µF capacitor at 500 Hz and at 700 Hz, respectively. Note the widening of lobe area as frequency is increased and using (21), the pinched loop is born at approximately 410 Hz.

Figure 11a shows the effect of connecting a $10H$ inductor in parallel with the inverse memristor. It is clear that the pinch point lies in the first quadrant. By

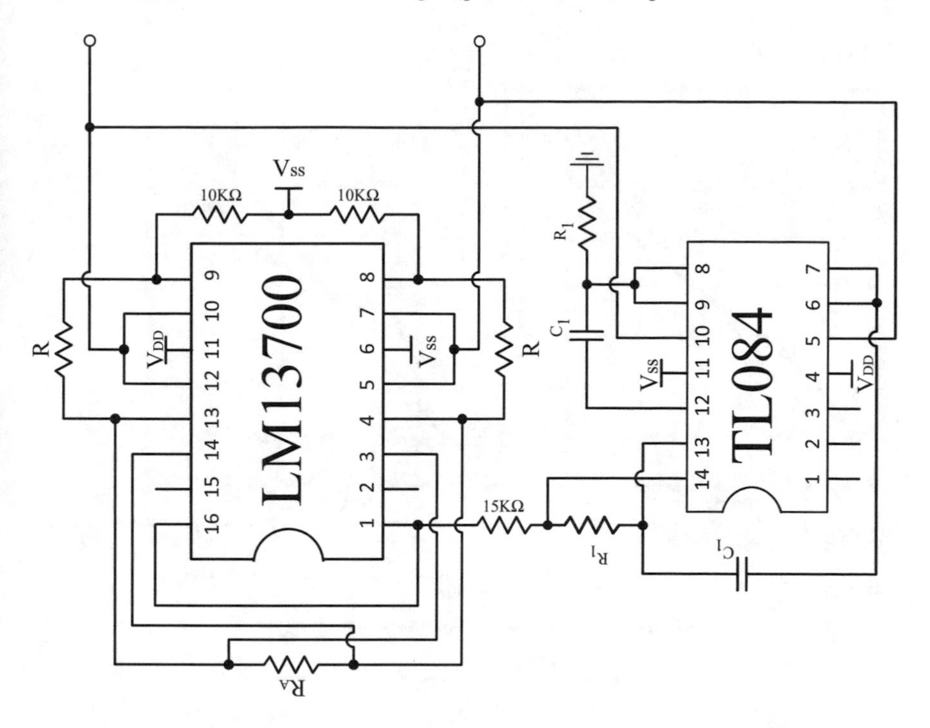

Fig. 8 Emulation circuit of voltage-controlled inverse memristor (Fouda et al. 2015)

increasing the value of the inductance, the pinched point moves up until the loop becomes elliptic without any intersection which means that the inductance dominates the behavior of the circuit. Unlike Fig. 11b where the pinched point lies in the third quadrant, by increasing the capacitance value, the pinch point moves down until it gets out of the boundaries and the loop becomes inclined elliptic where the capacitance behavior dominates the inverse memristor behavior.

5 Circuit Identification

Briefly, from a circuit point of view, (1) can represent two different types of circuits; based on the nature of x and y. When $x(t) = i(t)$ and $y(t) = v(t)$ then (1) can be translated into five series connected impedances, as shown in Fig. 12a. Alternatively, if $x(t) = v(t)$ and $y(t) = i(t)$ then (1) can be translated into five parallel connected admittances, as also shown in Fig. 12b. In Fig. 12a, the five impedances are identified respectively as

- a linear resistance R corresponding to the linear proportional coefficient a in (1) ($R = a$).

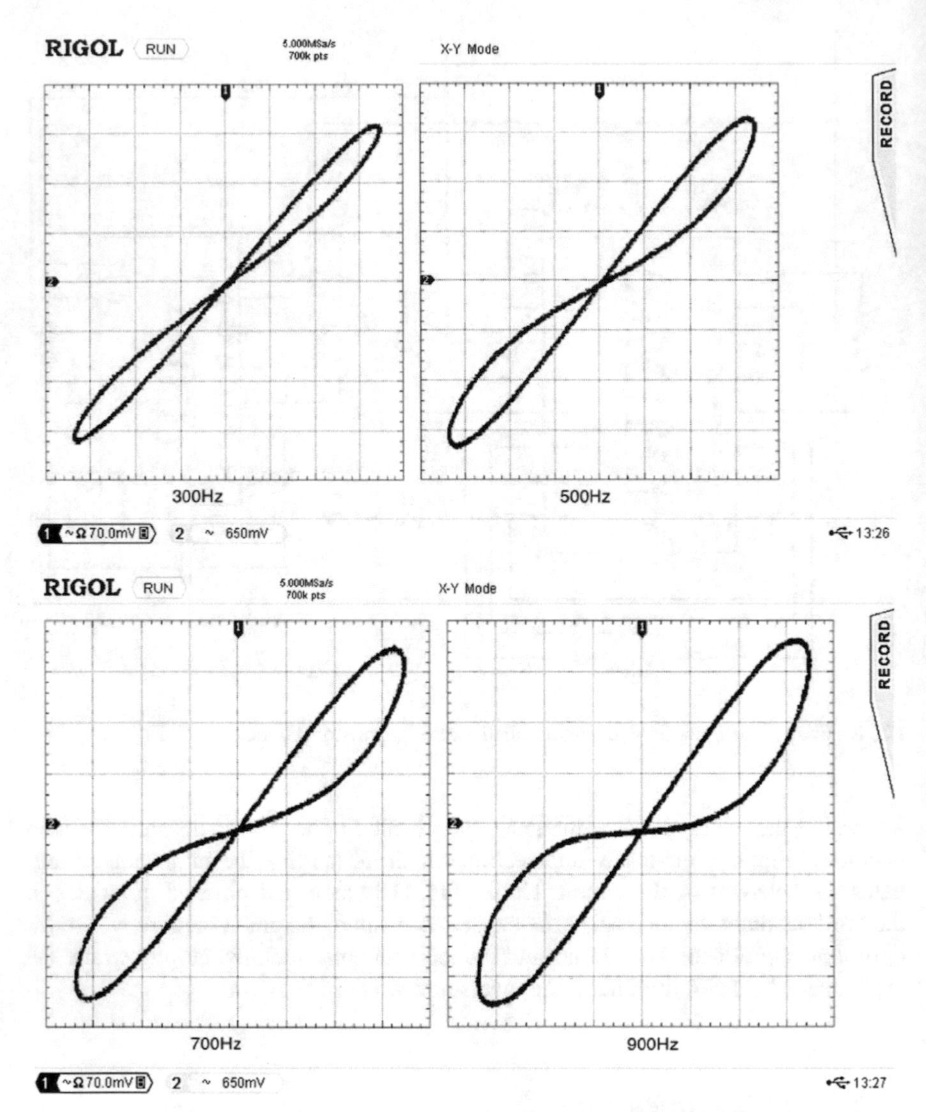

Fig. 9 Experimental verification of the circuit in Fig. 8 at 300, 500, 700 and 900 Hz. X-axis is $v(t)$ and Y-axis is $i(t) \times 56\,\text{k}\Omega$

- a linear inductance L corresponding to the linear derivative coefficient b in (1) $(L = b)$.
- a linear capacitance C corresponding to the linear integration coefficient d in (1) $(C = 1/d)$.
- a memristor M corresponding to the nonlinear integral term with coefficient c in (1). From (1), $M(q) = c \int i(\tau)d\tau$. Under sinusoidally exciting $i(t)$, the charge $q(t)$

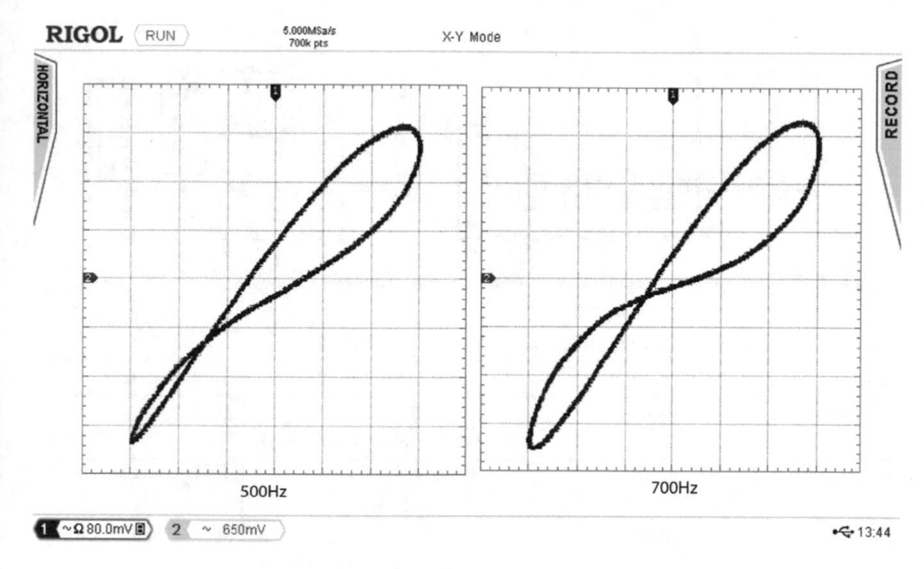

Fig. 10 Experimental results showing non-symmetrical loops at 500 and 700 Hz

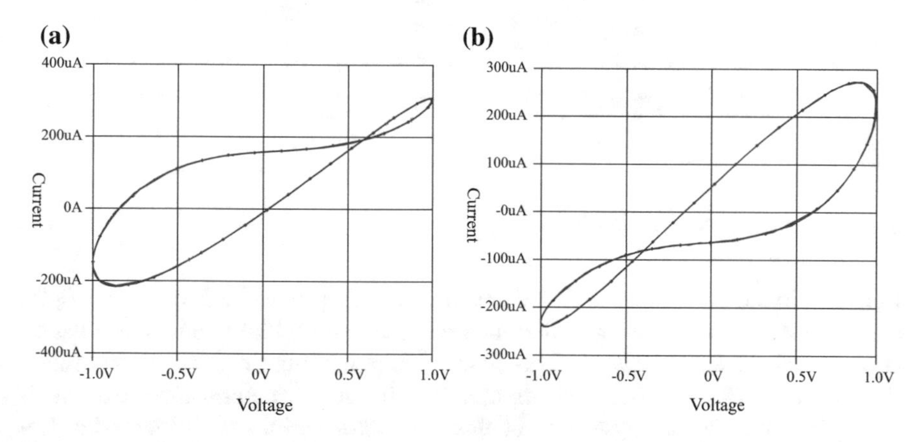

Fig. 11 SPICE simulation of parallel inverse memristor with **a** 10H inductor and **b** 50 nF capacitor

is inversely proportional to the frequency ω and hence the memristance decays with increasing frequency (Adhikari et al. 2013).

- a new element, which we term the inverse memristance $\overline{M}$, corresponding to the nonlinear derivative term with coefficient e in (1). From (1), $\overline{M}(q) = edi(t)/dt$. Under sinusoidal $i(t)$, $\overline{M}$ increases proportional to ω.

All series-connected cases are summarized in Table 1. Similarly, the same five impedances can be transformed into their admittance equivalents in Fig. 12b.

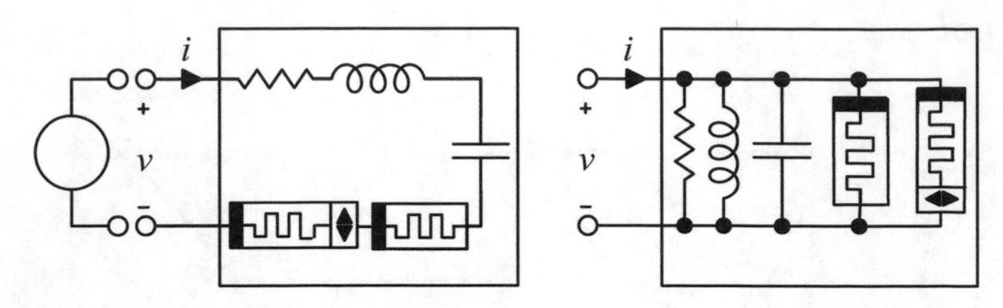

Fig. 12 Series and parallel connections of the five impedances $(R, L, C, M, \overline{M})$

Table 1 A summary of some special cases of the proposed Eq. (1)

Case	(x_p, y_p)	Boundary points	Double-loop
$b = d = e = 0$	$(0, 0)$	$(0, 0)$ $(\pm k, \pm k(a + ck\frac{cos(\phi)}{\omega}))$	$\Delta(\frac{y}{x})_{max} = \frac{ck}{\omega}$
$b = c = d = 0$	$(0, 0)$	$(0, 0)$ $(\pm k, \pm ak)$	$\Delta(\frac{y}{x})_{max} = ek\omega$
$b = d = 0$	$(0, 0)$	$(0, 0)$ $(\pm k, \pm k(a + ck\frac{cos(\phi)}{\omega}))$	$\Delta(\frac{y}{x})_{max} = k(e\omega - \frac{c}{\omega})$
$d = e = 0$	$x_p = \frac{b\omega^2}{c}$ $y_p = ax_p + \frac{kdcos(\phi)}{\omega}$	$(0, \pm kb\omega)$ $(\pm k, \pm k(a + ck\frac{cos(\phi)}{\omega}))$	Non-symmetrical $y \approx (a + \frac{ckcos(\phi)}{\omega})x$
$b = c = 0$	$x_p = \frac{d}{e\omega^2}$ $y_p = ax_p + \frac{kdcos(\phi)}{\omega}$	$(0, \frac{kd}{\omega}(cos(\phi) \pm 1))$ $(\pm k, \pm k(a + ck\frac{cos(\phi)}{\omega}))$	Non-symmetrical $y \approx ax + \frac{kdcos(\phi)}{\omega}$

5.1 Impedance Analysis

Referring to the first row in Table 1 where $a \neq 0$ and $c \neq 0$, while $b = d = e = 0$, (1) then represents a current-controlled memristor M with initial memristance equals $cq(0)$ in series with a resistor a. This connection represents a memristor with memristance $R_m = R_i + cq(t)$ and R_i representing the initial memristance and equals $a - cq(0)$. The pinched hysteresis of this memristor shrinks with increasing frequency and eventually disappears since it can be shown that $\triangle(y/x)_{max} = ck/\omega$. Therefore, maximizing the hysteresis behavior requires minimizing ω.

The second row in Table 1, where $a \neq 0, e \neq 0$ and $b = c = d = 0$ corresponds to the inverse memristor $\overline{M}$. A symmetric pinched hysteresis loop is also observed and is stimulated with increasing the frequency and vanishes as $\omega \to 0$ since it can be shown that $\triangle(y/x)_{max} = ek\omega$. Hence, maximizing the hysteresis loop requires increasing ω.

Row 3 in Table 1 shows the case of a memristor and inverse memristor connected in series; in which case only $b = d = 0$ and $\triangle(y/x)_{max} = k(e\omega - \frac{c}{\omega})$. Note the existence of a critical frequency $\omega_o = \sqrt{c/e}$ at which the hysteresis loop disappears and reduces to a straight line. Higher or lower than this critical frequency, the area of the hysteresis loop increases. This is illustrated in Figs. 13a, b respectively for

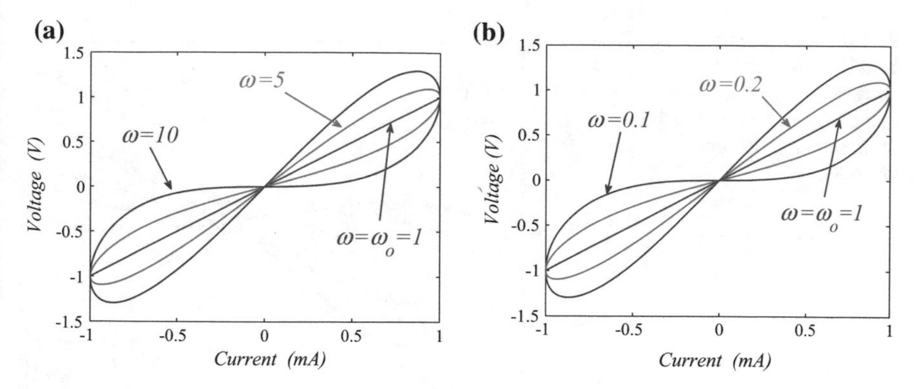

Fig. 13 Behaviour of an $M - \overline{M}$ series connection **a** increasing ω above ω_o and **b** reducing ω below ω_o

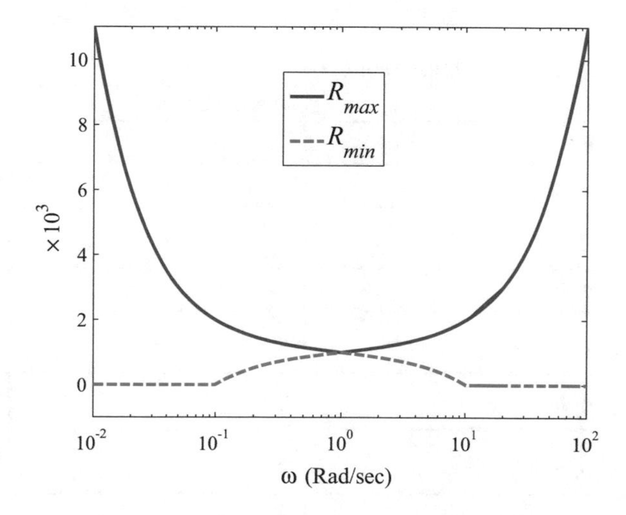

Fig. 14 Maximum and minimum resistances for a series $M - \overline{M}$ connection

$a = 10^3, c = e = 10^5, k = 10^{-3}$ and $\phi = \pi/2$; which yields $\omega_o = 1$. The maximum and minimum resistance (R_{max}, R_{min}) with this series $M - \overline{M}$ connection is shown in Fig. 14.

Adding a reactive element (capacitor or inductor) in series with M or $\overline{M}$ makes the hysteresis loops asymmetric. Figure 5 shows the effect of varying a when $b = c = 1$ and $\omega = 0.2\pi$ on the case represented in row 4 of Table 1. This case corresponds to an inductor in series with a memristor (series $L - M$). Note that all loops in Fig. 15 are asymmetric but with the same pinch point. Finally, the case in row 5 of Table 1 is that of a capacitor in series with a memristor (series $C - M$). Table 2 represents a summary of the complex impedances that can be obtained from (1) along with their circuit representations.

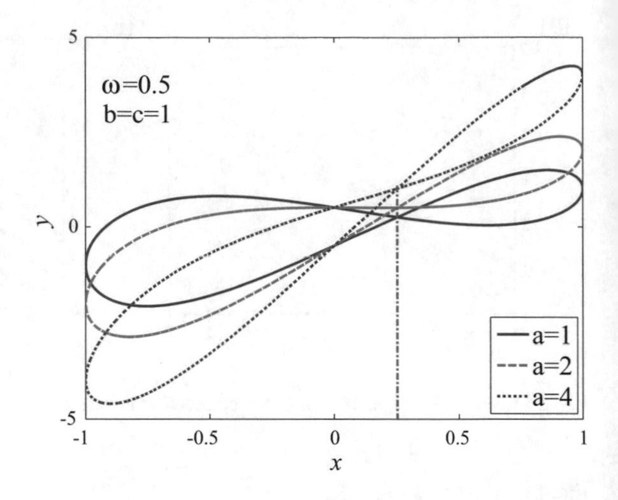

Fig. 15 Asymmetric double-loop hysteresis corresponding to row 4 of Table 1 (an inductor in series with a memristor) for variable *a*

Table 2 Impedances from (1)

Non-zero	Element(s)	Circuit Model
a	*R*	
b	*L*	
d	*C*	
a, b	*RL*	
a, d	*RC*	
a, b d	*RLC*	
a, c	*M*	
a, e	$\overline{M}$	
a, b, c	*ML*	
a, c, d	*MC*	
a, b, c, d	*MLC*	
a, b, e	$\overline{M}L$	
a, b, d, e	$\overline{M}LC$	
a, b,, c, d, e	$M\overline{M}LC$	

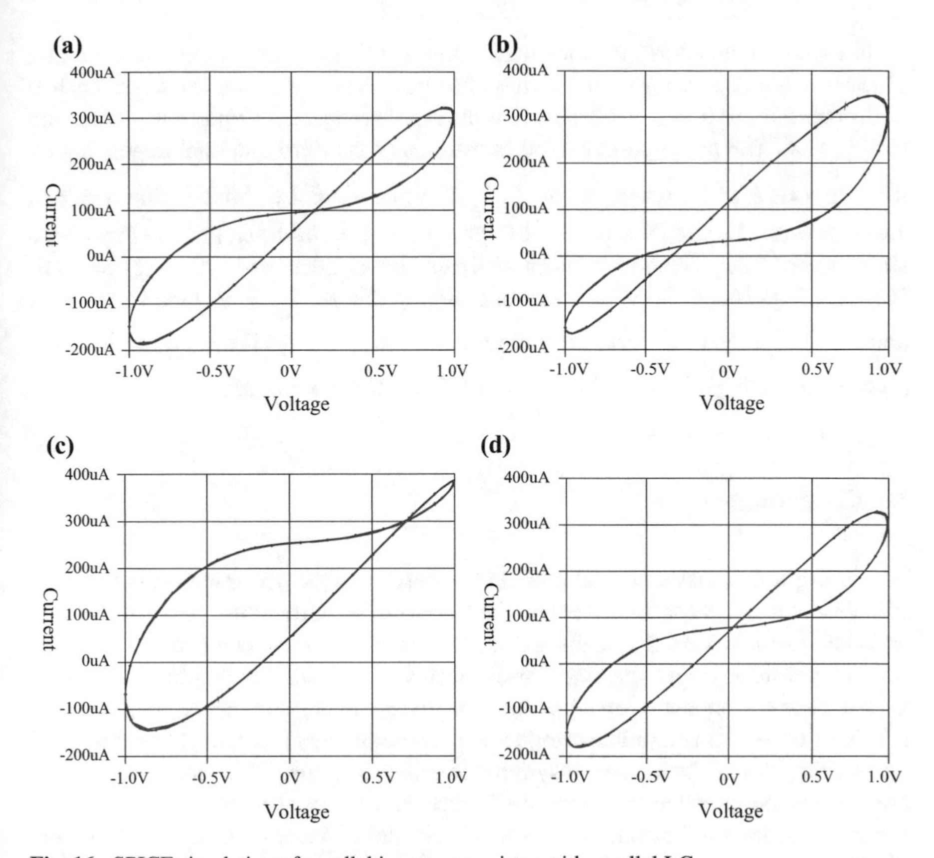

Fig. 16 SPICE simulation of parallel inverse memristor with parallel LC

5.2 *Admittance Analysis*

By setting $y = i(t)$ and $x = v(t)$, (1) reads as

$$i(t) = [a + c \int_0^t v(\tau)d\tau + e\frac{dv}{dt}]v(t) + b\frac{dv}{dt} + d \int_0^t v(\tau)d\tau \tag{25}$$

where the coefficient a has transconductance unit (℧), b has capacitance unit (F), d has inverse inductance unit (H^{-1}), c has unit $(sec \cdot V \cdot \Omega)^{-1}$, and e has the unit of $(sec/V \cdot \Omega)$. Similar to the impedance formation, different complex admittance can be obtained. For example, when $a \neq 0, c \neq 0$ and $b = d = e = 0$, the previous equation represents a voltage-controlled memristor with initial memductance equal to $a - c\phi(0)$. While when $a \neq 0, e \neq 0$ and $b = c = d = 0$, it represents a voltage-controlled inverse memristor. It is straightforward to build a table similar to Table 2 for all cases.

In case of connecting parallel $\overline{M}$LC circuit, the hysteresis loop can be either pinched or not depending on the values of b and d. The hysteresis would be pinched with double loops if $|x_p| < A$ where A is the input voltage amplitude and single loop for $|x_p| > A$. The pinch point can be in the first or the third quadrant depending on the values of b and d where for $\omega < \sqrt{\frac{d}{b}}$, the pinch point lies in the first quadrant and vice versa. Figure 16 is plotted by applying $v(t) = sin(400\pi t)$ where Fig. 16a, b showing the effect of changing the parallel capacitance 50 nF and 100 nF at $L = 10$ H. However, Fig. 16c, d show the effect of changing the parallel inductance from 10 H and 5 H at $C = 50$ nF. At the parallel resonance $\omega_o = \sqrt{\frac{1}{LC}}$, the coordinates of the pinch point is $(0, \frac{1}{L\omega})$ at $\omega = 400\pi, L = 10$ H and $C = 63.357$ nF.

6 Conclusion

In this chapter, A mathematical model to represent all the linear elements has been discussed. Different special cases have been introduced and verified using numerical, and circuit simulations. As we discussed, the statement "if it is pinched, it is memristor" is not valid anymore since we proved that inverse memristor has pinched hysteresis but it has the opposite behavior. Also the inverse memristor can be obtained by a nonlinear capacitor or nonlinear inductor. Connecting any reactive element to ideal symmetric pinched device gives asymmetric behavior. Thus, the asymmetric pinched devices can be modeled as symmetric devices with a reactive element. Moreover, according to the discussion, the series and parallel connection of the conventional elements in addition to the memristor and inverse memristor especially memristor-inverse memristor connection give new properties. These properties are new to the circuit theory.

References

Adamatzky, A., & Chua, L. (2013). *Memristor Networks*. Springer Science & Business Media.

Adhikari, S. P., Sah, M. P., Kim, H., & Chua, L. O. (2013). Three fingerprints of memristor. *IEEE Transactions on Circuits and Systems I: Regular Papers*, *60*(11), 3008–3021.

Alharbi, A. G., Fouda, M. E., & Chowdhury, M. H. (2015a). Memristor emulator based on practical current controlled model. In *2015 IEEE 58th International Midwest Symposium on Circuits and Systems (MWSCAS)* (pp. 1–4). IEEE.

Alharbi, A. G., Fouda, M. E., & Chowdhury, M. H. (2015b). A novel memristor emulator based only on an exponential amplifier and ccii+. In *2015 IEEE International Conference on Electronics, Circuits, and Systems (ICECS)* (pp. 376–379). IEEE.

Alharbi, A. G., Khalifa, Z. J., Fouda, M. E., & Chowdhury, M. H. (2015c). Memristor emulator based on single ccii. In *2015 27th International Conference on Microelectronics (ICM)* (pp. 174–177). IEEE.

Biolek, D., Biolek, Z., & Biolkova, V. (2011). Pinched hysteretic loops of ideal memristors, memcapacitors and meminductors must be 'self-crossing'. *Electronics Letters*, *47*(25), 1385–1387.

Biolek, Z., Biolek, D., & Biolková, V. (2015). Specification of one classical fingerprint of ideal memristor. *Microelectronics Journal*, *46*(4), 298–300.

Chua, L. (1971). Memristor-the missing circuit element. *IEEE Transactions on Circuit Theory*, *18*(5), 507–519.

ElSlehdar, A., Fouad, A., & Radwan, A. (2015). Memristor based n-bits redundant binary adder. *Microelectronics Journal*, *46*, 207–213.

Elwakil, A. S., Fouda, M. E., & Radwan, A. G. (2013). A simple model of double-loop hysteresis behavior in memristive elements. *IEEE Transactions on Circuits and Systems*, *60*(8), 487–491.

Fouda, M., Elwakil, A., & Radwan, A. (2015). Pinched hysteresis with inverse-memristor frequency characteristics in some nonlinear circuit elements. *Microelectronics Journal*, *46*(9), 834–838.

Fouda, M. E., Khatib, M. A., Mosad, A. G., & Radwan, A. G. (2013). Generalized analysis of symmetric and asymmetric memristive two-gate relaxation oscillators. *IEEE Transactions on Circuits and Systems I: Regular Papers*, *60*(10), 2701–2708.

Fouda, M. E., & Radwan, A. G. (2015a). Power dissipation of memristor-based relaxation oscillators. *Radioengineering*.

Fouda, M. E., & Radwan, A. G. (2015b). Fractional-order memristor response under dc and periodic signals. *Circuits, Systems, and Signal Processing*, *34*(3), 961–970.

Fouda, M. E., & Radwan, A. G. (2015c). Resistive-less memcapacitor-based relaxation oscillator. *International Journal of Circuit Theory and Applications*, *43*(7), 959–965.

Gambuzza, L. V., Fortuna, L., Frasca, M., & Gale, E. (2015). Experimental evidence of chaos from memristors. *International Journal of Bifurcation and Chaos*, *25*(08), 1550101.

Hussein, A. I. & Fouda, M. E. (2013). A simple mos realization of current controlled memristor emulator. In *2013 25th International Conference on Microelectronics (ICM)* (pp. 1–4). IEEE.

Joglekar, Y. N., & Wolf, S. J. (2009). The elusive memristor: properties of basic electrical circuits. *European Journal of Physics*, *30*(4), 661.

Khatib, M. A., Fouda, M. E., Mosad, A. G., Salama, K. N., & Radwan, A. (2012). Memristor-based relaxation oscillators using digital gates. In *2012 Seventh International Conference on Computer Engineering & Systems (ICCES)* (pp. 98–102). IEEE.

Kozma, R., Pino, R. E., & Pazienza, G. E. (2012). *Advances in neuromorphic memristor science and applications* (Vol. 4). Springer Science & Business Media.

Pickett, M. D., Strukov, D. B., Borghetti, J. L., Yang, J. J., Snider, G. S., Stewart, D. R., et al. (2009). Switching dynamics in titanium dioxide memristive devices. *Journal of Applied Physics*, *106*(7), 074508.

Radwan, A., & Fouda, M. (2014). Simple floating voltage-controlled memductor emulator for analog applications. *Radioengineering*.

Radwan, A. G. & Fouda, M. E. (2015). *On the mathematical modeling of memristor, memcapacitor, and meminductor* (Vol. 26). Springer.

Radwan, A. G., Moaddy, K., & Momani, S. (2011). Stability and nonstandard finite difference method of the generalized chua's circuit. *Computers and Mathematics with Applications*, *62*, 961–970.

Radwan, A. G., Zidan, M., & Salama, K. N. (2010a). Hp memristor mathematical model for periodic signals and dc. In *53rd IEEE International Midwest Symposium on Circuits and Systems (MWSCAS)* (pp. 861–864).

Radwan, A. G., Zidan, M., & Salama K. N. (2010b). On the mathematical modeling of memristors. In *22nd IEEE International Conference on Microelectronics (ICM)* (pp. 284–287).

Semiconductor, N. (2000). Lm13700 dual operational transconductance amplifiers with linearizing diodes and buffers.

Talukdar, A., Radwan, A., & Salama, K. N. (2011a). Generalized model for memristor-based wien-family oscillators. *Microelectronics Journal*, 1032–1038.

Talukdar, A., Radwan, A., & Salama, K. N. (2011b). Memristor-based third order oscillator: beyond oscillation. *Applied Nanoscience*, *1*, 143–145.

Talukdar, A., Radwan, A., & Salama, K. N. (2012). Nonlinear dynamics of memristor based 3rd order oscillatory system. *Microelectronics Journal*, 169–175.

Vaidyanathan, S., & Volos, C. (2016a). *Advances and applications in chaotic systems* (Vol. 636). Springer.

Vaidyanathan, S., & Volos, C. (2016b). *Advances and applications in nonlinear control systems* (Vol. 635). Springer.

Zidan, M., Omran, H., Radwan, A. G., & Salama, K. N. (2011). Memristor-based reactance-less oscillator. *Electronics Letters*, *47*(22), 1220–1221.

Zidan, M., Omran, H., Smith, C., Syed, A., Radwan, A. G., & Salama, K. N. (2014). A family of memristor-based reactance-less oscillators. *International Journal Circuit Theory and Applications*, *42*(22), 1103–1122.

A Conservative Hyperchaotic Hyperjerk System Based on Memristive Device

Sundarapandian Vaidyanathan

Abstract Memristor-based systems and their potential applications, in which memristor is both a nonlinear element and a memory element, have been received significant attention in the control literature. In this work, we propose a conservative memristor-based hyperchaotic hyperjerk system with infinite number of equilibrium points. In classical mechanics, the third-order time-derivative of displacement is called *jerk*, while the fourth-order time-derivative of displacement is called *snap*. As a result, a dynamical system which is represented by an nth order ordinary differential equation with $n > 3$ is considered as a *hyperjerk* system. Hyperjerk systems have received significant attention in the control literature. In this research work, a conservative memristor-based hyperjerk system has been designed which displays rich, *hyperchaotic* behavior. Interestingly, this hyperjerk system displays an infinite number of equilibrium points because of the presence of a memristive device. In this work, we obtain the Lyapunov exponents of the memristor-based system as $L_1 = 0.2098$, $L_2 = 0.2035$, $L_3 = 0$ and $L_4 = -0.4133$. Since there are two positive Lyapunov exponents, the memristor-based system is hyperchaotic. Also, the Kaplan-Yorke dimension of the memristor-based hyperchaotic system is obtained as $D_{KY} = 4$. Next, we design adaptive control and synchronization schemes for the memristor-based hyperjerk system with unknown parameters via backstepping control method. The main adaptive control and synchronization results are established using Lyapunov stability theory. MATLAB simulations are shown to illustrate all the main results of this work.

Keywords Memristor · Hyperjerk system · Conservative system · Chaos · Hyperchaos · Adaptive control · Synchronization · Backstepping control

S. Vaidyanathan (✉)
Research and Development Centre, Vel Tech University,
Avadi, Chennai 600062, Tamil Nadu, India
e-mail: sundarcontrol@gmail.com

© Springer International Publishing AG 2017

S. Vaidyanathan and C. Volos (eds.), *Advances in Memristors, Memristive Devices and Systems*, Studies in Computational Intelligence 701,
DOI 10.1007/978-3-319-51724-7_16

1 Introduction

Chua's circuit (Matsumoto 1984), the Cellular Neural Networks (CNNs) (Chua and Yang 1988a, b) and the memristor (Chua 1971) are three attractive inventions of Prof. Leon O. Chua and these inventions are widely regarded as the major breakthroughs in the literature of the nonlinear control systems. Chua's circuit has been applied in various areas in engineering (Liu et al. 2004; Fortuna et al. 2009; Chua 1994; Albuquerque et al. 2008; Tang and Wang 2005). Cellular Neural Networks have been applied in various areas such as chaos (Vaidyanathan 2016e), secure communications (Wang et al. 2012b), cryptosystem (Cheng and Cheng 2013), etc. The studies on memristor (Joglekar and Wolf 2009; Shin et al. 2011; Wang et al. 2012a; Shang et al. 2012; Adhikari et al. 2012, 2013; Yang et al. 2013) have received significant attention only recently after the realization of a solid-state thin film two-terminal memristor at Hewlett-Packard Laboratories (Strukov et al. 2008).

Memristor was proposed by L.O. Chua as the fourth basic circuit element besides the three conventional ones (resistor, inductor and capacitor) (Tetzlaff 2014).

Memristor depicts the relationship between two fundamental circuit variables, *viz.* the charge (q) and the flux (φ). Hence, there are two kinds of memristors: (1) *charge-controlled memristor*, and (2) *flux-controlled memristor*.

A charge-controlled memristor is described by

$$v_M = M(q) i_M \tag{1}$$

where v_M is the *voltage* across the memristor and i_M is the *current* through the memristor. Here, the *memristance* (M) is defined by

$$M(q) = \frac{d\varphi(q)}{dq} \tag{2}$$

A flux-controlled memristor is given by

$$i_M = W(\varphi) v_M \tag{3}$$

where $W(\varphi)$ is the *memductance*, which is defined by

$$W(q) = \frac{dq(\varphi)}{d\varphi} \tag{4}$$

By generalizing the original definition of a memristor (Chua 1971; Tetzlaff 2014), a *memristive system* is defined as

$$\begin{cases} \dot{x} = f(x, u, t) \\ y = g(x, u, t)u \end{cases} \tag{5}$$

where x is the *state*, u is the *input* and y is the *output* of the system (5). We assume that the function f is a continuously differentiable, n-dimensional vector field and g is a continuous scalar function.

Based on the definition of memristive system (Bao et al. 2013; Pershin et al. 2009; Tetzlaff 2014), a memristive device is introduced in this section and used in our whole chapter.

The memristive device is described by the following form:

$$\begin{cases} \dot{x}_1 = x_2 \\ \dot{x}_2 = W(x_1, x_2) = (1 - x_1)x_2 \end{cases} \tag{6}$$

Here x_2, y and x_1 are the input, output and state of the memristive device, respectively.

The intrinsic nonlinear characteristic of memristor has applications in implementing chaotic systems with complex dynamics as well as special features (Itoh and Chua 2008; Muthuswamy and Kokate 2009). For example, a simple memristor-based chaotic system including only three elements (an inductor, a capacitor and a memristor) was introduced in Muthuswamy and Chua (2010). Also, a system containing an HP memristor model and triangular wave sequence can generate multi-scroll chaotic attractors (Li et al. 2014). Moreover, a four-dimensional hyperchaotic memristive system with a line equilibrium was presented by Li et al. (2013).

Chaos theory deals with the qualitative study of chaotic dynamical systems and their applications in science and engineering. A dynamical system is called *chaotic* if it satisfies the three properties: boundedness, infinite recurrence and sensitive dependence on initial conditions (Azar and Vaidyanathan 2015).

Some classical paradigms of 3-D chaotic systems in the literature are Lorenz system (1963), Rössler system (1976), ACT system (Arneodo et al. 1981), Sprott systems (1994), Chen system (1999), Lü system (2002), Cai system (2007), Tigan system (2008), etc.

Many new chaotic systems have been discovered in the recent years such as Zhou system (2008), Zhu system (2010), Li system (2008), Wei-Yang system (2010), Sundarapandian systems (2012; 2013), Vaidyanathan systems (2013a, b, 2014a, b, c, d, 2015b, g, m, n, 2013, 2015b, 2015, 2014c, 2015b, d, f, 2015, 2016, 2016a, c, j, 2016), Pehlivan system (2014), Sampath system (2015), Pham system (2014), etc.

Chaos theory has many applications in science and engineering such as chemical systems (Vaidyanathan 2015i, g, s, o, t, c, k, u), biological systems (Vaidyanathan 2015d, e, a, j, p, f, x, q, y, r, z, h, v, l, w), memristors (Pham et al. 2015; Volos et al. 2015; Abdurrahman et al. 2015), etc.

The study of control of a chaotic system investigates feedback control methods that globally or locally asymptotically stabilize or regulate the outputs of a chaotic system. Many methods have been designed for control and regulation of chaotic systems such as active control (Sundarapandian 2010, 2011; Vaidyanathan 2011b), adaptive control (Vaidyanathan et al. 2014a, 2015e, h), backstepping control (Li et al. 2007; Wang and Ge 2008), sliding mode control (Vaidyanathan 2012c, e, 2016k, f, i, h), etc.

Synchronization of chaotic systems is a phenomenon that occurs when two or more chaotic systems are coupled or when a chaotic system drives another chaotic system. Because of the butterfly effect which causes exponential divergence of the trajectories of two identical chaotic systems started with nearly the same initial conditions, the synchronization of chaotic systems is a challenging research problem in the chaos literature (Azar and Vaidyanathan 2015, 2016; Azar et al. 2017; Vaidyanathan and Volos 2016a, b).

Pecora and Carroll pioneered the research on synchronization of chaotic systems with their seminal papers (Carroll and Pecora 1991; Pecora and Carroll 1990). The active control method (Karthikeyan and Sundarapandian 2014; Sarasu and Sundarapandian 2011a, b; Sundarapandian and Karthikeyan 2012b; Vaidyanathan 2011a, 2012d; Vaidyanathan and Rajagopal 2011a, b; Vaidyanathan and Rasappan 2011) is typically used when the system parameters are available for measurement. Adaptive control method (Sarasu and Sundarapandian 2012a, b, c; Sundarapandian and Karthikeyan 2011a, b, 2012a; Vaidyanathan 2012b, 2013c; Vaidyanathan and Azar 2015a; Vaidyanathan and Pakiriswamy 2013; Vaidyanathan and Rajagopal 2011c, 2012; Vaidyanathan et al. 2014b, 2015c) is typically used when some or all the system parameters are not available for measurement and estimates for the uncertain parameters of the systems.

Sampled-data feedback control method (Gan and Liang 2012; Xiao et al. 2014) and time-delay feedback control method (Chen et al. 2014; Jiang et al. 2004) are also used for synchronization of chaotic systems. Backstepping control method (Rasappan and Vaidyanathan 2012a, b, c, 2013, 2014; Suresh and Sundarapandian 2013; Vaidyanathan and Rasappan 2014; Vaidyanathan et al. 2015a, i; Vaidyanathan 2016g; Vaidyanathan et al. 2016) is also applied for the synchronization of chaotic systems. Backstepping control is a recursive method for stabilizing the origin of a control system in strict-feedback form Khalil (2001). In this research work, we apply backstepping control method for the adaptive control and synchronization of the novel hyperjerk system.

Sliding mode control method (Sundarapandian and Sivaperumal 2011; Vaidyanathan 2012a, 2014e; Vaidyanathan and Azar 2015c, d; Vaidyanathan and Sampath 2011, 2012; Sampath and Vaidyanathan 2016) is also a popular method for the synchronization of chaotic systems.

In the control literature, there is significant interest in investigating novel jerk chaotic systems (Azar and Vaidyanathan 2015, 2016; Azar et al. 2017; Vaidyanathan and Volos 2016a, b). It is well-known that a jerk system is represented by an explicit third-order differential equation which describes the time evolution of a single scalar variable. Thus, a jerk system can be described as

$$\frac{d^3x}{dt^3} = f\left(\frac{d^2x}{dt^2}, \frac{dx}{dt}, x\right) \tag{7}$$

In classical mechanics, the differential equation (7) is called a *jerk system*, because the successive derivatives of the displacement in a mechanical system are velocity, acceleration and jerk.

A generalization of the jerk dynamics is given by the dynamics

$$\frac{d^{(n)}x}{dt^n} = f\left(\frac{d^{(n-1)}x}{dt^{n-1}}, \ldots, \frac{dx}{dt}, x\right), \quad (n \geq 4) \tag{8}$$

An ordinary differential equation of the form (8) is called a *hyperjerk* system since it involves time derivatives of a jerk function (Vaidyanathan 2016b).

In Chlouverakis and Sprott (2006), Chlouverakis and Sprott discovered a simple hyperchaotic hyperjerk system given by the dynamics

$$\frac{d^4x}{dt^4} + \frac{d^3x}{dt^3}x^4 + A\frac{d^2x}{dt^2} + \frac{dx}{dt} + x = 0 \tag{9}$$

In system form, the differential equation (9) can be expressed as

$$\begin{cases} \dot{x}_1 = x_2 \\ \dot{x}_2 = x_3 \\ \dot{x}_3 = x_4 \\ \dot{x}_4 = -x_1 - x_2 - Ax_3 - x_1^4 x_4 \end{cases} \tag{10}$$

When $A = 3.6$, the hyperjerk system (10) exhibits *hyperchaos* with the Lyapunov exponents

$$L_1 = 0.132, \;\; L_2 = 0.035, \;\; L_3 = 0, \;\; L_4 = -1.25 \tag{11}$$

Thus, the maximum Lyapunov exponent (MLE) of the Chlouverakis-Sprott hyperchaotic hyperjerk system (10) is $L_1 = 0.132$ and the Kaplan-Yorke dimension of this hyperjerk system is easily calculated a $D_{KY} = 3.13$.

In Vaidyanathan (2016d), Vaidyanathan derived a novel hyperjerk system by adding a hyperbolic sinusoidal nonlinearity to the Chlouverakis-Sprott hyperjerk system (4) and with a different set of values for the system parameters.

Vaidyanathan hyperjerk system is given in system form as

$$\begin{cases} \dot{x}_1 = x_2 \\ \dot{x}_2 = x_3 \\ \dot{x}_3 = x_4 \\ \dot{x}_4 = -x_1 - x_2 - ax_3 - b\sinh(x_2) - x_1^4 x_4 \end{cases} \tag{12}$$

where the parameter values are taken as

$$a = 3.7, \;\; b = 0.05, \;\; c = 1.3 \tag{13}$$

For the parameter values in (13), the Lyapunov exponents of the Vaidyanathan hyperjerk system (12) are obtained as

$$L_1 = 0.14219, \;\; L_2 = 0.04605, \;\; L_3 = 0, \;\; L_4 = -1.39267 \tag{14}$$

Thus, the maximal Lyapunov exponent (MLE) of the Vaidyanathan hyperjerk system (12) is seen as $L_1 = 0.14219$, and the Kaplan-Yorke dimension of the Vaidyanathan hyperjerk system (12) is derived as $D_{KY} = 3.1348$.

Conservative chaotic systems are characterized by the property that they are volume preserving. Classical examples of conservative chaotic systems are Hénon-Heiles system (1964), Nosé-Hoover system (1995), Sprott system (1997), etc. Recently, many conservative chaotic systems have been reported such as Vaidyanathan-Volos system (2015), Vaidyanathan-Pakiriswamy systems (2015; 2016), etc.

In this work, we propose a conservative memristor-based hyperchaotic hyperjerk system with infinite number of equilibrium points.

In this work, we obtain the Lyapunov exponents of the memristor-based system as $L_1 = 0.2098$, $L_2 = 0.2035$, $L_3 = 0$ and $L_4 = -0.4133$. Since there are two positive Lyapunov exponents, the memristor-based system is hyperchaotic. Also, the Kaplan-Yorke dimension of the memristor-based hyperchaotic system is obtained as $D_{KY} = 4$.

Next, we design adaptive control and synchronization schemes for the memristor-based hyperjerk system with unknown parameters via backstepping control method. The main adaptive control and synchronization results are established using Lyapunov stability theory. MATLAB simulations are shown to illustrate all the main results of this work.

2 A 4-D Novel Conservative Hyperchaotic Memristive Hyperjerk System

In this chapter, a novel 4-D conservative hyperchaotic memristive hyperjerk system is proposed by using the memristive device (6) and the reported approach in Bao et al. (2013).

The 4-D novel memristive system is given in system form as follows:

$$\begin{cases} \dot{x}_1 = x_2 \\ \dot{x}_2 = x_3 \\ \dot{x}_3 = x_4 \\ \dot{x}_4 = -x_3 - ax_4 - bx_3x_4 - c\sin x_3 - y \end{cases} \tag{15}$$

where a, b, c are positive parameters and $y = W(x_1, x_2) = (1 - x_1)x_2$ is the output of the memristive device (6).

Next, we take the parameters of the memristor-based system (15) as

$$a = 0.5, \quad b = 0.4, \quad c = 0.1 \tag{16}$$

We choose the initial conditions of the system (15) as

$$x_1(0) = 0.1, \quad x_2(0) = 0.1, \quad x_3(0) = 0, \quad x_4(0) = 0 \tag{17}$$

For the parameter values (16) and the initial values (17), the Lyapunov exponents of the memristor-based system (15) are obtained as

$$L_1 = 0.2098, \quad L_2 = 0.2035, \quad L_3 = 0, \quad L_4 = -0.4133 \tag{18}$$

Thus, the memristor-based system (15) is a hyperchaotic system because it has two positive Lyapunov exponents (Azar and Vaidyanathan 2016; Azar et al. 2017; Vaidyanathan and Volos 2016a, b).

Since the sum of the Lyapunov exponents in (18), it follows that the hyperchaotic memristive hyperjerk system (15) is *conservative*.

The Kaplan-Yorke dimension describes the complexity of a chaotic attractor (Frederickson et al. 1983). Suppose that a chaotic system of order n has n Lyapunov exponents $L_1, L_2, \ldots, L_n$, which are arranged in decreasing order, *i.e.*

$$L_1 \geq L_2 \geq \cdots \geq L_n \tag{19}$$

Then the Kaplan-Yorke dimension of the chaotic system of order n is defined by

$$D_{KY} = j + \frac{1}{|L_{j+1}|} \sum_{i=1}^{j} L_j \tag{20}$$

where j is the largest integer satisfying $\sum_{i=1}^{j} L_i \geq 0$ and $\sum_{i=1}^{j+1} L_i < 0$.

The Kaplan-Yorke dimension of the hyperchaotic memristive system (15) is calculated as

$$D_{KY} = 3 + \frac{L_1 + L_2 + L_3}{|L_4|} = 4 \tag{21}$$

The high value of the Kaplan-Yorke dimension of the hyperchaotic memristive system (15) shows the high complexity of the system.

The equilibrium points of the hyperchaotic memristive system (15) are obtained by solving the system of equations

$$
\begin{cases}
x_2 = 0 \\
x_3 = 0 \\
x_4 = 0 \\
-x_3 - ax_4 - bx_3x_4 - c\sin x_3 - (1 - x_1)x_2 = 0
\end{cases}
\tag{22}
$$

Solving the system (22), we obtain infinite number of equilibrium points of the system (15) given by

$$
E_k = \begin{bmatrix} k \\ 0 \\ 0 \\ 0 \end{bmatrix}, \ (k \in \mathbf{R})
\tag{23}
$$

which are saddle-foci.

Figures 1, 2, 3 and 4 show the 3-D projections of the hyperchaotic memristive system (15) in (x_1, x_2, x_3), (x_1, x_2, x_4), (x_1, x_3, x_4), and (x_2, x_3, x_4), spaces, respectively. Figure 5 shows the Lyapunov exponents of the hyperchaotic memristive system (15).

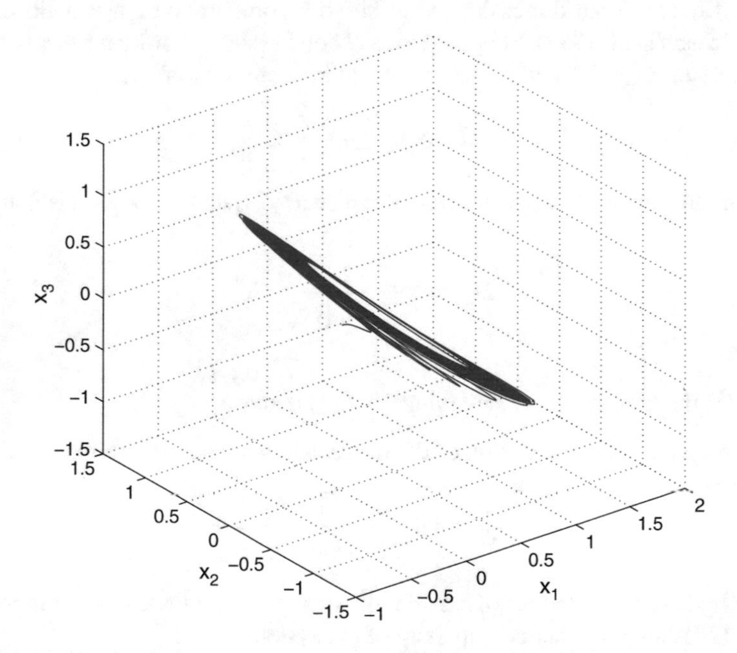

Fig. 1 3-D phase portrait of the hyperchaotic memristive hyperjerk system in (x_1, x_2, x_3) space

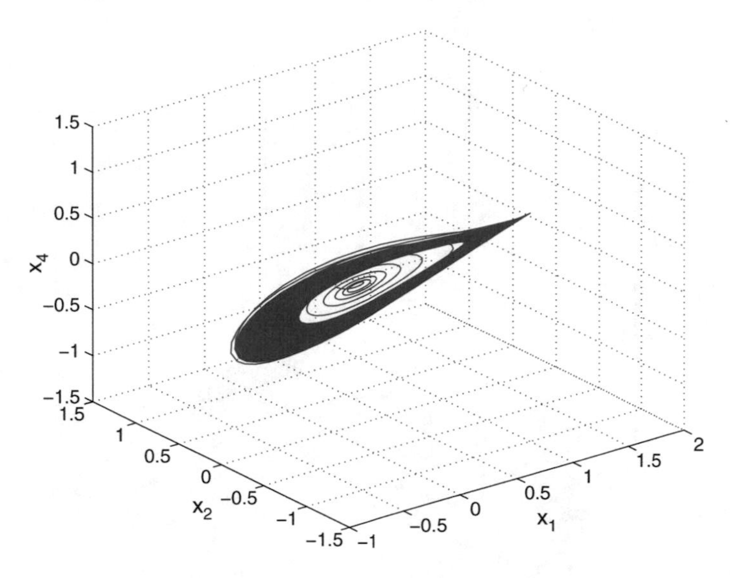

Fig. 2 3-D phase portrait of the hyperchaotic memristive hyperjerk system in (x_1, x_2, x_4) space

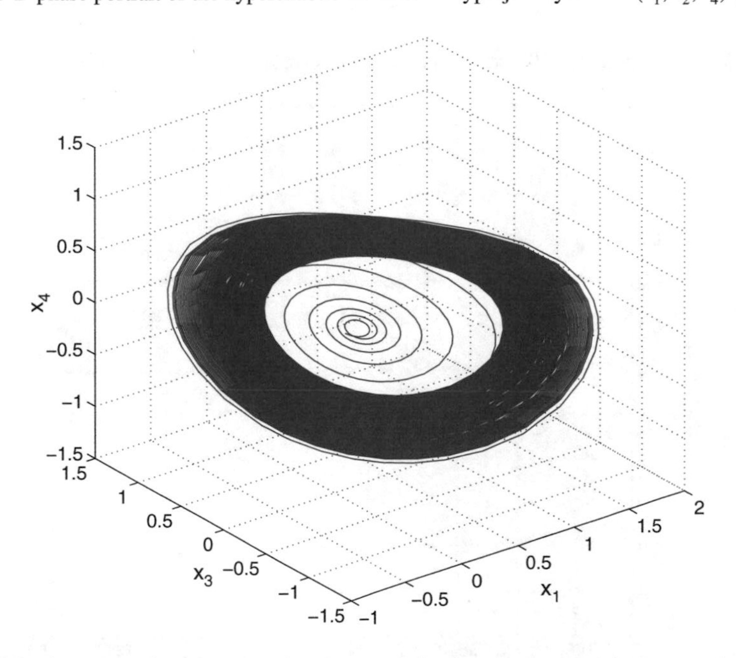

Fig. 3 3-D phase portrait of the hyperchaotic memristive hyperjerk system in (x_1, x_3, x_4) space

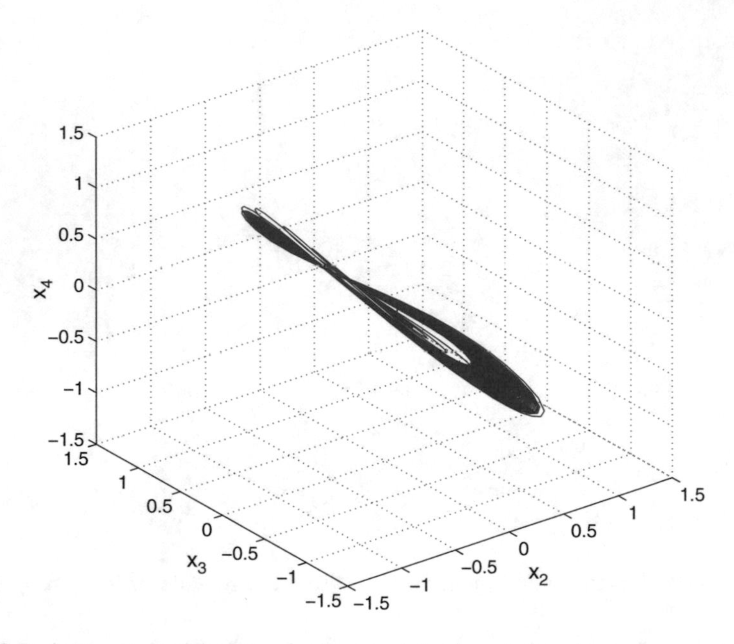

Fig. 4 3-D phase portrait of the hyperchaotic memristive hyperjerk system in (x_2, x_3, x_4) space

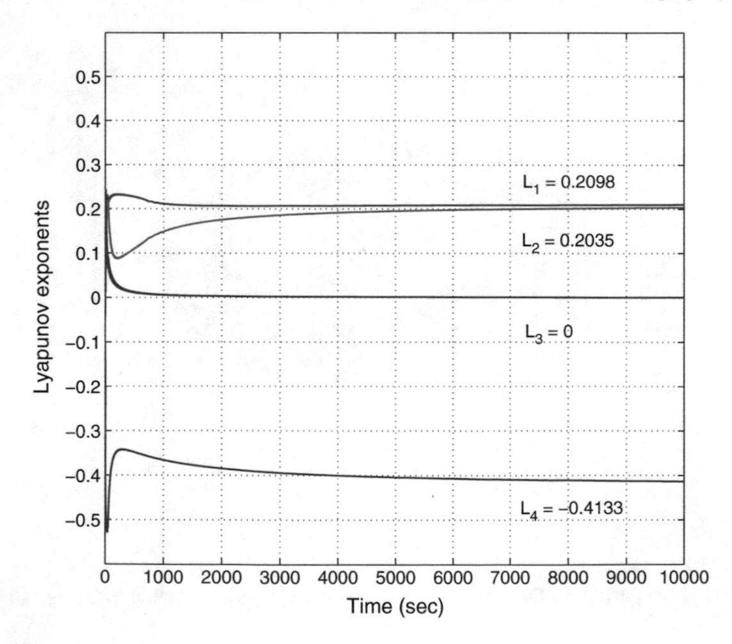

Fig. 5 Lyapunov exponents of the hyperchaotic memristive hyperjerk system

3 Adaptive Control of the Novel Hyperchaotic Memristive Hyperjerk System

In this section, we use backstepping control method to derive an adaptive feedback control law for globally stabilizing the 4-D novel hyperchaotic memristive hyperjerk system with unknown parameters.

Thus, we consider the 4-D novel hyperjerk system given by

$$\begin{cases} \dot{x}_1 = x_2 \\ \dot{x}_2 = x_3 \\ \dot{x}_3 = x_4 \\ \dot{x}_4 = -x_3 - ax_4 - bx_3x_4 - c\sin x_3 - y + u \end{cases} \tag{24}$$

where a, b, c are unknown constant parameters, $y = (1 - x_1)x_2$ is the output of the memristive device, and u is a backstepping control law to be determined using estimates $\hat{a}(t), \hat{b}(t)$ and $\hat{c}(t)$ for a, b and c, respectively.

The parameter estimation errors are defined as:

$$\begin{cases} e_a(t) = a - \hat{a}(t) \\ e_b(t) = b - \hat{b}(t) \\ e_c(t) = c - \hat{c}(t) \end{cases} \tag{25}$$

Differentiating (25) with respect to t, we obtain the following equations:

$$\begin{cases} \dot{e}_a(t) = -\dot{\hat{a}}(t) \\ \dot{e}_b(t) = -\dot{\hat{b}}(t) \\ \dot{e}_c(t) = -\dot{\hat{c}}(t) \end{cases} \tag{26}$$

Next, we shall state and prove the main result of this section.

Theorem 1 *The 4-D novel hyperchaotic memristive hyperjerk system (24), with unknown parameters a, b and c, is globally and exponentially stabilized by the adaptive feedback control law,*

$$u(t) = -5x_1 - 9x_2 - 8x_3 - [4 - \hat{a}(t)]x_4 - x_1x_2 + \hat{b}(t)x_3x_4 + \hat{c}(t)\sin x_3 - kz_4 \tag{27}$$

where $k > 0$ is a gain constant,

$$z_4 = 3x_1 + 5x_2 + 3x_3 + x_4 \tag{28}$$

and the update law for the parameter estimates $\hat{a}(t), \hat{b}(t), \hat{c}(t)$ *is given by*

$$\begin{cases} \dot{\hat{a}}(t) = -z_4 x_4 \\ \dot{\hat{b}}(t) = -z_4 x_3 x_4 \\ \dot{\hat{c}}(t) = -z_4 \sin x_3 \end{cases} \tag{29}$$

Proof We prove this result via Lyapunov stability theory (Khalil 2001).
First, we define a quadratic Lyapunov function

$$V_1(z_1) = \frac{1}{2} z_1^2 \tag{30}$$

where

$$z_1 = x_1 \tag{31}$$

Differentiating V_1 along the dynamics (24), we get

$$\dot{V}_1 = z_1 \dot{z}_1 = x_1 x_2 = -z_1^2 + z_1(x_1 + x_2) \tag{32}$$

Now, we define

$$z_2 = x_1 + x_2 \tag{33}$$

Using (33), we can simplify the Eq. (32) as

$$\dot{V}_1 = -z_1^2 + z_1 z_2 \tag{34}$$

Secondly, we define a quadratic Lyapunov function

$$V_2(z_1, z_2) = V_1(z_1) + \frac{1}{2} z_2^2 = \frac{1}{2} \left(z_1^2 + z_2^2 \right) \tag{35}$$

Differentiating V_2 along the dynamics (24), we get

$$\dot{V}_2 = -z_1^2 - z_2^2 + z_2(2x_1 + 2x_2 + x_3) \tag{36}$$

Now, we define

$$z_3 = 2x_1 + 2x_2 + x_3 \tag{37}$$

Using (37), we can simplify the Eq. (36) as

$$\dot{V}_2 = -z_1^2 - z_2^2 + z_2 z_3 \tag{38}$$

Thirdly, we define a quadratic Lyapunov function

$$V_3(z_1, z_2, x_3) = V_2(z_1, z_2) + \frac{1}{2}z_3^2 = \frac{1}{2}\left(z_1^2 + z_2^2 + z_3^2\right) \tag{39}$$

Differentiating V_3 along the dynamics (24), we get

$$\dot{V}_3 = -z_1^2 - z_2^2 - z_3^2 + z_3(3x_1 + 5x_2 + 3x_3 + x_4) \tag{40}$$

Now, we define

$$z_4 = 3x_1 + 5x_2 + 3x_3 + x_4 \tag{41}$$

Using (41), we can simplify the Eq. (40) as

$$\dot{V}_2 = -z_1^2 - z_2^2 - z_3^2 + z_3 z_4 \tag{42}$$

Finally, we define a quadratic Lyapunov function

$$V(z_1, z_2, z_3, z_4, e_a, e_b, e_c) = V_3(z_1, z_2, z_3) + \frac{1}{2}z_4^2 + \frac{1}{2}e_a^2 + \frac{1}{2}e_b^2 + \frac{1}{2}e_c^2 \tag{43}$$

which is a positive definite function on $\mathbf{R}^7$.

Differentiating V along the dynamics (24), we get

$$\dot{V} = -z_1^2 - z_2^2 - z_3^2 - z_4^2 + z_4(z_4 + z_3 + \dot{z}_4) - e_a\dot{\hat{a}} - e_b\dot{\hat{b}} - e_c\dot{\hat{c}} \tag{44}$$

Equation (44) can be written compactly as

$$\dot{V} = -z_1^2 - z_2^2 - z_3^2 - z_4^2 + z_4 S - e_a\dot{\hat{a}} - e_b\dot{\hat{b}} - e_c\dot{\hat{c}} \tag{45}$$

where

$$S = z_4 + z_3 + \dot{z}_4 = z_4 + z_3 + 3\dot{x}_1 + 5\dot{x}_2 + 3\dot{x}_3 + \dot{x}_4 \tag{46}$$

A simple calculation gives

$$S = 5x_1 + 9x_2 + 8x_3 + (4 - a)x_4 + x_1x_2 - bx_3x_4 - c\sin x_3 + u \tag{47}$$

Substituting the adaptive control law (27) into (47), we obtain

$$S = -\left[a - \hat{a}(t)\right]x_4 - \left[b - \hat{b}(t)\right]x_3x_4 - \left[c - \hat{c}(t)\right]\sin x_3 - kz_4 \tag{48}$$

Using the definitions (26), we can simplify (48) as

$$S = -e_a x_1 - e_b x_3 - e_c x_4 - kz_4 \tag{49}$$

Substituting the value of S from (49) into (45), we obtain

$$\begin{aligned}\dot{V} = -z_1^2 - z_2^2 - z_3^2 - (1+k)z_4^2 + e_a\left[-z_4x_4 - \dot{\hat{a}}\right] \\ +e_b\left[-z_4x_3x_4 - \dot{\hat{b}}\right] + e_c\left[-z_4\sin x_3 - \dot{\hat{c}}\right]\end{aligned} \tag{50}$$

Substituting the update law (29) into (50), we get

$$\dot{V} = -z_1^2 - z_2^2 - z_3^2 - (1+k)z_4^2, \tag{51}$$

which is a negative semi-definite function on $\mathbf{R}^7$.

From (51), it follows that the vector $\mathbf{z}(t) = (z_1(t), z_2(t), z_3(t), z_4(t))$ and the parameter estimation error $(e_a(t), e_b(t), e_c(t))$ are globally bounded, i.e.

$$\left[z_1(t)\ z_2(t)\ z_3(t)\ z_4(t)\ e_a(t)\ e_b(t)\ e_c(t)\right] \in \mathbf{L}_\infty \tag{52}$$

Also, it follows from (51) that

$$\dot{V} \le -z_1^2 - z_2^2 - z_3^2 - z_4^2 = -\|\mathbf{z}\|^2 \tag{53}$$

That is,

$$\|\mathbf{z}\|^2 \le -\dot{V} \tag{54}$$

Integrating the inequality (54) from 0 to t, we get

$$\int_0^t \|\mathbf{z}(\tau)\|^2\, d\tau \le V(0) - V(t) \tag{55}$$

From (55), it follows that $\mathbf{z}(t) \in \mathbf{L}_2$.

From Eq. (24), it can be deduced that $\dot{\mathbf{z}}(t) \in \mathbf{L}_\infty$.

Thus, using Barbalat's lemma (Khalil 2001), we conclude that $\mathbf{z}(t) \to \mathbf{0}$ exponentially as $t \to \infty$ for all initial conditions $\mathbf{z}(0) \in \mathbf{R}^4$.

Hence, it is immediate that $\mathbf{x}(t) \to \mathbf{0}$ exponentially as $t \to \infty$ for all initial conditions $\mathbf{x}(0) \in \mathbf{R}^4$.

This completes the proof. ■

For the numerical simulations, the classical fourth-order Runge-Kutta method with step size $h = 10^{-8}$ is used to solve the system of differential equations (24) and (29), when the adaptive control law (27) and the parameter update law (29) are applied.

The parameter values of the novel hyperjerk system (24) are taken as in the hyperchaotic case (16), *i.e.*

$$a = 0.5, \quad b = 0.4, \quad c = 0.1 \tag{56}$$

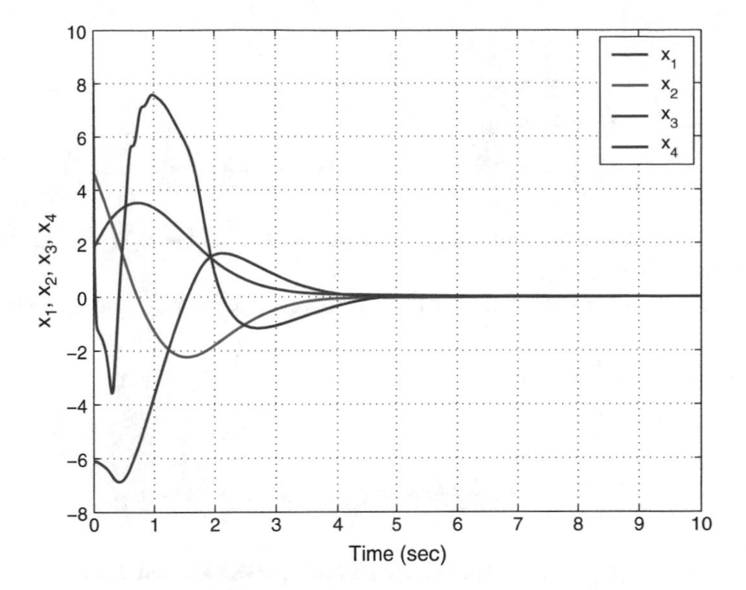

Fig. 6 Time-history of the controlled states x_1, x_2, x_3, x_4 of the novel hyperjerk system

We take the positive gain constant as $k = 10$.

As initial conditions of the hyperjerk system (24), we take

$$x_1(0) = 1.8, \quad x_2(0) = 4.7, \quad x_3(0) = -6.2, \quad x_4(0) = 8.1 \tag{57}$$

Also, as initial conditions of the parameter estimates $\hat{a}(t)$ and $\hat{b}(t)$, we take

$$\hat{a}(0) = 5.4, \quad \hat{b}(0) = 2.9, \quad \hat{c}(0) = 3.5 \tag{58}$$

In Fig. 6, the exponential convergence of the controlled states $x_1(t)$, $x_2(t)$, $x_3(t)$, $x_4(t)$ is depicted, when the adaptive control law (27) and the parameter update law (29) are implemented.

4 Adaptive Synchronization of the Identical Novel Hyperjerk Systems

In this section, we use backstepping control to derive an adaptive control law for globally and exponentially synchronizing the identical novel hyperchaotic memristive hyperjerk systems with unknown parameters.

As the master system, we consider the 4-D novel memristive hyperjerk system given by

$$\begin{cases} \dot{x}_1 = x_2 \\ \dot{x}_2 = x_3 \\ \dot{x}_3 = x_4 \\ \dot{x}_4 = -x_3 - ax_4 - bx_3x_4 - c\sin x_3 - \varphi \end{cases} \tag{59}$$

where a, b, c are unknown constant parameters and $\varphi = (1 - x_1)x_2$ is the output of the memristive device.

As the slave system, we consider the 4-D novel hyperjerk system given by

$$\begin{cases} \dot{y}_1 = y_2 \\ \dot{y}_2 = y_3 \\ \dot{y}_3 = y_4 \\ \dot{y}_4 = -y_3 - ay_4 - by_3y_4 - c\sin y_3 - \hat{\varphi} + u \end{cases} \tag{60}$$

where $\hat{\varphi} = (1 - y_1)y_2$ is the output of the memristive device and u is a backstepping control to be determined using estimates $\hat{a}(t)$, $\hat{b}(t)$ and $\hat{c}(t)$ for a, b and c, respectively.

We define the synchronization errors between the states of the master system (59) and the slave system (60) as

$$e_i = y_i - x_i, \quad (i = 1, 2, 3, 4) \tag{61}$$

Then the error dynamics is easily obtained as

$$\begin{cases} \dot{e}_1 = e_2 \\ \dot{e}_2 = e_3 \\ \dot{e}_3 = e_4 \\ \dot{e}_4 = -e_2 - e_3 - ae_4 - b(y_3y_4 - x_3x_4) - c(\sin y_3 - \sin x_3) \\ \qquad +y_1y_2 - x_1x_2 + u \end{cases} \tag{62}$$

The parameter estimation errors are defined as:

$$\begin{cases} e_a(t) = a - \hat{a}(t) \\ e_b(t) = b - \hat{b}(t) \\ e_c(t) = c - \hat{c}(t) \end{cases} \tag{63}$$

Differentiating (63) with respect to t, we obtain the following equations:

$$\begin{cases} \dot{e}_a(t) = -\dot{\hat{a}}(t) \\ \dot{e}_b(t) = -\dot{\hat{b}}(t) \\ \dot{e}_c(t) = -\dot{\hat{c}}(t) \end{cases} \tag{64}$$

Next, we shall state and prove the main result of this section.

Theorem 2 *The identical 4-D hyperjerk systems (59) and (60) with unknown parameters a, b and c are globally and exponentially synchronized by the adaptive control law*

$$\begin{aligned} u = -5e_1 - 9e_2 - 8e_3 - [4 - \hat{a}(t)]e_4 + \hat{b}(t)(y_3y_4 - x_3x_4) \\ +\hat{c}(t)(\sin y_3 - \sin x_3) - y_1y_2 + x_1x_2 - kz_4 \end{aligned} \tag{65}$$

where $k > 0$ is a gain constant,

$$z_4 = 3e_1 + 5e_2 + 3e_3 + e_4, \tag{66}$$

and the update law for the parameter estimates $\hat{a}(t), \hat{b}(t), \hat{c}(t)$ is given by

$$\begin{cases} \dot{\hat{a}}(t) = -z_4e_4 \\ \dot{\hat{b}}(t) = -z_4(y_3y_4 - x_3x_4) \\ \dot{\hat{c}}(t) = -z_4(\sin y_3 - \sin x_3) \end{cases} \tag{67}$$

Proof We prove this result via backstepping control method and Lyapunov stability theory (Khalil 2001).

First, we define a quadratic Lyapunov function

$$V_1(z_1) = \frac{1}{2}z_1^2 \tag{68}$$

where

$$z_1 = e_1 \tag{69}$$

Differentiating V_1 along the error dynamics (62), we get

$$\dot{V}_1 = z_1\dot{z}_1 = e_1e_2 = -z_1^2 + z_1(e_1 + e_2) \tag{70}$$

Now, we define

$$z_2 = e_1 + e_2 \tag{71}$$

Using (71), we can simplify the Eq. (70) as

$$\dot{V}_1 = -z_1^2 + z_1z_2 \tag{72}$$

Secondly, we define a quadratic Lyapunov function

$$V_2(z_1, z_2) = V_1(z_1) + \frac{1}{2}z_2^2 = \frac{1}{2}\left(z_1^2 + z_2^2\right) \tag{73}$$

Differentiating V_2 along the error dynamics (62), we get

$$\dot{V}_2 = -z_1^2 - z_2^2 + z_2(2e_1 + 2e_2 + e_3) \quad (74)$$

Now, we define

$$z_3 = 2e_1 + 2e_2 + e_3 \quad (75)$$

Using (75), we can simplify the Eq. (74) as

$$\dot{V}_2 = -z_1^2 - z_2^2 + z_2 z_3 \quad (76)$$

Thirdly, we define a quadratic Lyapunov function

$$V_3(z_1, z_2, x_3) = V_2(z_1, z_2) + \frac{1}{2} z_3^2 = \frac{1}{2} \left(z_1^2 + z_2^2 + z_3^2 \right) \quad (77)$$

Differentiating V_3 along the error dynamics (62), we get

$$\dot{V}_3 = -z_1^2 - z_2^2 - z_3^2 + z_3(3e_1 + 5e_2 + 3e_3 + e_4) \quad (78)$$

Now, we define

$$z_4 = 3e_1 + 5e_2 + 3e_3 + e_4 \quad (79)$$

Using (79), we can simplify the Eq. (78) as

$$\dot{V}_2 = -z_1^2 - z_2^2 - z_3^2 + z_3 z_4 \quad (80)$$

Finally, we define a quadratic Lyapunov function

$$V(z_1, z_2, z_3, z_4, e_a, e_b, e_c) = V_3(z_1, z_2, z_3) + \frac{1}{2} z_4^2 + \frac{1}{2} e_a^2 + \frac{1}{2} e_b^2 + \frac{1}{2} e_c^2 \quad (81)$$

Differentiating V along the error dynamics (62), we get

$$\dot{V} = -z_1^2 - z_2^2 - z_3^2 - z_4^2 + z_4(z_4 + z_3 + \dot{z}_4) - e_a \dot{\hat{a}} - e_b \dot{\hat{b}} - e_c \dot{\hat{c}} \quad (82)$$

Equation (82) can be written compactly as

$$\dot{V} = -z_1^2 - z_2^2 - z_3^2 - z_4^2 + z_4 S - e_a \dot{\hat{a}} - e_b \dot{\hat{b}} - e_c \dot{\hat{c}} \quad (83)$$

where

$$S = z_4 + z_3 + \dot{z}_4 = z_4 + z_3 + 3\dot{e}_1 + 5\dot{e}_2 + 3\dot{e}_3 + \dot{e}_4 \quad (84)$$

A simple calculation gives

$$S = 5e_1 + 9e_2 + 8e_3 + (4-a)e_4 - b(y_3y_4 - x_3x_4) - c(\sin y_3 - \sin x_3) \\ + y_1y_2 - x_1x_2 + u \tag{85}$$

Substituting the adaptive control law (65) into (85), we obtain

$$S = -[a - \hat{a}(t)]e_4 - [b - \hat{b}(t)](y_3y_4 - x_3x_4) - [c - \hat{c}(t)](\sin y_3 - \sin x_3) - kz_4 \tag{86}$$

Using the definitions (64), we can simplify (86) as

$$S = -e_a e_4 - e_b(y_3y_4 - x_3x_4) - e_c(\sin y_3 - \sin x_3) - kz_4 \tag{87}$$

Substituting the value of S from (87) into (83), we obtain

$$\begin{cases} \dot{V} = -z_1^2 - z_2^2 - z_3^2 - (1+k)z_4^2 + e_a\left[-z_4e_4 - \dot{\hat{a}}\right] \\ \quad + e_b\left[-z_4(y_3y_4 - x_3x_4) - \dot{\hat{b}}\right] + e_c\left[-z_4(\sin y_3 - \sin x_3) - \dot{\hat{c}}\right] \end{cases} \tag{88}$$

Substituting the update law (67) into (88), we get

$$\dot{V} = -z_1^2 - z_2^2 - z_3^2 - (1+k)z_4^2, \tag{89}$$

which is a negative semi-definite function on $\mathbf{R}^7$.

From (89), it follows that the vector $\mathbf{z}(t) = (z_1(t), z_2(t), z_3(t), z_4(t))$ and the parameter estimation error $(e_a(t), e_b(t), e_c(t))$ are globally bounded, i.e.

$$\left[z_1(t)\ z_2(t)\ z_3(t)\ z_4(t)\ e_a(t)\ e_b(t)\ e_c(t)\right] \in \mathbf{L}_\infty \tag{90}$$

Also, it follows from (89) that

$$\dot{V} \le -z_1^2 - z_2^2 - z_3^2 - z_4^2 = -\|\mathbf{z}\|^2 \tag{91}$$

That is,

$$\|\mathbf{z}\|^2 \le -\dot{V} \tag{92}$$

Integrating the inequality (92) from 0 to t, we get

$$\int_0^t |\mathbf{z}(\tau)|^2\, d\tau \le V(0) - V(t) \tag{93}$$

From (93), it follows that $\mathbf{z}(t) \in \mathbf{L}_2$.

From Eq. (62), it can be deduced that $\dot{\mathbf{z}}(t) \in \mathbf{L}_\infty$.

Thus, using Barbalat's lemma (Khalil 2001), we conclude that $\mathbf{z}(t) \to \mathbf{0}$ exponentially as $t \to \infty$ for all initial conditions $\mathbf{z}(0) \in \mathbf{R}^4$.

Hence, it is immediate that $\mathbf{e}(t) \to \mathbf{0}$ exponentially as $t \to \infty$ for all initial conditions $\mathbf{e}(0) \in \mathbf{R}^4$.

This completes the proof. ■

For the numerical simulations, the classical fourth-order Runge-Kutta method with step size $h = 10^{-8}$ is used to solve the system of differential equations (59) and (60).

The parameter values of the novel hyperjerk system are taken as in the hyperchaotic case (16), *viz.*

$$a = 0.5, \quad b = 0.4, \quad c = 0.1 \tag{94}$$

Also, as initial conditions of the master system (59), we take

$$x_1(0) = 1.7, \quad x_2(0) = -1.5, \quad x_3(0) = -0.2, \quad x_4(0) = 1.9 \tag{95}$$

As initial conditions of the slave system (60), we take

$$y_1(0) = -0.3, \quad y_2(0) = 2.4, \quad y_3(0) = 3.1, \quad y_4(0) = -1.2 \tag{96}$$

Furthermore, as initial conditions of the parameter estimates $\hat{a}(t)$, $\hat{b}(t)$ and $\hat{c}(t)$, we take

$$\hat{a}(0) = 2.1, \quad \hat{b}(0) = 4.9, \quad \hat{c}(0) = 7.5 \tag{97}$$

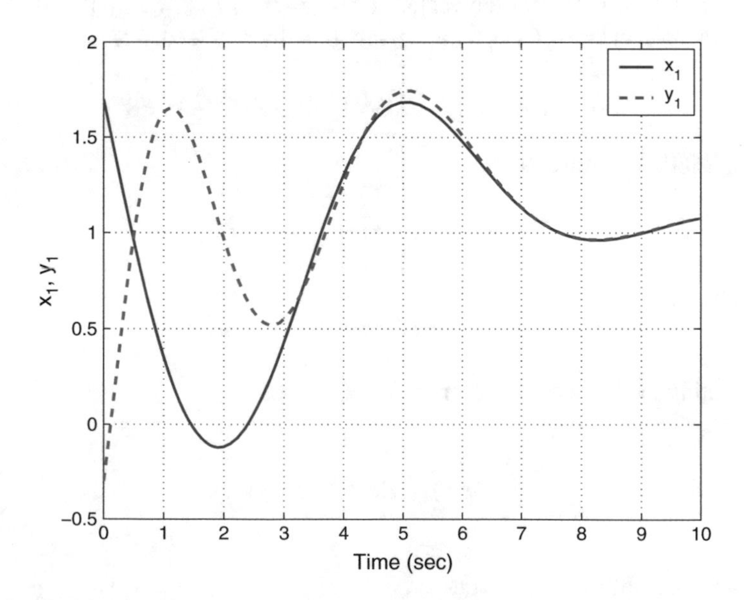

Fig. 7 Synchronization of the states x_1 and y_1 of the novel hyperjerk systems

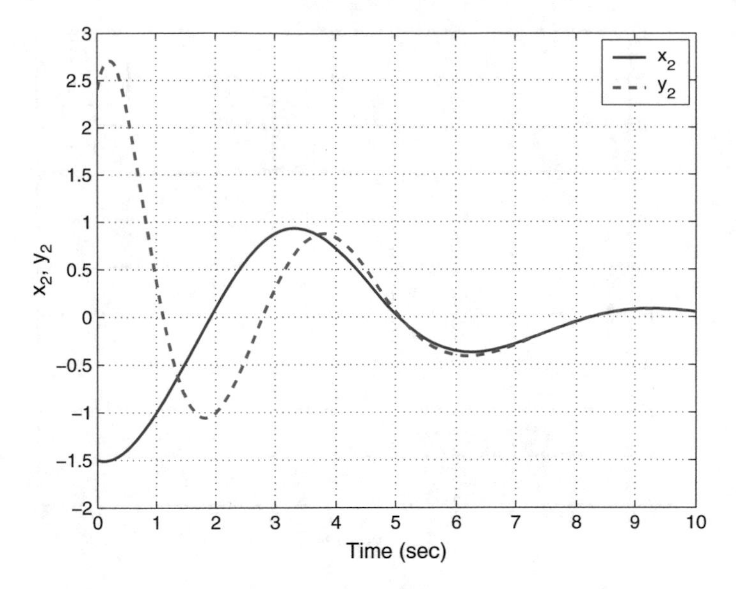

Fig. 8 Synchronization of the states x_2 and y_2 of the novel hyperjerk systems

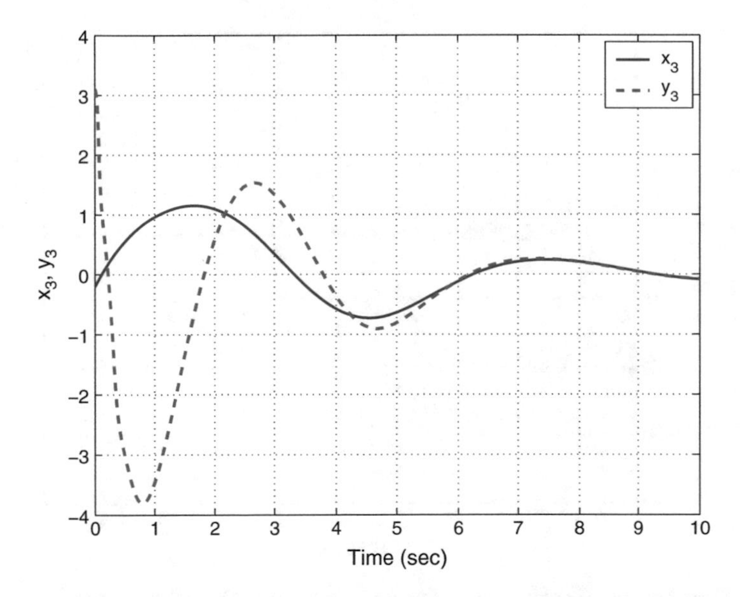

Fig. 9 Synchronization of the states x_3 and y_3 of the novel hyperjerk systems

Figures 7, 8, 9 and 10 depict the complete synchronization of the identical 4-D hyperjerk systems (59) and (60).

Figure 11 shows the time-history of the complete synchronization errors.

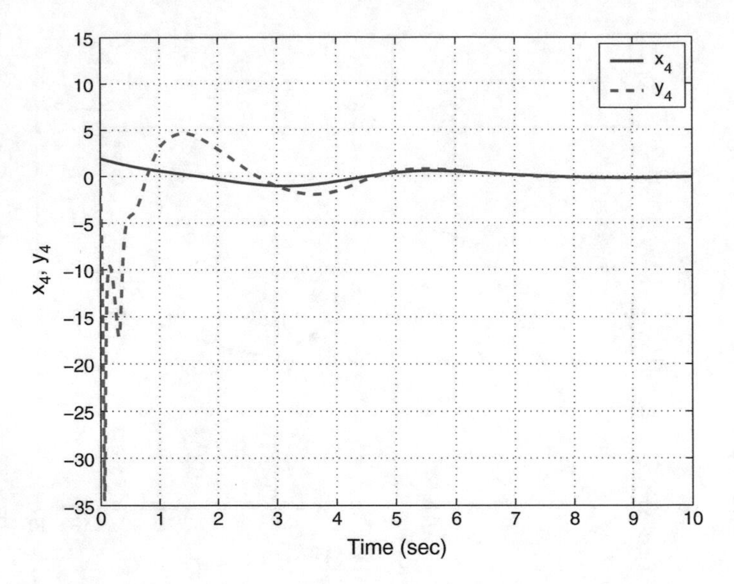

Fig. 10 Synchronization of the states x_4 and y_4 of the novel hyperjerk systems

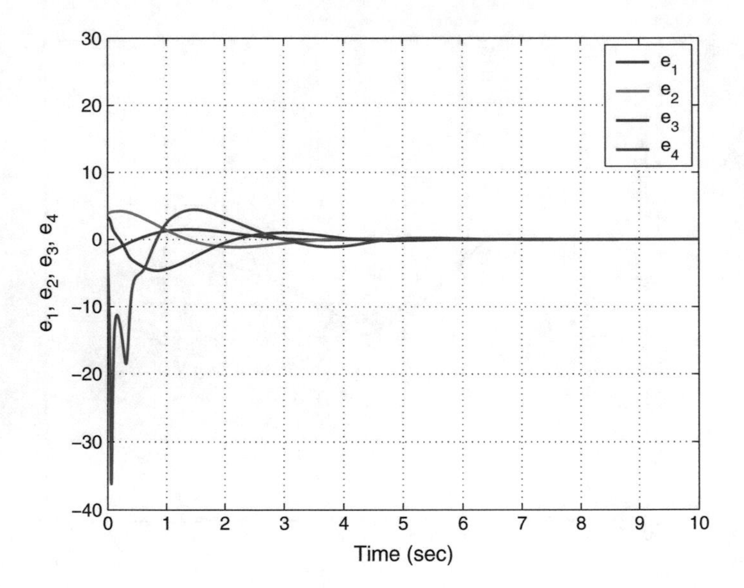

Fig. 11 Time-history of the synchronization errors of the novel hyperjerk systems

5 Conclusions

In this work, we have proposed a conservative memristor-based hyperchaotic hyperjerk system with infinite number of equilibrium points. There is great interest in the chaos literature in finding novel hyperchaotic hyperjerk systems. In this chapter, we derived a conservative hyperchaotic hyperjerk system with the help of a memristive device. In this work, we derived the Lyapunov exponents of the novel memristive hyperjerk system as $L_1 = 0.2098$, $L_2 = 0.2035$, $L_3 = 0$ and $L_4 = -0.4133$. Since there are two positive Lyapunov exponents, the memristor-based hyperjerk system is hyperchaotic. Also, the Kaplan-Yorke dimension of the memristor-based hyperchaotic system has been found as $D_{KY} = 4$. Next, we designed adaptive control and synchronization schemes for the memristor-based hyperjerk system with unknown parameters via backstepping control method. The main adaptive control and synchronization results were established using Lyapunov stability theory. MATLAB simulations have been shown to illustrate all the main results of this work. Also, it is well-known that hyperchaotic system, which is characterized by more than one positive Lyapunov exponent, exhibits a higher level of complexity than a conventional chaotic system. Hence, we can apply this memristor-based hyperchaotic system in practical applications like encryption, cryptosystems, neural networks and secure communications.

References

Abdurrahman, A., Jiang, H., & Teng, Z. (2015). Finite-time synchronization for memristor-based neural networks with time-varying delays. *Neural Networks*, *69*, 20–28.

Adhikari, S. P., Yang, C., Kim, H., & Chua, L. O. (2012). Memristor bridge synapse-based neural network and its learning. *IEEE Transactions on Neural Networks and Learning Systems*, *23*, 1426–1435.

Adhikari, S. P., Sad, M. P., Kim, H., & Chua, L. O. (2013). Three fingerprints of memristor. *IEEE Transactions on Circuits and Systems I: Regular Papers*, *60*(11), 3008–3021.

Albuquerque, H. A., Rubinger, R. M., & Rech, P. C. (2008). Self-similar structures in a 2D parameter-space of an inductorless Chua's circuit. *Physics Letters A*, *372*, 4793–4798.

Arneodo, A., Coullet, P., & Tresser, C. (1981). Possible new strange attractors with spiral structure. *Commonications in Mathematical Physics*, *79*(4), 573–576.

Azar, A. T., & Vaidyanathan, S. (2015). *Chaos Modeling and Control Systems Design* (Vol. 581). Germany: Springer.

Azar, A. T., & Vaidyanathan, S. (2016). *Advances in Chaos Theory and Intelligent Control*. Berlin, Germany: Springer.

Azar, A. T., Vaidyanathan, S., & Ouannas, A. (2017). *Fractional Order Control and Synchronization of Chaotic Systems*. Berlin, Germany: Springer.

Bao, B., Zou, X., Liu, Z., & Hu, F. (2013). Generalized memory element and chaotic memory system. *International Journal of Bifurcation and Chaos*, *23*(1350), 135.

Cai, G., & Tan, Z. (2007). Chaos synchronization of a new chaotic system via nonlinear control. *Journal of Uncertain Systems*, *1*(3), 235–240.

Carroll, T. L., & Pecora, L. M. (1991). Synchronizing chaotic circuits. *IEEE Transactions on Circuits and Systems*, *38*(4), 453–456.

Chen, G., & Ueta, T. (1999). Yet another chaotic attractor. *International Journal of Bifurcation and Chaos*, *9*(7), 1465–1466.
Chen, W. H., Wei, D., & Lu, X. (2014). Global exponential synchronization of nonlinear time-delay Lur'e systems via delayed impulsive control. *Communications in Nonlinear Science and Numerical Simulation*, *19*(9), 3298–3312.
Cheng, C. J., & Cheng, C. B. (2013). An asymmetric image cryptosystem based on the adaptive synchronization of an uncertain unified chaotic system and a cellular neural network. *Communications in Nonlinear Science and Numerical Simulation*, *18*(10), 2825–2837.
Chlouverakis, K. E., & Sprott, J. C. (2006). Chaotic hyperjerk systems. *Chaos, Solitons and Fractals*, *28*, 739–746.
Chua, L. O. (1971). Memristor-the missing circuit element. *IEEE Transactions on Circuit Theory*, *18*(5), 507–519.
Chua, L. O. (1994). Chua's circuit: An overview ten years later. *Journal of Circuits, Systems and Computers*, *04*, 117–159.
Chua, L. O., & Yang, L. (1988a). Cellular neural networks: Applications. *IEEE Transactions on Circuits and Systems*, *35*, 1273–1290.
Chua, L. O., & Yang, L. (1988b). Cellular neural networks: Theory. *IEEE Transactions on Circuits and Systems*, *35*, 1257–1272.
Fortuna, L., Frasca, M., & Xibilia, M. G. (2009). *Chua's circuit implementations: yesterday, today and tomorrow*. Singapore: World Scientific.
Frederickson, P., Kaplan, J. L., Yorke, E. D., & York, J. A. (1983). The Lyapunov dimension of strange attractors. *Journal of Differential Equations*, *49*, 185–207.
Gan, Q., & Liang, Y. (2012). Synchronization of chaotic neural networks with time delay in the leakage term and parametric uncertainties based on sampled-data control. *Journal of the Franklin Institute*, *349*(6), 1955–1971.
Hénon, M., & Heiles, C. (1964). The applicability of the third integral of motion: Some numerical experiments. *Astrophysical Journal*, *69*, 73–79.
Hoover, W. G. (1995). Remark on 'Some simple chaotic flows'. *Physical Review E*, *51*, 759–760.
Itoh, M., & Chua, L. O. (2008). Memristor oscillators. *International Journal of Bifurcation and Chaos*, *18*(11), 3183–3206.
Jiang, G. P., Zheng, W. X., & Chen, G. (2004). Global chaos synchronization with channel time-delay. *Chaos, Solitons & Fractals*, *20*(2), 267–275.
Joglekar, Y. N., & Wolf, S. J. (2009). The elusive memristor: Properties of basic electrical circuits. *European Journal of Physics*, *30*(4), 661–675.
Karthikeyan, R., & Sundarapandian, V. (2014). Hybrid chaos synchronization of four-scroll systems via active control. *Journal of Electrical Engineering*, *65*(2), 97–103.
Khalil, H. K. (2001). *Nonlinear systems* (3rd ed.). New Jersey: Prentice Hall.
Li, D. (2008). A three-scroll chaotic attractor. *Physics Letters A*, *372*(4), 387–393.
Li, G. H., Zhou, S. P., & Yang, K. (2007). Controlling chaos in Colpitts oscillator. *Chaos, Solitons and Fractals*, *33*, 582–587.
Li, H., Wang, L., Duan, S. (2014). A memristor-mased scroll chaotic system—design, analysis and circuit implementation. *International Journal of Bifurcation and Chaos*, *24*(07),1450,099
Li, Q., Hu, S., Tang, S., & Zeng, G. (2013). Hyperchaos and horseshoe in a 4D memristive system with a line of equilibria and its implementation. *International Journal of Circuit Theory and Applications*, *42*(11), 1172–1188.
Liu, L., Wu, X., & Hu, H. (2004). Estimating system parameters of Chua's circuit from synchronizing signal. *Physics Letters A*, *324*(1), 36–41.
Lorenz, E. N. (1963). Deterministic periodic flow. *Journal of the Atmospheric Sciences*, *20*(2), 130–141.
Lü, J., & Chen, G. (2002). A new chaotic attractor coined. *International Journal of Bifurcation and Chaos*, *12*(3), 659–661.
Matsumoto, T. (1984). A chaotic attractor from Chua's circuit. *IEEE Transactions on Circuits and Systems*, *31*, 1055–1058.

Muthuswamy, B., & Chua, L. O. (2010). Simplest chaotic circuit. *International Journal of Bifurcation and Chaos*, *20*(5), 1567–1580.
Muthuswamy, B., & Kokate, P. (2009). Memristor based chaotic circuits. *IETE Technical Review*, *26*(6), 417–429.
Pecora, L. M., & Carroll, T. L. (1990). Synchronization in chaotic systems. *Physical Review Letters*, *64*(8), 821–824.
Pehlivan, I., Moroz, I. M., & Vaidyanathan, S. (2014). Analysis, synchronization and circuit design of a novel butterfly attractor. *Journal of Sound and Vibration*, *333*(20), 5077–5096.
Pershin, Y. V., Fontaine, S. L., & Ventra, M. D. (2009). Memristive model of amoeba learning. *Physical Review E*, *80*(021), 926.
Pham, V. T., Volos, C., Jafari, S., & Wang, X. (2014). Generating a novel hyperchaotic system out of equilibrium. *Optoelectronics and Advanced Materials-Rapid Communications*, *8*, 535–539.
Pham, V. T., Volos, C. K., Vaidyanathan, S., Le, T. P., & Vu, V. Y. (2015). A memristor-based hyperchaotic system with hidden attractors: Dynamics, synchronization and circuital emulating. *Journal of Engineering Science and Technology Review*, *8*(2), 205–214.
Rasappan, S., & Vaidyanathan, S. (2012a). Global chaos synchronization of WINDMI and Coullet chaotic systems by backstepping control. *Far East Journal of Mathematical Sciences*, *67*(2), 265–287.
Rasappan, S., & Vaidyanathan, S. (2012b). Hybrid synchronization of n-scroll Chua and Lur'e chaotic systems via backstepping control with novel feedback. *Archives of Control Sciences*, *22*(3), 343–365.
Rasappan, S., & Vaidyanathan, S. (2012c). Synchronization of hyperchaotic Liu system via backstepping control with recursive feedback. *Communications in Computer and Information Science*, *305*, 212–221.
Rasappan, S., & Vaidyanathan, S. (2013). Hybrid synchronization of *n*-scroll chaotic Chua circuits using adaptive backstepping control design with recursive feedback. *Malaysian Journal of Mathematical Sciences*, *7*(2), 219–246.
Rasappan, S., & Vaidyanathan, S. (2014). Global chaos synchronization of WINDMI and Coullet chaotic systems using adaptive backstepping control design. *Kyungpook Mathematical Journal*, *54*(1), 293–320.
Rössler, O. E. (1976). An equation for continuous chaos. *Physics Letters A*, *57*(5), 397–398.
Sampath, S., & Vaidyanathan, S. (2016). Hybrid synchronization of identical chaotic systems via novel sliding control method with application to Sampath four-scroll chaotic system. *International Journal of Control Theory and Applications*, *9*(1), 221–235.
Sampath, S., Vaidyanathan, S., Volos, C. K., & Pham, V. T. (2015). An eight-term novel four-scroll chaotic System with cubic nonlinearity and its circuit simulation. *Journal of Engineering Science and Technology Review*, *8*(2), 1–6.
Sarasu, P., & Sundarapandian, V. (2011a). Active controller design for the generalized projective synchronization of four-scroll chaotic systems. *International Journal of Systems Signal Control and Engineering Application*, *4*(2), 26–33.
Sarasu, P., & Sundarapandian, V. (2011b). The generalized projective synchronization of hyperchaotic Lorenz and hyperchaotic Qi systems via active control. *International Journal of Soft Computing*, *6*(5), 216–223.
Sarasu, P., & Sundarapandian, V. (2012a). Adaptive controller design for the generalized projective synchronization of 4-scroll systems. *International Journal of Systems Signal Control and Engineering Application*, *5*(2), 21–30.
Sarasu, P., & Sundarapandian, V. (2012b). Generalized projective synchronization of three-scroll chaotic systems via adaptive control. *European Journal of Scientific Research*, *72*(4), 504–522.
Sarasu, P., & Sundarapandian, V. (2012c). Generalized projective synchronization of two-scroll systems via adaptive control. *International Journal of Soft Computing*, *7*(4), 146–156.
Shang, Y., Fei, W., & Yu, H. (2012). Analysis and modeling of internal state variables for dynamic effects of nonvolatile memory devices. *IEEE Transactions on Circuits and Systems I: Regular Papers*, *59*, 1906–1918.

Shin, S., Kim, K., & Kang, S. M. (2011). Memristor applications for programmable analog ICs. *IEEE Transactions on Nanotechnology*, *410*, 266–274.
Sprott, J. C. (1994). Some simple chaotic flows. *Physical Review E*, *50*(2), 647–650.
Sprott, J. C. (1997). Some simple chaotic jerk functions. *American Journal of Physics*, *65*, 537–543.
Strukov, D., Snider, G., Stewart, G., & Williams, R. (2008). The missing memristor found. *Nature*, *453*, 80–83.
Sundarapandian, V. (2010). Output regulation of the Lorenz attractor. *Far East Journal of Mathematical Sciences*, *42*(2), 289–299.
Sundarapandian, V. (2011). Output regulation of the Arneodo-Coullet chaotic system. *Communications in Computer and Information Science*, *133*, 98–107.
Sundarapandian, V. (2013). Analysis and anti-synchronization of a novel chaotic system via active and adaptive controllers. *Journal of Engineering Science and Technology Review*, *6*(4), 45–52.
Sundarapandian, V., & Karthikeyan, R. (2011a). Anti-synchronization of hyperchaotic Lorenz and hyperchaotic Chen systems by adaptive control. *International Journal of Systmes Signal Control and Engineering Application*, *4*(2), 18–25.
Sundarapandian, V., & Karthikeyan, R. (2011b). Anti-synchronization of Lü and Pan chaotic systems by adaptive nonlinear control. *European Journal of Scientific Research*, *64*(1), 94–106.
Sundarapandian, V., & Karthikeyan, R. (2012a). Adaptive anti-synchronization of uncertain Tigan and Li systems. *Journal of Engineering and Applied Sciences*, *7*(1), 45–52.
Sundarapandian, V., & Karthikeyan, R. (2012b). Hybrid synchronization of hyperchaotic Lorenz and hyperchaotic Chen systems via active control. *Journal of Engineering and Applied Sciences*, *7*(3), 254–264.
Sundarapandian, V., & Pehlivan, I. (2012). Analysis, control, synchronization and circuit design of a novel chaotic system. *Mathematical and Computer Modelling*, *55*(7–8), 1904–1915.
Sundarapandian, V., & Sivaperumal, S. (2011). Sliding controller design of hybrid synchronization of four-wing chaotic systems. *International Journal of Soft Computing*, *6*(5), 224–231.
Suresh, R., & Sundarapandian, V. (2013). Global chaos synchronization of a family of *n*-scroll hyperchaotic Chua circuits using backstepping control with recursive feedback. *Far East Journal of Mathematical Sciences*, *73*(1), 73–95.
Tang, F., & Wang, L. (2005). An adaptive active control for the modified Chua's circuit. *Physics Letters A*, *346*, 342–346.
Tetzlaff, R. (2014). *Memristors and memristive systems*. Berlin: Springer.
Tigan, G., & Opris, D. (2008). Analysis of a 3D chaotic system. *Chaos, Solitons and Fractals*, *36*, 1315–1319.
Vaidyanathan, S. (2011a). Hybrid chaos synchronization of Liu and Lü systems by active nonlinear control. *Communications in Computer and Information Science*, *204*, 1–10.
Vaidyanathan, S. (2011b). Output regulation of the unified chaotic system. *Communications in Computer and Information Science*, *204*, 84–93.
Vaidyanathan, S. (2012a). Analysis and synchronization of the hyperchaotic Yujun systems via sliding mode control. *Advances in Intelligent Systems and Computing*, *176*, 329–337.
Vaidyanathan, S. (2012b). Anti-synchronization of Sprott-L and Sprott-M chaotic systems via adaptive control. *International Journal of Control Theory and Applications*, *5*(1), 41–59.
Vaidyanathan, S. (2012c). Global chaos control of hyperchaotic Liu system via sliding control method. *International Journal of Control Theory and Applications*, *5*(2), 117–123.
Vaidyanathan, S. (2012d). Output regulation of the Liu chaotic system. *Applied Mechanics and Materials*, *110–116*, 3982–3989.
Vaidyanathan, S. (2012e). Sliding mode control based global chaos control of Liu-Liu-Liu-Su chaotic system. *International Journal of Control Theory and Applications*, *5*(1), 15–20.
Vaidyanathan, S. (2013a). A new six-term 3-D chaotic system with an exponential nonlinearity. *Far East Journal of Mathematical Sciences*, *79*(1), 135–143.
Vaidyanathan, S. (2013b). Analysis and adaptive synchronization of two novel chaotic systems with hyperbolic sinusoidal and cosinusoidal nonlinearity and unknown parameters. *Journal of Engineering Science and Technology Review*, *6*(4), 53–65.

Vaidyanathan, S. (2013c). Analysis, control and synchronization of hyperchaotic Zhou system via adaptive control. *Advances in Intelligent Systems and Computing*, *177*, 1–10.
Vaidyanathan, S. (2014a). A new eight-term 3-D polynomial chaotic system with three quadratic nonlinearities. *Far East Journal of Mathematical Sciences*, *84*(2), 219–226.
Vaidyanathan, S. (2014b). Analysis and adaptive synchronization of eight-term 3-D polynomial chaotic systems with three quadratic nonlinearities. *European Physical Journal: Special Topics*, *223*(8), 1519–1529.
Vaidyanathan, S. (2014c). Analysis, control and synchronisation of a six-term novel chaotic system with three quadratic nonlinearities. *International Journal of Modelling, Identification and Control*, *22*(1), 41–53.
Vaidyanathan, S. (2014d). Generalized projective synchronisation of novel 3-D chaotic systems with an exponential non-linearity via active and adaptive control. *International Journal of Modelling, Identification and Control*, *22*(3), 207–217.
Vaidyanathan, S. (2014e). Global chaos synchronization of identical Li-Wu chaotic systems via sliding mode control. *International Journal of Modelling, Identification and Control*, *22*(2), 170–177.
Vaidyanathan, S. (2015a). 3-cells Cellular Neural Network (CNN) attractor and its adaptive biological control. *International Journal of PharmTech Research*, *8*(4), 632–640.
Vaidyanathan, S. (2015b). A 3-D novel highly chaotic system with four quadratic nonlinearities, its adaptive control and anti-synchronization with unknown parameters. *Journal of Engineering Science and Technology Review*, *8*(2), 106–115.
Vaidyanathan, S. (2015c). A novel chemical chaotic reactor system and its adaptive control. *International Journal of ChemTech Research*, *8*(7), 146–158.
Vaidyanathan, S. (2015d). Adaptive backstepping control of enzymes-substrates system with ferroelectric behaviour in brain waves. *International Journal of PharmTech Research*, *8*(2), 256–261.
Vaidyanathan, S. (2015e). Adaptive biological control of generalized Lotka-Volterra three-species biological system. *International Journal of PharmTech Research*, *8*(4), 622–631.
Vaidyanathan, S. (2015f). Adaptive chaotic synchronization of enzymes-substrates system with ferroelectric behaviour in brain waves. *International Journal of PharmTech Research*, *8*(5), 964–973.
Vaidyanathan, S. (2015g). Adaptive control of a chemical chaotic reactor. *International Journal of PharmTech Research*, *8*(3), 377–382.
Vaidyanathan, S. (2015h). Adaptive control of the FitzHugh-Nagumo chaotic neuron model. *International Journal of PharmTech Research*, *8*(6), 117–127.
Vaidyanathan, S. (2015i). Adaptive synchronization of chemical chaotic reactors. *International Journal of ChemTech Research*, *8*(2), 612–621.
Vaidyanathan, S. (2015j). Adaptive synchronization of generalized Lotka-Volterra three-species biological systems. *International Journal of PharmTech Research*, *8*(5), 928–937.
Vaidyanathan, S. (2015k). Adaptive synchronization of novel 3-D chemical chaotic reactor systems. *International Journal of ChemTech Research*, *8*(7), 159–171.
Vaidyanathan, S. (2015l). Adaptive synchronization of the identical FitzHugh-Nagumo chaotic neuron models. *International Journal of PharmTech Research*, *8*(6), 167–177.
Vaidyanathan, S. (2015m). Analysis, control and synchronization of a 3-D novel jerk chaotic system with two quadratic nonlinearities. *Kyungpook Mathematical Journal*, *55*, 563–586.
Vaidyanathan, S. (2015n). Analysis, properties and control of an eight-term 3-D chaotic system with an exponential nonlinearity. *International Journal of Modelling, Identification and Control*, *23*(2), 164–172.
Vaidyanathan, S. (2015o). Anti-synchronization of brusselator chemical reaction systems via adaptive control. *International Journal of ChemTech Research*, *8*(6), 759–768.
Vaidyanathan, S. (2015p). Chaos in neurons and adaptive control of Birkhoff-Shaw strange chaotic attractor. *International Journal of PharmTech Research*, *8*(5), 956–963.
Vaidyanathan, S. (2015q). Chaos in neurons and synchronization of Birkhoff-Shaw strange chaotic attractors via adaptive control. *International Journal of PharmTech Research*, *8*(6), 1–11.

Vaidyanathan, S. (2015r). Coleman-Gomatam logarithmic competitive biology models and their ecological monitoring. *International Journal of PharmTech Research*, *8*(6), 94–105.
Vaidyanathan, S. (2015s). Dynamics and control of brusselator chemical reaction. *International Journal of ChemTech Research*, *8*(6), 740–749.
Vaidyanathan, S. (2015t). Dynamics and control of tokamak system with symmetric and magnetically confined plasma. *International Journal of ChemTech Research*, *8*(6), 795–803.
Vaidyanathan, S. (2015u). Global chaos synchronization of chemical chaotic reactors via novel sliding mode control method. *International Journal of ChemTech Research*, *8*(7), 209–221.
Vaidyanathan, S. (2015v). Global chaos synchronization of the forced Van der Pol chaotic oscillators via adaptive control method. *International Journal of PharmTech Research*, *8*(6), 156–166.
Vaidyanathan, S. (2015w). Global chaos synchronization of the Lotka-Volterra biological systems with four competitive species via active control. *International Journal of PharmTech Research*, *8*(6), 206–217.
Vaidyanathan, S. (2015x). Lotka-Volterra population biology models with negative feedback and their ecological monitoring. *International Journal of PharmTech Research*, *8*(5), 974–981.
Vaidyanathan, S. (2015y). Lotka-Volterra two species competitive biology models and their ecological monitoring. *International Journal of PharmTech Research*, *8*(6), 32–44.
Vaidyanathan, S. (2015z). Output regulation of the forced Van der Pol chaotic oscillator via adaptive control method. *International Journal of PharmTech Research*, *8*(6), 106–116.
Vaidyanathan, S. (2016a). A novel 3-D conservative chaotic system with a sinusoidal nonlinearity and its adaptive control. *International Journal of Control Theory and Applications*, *9*(1), 115–132.
Vaidyanathan, S. (2016b). A novel hyperchaotic hyperjerk system with two nonlinearities, its analysis, adaptive control and synchronization via backstepping control method. *International Journal of Control Theory and Applications*, *9*(1), 257–278.
Vaidyanathan, S. (2016c). An eleven-term novel 4-D hyperchaotic system with three quadratic nonlinearities, analysis, control and synchronization via adaptive control method. *International Journal of Control Theory and Applications*, *9*(1), 21–43.
Vaidyanathan, S. (2016d). Analysis, adaptive control and synchronization of a novel 4-D hyperchaotic hyperjerk system via backstepping control method. *Archives of Control Sciences*, *26*(3), 311–338.
Vaidyanathan, S. (2016e). Anti-synchronization of 3-cells Cellular Neural Network attractors via integral sliding mode control. *International Journal of PharmTech Research*, *9*(1), 193–205.
Vaidyanathan, S. (2016f). Anti-synchronization of Duffing double-well chaotic oscillators via integral sliding mode control. *International Journal of ChemTech Research*, *9*(2), 297–304.
Vaidyanathan, S. (2016g). Anti-synchronization of enzymes-substrates biological systems via adaptive backstepping control. *International Journal of PharmTech Research*, *9*(2), 193–205.
Vaidyanathan, S. (2016h). Global chaos control of the FitzHugh-Nagumo chaotic neuron model via integral sliding mode control. *International Journal of PharmTech Research*, *9*(4), 413–425.
Vaidyanathan, S. (2016i). Global chaos control of the generalized Lotka-Volterra three-species system via integral sliding mode control. *International Journal of PharmTech Research*, *9*(4), 399–412.
Vaidyanathan, S. (2016j). Mathematical analysis, adaptive control and synchronization of a ten-term novel three-scroll chaotic system with four quadratic nonlinearities. *International Journal of Control Theory and Applications*, *9*(1), 1–20.
Vaidyanathan, S. (2016k). Sliding mode controller design for the global stabilization of chaotic systems and its application to Vaidyanathan jerk system. *Studies in Computational Intelligence*, *636*, 537–552.
Vaidyanathan, S., & Azar, A. T. (2015a). Analysis and control of a 4-D novel hyperchaotic system. In A. T. Azar & S. Vaidyanathan (Eds.), *Chaos modeling and control systems design*, Studies in Computational Intelligence, (Vol. 581, pp. 19–38). Germany: Springer.

Vaidyanathan, S., & Azar, A. T. (2015b). Analysis, control and synchronization of a nine-term 3-D novel chaotic system. In A. T. Azar & S. Vaidyanathan (Eds.), *Chaos modelling and control systems design*, Studies in Computational Intelligence, (Vol. 581, pp. 19–38). Germany: Springer.
Vaidyanathan, S., & Azar, A. T. (2015c). Anti-synchronization of identical chaotic systems using sliding mode control and an application to Vaidhyanathan-Madhavan chaotic systems. *Studies in Computational Intelligence*, *576*, 527–547.
Vaidyanathan, S., & Azar, A. T. (2015d). Hybrid synchronization of identical chaotic systems using sliding mode control and an application to Vaidhyanathan chaotic systems. *Studies in Computational Intelligence*, *576*, 549–569.
Vaidyanathan, S., & Boulkroune, A. (2016). A novel hyperchaotic system with two quadratic nonlinearities, its analysis and synchronization via integral sliding mode control. *International Journal of Control Theory and Applications*, *9*(1), 321–337.
Vaidyanathan, S., & Madhavan, K. (2013). Analysis, adaptive control and synchronization of a seven-term novel 3-D chaotic system. *International Journal of Control Theory and Applications*, *6*(2), 121–137.
Vaidyanathan, S., & Pakiriswamy, S. (2013). Generalized projective synchronization of six-term Sundarapandian chaotic systems by adaptive control. *International Journal of Control Theory and Applications*, *6*(2), 153–163.
Vaidyanathan, S., & Pakiriswamy, S. (2015). A 3-D novel conservative chaotic System and its generalized projective synchronization via adaptive control. *Journal of Engineering Science and Technology Review*, *8*(2), 52–60.
Vaidyanathan, S., & Pakiriswamy, S. (2016). A five-term 3-D novel conservative chaotic system and its generalized projective synchronization via adaptive control method. *International Journal of Control Theory and Applications*, *9*(1), 61–78.
Vaidyanathan, S., & Rajagopal, K. (2011a). Anti-synchronization of Li and T chaotic systems by active nonlinear control. *Communications in Computer and Information Science*, *198*, 175–184.
Vaidyanathan, S., & Rajagopal, K. (2011b). Global chaos synchronization of hyperchaotic Pang and Wang systems by active nonlinear control. *Communications in Computer and Information Science*, *204*, 84–93.
Vaidyanathan, S., & Rajagopal, K. (2011c). Global chaos synchronization of Lü and Pan systems by adaptive nonlinear control. *Communications in Computer and Information Science*, *205*, 193–202.
Vaidyanathan, S., & Rajagopal, K. (2012). Global chaos synchronization of hyperchaotic Pang and hyperchaotic Wang systems via adaptive control. *International Journal of Soft Computing*, *7*(1), 28–37.
Vaidyanathan, S., & Rajagopal, K. (2016). Analysis, control, synchronization and LabVIEW implementation of a seven-term novel chaotic system. *International Journal of Control Theory and Applications*, *9*(1), 151–174.
Vaidyanathan, S., & Rasappan, S. (2011). Global chaos synchronization of hyperchaotic Bao and Xu systems by active nonlinear control. *Communications in Computer and Information Science*, *198*, 10–17.
Vaidyanathan, S., & Rasappan, S. (2014). Global chaos synchronization of *n*-scroll Chua circuit and Lur'e system using backstepping control design with recursive feedback. *Arabian Journal for Science and Engineering*, *39*(4), 3351–3364.
Vaidyanathan, S., & Sampath, S. (2011). Global chaos synchronization of hyperchaotic Lorenz systems by sliding mode control. *Communications in Computer and Information Science*, *205*, 156–164.
Vaidyanathan, S., & Sampath, S. (2012). Anti-synchronization of four-wing chaotic systems via sliding mode control. *International Journal of Automation and Computing*, *9*(3), 274–279.
Vaidyanathan, S., & Volos, C. (2015). Analysis and adaptive control of a novel 3-D conservative no-equilibrium chaotic system. *Archives of Control Sciences*, *25*(3), 333–353.
Vaidyanathan, S., & Volos, C. (2016a). *Advances and applications in chaotic systems*. Berlin: Springer.

Vaidyanathan, S., & Volos, C. (2016b). *Advances and applications in nonlinear control systems*. Berlin: Springer.

Vaidyanathan, S., Volos, C., & Pham, V. T. (2014a). Hyperchaos, adaptive control and synchronization of a novel 5-D hyperchaotic system with three positive Lyapunov exponents and its SPICE implementation. *Archies of Control Sciences*, *24*(4), 409–446.

Vaidyanathan, S., Volos, C., & Pham, V. T. (2014b). Hyperchaos, adaptive control and synchronization of a novel 5-D hyperchaotic system with three positive Lyapunov exponents and its SPICE implementation. *Archives of Control Sciences*, *24*(4), 409–446.

Vaidyanathan, S., Volos, C., Pham, V. T., Madhavan, K., & Idowu, B. A. (2014c). Adaptive backstepping control, synchronization and circuit simulation of a 3-D novel jerk chaotic system with two hyperbolic sinusoidal nonlinearities. *Archives of Control Sciences*, *24*(3), 375–403.

Vaidyanathan, S., Idowu, B. A., & Azar, A. T. (2015a). Backstepping controller design for the global chaos synchronization of Sprott's jerk systems. *Studies in Computational Intelligence*, *581*, 39–58.

Vaidyanathan, S., Rajagopal, K., Volos, C. K., Kyprianidis, I. M., & Stouboulos, I. N. (2015b). Analysis, adaptive control and synchronization of a seven-term novel 3-D chaotic system with three quadratic nonlinearities and its digital implementation in LabVIEW. *Journal of Engineering Science and Technology Review*, *8*(2), 130–141.

Vaidyanathan, S., Volos, C., Pham, V. T., & Madhavan, K. (2015c). Analysis, adaptive control and synchronization of a novel 4-D hyperchaotic hyperjerk system and its SPICE implementation. *Archives of Control Sciences*, *25*(1), 5–28.

Vaidyanathan, S., Volos, C. K., Kyprianidis, I. M., Stouboulos, I. N., & Pham, V. T. (2015d). Analysis, adaptive control and anti-synchronization of a six-term novel jerk chaotic system with two exponential nonlinearities and its circuit simulation. *Journal of Engineering Science and Technology Review*, *8*(2), 24–36.

Vaidyanathan, S., Volos, C. K., & Madhavan, K. (2015e). Analysis, control, synchronization and SPICE implementation of a novel 4-D hyperchaotic Rikitake dynamo System without equilibrium. *Journal of Engineering Science and Technology Review*, *8*(2), 232–244.

Vaidyanathan, S., Volos, C. K., & Pham, V. T. (2015f). Analysis, adaptive control and adaptive synchronization of a nine-term novel 3-D chaotic system with four quadratic nonlinearities and its circuit simulation. *Journal of Engineering Science and Technology Review*, *8*(2), 181–191.

Vaidyanathan, S., Volos, C. K., & Pham, V. T. (2015g). Global chaos control of a novel nine-term chaotic system via sliding mode control. In A. T. Azar & Q. Zhu (Eds.), *Advances and applications in sliding mode control systems*, Studies in Computational Intelligence, (Vol. 576, pp. 571–590). Germany: Springer.

Vaidyanathan, S., Volos, C. K., Pham, V. T., & Madhavan, K. (2015h). Analysis, adaptive control and synchronization of a novel 4-D hyperchaotic hyperjerk system and its SPICE implementation. *Archives of Control Sciences*, *25*(1), 135–158.

Vaidyanathan, S., Volos, C. K., Rajagopal, K., Kyprianidis, I. M., & Stouboulos, I. N. (2015i). Adaptive backstepping controller design for the anti-synchronization of identical WINDMI chaotic systems with unknown parameters and its SPICE implementation. *Journal of Engineering Science and Technology Review*, *8*(2), 74–82.

Vaidyanathan, S., Madhavan, K., & Idowu, B. A. (2016). Backstepping control design for the adaptive stabilization and synchronization of the Pandey jerk chaotic system with unknown parameters. *International Journal of Control Theory and Applications*, *9*(1), 299–319.

Volos, C. K., Kyprianidis, I. M., Stouboulos, I. N., Tlelo-Cuautle, E., & Vaidyanathan, S. (2015). Memristor: A new concept in synchronization of coupled neuromorphic circuits. *Journal of Engineering Science and Technology Review*, *8*(2), 157–173.

Wang, L., Zhang, C., Chen, L., Lai, J., & Tong, J. (2012a). A novel memristor-based rSRAM structure for multiple-bit upsets immunity. *IEICE Electronics Express*, *9*, 861–867.

Wang, X., & Ge, C. (2008). Controlling and tracking of Newton-Leipnik system via backstepping design. *International Journal of Nonlinear Science*, *5*(2), 133–139.

Wang, X., Xu, B., & Luo, C. (2012b). An asynchronous communication system based on the hyperchaotic system of 6th-order cellular neural network. *Optics Communications*, *285*(24), 5401–5405.

Wei, Z., & Yang, Q. (2010). Anti-control of Hopf bifurcation in the new chaotic system with two stable node-foci. *Applied Mathematics and Computation*, *217*(1), 422–429.

Xiao, X., Zhou, L., & Zhang, Z. (2014). Synchronization of chaotic Lur'e systems with quantized sampled-data controller. *Communications in Nonlinear Science and Numerical Simulation*, *19*(6), 2039–2047.

Yang, J. J., Strukov, D. B., & Stewart, D. R. (2013). Memristive devices for computing. *Nature Nanotechnology*, *8*, 13–24.

Zhou, W., Xu, Y., Lu, H., & Pan, L. (2008). On dynamics analysis of a new chaotic attractor. *Physics Letters A*, *372*(36), 5773–5777.

Zhu, C., Liu, Y., & Guo, Y. (2010). Theoretic and numerical study of a new chaotic system. *Intelligent Information Management*, *2*, 104–109.

Logic Synthesis for Majority Based In-Memory Computing

Saeideh Shirinzadeh, Mathias Soeken, Pierre-Emmanuel Gaillardon and Rolf Drechsler

Abstract The resistive switching property exhibited by many emerging memory technologies enables the execution of logic operations directly with memory arrays. This opens new horizons to a modern era of computer architectures beyond the traditional Von Neumann architectures which have separated memory and computing units. In this chapter, the memristive behavior of RRAM is abstracted as a majority based logic operation for efficient synthesis of logic-in-memory circuits and systems. A majority based *Programmable Logic-in-Memory* (PLiM) architecture is also introduced and compiled addressing the latency and area issues.

Keywords Majority-inverter graph · In-memory computing · Logic synthesis · PLiM architecture

1 Introduction

Resistive Random Access Memories (RRAMs) have gained high research attention to the design and synthesis of in-memory computing circuits and systems. The majority of approaches proposed so far with this respect, exploit *Material Implication* (IMP) operated by RRAMs to synthesize Boolean functions. However, the high number of

S. Shirinzadeh (✉) · R. Drechsler
Group of Computer Architecture, University of Bremen, Bremen, Germany
e-mail: s.shirinzadeh@cs.uni-bremen.de

M. Soeken
Integrated Systems Laboratory, EPFL, Lausanne, Switzerland
e-mail: mathias.soeken@epfl.ch

P.-E. Gaillardon
Electrical and Computer Engineering Department, University of Utah, Salt Lake City, UT, USA
e-mail: pierre-emmanuel.gaillardon@utah.edu

R. Drechsler
Cyber-Physical Systems, DFKI GmbH, Bremen, Germany
e-mail: drechsler@uni-bremen.de

© Springer International Publishing AG 2017

S. Vaidyanathan and C. Volos (eds.), *Advances in Memristors, Memristive Devices and Systems*, Studies in Computational Intelligence 701,
DOI 10.1007/978-3-319-51724-7_17

instructions that is required to implement a complete system is a serious drawback of implication logic. Using data structures such as *Binary Decision Diagrams* (BDDs) (Chakraborti et al. 2014), and *And-Inverter Graphs* (AIGs) (Bürger et al. 2013) has been previously proposed for optimization of memristive in-memory computing circuits. Nevertheless, both approaches still require a high number of instructions.

A *Majority-Inverter Graph* (MIG) is a logic representation, proposed in Amarù et al. (2014), which has a high flexibility in depth optimization and therefore enables design of high speed logic circuits and FPGA implementations (Amarù et al. 2015a). In comparison with the well-known data structures BDDs and AIGs, MIGs have experimentally shown better results in logic optimization, especially in propagation delay (Amarù et al. 2014). In Gaillardon et al. (2016), it was shown that MIGs are highly qualified for logic synthesis of RRAM-based circuits since they can efficiently execute the intrinsic *resistive majority* operation in RRAMs indicated by RM_3. This enables more efficient in-memory computing exploiting a majority oriented paradigm.

Logic synthesis for majority based in-memory computing also can vary with respect to differently specified MIG implementation methodologies. An MIG can be mapped to a netlist of RRAM devices in a fully customized manner such that a set of MIG features determines the requirements of the final implementation. Obviously, such a customized synthesis approach results in customized cost metrics, which need specifically designed optimization schemes. On the other hand, in-memory computing can be also performed in an instruction based manner. This way, a given MIG is translated instruction by instruction, which can be easily implemented upon a standard resistive crossbar.

An instruction based architecture based on memristive arrays called *Programmable Logic-in-Memory* (PLiM) was proposed in Gaillardon et al. (2016). For programs executed on this architecture, the number of instructions and the required number of RRAMs are important cost metrics to measure the quality. The instruction set for the PLiM architecture is based on the RM_3 operation, which computes the majority-of-three operands when one input is inverted. This corresponds directly to the physical implementation of the RRAM proposed in Linn et al. (2012). Consequently, MIGs are the natural abstraction to derive PLiM programs, i.e., sequences of RM_3 instructions.

An arbitrary function can be represented by several structurally different but functionally equivalent MIGs, which leads to PLiM programs of different quality. Further, even for the same MIG representation, there exists several ways to translate a graph into PLiM programs that vary in the number of RRAM devices and instructions. Hence, the PLiM architecture can highly benefit from MIG optimization and efficient node translation schemes.

This chapter presents logic synthesis for majority based in-memory computing exploiting the memristive behavior of RRAMs. In Sect. 2, MIGs are introduced and their Boolean algebra is explained. The section also provides an overview of majority-based in-memory computing, and the state-of-the-art. In Sect. 3, a customized MIG-based approach is presented for synthesis of logic-in-memory circuits. Section 4 studies instruction based in-memory computing for the PLiM architecture and also presents an efficient and fully automated compilation procedure for it. Experimental evaluations for the customized and instruction based approaches are given in the corresponding sections presenting them. The concluding remarks of the chapter are presented in Sect. 5.

2 Background

2.1 *Majority-Inverter Graphs*

An MIG is a data structure for Boolean function representation and optimization. An MIG is defined as a logic network that consists of 3-input majority nodes and regular/complemented edges (Amarù et al. 2014, 2015b, 2016).

MIGs can efficiently represent Boolean functions thanks to the expressive power of the *majority operator* (MAJ) $M(a, b, c) = a \cdot b + a \cdot c + b \cdot c = (a + b) \cdot (a + c) \cdot (b + c)$. Indeed, a majority operator can be configured to behave as a traditional conjunction (AND) or disjunction (OR) operator. In the case of 3-input majority operator, fixing one input to 0 realizes an AND while fixing one input to 1 realizes an OR. As a consequence of the AND/OR inclusion by MAJ, traditional *And-Or-Inverter Graphs* (AOIGs) are a special case of MIGs and MIGs can be easily derived from AOIGs. An example MIG representation derived from its optimal AOIG is depicted in Fig. 1a. AND/OR operators are replaced node-wise by MAJ operators with a constant input, and the inverters are shown by black dots on the branches.

Intuitively, MIGs are at least as compact as AOIGs. However, even smaller MIG representation arise when fully exploiting the majority functionality, i.e., with non-constant inputs (Amarù et al. 2016). We are interested in compact MIG representations because they translate into smaller and faster physical implementations. In order to manipulate MIGs and reach advantageous MIG representations, a dedicated Boolean algebra was introduced in Amarù et al. (2014). The axiomatic system for the MIG Boolean algebra, referred to as Ω, is defined by the five following primitive axioms.

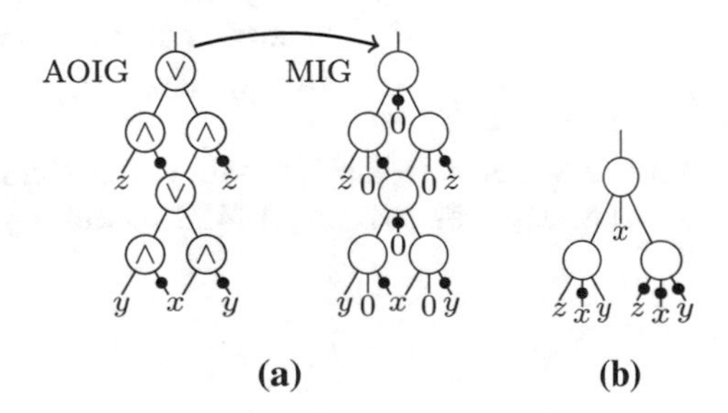

Fig. 1 Example MIG representation **a** derived by transposing its optimal AOIG representation, and **b** after optimization

$$\Omega \begin{cases} \textbf{Commutativity} - \Omega.C \\ M(x,y,z) = M(y,x,z) = M(z,y,x) \\ \textbf{Majority} - \Omega.M \\ M(x,x,z) = x \\ M(x,\bar{x},z) = z \\ \textbf{Associativity} - \Omega.A \\ M(x,u,,M(y,u,z)) = M(z,u,M(y,u,x)) \\ \textbf{Distributivity} - \Omega.D \\ M(x,y,M(u,v,z)) = M(M(x,y,u),M(x,y,v),z) \\ \textbf{Inverter Propagation} - \Omega.I \\ \overline{M}(x,y,z) = M(\bar{x},\bar{y},\bar{z}) \end{cases}$$

The axioms are inspired from median algebra (Isbell 1980) and the properties of the median operator in a distributive lattice (Birkhoff and Kiss 1947). A strong property of MIGs and their algebraic framework concerns reachability. It has been proved that by using a sequence of transformations drawn from Ω it is possible to traverse the entire MIG representation space (Amarù et al. 2016). In other words, given two equivalent MIG representations, it is possible to transform one into the other by just using axioms in Ω. This results is of paramount interest to logic synthesis because it guarantees that the best MIG can always be reached. Unfortunately, deriving a sequence of Ω transformations is an intractable problem. As for traditional logic optimization, heuristic techniques provide here fast solutions with reasonable quality (De Micheli 1994).

By using the MIG algebraic framework it is possible to obtain a better MIG for the example in Fig. 1a. Figure 1b shows the MIG structure, which is optimized in both depth (number of levels) and size (number of nodes). Such MIGs can be reached using a sequence of Ω axioms starting from their unoptimized structures.

Although Ω axioms are sufficient to transform a given MIG to any equivalent one, the length of the transformation sequence might be impractical to execute. To solve this problem, a more advanced set of transformations derived from the basic rules in Ω was proposed in Amarù et al. (2014) which is denoted by Ψ. The following set includes those axioms of Ψ that are used in this chapter.

$$\Psi \begin{cases} \textbf{Relevance} - \Psi.R \\ M(x,y,z) = M(x,z,z_{x/\bar{y}}) \\ \textbf{Complementary Associativity} - \Psi.C \\ M(x,u,M(y,\bar{u},z)) = M(x,u,M(y,x,z)) \end{cases}$$

where $z_{x/\bar{y}}$ means replacing x with $\bar{y}$. We refer the reader to paper (Amarù et al. 2016) for an in-depth discussion on MIG optimization recipes.

Fig. 2 The intrinsic majority operation within an RRAM

P	Q	R	R'
0	0	0	0
0	1	0	0
1	0	0	1
1	1	0	0

$R' = P \cdot \overline{Q}$

P	Q	R	R'
0	0	1	1
0	1	1	0
1	0	1	1
1	1	1	1

$R' = P + \overline{Q}$

2.2 *Majority for Logic-in-Memory*

RRAMs are two-terminal devices which internal resistance R can be switched between two logic states 0 and 1 designating high and low resistance values, respectively. Denoting the top and bottom terminals by P and Q, the memory can be switched with a negative or positive voltage V_{PQ} based on the device polarity. The truth tables in Fig. 2 show how the next state (R') of the switch is resulted from P, Q, and the current state (R). The built-in majority operation described in Fig. 2 can be formally expressed as the following (Gaillardon et al. 2016):

$$\begin{aligned} R' = \mathrm{RM}_3(P, Q, R) &= (P \cdot \overline{Q}) \cdot \overline{R} + (P + \overline{Q}) \cdot R \\ &= P \cdot R + \overline{Q} \cdot R + P \cdot \overline{Q} \cdot \overline{R} \\ &= P \cdot R + \overline{Q} \cdot R + P \cdot \overline{Q} \cdot R + P \cdot \overline{Q} \cdot \overline{R} \\ &= P \cdot R + \overline{Q} \cdot R + P \cdot \overline{Q} \\ &= M(P, \overline{Q}, R) \end{aligned}$$

The operation above is referred to 3-input resistive majority $\mathrm{RM}_3(x, y, z)$, with $\mathrm{RM}_3(x, y, z) = M(x, \bar{y}, z)$ (Gaillardon et al. 2016). According to RM_3, the next state of a resistive switch is equal to a result of a built-in majority gate when one of the three variables x, y, and z is already preloaded and the variable corresponding to the logic state of the bottom electrode is inverted.

2.3 *State-of-the-Art of In-Memory Computing*

So far, few synthesis approaches using logic representations have been proposed for logic synthesis of in-memory computing circuits. Most of the existing approaches in this area exploit material implication for realization of the nodes of their employed graph based data structures.

In Chattopadhyay and Rakosi (2011), material implication was used to synthesize combinational logic circuits with resistive memories using *Or-Inverter Graphs* (OIGs). The approach applies an extension of the delay minimization algorithm proposed in Beatty (1972) to the OIGs and also uses an area minimization to lower the costs of the equivalent circuits constructed with resistive memories.

Another approach presented in Bürger et al. (2013) exploits AIGs for synthesis of in-memory computing logic circuits. The approach maps an arbitrary Boolean function to an AIG and optimizes it. An optimized AIG representing a given function is then mapped to an equivalent network of implication gates. The approach executes a given Boolean function with $N + 2$ RRAMs, where N is the number of input RRAM devices, which keep their initial values until the target function is executed, and 2 is the number of work RRAMs, which internal states are changed during the operations. Nevertheless, some extra RRAMs are also considered to maintain values of the implication gates possessing more than one fanout.

A BDD-based approach has been proposed in Chakraborti et al. (2014) for synthesis of Boolean functions with resistive switches. Two implication based realizations are proposed for a 2-to-1 multiplexer (MUX) one for a minimum number of resistive switches and the other for a minimum number of operations when lower latency is of higher importance than area. It has not been referred to any BDD optimization method in Chakraborti et al. (2014) to lower either the number of RRAM devices or operations. For a given Boolean function, the approach maps the nodes of the corresponding BDD representation to a netlist of RRAM devices using one of the aforementioned implication based MUX realizations. Two methods for serial and parallel evaluation of BDDs are proposed in Chakraborti et al. (2014) which lead to different cost metrics. In the serial method, the BDD is traversed in depth-first-manner and each node corresponding to a MUX realization is computed in order. This method decreases the number of required RRAM devices but as a consequence the length of computational operations increases significantly with respect to the size of Boolean function which might not be of interest for many applications. The parallel method evaluates all of BDD nodes in a level at the same time but still requires more RRAMs regarding the number of complemented edges and fanouts.

3 Customized In-Memory Computing

This section presents a full custom approach for MIG-based synthesis of logic-in-memory computing circuits. The presented customized approach is based on a realization proposed for the majority gate with RRAM devices and its corresponding MIG mapping methodology. Then, several MIG optimization algorithms are presented with respect to area and depth, i.e., the number of RRAMs and instructions, respectively (Shirinzadeh et al. 2016).

3.1 Realization of Majority Gate with RRAMs

Exploiting RM_3 (Gaillardon et al. 2016), the majority gate can be realized using four RRAMs within only three instructions:

$$\mathbf{01}: X = x, Y = y, Z = z, A = 0$$
$$\mathbf{02}: P_A = 1, Q_A = y, R_A = 0 \rightarrow R'_A = \bar{y}$$
$$\mathbf{03}: P_Z = x, Q_Z = \bar{y}, R_Z = z \rightarrow R'_Z = M(x, y, z)$$

In the first step, the initial values of input variables as well as an additional RRAM are loaded by applying appropriate voltage levels to the top and bottom electrodes of RRAM devices. Step 2 executes the required NOT operation in RRAM device A. In the last step, the majority function is executed by use of RM_3 at RRAM device Z.

3.2 Design Methodology

Although the realization imposes sequential circuit implementation, it allows a reduction in area by reusing RRAMs released from previous computations. The presented synthesis approach considers one MIG level at each time, such that the employed RRAMs to evaluate the level can be used for other levels. Starting from the input of the graph, the RRAMs in a level are released when all the required instructions are executed. Then, the RRAMs are reused for the upper level and this procedure is continued until the target function is evaluated. Such an implementation requires as many majority gates as the maximum number of nodes in any level of the MIG. Hence, depending on the use of IMP or MAJ in the realization, the corresponding number of RRAMs and steps for synthesizing the MIG is four times the number of required majority gates and three times the number of levels, respectively. However, still some additional RRAMs are needed in the presence of complemented edges. Table 1 shows the number of RRAMs and instructions of the resulting in-memory computing circuits.

For every complemented edge in the graph a NOT gate is required. The negation can be executed by an RM_3 operation with 0 as one input, as shown in second step of the realization. This will require one extra RRAM to be loaded by 0 that can be done in parallel with the data loading step and an additional instruction. For a correct evaluation, the ingoing complemented edges of any level should be first inverted. It is obvious that the required instructions for all complemented edges in a level can be executed simultaneously. In other words, the additional steps required for complemented edges are equal to the number of MIG levels with ingoing complemented edges. Similarly, the total number of RRAMs required for the synthesis of the whole graph is equal to the maximum of four times the number of nodes in the level plus the number of ingoing complemented edges over all MIG levels.

3.3 MIG Rewriting for Customized In-Memory Computing

Having the cost metrics shown in Table 1, an MIG representing a given Boolean function can be optimized by applying a set of valid transformations to find an

Table 1 The cost metrics of MIG implementation using RRAMs

Symbol	Definition	Value
N_i	No. of nodes in the *i*th level of the MIG	Given
C_i	No. of ingoing complemented edges in the *i*th level of the MIG	Given
D	The depth of the MIG	Given
L	No. of MIG levels with ingoing complemented edges	Given
R	No. of RRAMs	$\max_{0 \le i \le D} (4N_i + C_i)$
I	No. of instructions	$3D + L$

equivalent but more efficient MIG. MIG optimization in terms of delay and area aims at finding the smallest depth, i.e., the number of levels, or the size of the graph, i.e., the number of nodes. Using RRAMs for implementation, the metrics determining area and delay depend on a combination of MIG features that some of them are not intended in conventional area and depth optimization. However, a reduction in area and especially depth might lower costs of an RRAM-based implementation. Thus, specific optimization techniques are required to find an optimum MIG with respect to the number of RRAMs and computational steps.

Two MIG rewriting algorithms for logic synthesis of in-memory computing circuits are presented in this section. The first algorithm optimizes MIGs with respect to both objectives simultaneously, while the other one aims at reducing the number of instructions, which is often regarded to be more important compared to the number of RRAM devices. For a better understanding of the MIG optimization algorithms tailored for in-memory computing, conventional area and depth optimization algorithms for standard implementation of MIGs proposed in Amarù et al. (2014) are introduced first.

1 **for** *(cycles = 0; cycles < effort; cycles++)* **do**
2 $\quad \Omega.M;\ \Omega.D_{R\rightarrow L};$
3 $\quad \Omega.A;\ \Psi.C;$ } eliminate
4 $\quad \Omega.M;\ \Omega.D_{R\rightarrow L};$
5 **end**

Algorithm 1: Conventional MIG area optimization (based on (Amarù et al. 2014))

The framework for area optimization given in Algorithm 1 is based on conventional MIG area optimization algorithm proposed in Amarù et al. (2014). Using *eliminate* ($\Omega.M$; $\Omega.D_{R\rightarrow L}$) some of the MIG nodes can be removed by repeatedly applying majority rule ($\Omega.M$) and distributivity from right to left ($\Omega.D_{R\rightarrow L}$) to the entire MIG. Assuming x, y, z, u, and v as input variables $\Omega.D_{R\rightarrow L}$ transforms $M(M(x, y, u), M(x, y, v), z)$ to $M(x, y, M(u, v, z))$ which means the total number of nodes has decreased from three to two. In order to enable further reduction in the number of nodes, the MIG is reshaped by use of associativity axioms $\Omega.A$, $\Psi.C$, which allow to move the variables between adjacent levels. Then, *eliminate* is applied again to optimize the size of the newly arranged MIG. The area optimization algorithm can be iterated for a maximum number of cycles called *effort*. From the point of area in an RRAM-based circuit, although Algorithm 1 can reduce the number of physical RRAMs by removing unnecessary nodes, it does not address the issue of complemented edges that are important in both aforementioned cost metrics.

```
1 for (cycles = 0; cycles < effort; cycles++) do
2     Ω.M; Ω.D_{L→R}; Ω.A; Ψ.C;  ⎫
3     Ψ.R;                         ⎬ push-up
4     Ω.M; Ω.D_{L→R}; Ω.A; Ψ.C;  ⎭
5 end
```

Algorithm 2: Conventional MIG depth optimization (based on (Amarù et al. 2014))

Algorithm 2 is structurally similar to the MIG depth optimization procedure proposed in Amarù et al. (2014) with slightly shorter iterations. In general, the depth of the graph is of high importance in MIG optimization to lower the latency of the resulting circuits. The depth of the MIG can be reduced by pushing the critical variable with the longest arrival time to upper levels. This can be possible by the process *push-up* shown in Algorithm 2. *Push-up* includes majority, distributivity, and associativity axioms. It is obvious that the majority rule may reduce depth by removing unnecessary nodes. Applying distributivity from left to right ($\Omega.D_{L\rightarrow R}$) such that $M(x, y, M(u, v, z))$ is transformed to $M(M(x, y, u), M(x, y, v), z)$ may also result in an MIG with smaller depth. If either x or y is the critical variable with the latest arrival, distributivity cannot reduce the depth of $M(x, y, M(u, v, z))$. However, if z is the critical variable, applying $\Omega.D_{L\rightarrow R}$ will reduce the depth of MIG by pushing z one level up. In the cases that the associativity rules ($\Omega.A$, $\Psi.C$) are applicable, the depth can be reduced by one if the axioms move the critical variable to the upper level. After performing *push-up*, the relevance axiom ($\Psi.R$) is applied to replace the reconvergent variables that might provide further possibility of depth reduction for another *push-up*.

Although Algorithm 2 decreases the number of instructions in an in-memory computing circuit, it does not consider the issue of complemented edges. Moreover, the depth reduction by Algorithm 2 is only possible at a cost in area, since $\Omega.D_{L\rightarrow R}$ adds one extra node to the graph. This may increase the area of the resulting circuit if the size of the critical level, i.e., the level with the maximum number of required RRAMs, is increased. $\Omega.A$ and $\Psi.C$ can also have a similar effect on the maximum level size by moving one node to the critical level. A simple example for this is applying $\Omega.A$ to $M(x, u, M(y, u, M(p, q, r)))$ that has a depth of three and one node in each level. The transformation results in $M(M(p, q, r), u, M(y, u, x))$ of depth two and two nodes in the lower level. Although the late arrival variable ($M(p, q, r)$) is pushed up, the number of nodes in one level, that might be the critical level, has increased from one to two. This effect is not of interest for implementation of MIGs with resistive arrays, however using $\Psi.C$ might be with a positive spin in this case because of the possibility of reducing the number of complemented edges.

```
1 for (cycles = 0; cycles < effort; cycles++) do
2     Ω.M_{L→R}; Ω.D_{L→R}; Ω.A; Ψ.C;    \
3     Ω.I_{R→L(1−3)};                      } push-up
4     Ω.M_{L→R}; Ω.D_{L→R}; Ω.A; Ψ.C;    /
5     Ω.A; Ω.D_{R→L};
6 end
```

Algorithm 3: Multi-objective optimization for in-memory computing costs

None of the algorithms explained above suggest a solution for the issue of complemented edges that contain an important part of both cost metrics of in-memory computing circuits. Moreover, a single-objective MIG optimization algorithm considers either area or delay that leads to circuits worsened with respect to the other objective. Hence, a multi-objective MIG optimization algorithm is presented to obtain efficient RRAM-based logic circuits with a good trade-off between both cost metrics. The algorithm includes a combination of conventional area and depth rewritings besides techniques tackling complemented edges from both aspects of area and delay.

As shown in Algorithm 3, the presented MIG rewriting for in-memory computing costs starts with applying *push-up* to obtain a smaller depth. Then, the complemented edges are aimed by applying an extension of inverter propagation axiom from right to left ($\Omega.I_{R\rightarrow L}$) for the condition that the considered node has at least two outgoing complemented edges. The three cases satisfying this condition and their equivalent majority gates are shown below and discussed in the following considering their effect on both cost metrics.

$$M(\bar{x}, \bar{y}, \bar{z}) = \overline{M}(x, y, z) \tag{1}$$

$$\overline{M}(\bar{x}, \bar{y}, z) = M(x, y, \bar{z}) \tag{2}$$

$$M(\bar{x}, \bar{y}, z) = \overline{M}(x, y, \bar{z}) \tag{3}$$

In the first case, the ingoing complemented edges of the gate are decreased from three to zero, while one complement attribute is moved to the upper level, i.e., the level including the output of the gate. Assuming that the current level, i.e., the level including the ingoing edges, is the critical level with the maximum number of required RRAMs, this case is favorable for area optimization. However, if the upper level is the critical level, the number of required RRAMs will increase by only one. Similar scenarios exist for the two other cases, although the last case might be less interesting because the number of complemented edges in both levels is changed equally by one. That means a penalty of one is possible as the cost for a reduction of one, while transformations (1) and (2) may result in RRAM reductions of three and two, respectively.

To reduce the number of instructions, the number of levels possessing complemented edges should be reduced. Depending on the presence of complemented edges by other gates in both levels, the two first transformations given above might reduce or increase the number of instructions or even leave it unchanged. Case (1) is

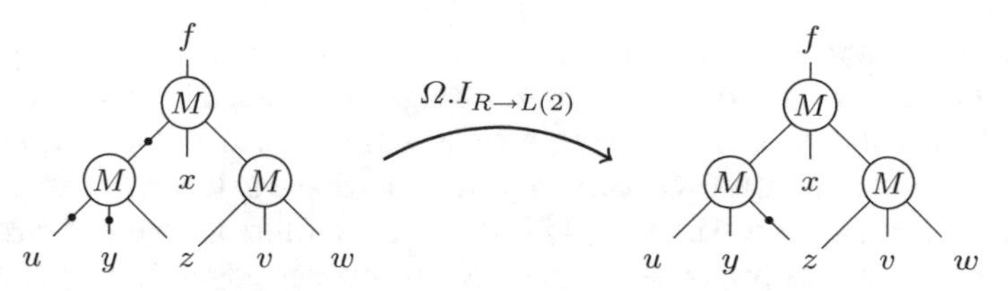

Fig. 3 Applying an extension of $\Omega.I_{R\to L}$ to reduce the extra RRAMs and steps caused by complemented edges

beneficial if the upper level already has complement edges and also the transformation removes all the complemented edges from the current level. It might be also neutral if none of the levels are going to be improved to a complement-free level. The worst case occurs when moving the complement attribute to the upper level which increments the number of levels with complement edges. Similar arguments can be made for the remaining cases. However, case (2) is more favorable because it never adds a level with complemented edges and case (3) can not be advantageous because it can never release a level from complemented edges.

Figure 3 shows a simple MIG that is applicable to transformation (2) ($\Omega.I_{R\to L(2)}$). The transformation has released one level of the MIG from the complement attribute (indicated by a black dot on the edge), which results in a smaller number of instructions. Furthermore, as a result of removing one complemented edge from the critical level, the required number of RRAMs is decreased by one.

After applying inverter propagation for the aforementioned conditions ($\Omega.I_{R\to L(1-3)}$), the MIG is also reshaped and more chances for reducing the depth might be created. Thus, *push-up* is applied to the entire MIG again to reduce the number of instructions as much as possible. In the last step, the number of RRAMs are reduced to make a trade-off between both cost metrics. Applying $\Omega.A$, some of changes by *push-up* that have increased the maximum level size can be undone. Finally, distributivity from right to left ($\Omega.D_{R\to L}$) is applied to the graph to reduce the number of nodes in levels.

1 **for** *(cycles = 0; cycles < effort; cycles++)* **do**
2 $\quad \Omega.M_{L\to R}; \Omega.D_{L\to R}; \Omega.A; \Psi.C;$ (push-up)
3 $\quad \Omega.I_{R\to L}$ (push-up)
4 $\quad \Omega.I_{R\to L(1-3)}$ (push-up)
5 $\quad \Omega.M_{L\to R}; \Omega.D_{L\to R}; \Omega.A; \Psi.C;$ (push-up)
6 **end**

Algorithm 4: Instruction count optimization

Due to the importance of latency in logic synthesis, and the issue of sequential implementation, another MIG optimization algorithm is also presented in this chapter to reduce the number of instructions of logic-in-memory computing circuits. In the presented instruction count optimization algorithm, two axioms of inverter propagation are applied to the MIG after *push-up*. First, only the axiom presented by case (1), i.e., the base rule of inverter propagation from right to left ($\Omega.I_{R\rightarrow L}$), is applied to the entire MIG to lower the number of levels with complemented edges. Since the transformation moves one complement attribute to the upper level, it might create new inverter propagation candidates for the all three discussed cases if the upper level already has one or two ingoing complemented edges. Hence, $\Omega.I_{R\rightarrow L(1-3)}$ is applied again to ensure maximum coverage of complemented edges. Although case (3) can not reduce the number of instructions, it is not excluded from $\Omega.I_{R\rightarrow L(1-3)}$ due to its effect on balancing the levels' sizes. Finally, *push-up* is applied to the MIG to reduce the depth more if new opportunities are generated. It should be noted that the number of instructions is mainly determined by the MIG depth. In fact, in the worst case caused by complemented edges, the total number of instructions would be equal to four times the number of levels, i.e., the MIG depth.

3.4 Experimental Evaluation of Customized In-Memory Computing

To assess the performance of the presented customized logic-in-memory computing approach, experiments are carried out over a benchmark set including 25 Boolean functions from LGsynth91 (Yang 1991) with a number of input variables from 7 to 135. The number of cycles (effort) is set to 4 in all experiments.

The comparison of results obtained by the conventional MIG area and depth rewritings and the presented rewritings for the customized in-memory computing approach are shown in Table 2. The instruction count optimization algorithm has resulted in MIGs with the smallest number of instructions that is reduced by 29.71% on average in comparison with the conventional depth optimization algorithm. This shows that the employed techniques to reduce the complemented edges have been effective. The results in Table 2 also show a trade-off between both cost metrics obtained by the presented multi-objective algorithm. It achieves an average reduction of 32.55% in the number of instructions at a fair cost of 19.77% increase in the number of RRAMs compared to the conventional area optimization algorithm.

4 Instruction Based In-Memory Computing

This section studies PLiM architecture (Gaillardon et al. 2016) and presents an optimization procedure for it, including MIG rewriting and compilation, to reduce the the length of instructions and the number of required RRAM devices (Soeken et al. 2016).

Table 2 Experimental results of customized in-memory computing

Benchmark	PI/PO	Area optimization		Depth optimization		Multi-objective optimization		Instruction count optimization	
		#*I*	#*R*	#*I*	#*R*	#*I*	#*R*	#*I*	#*R*
5xp1	7/10	40	121	40	153	36	149	28	182
alu4	14/8	104	1111	88	1324	72	1370	56	1717
apex1	45/45	92	1854	68	2373	56	2343	44	2972
apex2	39/3	104	249	84	423	56	358	47	435
apex4	9/19	76	2703	64	2894	64	2820	48	3602
apex5	117/88	100	861	52	1227	47	1053	35	1286
apex6	135/99	72	763	52	891	44	1018	36	1191
apex7	49/37	64	200	52	275	48	277	44	348
b9	41/21	36	168	32	168	32	168	28	168
clip	9/5	52	184	44	198	40	217	36	275
cm150a	21/1	36	88	36	88	32	95	32	90
cm162a	14/5	36	60	28	60	30	60	24	65
cm163a	16/5	28	68	28	68	27	68	24	68
cordic	23/2	56	143	48	174	48	134	39	162
misex1	8/7	28	73	24	92	24	76	20	94
misex3	14/14	92	1112	84	1488	67	1444	52	1762
parity	16/1	64	160	64	160	53	152	48	152
seq	41/35	112	1457	88	1798	64	1970	60	2498
t481	16/1	76	71	48	120	52	90	40	123
table 5	17/15	104	1126	84	1881	64	1723	52	2252
too_large	38/3	124	213	96	370	64	322	48	392
x1	51/35	64	309	40	528	36	435	28	509
x2	10/7	35	45	28	66	26	46	24	68
x3	135/99	72	750	52	951	44	1008	36	1201
x4	94/71	44	380	32	409	28	391	24	563
AVG		68.44	570.76	54.24	727.16	46.16	711.48	38.12	887

#*I*: number of RM_3 instructions, #*R*: number of RRAMs

4.1 PLiM Architecture

The Programmable Logic-in-Memory (PLiM) architecture aims at enabling logic operations on a regular RRAM array. While every memory node can implement basic operations, the difficulty of operating logic on a memory array lies in the distribution of signals and the scheduling of operations. The PLiM controller consists of a wrapper of the RRAM array (Fig. 4) and works as a simple processor core, reading instructions from the memory array and performing computing operations (majority) within the memory array. As a wrapper, PLiM uses the addressing and read/write peripheral circuitries of the RRAM array. When LiM = 0, the controller is *off* and the whole array works as a standard RAM system. When LiM = 1, the circuit starts performing computation. The controller consists of a simple finite state machine and few work registers, in order to operate the RM_3 instruction, as detailed

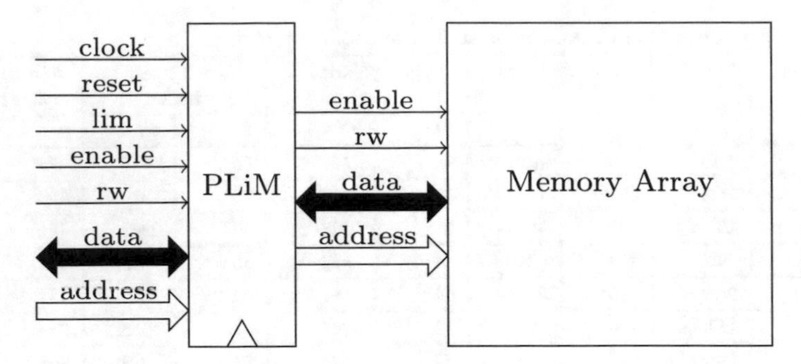

Fig. 4 The PLiM architecture: the PLiM controller operates as a wrapper over the RRAM array and schedules the RM_3 operations (Gaillardon et al. 2016)

in Gaillardon et al. (2016). The instruction format consists of the first operand A, the second operand B, and the destination address Z of the results. Single-bit operands A and B are then read from constants or from the memory array, and logic operation is performed during the write operation to the memory location Z by setting P to A and Q to B. The new value stored in the node Z is then $Z \leftarrow \langle A\bar{B}Z \rangle$. When the write operation is completed, a program counter is incremented, and a new cycle of operation is triggered.

4.2 *Motivation*

The main idea of a compilation procedure for PLiM computer is leveraging MIGs in order to derive RM_3 instruction sequences, which can run as programs on the PLiM architecture. In its current form, the PLiM architecture can only handle serial operations (Gaillardon et al. 2016). Therefore, only one MIG node might be computed each time and the total number of instructions is equal to the sum of instructions required to compute each MIG node. Accordingly, reducing the size of the MIG is considered to have a significant impact on the PLiM program with respect to the number of instructions. However, still further MIG optimization is possible to lower the costs caused by complemented edges. While the presence of a single complemented edge in an MIG node is of interest for benefiting from the intrinsic majority operation inside an RRAM, a second or third complemented edge imposes extra costs in both number of instructions and required RRAMs. Hence, MIG area rewriting, besides reducing number of nodes with multiple complemented edges, can be highly effective for optimizing the number of instructions, while the latter can also lower the number of required RRAMs.

As an example, consider the two equivalent MIGs in Fig. 5a, before optimization on the left and after optimization on the right. Translating them into RM_3 instructions yields:

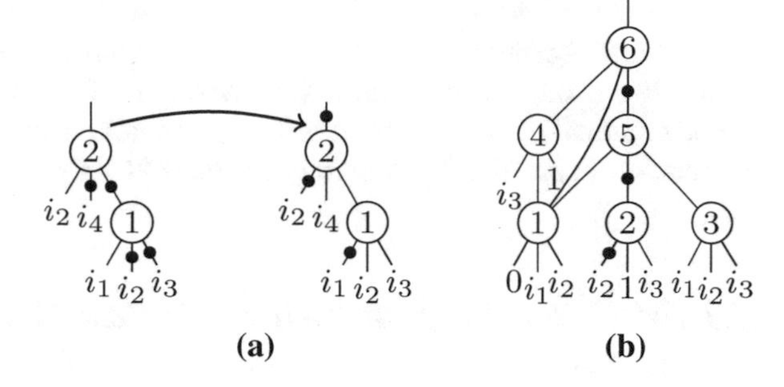

Fig. 5 Reducing the number of instructions and RRAMs, after **a** MIG rewriting and **b** node selection and translation

Before MIG optimization	
01: 0, 1, $@X_1$	$X_1 \leftarrow 0$
02: 1, i_3, $@X_1$	$X_1 \leftarrow \bar{i}_3$
03: i_1, i_2, $@X_1$	$X_1 \leftarrow N_1$
04: 0, 1, $@X_2$	$X_2 \leftarrow 0$
05: 1, $@X_1$, $@X_2$	$X_2 \leftarrow \bar{N}_1$
06: i_2, i_4, $@X_2$	$X_2 \leftarrow N_2$

After MIG optimization	
01: 0, 1, $@X_1$	$X_1 \leftarrow 0$
02: i_3, 0, $@X_1$	$X_1 \leftarrow i_3$
03: i_2, i_1, $@X_1$	$X_1 \leftarrow N_1$
04: i_4, i_2, $@X_1$	$X_1 \leftarrow N_2$

Here $@X_i$ refers to the address of RRAM X_i and N_j refers to the result of the MIG node j. Program addresses are bold in front of the RM_3 instruction A, B, C, and the second and fourth column comment the action of the instruction.

It can be seen that after optimization both the number of instructions and RRAMs are decreased, from 6 to 4 and from 2 to 1, respectively. The effect of multiple complement edge elimination is much larger when translating a large MIG.

Not only the MIG structure has an effect on the PLiM program, but also the order in which nodes are translated and which of the node's children are selected as operands A, B, and destination Z in the RM_3 instruction. As example, consider the MIG in Fig. 5b. Translating it in a naïve way, i.e., in order of their node indexes and selecting the RM_3 operands and destination in order of their children (from left to right), will result in the following program:

01: 0, 1, $@X_1$	$X1 \leftarrow 0$
02: 1, i_1, $@X_1$	$X_1 \leftarrow \bar{i}_1$
03: 0, 1, $@X_2$	$X_2 \leftarrow 0$
04: i_2, 0, $@X_2$	$X_2 \leftarrow i_2$
05: 0, $@X_1$, $@X_2$	$X_2 \leftarrow N_1$
06: 0, 1, $@X_3$	$X_3 \leftarrow 0$
07: i_3, 0, $@X_3$	$X_3 \leftarrow i_3$
08: 1, i_2, $@X_3$	$X_3 \leftarrow N_2$
09: 0, 1, $@X_4$	$X_4 \leftarrow 0$
10: 1, i_2, $@X_4$	$X_4 \leftarrow \bar{i}_2$
11: 0, 1, $@X_5$	$X_5 \leftarrow 0$
12: i_3, 0, $@X_5$	$X_5 \leftarrow i_3$
13: i_1, $@X_4$, $@X_5$	$X_5 \leftarrow N_3$
14: 0, 1, $@X_6$	$X_6 \leftarrow 0$
15: 1, i_3, $@X_6$	$X_6 \leftarrow \bar{i}_3$
16: 1, 0, $@X_7$	$X_7 \leftarrow 1$
17: $@X_2$, $@X_6$, $@X_7$	$X_7 \leftarrow N_4$
18: $@X_2$, $@X_3$, $@X_5$	$X_5 \leftarrow N_5$
19: $@X_7$, $@X_5$, $@X_2$	$X_2 \leftarrow N_6$

By changing the order in which the nodes are translated and also the order in which children are selected as operands and destination for the RM_3 instructions, a shorter program can be found (for the same MIG representation):

01: 0, 1, $@X_1$	$X_1 \leftarrow 0$
02: i_2, 0, $@X_1$	$X_1 \leftarrow i_2$
03: i_1, 1, $@X_1$	$X_2 \leftarrow N_1$
04: 1, 0, $@X_2$	$X_2 \leftarrow 1$
05: i_3, i_2, $@X_2$	$X_2 \leftarrow N_2$
06: 0, 1, $@X_3$	$X_3 \leftarrow 0$
07: 1, i_2, $@X_3$	$X_3 \leftarrow \bar{i}_2$
08: 0, 1, $@X_4$	$X_4 \leftarrow 0$
09: i_3, 0, $@X_4$	$X_5 \leftarrow i_3$
10: i_1, $@X_3$, $@X_4$	$X_4 \leftarrow N_3$
11: $@X_1$, $@X_2$, $@X_4$	$X_4 \leftarrow N_5$
12: 0, 1, $@X_2$	$X_2 \leftarrow 0$
13: i_3, 0, $@X_2$	$X_2 \leftarrow i_3$
14: $@X_1$, 0, $@X_2$	$X_2 \leftarrow N_4$
15: $@X_1$, $@X_4$, $@X_2$	$X_2 \leftarrow N_6$

It can be seen that by translating the nodes with a different order, the number of instructions is reduced from 19 to 15, and the number of RRAMs is reduced from 7 to 5. Based on this observations, the next part of this section describes algorithms for automatically finding a good MIG representation and for translating an MIG representation in an effective way to get a small PLiM program.

4.3 MIG Rewriting for Instruction Based In-Memory Computing

As discussed before, the first requirement for an efficient compilation process is a dedicated MIG rewriting algorithm such that optimizes MIG structures to be more convenient for compiling into RM_3 instructions. It was already shown that both the size of an MIG and the distribution of complemented edges have an effect on the PLiM program in number of instructions and number of RRAMs. Hence, we are interested in an MIG rewriting algorithm that (i) reduces the size of the MIG, and (ii) reduces the number of MIG nodes with multiple complemented edges.

1 **for** *(cycles = 0; cycles < effort; cycles++)* **do**
2 $\quad \Omega.M; \Omega.D_{R\to L};$
3 $\quad \Omega.A; \Psi.C;$
4 $\quad \Omega.M; \Omega.D_{R\to L};$
5 $\quad \Omega.I_{R\to L(1-3)};$
6 $\quad \Omega.I_{R\to L};$
7 **end**

Algorithm 5: MIG rewriting for PLiM architecture

The MIG rewriting approach required for PLiM is given in Algorithm 5 and follows the rewriting idea of Amarù et al. (2014). It can be iterated for a certain number of times, controlled by *effort*. The first three lines of Algorithm 5 are based on the conventional MIG area rewriting approach proposed in Amarù et al. (2014). These node elimination techniques are repeated after reshaping the MIG by applying $\Omega.A$; $\Omega.C$, which may provide further size reduction opportunities. To reduce the number of nodes with multiple inverted edges, the extended inverter propagation axioms $\Omega.I_{R\to L(1-3)}$ shown before are applied to the MIG.

At the end, since the MIG might have been changed after the three aforementioned transformations, $\Omega.I_{R\to L}$ is applied again to ensure the most costly case is eliminated. In general, applying the last two lines of Algorithm 5 over the entire MIG repetitively can lead to much fewer instructions and RRAM cost.

4.4 *The PLiM Compiler*

After applying the rewriting algorithm, the optimized MIG is compiled into a PLiM program. Algorithm 6 gives an overview of the compilation procedure. The algorithm keeps track of a map `COMP[v]` that stores for each MIG node v whether it has been computed or not. Initially, all leafs, i.e., primary inputs and the constants, are set to be computed. A priority queue Q keeps track of all vertices that can possibly be translated, called candidates. A vertex is a *candidate* if all its children are computed.

Input : MIG M
Output : PLiM program $P = \{I_1, I_2, \ldots, I_k\}$
1 **foreach** *leaf in M* **do**
2 | set `COMP[v]` $\leftarrow$ T;
3 **end**
4 **foreach** *MIG node in M* **do**
5 | **if** *all children of v are computed* **then**
6 | | Q.enqueue(v);
7 | **end**
8 **end**
9 **while** *Q is not empty* **do**
10 | set $c \leftarrow Q$.pop();
11 | set $P \leftarrow P \cup$ translate(c);
12 | set `COMP[c]` $\leftarrow$ T;
13 | **foreach** *parent of c* **do**
14 | | **if** *all children of v are computed* **then**
15 | | | Q.enqueue(v);
16 | | **end**
17 | **end**
18 **end**

Algorithm 6: Outline of compilation algorithm

The main loop of the algorithm starts by popping the best candidate c from the priority queue and translating it into a sequence of PLiM instructions. Afterwards, for each parent, it is checked whether it is computable, and if this is the case, it is inserted into Q.

The remaining of this section describes the details of the sorting criteria for Q, and the node translation process are in order.

4.4.1 Candidate Selection

The candidate selection strategy for the PLiM compiler is based on two principles: (i) releasing the RRAMs in-use as early as possible, and (ii) allocating RRAMs at the right time such that they are blocked as short as possible. Two example MIGs are shown to clarify the principles. Figure 6a shows an MIG with two candidates u and v, for which all of their children nodes are already computed. Candidate u

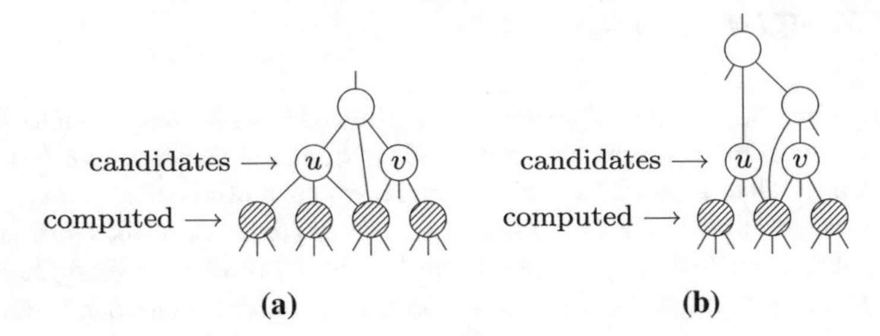

Fig. 6 Reducing the number of RRAMs by selecting the candidate with **a** more releasing children and **b** smaller fanout level index

has two *releasing children*, i.e., children who have single fan-out, while v has only one releasing child. In the case that u is selected for computation first, the RRAMs keeping its releasing children can be freed and reused for the next candidate.

Figure 6b shows a small MIG with two candidates u and v to illustrate the second principle. The output of u is only required when v is already computed. In other words, the number of RRAMs in use can increase if u is computed before v since the RRAM keeping u cannot be released before computing the root node of the MIG. This way, v is computed when an RRAM has been already allocated to retain the value of u. The number of additional RRAMs required in such condition can be considerable for large number of nodes.

In order to sort nodes in the priority queue in Algorithm 6, two nodes u and v are compared. Node u is preferred over v if (i) its number of releasing children is greater, or (ii) if u's parent with the largest level (ordered from PIs to POs) is on a lower level than v's parent with the smallest level. If no criteria is fulfilled, u and v are compared according to their node index.

4.4.2 Node Translation

To translate a node of the MIG into RM_3 instructions, three operands A and B, and destination Z should be selected in a way that the number of RRAMs and instructions are as low as possible. The operands A and B can be RRAMs or constants and the destination Z is a RRAM. Recall that the instruction computes $Z \leftarrow \langle A\bar{B}Z \rangle$. In the ideal case each MIG node can be translated into exactly one RM_3 instruction and can reuse one of its children's RRAMs as destination. In other cases additional RM_3 instructions and/or additional RRAMs are required.

Select Operand B

We first select which of the node's children should serve as operand B, i.e., the second operand of the RM_3 instruction. In total, four cases with subcases are checked in the

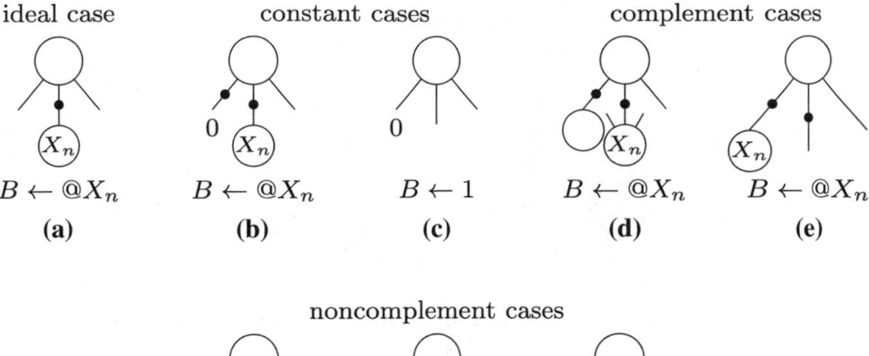

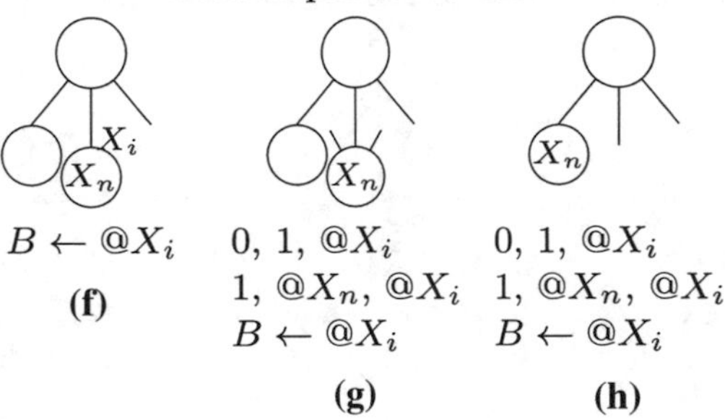

Fig. 7 Selecting operand *B*

given order which are illustrated in Fig. 7. Only the last two subcases require two additional instructions and one additional RRAM.

(a) *There is exactly one complemented child: B* is the RRAM storing this complemented child.

(b) *There is more than one complemented child, but also a constant child:* The non-constant complemented child is selected for *B*, since constants allow for more flexibility when selecting the remaining operands.

(c) *There is no complemented child, but there is a constant child: B* is assigned the inverse of the constant. Since we consider MIGs that only have the constant 0 child, *B* is assigned 1.

(d) *There is more than one complemented child, but at least one with multiple fan-out:* We select the RRAM of the child with multiple fan-out, as this excludes its use as destination.

(e) *There is more than one complemented child, none with multiple fan-out:* The RRAM of the first child is selected.

(f) *There is no complemented child, but for one child there exists a RRAM with its complemented value:* Each node is associated with an RRAM which holds or has held its computed result. In addition, if its inverted value is computed once and stored in an additional RRAM X_i, it is remembered for future use. In this case *B* can be assigned X_i.

(g) *There is no complemented child, but one child has multiple fan-out:* The child with multiple fan-out is selected with the same argumentation as above. Since it is

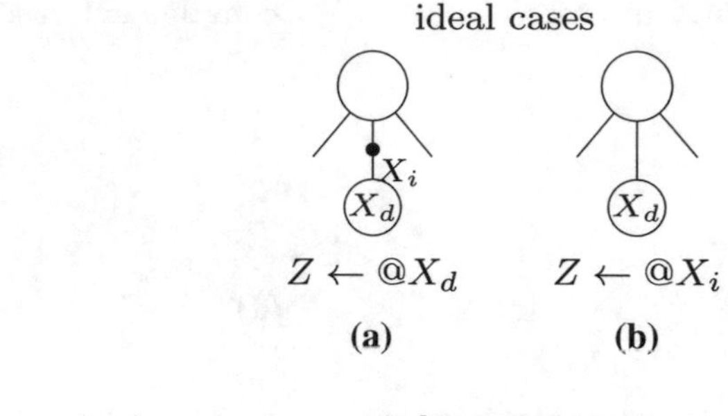

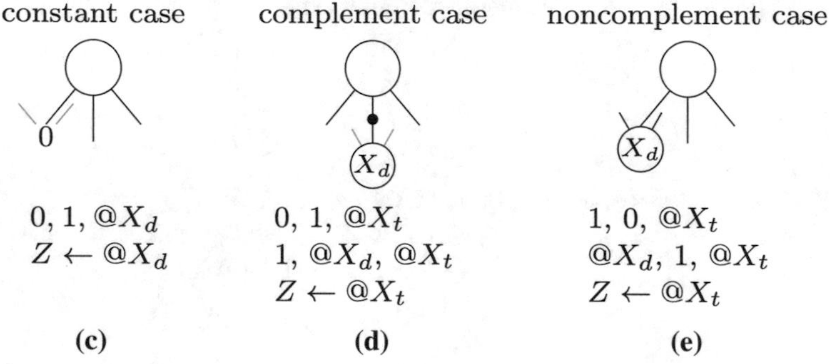

Fig. 8 Selecting destination Z

not inverted, an additional RRAM X_i needs to be allocated, loaded with 0, and then set to the complement of X_n. As described above, X_i is associated to the child for future use.
(h) *There is no inverted child and none has multiple fan-out:* The fist child is selected and an additional RRAM X_i is allocated to store the complement of X_n.

Select Destination Z

After the inverter selection, the destination RRAM, i.e., the third argument of the RM_3 instruction is selected. The aim is to reuse one of the children's RRAMs as work RRAM instead of creating a new one. In total, four cases (with subcases) are checked which are illustrated in Fig. 8. Only the first case allows to reuse an RRAM of the children for the destination, all the other cases require one or two additional instructions and one additional RRAM. Note that one of the children has already been selected as operand B and that this is implicitly considered in the following descriptions.
(a) *There is a complemented child with one fan-out, and there exists an RRAM with its complemented value:* The existing RRAM X_i for the complemented value can be used and it is safe to override it, since the child does not fan out to other parents.

(b) *There is a noncomplemented child with one fan-out:* The RRAM of this child can be used as destination and it is safe to override it. Note that case (a) is preferable compared to this one to avoid complemented children for operand A.
(c) *There is a constant child:* If there is a constant child (with or without multiple fan-out) a new RRAM is allocated and initialized to the constant value (considering complemented edges into account).
(d) *There is a complemented child:* If there is an inverted child X_d (with or without multiple fan-out), a new RRAM X_t is allocated and initialized to the complement of X_d using two RM_3 instructions.
(e) *There is a noncomplemented child with multiple fan-out:* The first child X_d is selected and it's value is copied into a new allocated RRAM X_t using two RM_3 instructions.

Select Operand A

The child that is selected as operand A is uniquely determined at this point since operand B and destination Z have been selected. Consequently, there is no case distinction w.r.t. to preference. However, there are still different actions to be taken depending on the child node.
(a) *The child node is constant:* A is set to the constant taking the complement edge into account.
(b) *The child node is noncomplemented:* A is set to the RRAM of the child node.
(c) *The child node is complemented, and there exists an RRAM with its complemented value:* A is set to the computed RRAM of the complemented value.
(d) *The child node is complemented, but there does not exist an RRAM with its complemented value:* A new RRAM X_i is allocated and assigned to the inverted value of the node. A is set to X_i.
At least one instruction is required and no additional RRAM needs to be allocated in order to translate one node. In the worst case, six additional instructions and three additional RRAMs are required, e.g., cases (h), (e), and (d) for selecting operand B, destination Z, and operand A, respectively.

We are interested in finding programs with a small number of instructions and a small number of RRAMs, i.e., optimized w.r.t. time and space. The presented MIG rewriting algorithm and the node translation scheme address and affect both optimization criteria whereas the candidate selection mainly targets the number of RRAMs.

In order to further reduce the number of RRAMs, an RRAM allocation approach is also employed. It implements an interface with two operations: (i) *request*, which returns an RRAM that is ready to use, and (ii) *release*, which releases an RRAM that is not required anymore. This interface is implemented by using a free list that is populated with released RRAMs. Whenever an RRAM is requested, first it is checked whether a free released RRAM exists that can be re-used, or a new fresh RRAM is allocated. RRAMs are requested whenever more than one instruction is required to translate a node (e.g., cases (g) and (h) for selecting operand B). RRAMs are released whenever all parents of a child have been computed.

4.5 Experimental Evaluation of Instruction Based In-Memory Computing

Table 3 evaluates the efficiency obtained by the presented PLiM compiler for the EPFL benchmarks.[1] The second column includes results for a naïve translation, where only the candidate selection scheme is disabled, based on the initial nonoptimized MIGs. The third and the forth columns represent results after MIG rewriting and both rewriting and compilation, respectively. The number of MIG nodes indicated by N is also provided to give a better understanding of the MIG before and after rewriting. It is clear that N also shows the number of MIG nodes for the compiled PLiM, since the same MIG after rewriting has been used.

As expected, the number of MIG nodes have been reduced or remained unchanged for a few cases after MIG rewriting. Although, the number of nodes after MIG rewriting does not show a significant reduction, the average of the number of instructions is reduced up to 20.09% compared to the naïve translation. This besides the 14.83% reduction achieved in the total number of RRAMs imply the effectiveness of the employed techniques for removing multiple inverted edges.

Performing both MIG rewriting and compilation, the number of required instructions and RRAMs reduces notably. The number of instructions and RRAM for the compiled PLiM are reduced on average by up to 19.95% and 61.4%, respectively in comparison with the corresponding values obtained for the naïve PLiM. This represents a significant reduction in both the latency and especially storage space metrics.

5 Conclusion

In this chapter, majority based in-memory computing was fully studied for two customized and instruction based approaches exploiting the memristive behavior of RRAMs. Both approaches benefit from the intrinsic majority operation of RRAMs. The customized synthesis approach utilizes MIG optimization algorithms to reduce the number of RRAMs and instructions, and maps the optimized MIGs to memristive arrays according to a level based implementation methodology. The instruction based approach presents PLiM computer architecture and its automatic compiler for translation of large Boolean functions into programs for the in-memory computing. It was observed that both the MIG representation and the way in which an MIG is compiled has a large impact on the resulting PLiM programs—in terms of required instructions as well as number of RRAMs.

[1] http://lsi.epfl.ch/benchmarks.

Table 3 Experimental results of PLiM architecture

Benchmark	PI/PO	Naïve			MIG rewriting					Rewriting and compilation			
		#*N*	#*I*	#*R*	#*N*	#*I*	impr. (%)	#*R*	impr. (%)	#*I*	impr.(%)	#*R*	impr. (%)
adder	256/129	1020	2844	512	1020	2037	28.38	386	24.61	1911	32.81	259	49.41
bar	135/128	3336	8136	523	3240	5895	27.54	371	29.06	6011	26.12	332	36.52
div	128/128	57247	146617	687	50841	147026	–0.03	771	–12.22	147608	–0.68	590	14.12
log2	32/32	32060	78885	1597	31419	60402	23.43	1487	6.89	60184	23.71	1256	21.35
max	512/130	2865	6731	1021	2845	5092	24.35	867	15.08	4996	25.78	579	43.29
multiplier	128/128	27062	76156	2798	26951	56428	25.91	1672	40.24	56009	26.45	419	85.03
sin	24/25	5416	12479	438	5344	10300	17.09	426	2.73	10223	18.08	402	8.22
sqrt	128/64	24618	60691	375	22351	47454	21.81	433	–15.46	49782	17.97	323	13.87
square	64/128	18484	54704	3272	18085	33625	38.53	3247	0.76	33369	39.00	452	86.19
cavlc	10/11	693	1919	262	691	1146	40.28	236	9.92	1124	41.43	102	61.07
ctrl	7/26	174	499	66	156	258	48.29	55	16.66	263	47.29	39	40.91
dec	8/256	304	822	257	304	783	4.74	257	0.00	777	5.47	258	–0.39
i2c	147/142	1342	3314	545	1311	2119	36.05	487	10.64	2028	38.81	234	57.06
int2float	11/7	260	648	99	257	432	33.33	83	16.16	428	33.95	41	58.59
mem_ctrl	1204/1231	46836	113244	8127	46519	85785	24.25	6708	17.46	84963	24.97	2223	72.65
priority	128/8	978	2461	315	977	2126	13.61	241	23.49	2147	12.76	149	52.70
router	60/30	257	503	117	257	407	19.09	112	4.27	401	20.28	64	45.30
voter	1001/1	13758	38002	1749	12992	25009	34.19	1544	11.72	24990	34.24	1063	39.22
AVG		9468.4	24346.2	910.4	9022.4	19452.96	20.09	775.32	14.83	19488.56	19.95	351.4	61.40

#*N*: number of MIG nodes, #*I*: number of RM_3 instructions, #*R*: number of RRAMS, improvement is calculated compared to naïve

Experimental results of the customized approach show that higher efficiency can be obtained with respect to the cost metrics of in-memory computing by specifically defined MIG optimization algorithms rather than conventional ones. Experiments for the instruction based approach show that compared to a naïve translation approach the number of instructions can be reduced and RRAMs can be reduced by up considerably. The presented compiler unlocks the potential of the PLiM architecture to process large scale computer programs using in-memory computing. This makes this promising emerging technology immediately available for nontrivial applications.

Acknowledgements This research work was partly supported by H2020-ERC-2014-ADG 669354 CyberCare, by the University of Bremen's graduate school SyDe, funded by the German Excellence Initiative, and by the Swiss National Science Foundation project number 200021 146600.

References

Amarù, L. G., Gaillardon, P.-E., & De Micheli, G. (2014). Majority-inverter graph: A novel data-structure and algorithms for efficient logic optimization. In *DAC* (pp. 194:1–194:6).

Amarù, L., Petkovska, A., Gaillardon, P.-E., Bruna, D. N., Ienne, P., & De Micheli, G. (2015a). Majority-inverter graph for FPGA synthesis. In *SASIMI*.

Amarù, L. G., Gaillardon, P.-E., & De Micheli, G. (2015b). Boolean logic optimization in majority-inverter graphs. In *DAC*.

Amarù, L. G., Gaillardon, P.-E., & De Micheli, G. (2016). Majority-inverter graph: A new paradigm for logic optimization. *IEEE Transactions on Computer-Aided Design of Integrated Circuits and Systems*, *35*(5), 806–819.

Beatty, J. C. (1972). An axiomatic approach to code optimization for expressions. *Journal of the ACM*, *19*(4), 613–640.

Birkhoff, G., & Kiss, S. A. (1947). A ternary operation in distributive lattices. *Bulletin of the American Mathematical Society*, *53*(8), 749–752.

Bürger, J., Teuscher, C., & Perkowski, M. (2013). Digital logic synthesis for memristors. In *Reed-Muller 2013*.

Chakraborti, S., Chowdhary, P., Datta, K., & Sengupta, I. (2014). BDD based synthesis of Boolean functions using memristors. In *IDT* (pp. 136–141).

Chattopadhyay, A., & Rakosi, Z. (2011). Combinational logic synthesis for material implication. In *VLSI-SoC* (pp. 200–203).

De Micheli, G. (1994). *Synthesis and optimization of digital circuits*. McGraw-Hill Higher Education.

Gaillardon, P.-E., Amarú, L., Siemon, A., Linn, E., Waser, R., & Chattopadhyay, A., et al. (2016). The programmable logic-in-memory (PLiM) computer. In *DATE* (pp. 427–432).

Isbell, J. R. (1980). Median algebra. *Transactions of the American Mathematical Society*, *260*(2), 319–362.

Linn, E., Rosezin, R., Tappertzhofen, S., Böttger, U., & Waser, R. (2012). Beyond von Neumann-logic operations in passive crossbar arrays alongside memory operations. *Nanotechnology*, *23*(305205).

Shirinzadeh, S., Soeken, M., Gaillardon, P.-E., & Drechsler, R. (2016). Fast logic synthesis for RRAM-based in-memory computing using majority-inverter graphs. In *DATE* (pp. 948–953).

Soeken, M., Shirinzadeh, S., Gaillardon, P.-E., Amarú, L. G., Drechsler, R., & De Micheli, G. (2016). An MIG-based compiler for programmable logic-in-memory architectures. In *DAC* (pp. 117:1–117:6).

Yang, S. (1991). *Logic synthesis and optimization benchmarks user guide: Version 3.0*. MCNC.

Analysis of Dynamic Linear Memristor Device Models

Balwinder Raj and Sundarapandian Vaidyanathan

Abstract The aim of this book chapter is to provide a comprehensive review report on the Memristor device. Development of linear model for memristor and analysis of memristor are the prime focus as its current requirement for high speed and low power circuits design. Detailed discussion about memristor device physics, structure, operation, mathematical modeling and TCAD simulations have been carried out for better understand of memristor. Moore's law, the semiconductor industry's obsession with the shrinking of transistors with the commensurate steady doubling on chip about every two years, has been a source of about 50 year technical and economic revolution. Numerous innovations by a large number of scientists and engineers have helped significantly to sustain Moore's law since the beginning of the Integrated Circuit (IC) era. As the cost of computer power to the consumer reduces, the cost of production for producers to sustain Moore's law follows an opposite trend, i.e. Research, Development, Manufacturing, and Test costs are increasing continuously with each new generation of chips. This had led to the reason for existence of Moore's second law, also called Rock's law, which is that the capital cost of a semiconductor fabrication also increases exponentially over time. The formation of memristor is a great achievement in semiconductor industry considering Moore's second law because of its very easy and less steps of fabrication which is the reason for memristor being so cheap, while its nano scale size is new direction to attain Moore's first law. Therefore, the modelling and simulation of memristor is essential to analyze more advanced features of memristor without spending a lot of money on fabrication and testing.

Keywords Memristor · Memristive device · Logic design · Linear model · Modeling · Simulation

B. Raj
Department of Electronics and Communication, National Institute of Technology, Jalandhar 144011, Punjab, India
e-mail: balwinderraj@gmail.com

S. Vaidyanathan (✉)
R & D Centre, Vel Tech University, #42, Avadi, Chennai 600062, Tamil Nadu, India
e-mail: sundarcontrol@gmail.com

© Springer International Publishing AG 2017

S. Vaidyanathan and C. Volos (eds.), *Advances in Memristors, Memristive Devices and Systems*, Studies in Computational Intelligence 701,
DOI 10.1007/978-3-319-51724-7_18

1 Introduction

Today, most electronic devices in technology inventions and all other applications use semiconductor components. The study of semiconductor devices is known as a branch of solid-state physics, whereas the designing and construction of the circuits to solve practical problems is included in electronics engineering (Raj et al. 2009; Sharma et al. 2014; Chua 1971). Moore's law, the semiconductor industry's obsession with the shrinking of transistors with the commensurate steady doubling on chip about every two years, has been a source of about 50 year technical and economic revolution. Whether this scaling paradigm will last for 10 or 15 years more, it will finally come to an end. The emphasis in electronic design will have to shift to devices that are not just increasingly infinitesimal but increasingly capable. Numerous innovations by a large number of scientists and engineers have helped significantly to sustain Moore's law since the beginning of the Integrated Circuit (IC) era (Raj et al. 2013; Pattanaik et al. 2012; Raj 2014; Gergel-Hackett et al. 2009; Bhushan et al. 2013).

Innovations listed below are examples of breakthroughs that have played a critical role in the advancement of integrated circuit technology by more than seven orders of magnitude in less than five decades:

i. The invention of the Integrated Circuit itself is the foremost contribution and the reason for existence of Moore's law, credited equally to Jack Kilby at Texas Instruments and Robert Noyce at Fairchild Semiconductor (Noyce 1961; Prodromakis and Papavassiliou 2011).
ii. The invention of the Complementary Metal–Oxide–Semiconductor (CMOS) in 1963 enabled extremely dense and high performance ICs (Wanlass 1967; Johnson 2010).
iii. Invention of the Dynamic Random Access Memory (DRAM) technology in 1967 made fabrication of single-transistor memory cells possible (Dennard 1968; Raj et al. 2009; Larrieu and Han 2013).
iv. The invention of deep UV excimer laser photolithography decreased the smallest features in ICs from 500 nm in 1990 to 32 nm in 2012. The trend is expected to reach smallest feature below 10 nm in next decades (Jain et al. 1982; Biolek et al. 2009).

Some of the new directions in research that may allow Moore's law to continue are:

i. Using deep-ultraviolet excimer laser photolithography- IBM researchers claimed to develop a technique to print circuitry only 29.9 nm wide using 193 nm Argon Fluoride excimer laser lithography (La Fontaine 2010; Raj et al. 2008; Joglekar and Wolf 2009).
ii. In April 2008, researchers at HP Labs successfully created a working memristor, whose existence had previously only been theorized. The memristor's unique properties permit the creation of smaller and better-performing electronic devices (Strukov et al. 2008; Raj et al. 2011).

iii. In February 2010 a breakthrough in transistors with the design and fabrication of the world's first Junctionless Transistor was announced. The researchers claim that the Junctionless transistors can be produced at 10 nm scale using existing fabrication techniques (Raj et al. 2008; Vishvakarma et al. 2007; Mohsin 2010).
iv. In April 2011, development of Single-Electron Transistor (SET) was announced, which was 1.5 nm in diameter and made out of oxide based materials (Fuechsle et al. 2012; Raj et al. 2011; Volos et al. 2015).
v. In February 2013, the development of the first working transistor consisting of a single atom placed precisely in a silicon crystal (not just picked from a large sample of random transistors) announced (Larrieu and Han 2013).

2 Background Study: Memristor Review

The analysis of Memristor fabricated on polymer sheet by simple techniques is done by Gergel-Hackett et al. (2009). The fabricated memristor was with ratio of ON/OFF resistance (R_{on}/R_{off} = 10000:1), life = ~14 days, and was able to show good characteristics even with >4000 flexes and requires V < 10 V for stability with thermal effects. The device was providing following advantages-

i. Both soft switching and hard switching compatible
ii. Behavior is not ambient dependent
iii. Suitable for flexible memory component
iv. Portable physically flexible device

Varghese and Gandhi (2009) have proposed a design for a low area differential pair which substantially reduces cubic distortion, provide better Total Harmonic Distortion and hence wide linear range with the help of memristor using in place of nonlinear resistor to provide better nonlinearity. The advantage of exploiting the nonlinearity was evident due to the reason that the memristor is 10^6 times more nonlinear at nanoscale as compared to micro scale. The possibilities of logic design using memristors are discussed in literature (Raja and Mourad 2010). Memristor as state element is analysed, also basic logic operation on two memristor with both inverting and non-inverting configuration is observed. Logic implementation on multiple memristor (wired AND) and memristor crossbar logic design for implementation of NAND gate is shown (Raja and Mourad 2010).

Figure 1 shows an example of Crossbar Array, where all possible combinations for state of a switch is shown and also the case if no switch is present. One switch is realized here with one memristor working as a memory cell. As a memristor does not consume area more than required for two perpendicular wires and power to save its state in standby mode, memristor is a best option for a switch based memory cell in Crossbar Architectures. Selection of a memory cell (memristor) is done by

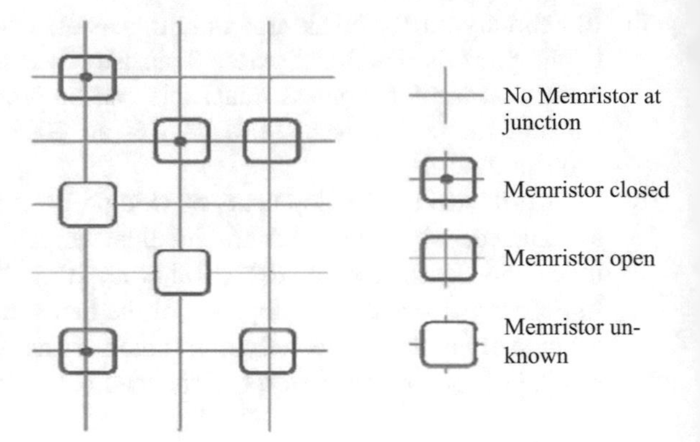

Fig. 1 Memristor crossbar array

applying voltage on those selected wire/lines. Wired AND, NAND, XOR, SOP implementation using memristor crossbar is given in Fig. 2a and future work includes-

i. Multiple output implementation
ii. Multi crossbar architecture
iii. Implication logic

Equivalent circuit for a NAND gate works in following three stages:

i. Latch input operation- In this stage, the input data from the three input terminals A, B, C is stored in the memristors M1–M3. Here, we have to apply a writing voltage (V_w) at the input terminal IN. Assuming that the inputs A, B, C are outputs of other memristors; the logic state of signals A, B, C is latched into M1–M3 respectively.
ii. Copy inputs & close AND- Here we are copying data from M1–M3 to M4–M6 and closing M7 by applying V = 0 at IN, K and V_w at AND. M7 will be used in next stage as this is inverting stage configuration.

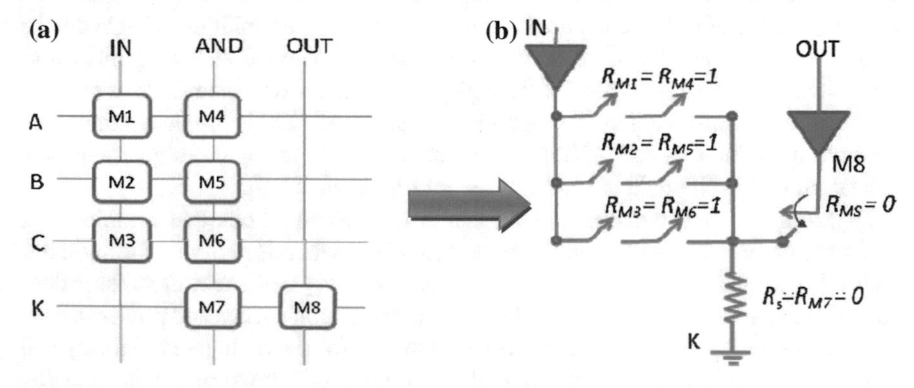

Fig. 2 Memristor crossbar with a 3-input NAND gate. **a** The crossbar with a 4 × 3 tile. **b** The equivalent circuit of NAND gate

iii. Evaluate and Capture- Here we have to apply read voltage (Vr) at IN and V_w at OUT.

In order to read the output of the memristor M8, we need to disable the memristor M7 by applying a large negative voltage on AND to open memristors M4–M7 first and then the memristor M8 can be read out by applying a voltage $V = 0$ at OUT and using the output terminal as an input to a subsequent crossbar logic gate. Figure 2b is showing Evaluate and Capture stage of 3-input NAND gate implementation. Memristor can have many resistance levels, we can use some discrete value as different logic levels and thus can use in multi value logic system as reported by Mohsin (2010). The memristor is basically a resistor whose resistance increases when current flows from one direction and decreases when current flows from opposite direction, So small pulses of current can be applied to change its resistance level to up or down depending on duration, direction and strength of current pulse. Both Hard switching and Soft Switching can be done on the device for binary or multilevel storage respectively. After storing values there is no need of refreshing as data is in resistance value form, therefore no external power source is required to save data for very large time duration.

Liu et al. (2010) propose a behavioral modeling method by constructing two different workable memristor models. The models proposed are based on experimental measured data of $Au/Ti_2O_5/Au$ and $Pt/TiO_2/Al$ memristor, fabricated in their lab. Curve fitting in MATLAB is also used. The proposed modeling method can be used efficiently to choose different memristor materials or fabrication technique, and the usefulness of which will be amplified by pursuing system-level analyzing and large scale design of memristor arrays. Compact models for current controlled and voltage controlled memristor implementable in SPICE, verilog-A and Spectre, which are suitable for frequency dependent memristive hysteresis behavior reported by Shin et al. (2010). This shows unique boundary assurance to simulate memristors whether they behave memristive or resistively. Parameter extraction, simulation results with macro model shown. Discussion given on- dependency of model on type of excitation signal and nonlinear dopant drift effect.

da Costa et al. (2012) were implemented the model in verilog- AMS and simulated in mentor graphics with AMS support. Switching between memristive and resistive states occurs when the potential barrier that separates the doped and undoped regions moves totally to one of sides and return only when the stimulus is inverted. The model can be used for mixed-signal or multi-domain simulation in circuit designs using memristor.

Figure 3 shows voltage and current output response for memristor model with input voltage frequency $f_0 = 1$ Hz, input voltage amplitude $V_0 = 1$ V, dopant ion mobility $\mu_v = 10^{-10}$ $cm^2s^{-1}V^{-1}$, device length D = 10 nm and off to on resistance ratio $R_{off}/R_{on} = 160$. It is clear from the Fig. 3 that if the input biasing is applied for a time period more than required for full length boundary movement, then the resistance of memristor will be saturated to one of its boundary value (either R_{off} or R_{on} depending upon biasing polarity) and will remain at it until the input bias polarity is reversed. The memristor will not show any nonlinearity in the case and

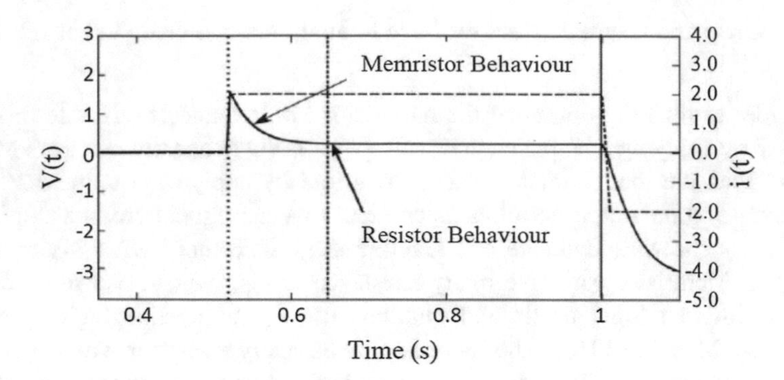

Fig. 3 Variation in stability memristor switching

will work as simple resistor. Changing the biasing will again change the resistance of memristor to other boundary value as shown in Fig. 3.

A preliminary SPICE macro model of memristor to develop models for the SPICE based analysis tools like HSPICE and Spectre was reported by author Mohanty (2013). An interpretation of the memristor device recalling the quasi-static expansion of Maxwell's equations and a review on Chua's argumentation about the memristor relating to the electromagnetic theory was also given. They have concluded that the Von-Neumann architecture, which is the base of all current computer systems, is not capable for carrying out computations with nano-devices and materials. There are lots options as different components but they are poor at mimicking the human brain. However, the memristor motivates future work in nano-electronics and nano-computing based on its capabilities.

3 Variables and Circuit Elements

Variables and circuit elements are important key components for any type of advanced system design. Fundamental variables and basic circuit elements had been evolution from many decades ago. New elements envisioning for the sake of completion of a physical system is not without scientific precedent. Indeed, the well known discovery of the periodic table for chemical elements by Mendeleef in 1869 is a case in point. From the circuit theory point of view, the relationship between two of four fundamental circuit variables; namely current (i), voltage (v), charge (q), flux-linkage (φ), define three basic two terminal elements (Sharma et al. 2015). Figure 4 shows the relationships between these variables and circuit elements.

Out of the six possible combinations of these variables, five are well known relationships. Two of these relationships are given by $q(t) = \int_{-\infty}^{t} i(\tau)d\tau$ and $\varphi(t) = \int_{-\infty}^{t} v(\tau)d\tau$. Other three relationships are given by axiomatic definition of the

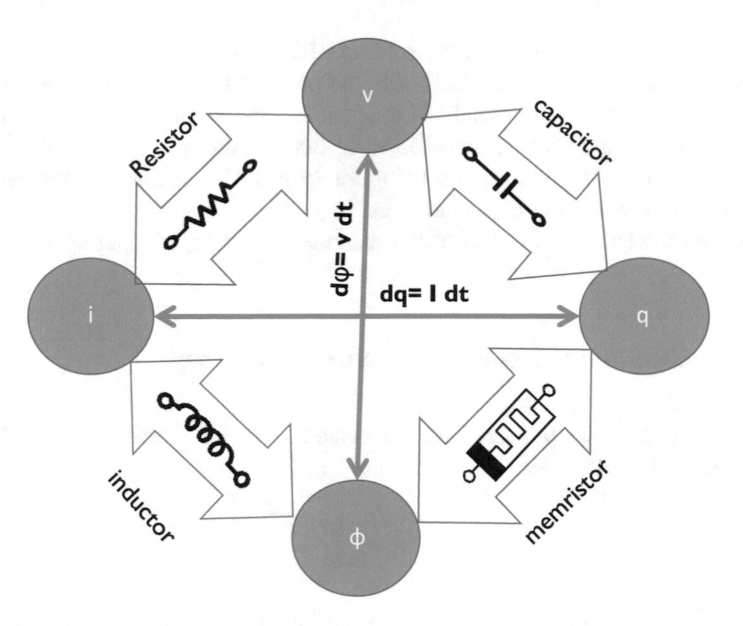

Fig. 4 Variables and circuit elements

Fig. 5 Memristor symbol

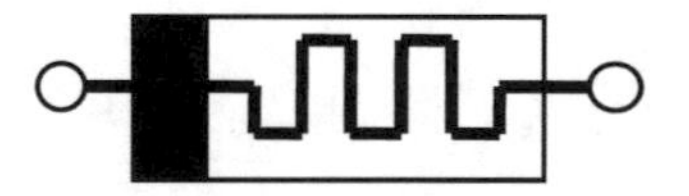

three fundamental circuit elements, namely, Resistor ($R = \frac{v}{i}$; relation between voltage and current), Capacitor $(v(t) = \frac{1}{C}q(t)$; relation between charge and voltage) and inductor $(i(t) = \frac{1}{L}q(t)$; relation between flux and current). For the sake of completeness from logical and axiomatic point of view, Prof. Leon O. Chua argued for the *Memristor* (a contraction for Memory resistor because it behaves somewhat like a nonlinear Resistor with memory) to set up a mathematical relationship between electric charge and magnetic flux (Chua 1971).

The memristor device is characterized by a nonlinear relation between charge and flux, i.e. time integrals of voltage and current. The symbol of memristor is shown in Fig. 5.

4 Linear Model of Memristor

There are two models of memristor, viz. linear model and non-linear model.

In this chapter, we shall discuss only the linear model of memristor in detail. The linear model of memristor is described for simple dopant ion drift kinetics in the memristor which is not including any type of nonlinearity issues for the case of simplicity. The modelling is done on MATLAB and results are presented in subsequent sessions. This chapter deals with the proposed linear model and discussion of its results. Further output analysis in terms of state variable has been done here with different driving voltage waveforms. In Sect. 5, the shortcomings of the linear model with output waveforms are being analysed. All the input parameters associated with the model are explained here.

First, the mathematical model of linear model is developed and explained below:

4.1 *Mathematical Linear Model of Memrisror*

In 1971, (Chua 1971) proposed memristance as the functional property of memristors; that correlates charge and flux, i.e.

$$M=\frac{d\varphi}{dq} \tag{1}$$

Since the flux is integration of voltage and charge is integration of current, the memristance has the same units as resistance. Later on, Chua and Kang (1976) generalized the concept to memristive systems, i.e.

$$v=R(x)i \tag{2}$$

$$\frac{dx}{dt}=f(x,i) \tag{3}$$

where v is the voltage, i is the current, and $R(x)$ is the instantaneous resistance that is dependent on the internal state variable x of the device. This state variable x is bounded within the interval [0, 1], and it is simply the normalized width of the doped region $x=\frac{w}{D}$ with D being the total thickness of the switching bilayer and w is width of doped region at the instant. At time t, the width of the doped region w depends on the amount of charge that has passed through the device; thus, the time derivative of w is a function of current, which can be described as

$$\frac{dw}{dt}=V_d=\mu E=\mu R_{on}\frac{i(t)}{D} \tag{4}$$

where V_d is the speed at which the boundary drifts between the doped and undoped regions, μ is the average dopant mobility, μE is the electric field across the doped region in the presence of current $i(t)$, R_{on} and R_{off} are the net resistances of the device when the active region is completely doped and undoped respectively.

The definition of memristance can be generalized as follows:

$$v(t) = M(t)i(t) \tag{5}$$

$$M(t) = R_{on}x(t) + R_{off}(1 - (x(t)) \tag{6}$$

$$v(t) = \left[R_{on}x(t) + R_{off}(1 - (x(t))\right]i(t) \tag{7}$$

$$\frac{dx}{dt} = \mu\frac{R_{on}}{D^2}i(t) \tag{8}$$

It is clear from above Eq. (8) that memristive effect is considerable in nanoscale devices due to the D^2 factor in denominator, which shows that memristive effect is 10^6 times better in nanoscale devices as compared to micro scale devices.

The MATLAB model simulation results, based on this mathematical model, are shown in Fig. 6 for sinusoidal Driving Voltage waveform.

The time axis is showing here the step number, as the time period $T = 1.5$ s is divided in 1000 steps due to the reason that MATLAB is a Digital Environment Programming Language and all the operations like integration and differentiation are performed here with difference equations and hence steps are required for the same. The driving voltage for this linear model is $v(t) = V_o \sin(2\pi ft)$ with $V_0 = 1.5$ V. As the signature characteristics of memristor (Chua 1971) is its pinched hysteresis loops in current-voltage characteristics, the output waveforms (I–V curves) in Fig. 6 are enough for validation of model by showing a pinched

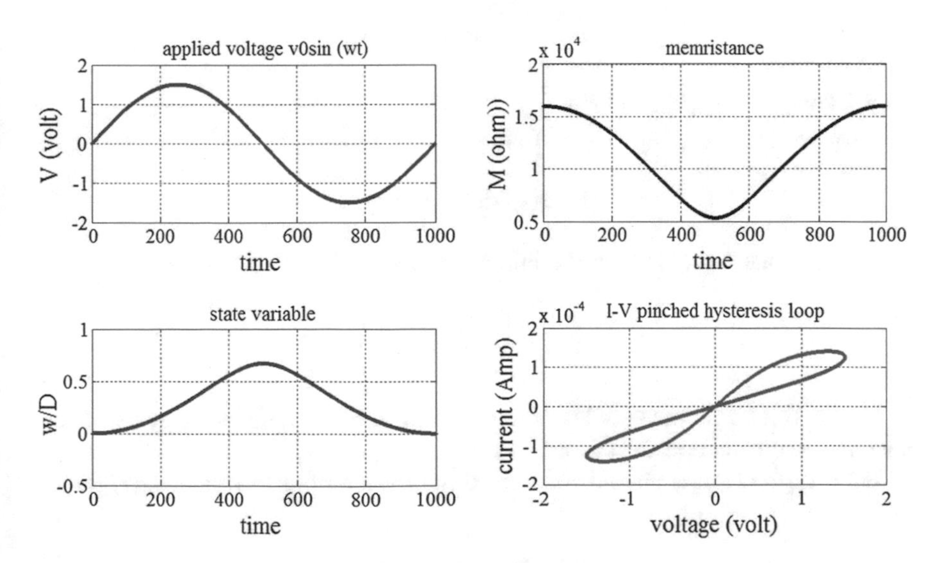

Fig. 6 Simulation curves for linear model with $V_0 = 1.5$ V, $T = 1.5$ s, $D = 10$ nm, $\mu = 10^{-14}\text{m}^2/\text{V.Sec}$, $R_{on} = 100\ \Omega$, $R_{off} = 16\ \text{K}\Omega$

hysteresis loop. The model is showing resistance switching between the predefined range, viz. $[R_{on}, R_{off}]$. Here the state variable x is having its maximum value ~ 0.65 and corresponding to this value the memristance of the device is $\sim 550\,\Omega$.

4.2 Boundary Movement Condition

The maximum value of state variable is not equal to 1. That is, the voltage amplitude is not sufficient for full length conductive layer, as it can be seen in Fig. 6. However, it also depends on the initial value of the state variable. We are assuming the initial value of the state to be zero, i.e. $x(0) = 0$.

Therefore, increasing the voltage amplitude increases the maximum value that can be obtained by the state variable keeping the time period of the driving voltage wave constant.

From the Eqs. (7) and (8), we have

$$\frac{dx}{dt} = \mu \frac{R_{on}}{D^2} \frac{v(t)}{\left[R_{on}x(t) + R_{off}(1 - x(t))\right]} \tag{9}$$

We know that $R_{off} \gg R_{on}$ and $0 < x(t) < 1$.

Thus, we can assume that $R_{on}x(t) + R_{off} \approx R_{off}$.

With this assumption, we can simplify Eq. (9) as follows:

$$\frac{dx}{dt} = \mu \frac{R_{on}}{D^2} \frac{v(t)}{R_{off}[1 - x(t)]} \tag{10}$$

We define $k = \mu \frac{R_{on}}{D^2 R_{off}}$ and $\beta = \frac{R_{off}}{R_{on}}$.

Then we can write Eq. (10) as follows:

$$[1 - x(t)]\, dx(t) = kv(t)\, dt \tag{11}$$

We integrate Eq. (11) with the initial condition $x(0) = 0$. Thus, we get

$$x(t) - \frac{[x(t)]^2}{2} = k\varphi(t) \tag{12}$$

For the full length travel of boundary between doped and undoped layers, we can take the final condition for the state variable as $x(T) = 1$.

From Eq. (12), it is clear that for maximum value of state variable $x(t) = 1$,

$$\text{flux } \varphi(t) = \frac{1}{2k} = \frac{\beta D^2}{2\mu} \tag{13}$$

Equation (13) is representing the amount of flux required for full length travel of the boundary between doped and undoped layers for any driving voltage.

In this work, we take $\beta = 160, D = 10^{-8}$ m, and $\mu = 10^{-14}$ m^2/V.sec. Also, the flux value $\varphi(t) = 0.8$ Weber for t being the time instant when $x(t) = 1$.

4.3 *Linear Model Results with Driving Voltage*

The Flux is defined as area under the voltage-time curve. We will now see the boundary movement for different types of driving voltage waveforms. For same Flux value the amplitude and shape of voltage wave changes. Correspondingly, the state variable value pattern also changes.

4.3.1 Sinusoidal Wave

For a sinusoidal voltage input, flux in positive half cycle is-

$$\int_{0}^{T/2} v(t)dt = \varphi\left(\frac{T}{2}\right) = V_0 \frac{T}{\pi} \tag{14}$$

with T = 2 s and $V_0 = 1.29$ V, the following wave output in Fig. 7 is obtained with the linear memristor model. The area under the driving voltage curve is obtained as $\varphi\left(\frac{T}{2}\right) = 0.821$ and the maximum value of state variable is obtained as $x\left(\frac{T}{2}\right) = 0.96$.

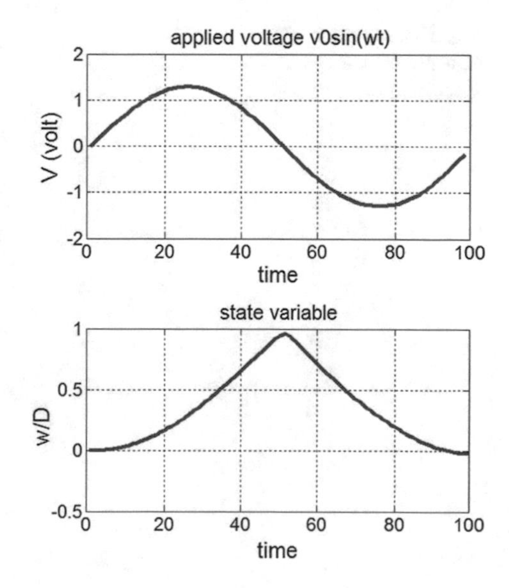

Fig. 7 The state variable response for sinusoidal voltage input

4.3.2 Triangular Wave

For a triangular voltage input, flux in positive half cycle is obtained as

$$\int_{0}^{T/2} v(t)dt = \varphi\left(\frac{T}{2}\right) = V_0\frac{T}{\pi} \quad (15)$$

with $T = 1.5$ s and $V_0 = 2.19$ V, the following wave output in Fig. 8 is obtained with the linear memristor model. The area under the driving voltage curve is obtained as $\varphi\left(\frac{T}{2}\right) = 0.821$ and the maximum value of state variable is obtained as $x\left(\frac{T}{2}\right) = 0.96$.

It is clear from Figs. 7 and 8 that the output response for both the triangular wave and sinusoidal wave input type is almost of same shape and so the maximum attainable value of state variable for the same flux (area under the curve) is also same. The case will be different for the input waves with abrupt changes (Fig. 9).

4.3.3 Square Pulse

- **Returning Zero type-** For a square pulse (RZ) voltage input, flux in positive half cycle is obtained as

$$\int_{0}^{T/2} v(t)dt = \varphi\left(\frac{T}{2}\right) = \int_{0}^{T/4} v(t)dt + \int_{T/4}^{T/2} v(t)dt = V_0\frac{T}{4} \quad (16)$$

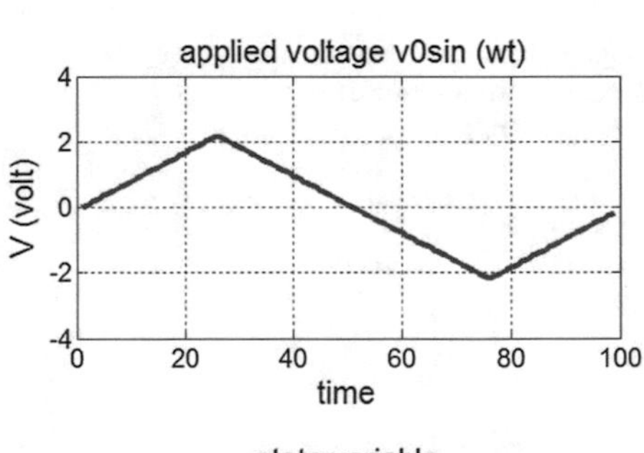

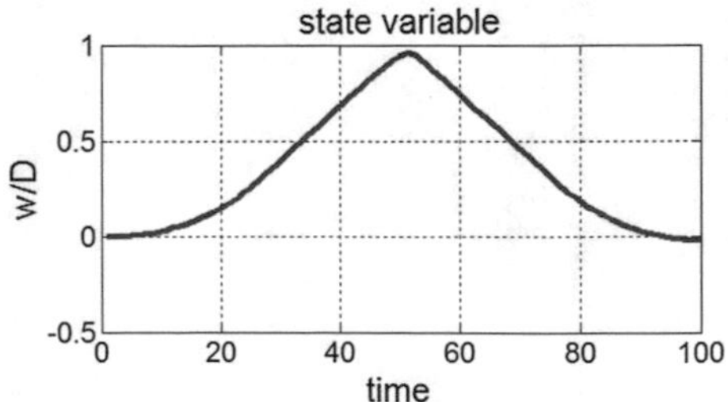

Fig. 8 The state variable response for triangular voltage input

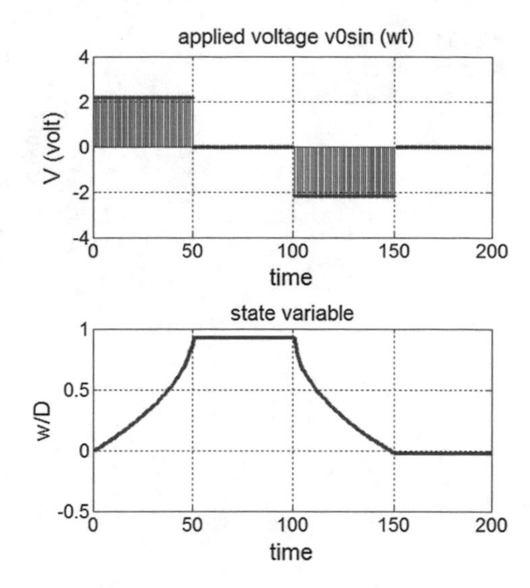

Fig. 9 The state variable response for square pulse (RZ) voltage input

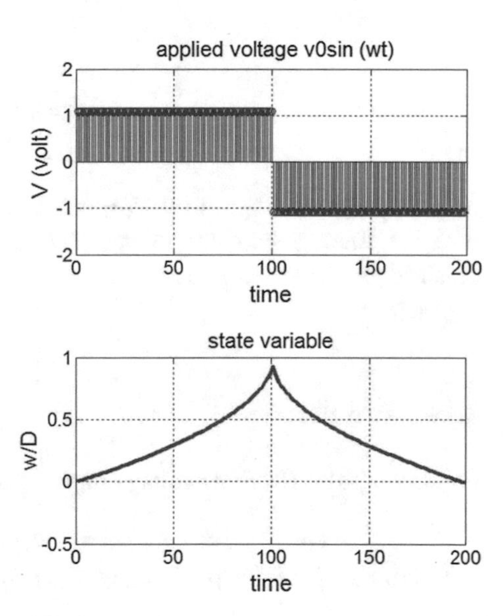

Fig. 10 The state variable response for square pulse (NRZ) voltage input

with $T = 1.5$ s and $V_0 = 2.183$ V, the following wave output in Fig. 10 is obtained with the linear memristor model. The area under the driving voltage curve is obtained as $\varphi\left(\frac{T}{2}\right) = 0.818$ and the maximum value of state variable is obtained as $x\left(\frac{T}{2}\right) = 0.87$.

Non-Returning Zero Type- For a square pulse (NRZ) voltage input, flux in positive half cycle is-

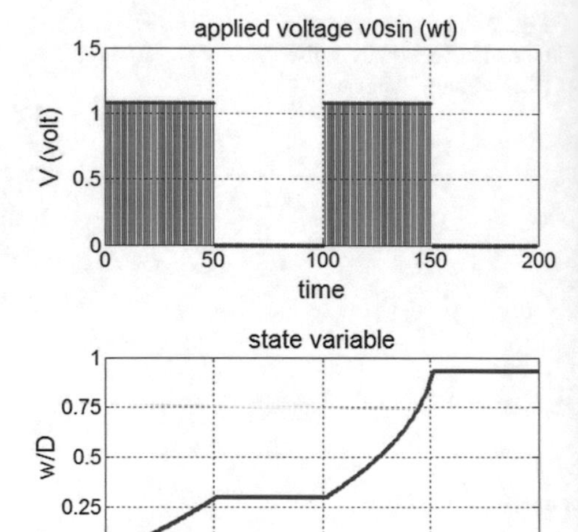

Fig. 11 The state variable response for two pulse (Digital) voltage input

$$\int_{0}^{T/2} v(t)dt = \varphi\left(\frac{T}{2}\right) = \frac{T}{2}V_0 \tag{17}$$

with $T = 1.5$ s and $V_0 = 1.09$ V, the following wave output in Fig. 11 is obtained with the linear memristor model. The area under the driving voltage curve is obtained as $\varphi\left(\frac{T}{2}\right) = 0.817$ and the maximum value of state variable is obtained as $x\left(\frac{T}{2}\right) = 0.91$.

4.3.4 Two Pulse

For a two pulse (Digital) voltage input, flux in positive voltage part is-

$$\int_{0}^{T} v(t)dt = \int_{0}^{T/4} v(t)dt + \int_{T/4}^{T/2} v(t)dt + \int_{T/2}^{3T/4} v(t)dt + \int_{3T/4}^{T} v(t)dt = \varphi(T) = \frac{T}{2}V_0 \tag{18}$$

with $T = 1.5$ s and $V_0 = 1.05$ V, the following wave output in Fig. 11 is obtained with the linear memristor model. The area under the driving voltage curve is obtained as $\varphi(T) = 0.788$ and the maximum value of state variable is obtained as $x(T) = 0.84$.

4.3.5 Four Pulse

For a Four pulse (Digital) voltage input, flux in positive voltage part is given by

$$\int_{0}^{T} v(t)dt = \varphi(T) = \frac{T}{2}V_0 \tag{19}$$

with $T = 1.5$ s and $V_0 = 1.05$ V, the following wave output in Fig. 12 is obtained with the linear memristor model. The area under the driving voltage curve is obtained as $\varphi(T) = 0.788$ and the maximum value of state variable is obtained as $x(T) = 0.84$.

This is clear from state variable responses for all above waveform types of Driving Voltage full length boundary movement can be achieved with a certain flux value. This Flux value can be calculated from the Eq. (13) for given set of parameter (β, μ, D) values. For abrupt changes in input voltage values there is significant nonlinear effect in the output response and for input types with continuous value changes the output response is not showing any nonlinearities and so the maximum attainable state variable value is more near to 1 in the same case.

The overview of results for all above types of waveforms is given in Table 1 below including maximum attainable state variable value and the Flux applied (Area under the Driving Voltage Curve) for the same.

The Table 1 clarifies that for maximum attainable state variable value, which is showing here full length boundary movement, the required Flux is almost same for all types of waveforms and this itself is almost equal to the theoretically calculated value of Flux for full length boundary Movement from (13).

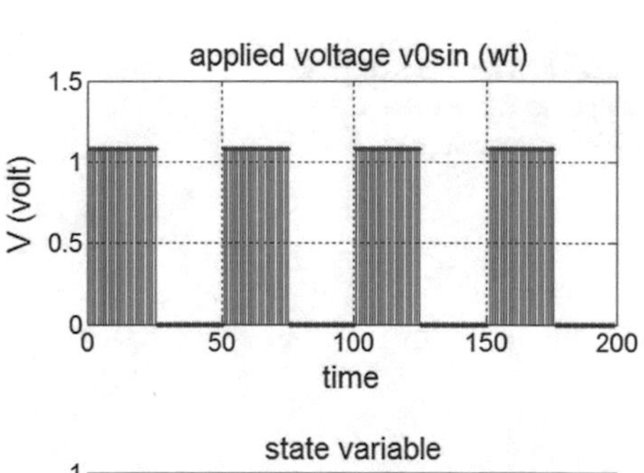

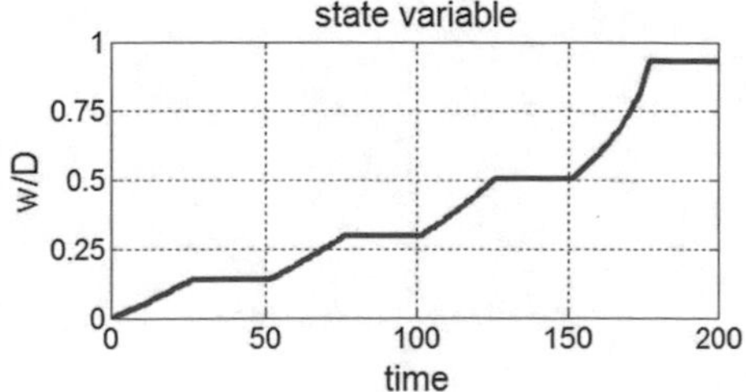

Fig. 12 The state variable response for Four pulse (Digital) voltage input

Table 1 Flux requirement for maximum attainable state variable value for different waveforms

Waveform shape	Area of positive cycle	Time period T (s)	Amplitude V_0(V)	Flux in positive cycle	$X_{max} = (w/D)_{max}$
Sinusoidal	$(1/\pi)V_0.T$	2	1.29	0.821	0.96
Triangular	$(¼)V_0.T$	1.5	2.19	0.821	0.96
Square NRZ	$(½)V_0.T$	1.5	1.09	0.817	0.91
Square RZ	$(¼)V_0.T$	1.5	2.183	0.818	0.87
2 Pulse	$(½)V_0.T$	1.5	1.05	0.787	0.84
4 Pulse	$(½)V_0.T$	1.5	1.05	0.787	0.84

The same calculation procedure can be applied for any certain value of state variable value. The only difference will be the value of $x(t)$ in (12) for which we are analyzing the model. We analyze the linear memristor model for the same set of parameter values, i.e. $\beta = 160$, $D = 10\,\text{nm}$, $\mu = 10^{-14}\ \text{m}^2/\text{V.Sec}$ with the final state variable value $x(t) = 0.27$. The corresponding flux value is obtained as

$$\varphi(t) = \frac{1}{k}\left[x(t) - \frac{x(t)^2}{2}\right] = 1.6\left[0.27 - \frac{(0.27)^2}{2}\right] = 0.374 \tag{20}$$

the driving voltage wave we are considering here is square pulse (RZ) with $V_0 = 1$ V and $T = 1.5$. For these values, the area under the curve is obtained as $V\left(\frac{T}{4}\right) = 0.375$, which is again equal to the theoretically calculated required flux value given in (20). The state variable response is shown in Fig. 13.

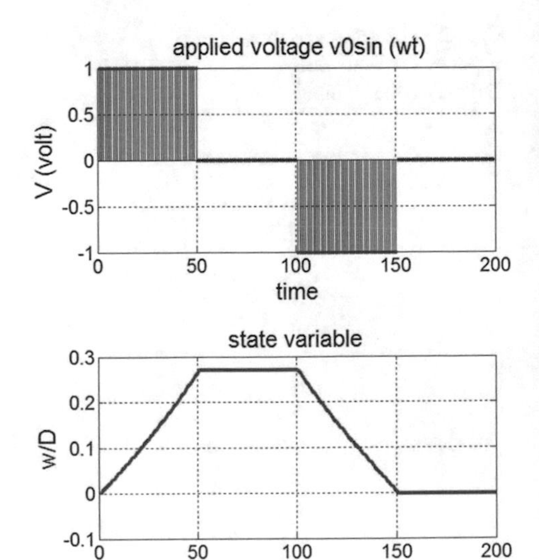

Fig. 13 The state variable response for square pulse (RZ) voltage input

Figure 13 clarifies that, except for the end positions, the nonlinear effects are insignificant for middle part of the device as the nonlinearities in output response are very less even for the input voltage wave with abrupt changes in the magnitude.

5 Shortcomings of the Linear Memristor Model

The shortcomings of the linear memristor models are its terminal state problem, boundary issues and nonlinear effects. These are as explained below:

5.1 *The Terminal State Problem*

After a limit of the voltage amplitude the state variable value goes out of the defined limit, i.e. [0,1]. This is named as '*The Terminal State Problem*' as after a limit the state variable cannot come to the range defined even if negative voltage is applied.

Figure 14 clarifies that in first positive half cycle the memristance is decreasing while the state variable is going towards maximum value and vice versa for the next negative half cycle. It can be seen that the maximum value of state variable is not exactly equal to 1. We expect that the voltage amplitude is not sufficient for full length conductive layer (moreover it also depends on the initial value of the state variable, but here we are assuming it to be zero).

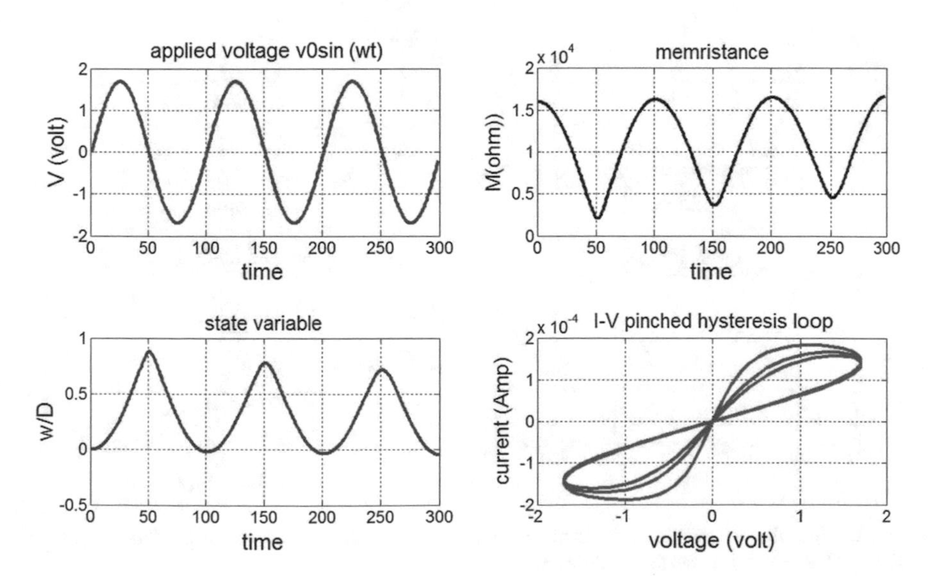

Fig. 14 Simulation curves for linear model with $V_0 = 1.69$ V, $T = 1.5$ s, Steps = 100/cycle

Table 2 First maximum and minimum value of state variable with increasing voltage amplitude

V_0	$X_{max \cdot 1}$	X_{min1}	X_{min2}
1.69	0.86	0	–0.03
1.7	0.9	0	–0.04
1.71	0.93	–0.02	–0.04
1.72	0.953	–0.02	–0.05
1.725	0.99	–0.025	–0.055
1.73	2	1.2	1.2

Increasing the voltage amplitude increases the maximum value that can be obtained by the state variable, but up to an extent only and after that terminal state problem arises.

Table 2 shows the corresponding maximum values of state variable. We can see from Table 2 that voltage amplitude value 1.73 is showing terminal state problem as the minimum value of state variable is also 1.2, which is not in the predefined range [0, 1]. Figure 15 clarifies the terminal state problem more clearly.

There are also other problems observed in Table 2 such as the negative minimum values of state variable. This is due to the programming in Digital Environment. Here application of sinusoidal driving voltage on memristor is done with 100 time steps and these steps are taking the value of $V_0 \sin(\omega t)$ for total step period which is at the starting instant of the step. So there is arising some asymmetry in wave in positive and negative half cycle.

Figure 16 depicts this phenomena with number of steps equal to 20 and 100.

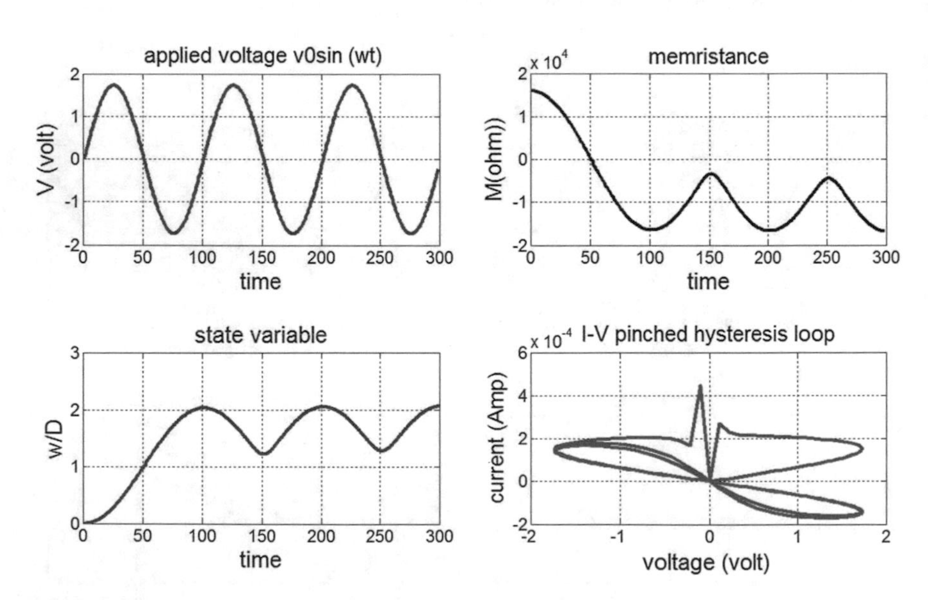

Fig. 15 Simulation curves for linear model showing The Terminal State Problem, with $V_0 = 1.73$ V, $T = 1.5$ s, Steps = 100/cycle

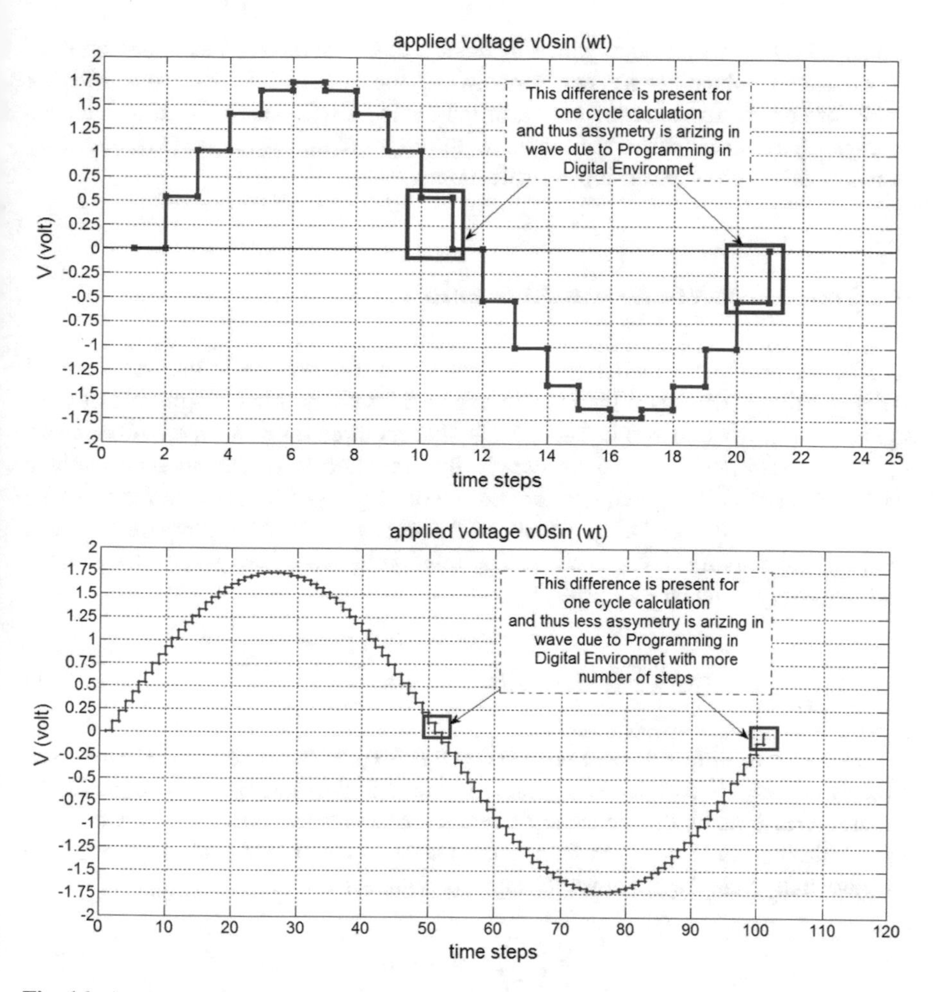

Fig. 16 Asymmetry in biasing voltage waveforms due to digital environment

5.2 *Boundary Issues and Nonlinear Effects*

When assuming that the generated electric field is small enough, the linear dopant drift model can approximate the dynamics of a memristor. However, this model is invalidated at boundaries when the boundary between doped and undoped regions is at either of end, i.e. when $w \gtrsim 0$ or $w \lesssim D$. This is due to the influence of a non-uniform electric field that significantly suppresses the drift of the dopants. The limitations of this model are revealed when, for example, driving a TiO_2/TiO_{2-x} memristor (*R*on = 100 Ω, *R*off = 16 kΩ, $w_0 = 5$ nm, $D = 10$ nm, and $\mu = 10^{-14}$ m^2/V s) into its extreme states, i.e., saturation ($w = D$) and depletion ($w = 0$). In the case of saturation, *w* exceeds the limit value of D (10 nm), whereas the device's memristance falls below the 100 Ω cut-off value (*R*on). Likewise, in depletion,

w can take negative values with the memristance exceeding the upper limit of 16 kΩ (Roff), which is clearly erroneous. At the edges of device some fringing fields also exists and over the length of the device also some nonlinearities exists. Applying a nonlinear drift over dopants at the edges of device to get rid of boundary issues is called Non-linear dopant drift model.

6 Frequency-Voltage Relationship

The Linear Memristor model analysis shows that the increase in the frequency of Driving Voltage decreases the Flux in one cycle with the same magnitude of voltage. As we have discussed in Table 1, the flux required for movement of boundary up to a certain position or more clearly flux required for a certain state variable value is constant if the parameters are taken not varying their value. So for applying same Flux with higher Driving Voltage frequency we have to increase the peak value of the Driving Voltage wave. The relation between this peak voltage value and frequency is linear and given by

$$x(t) - \frac{[x(t)]^2}{2} = k\varphi(t), \quad k = \mu \frac{R_{on}}{D^2 R_{off}}, \; \beta = \frac{R_{off}}{R_{on}} \tag{21}$$

We take two different cases with $x(t) = 1$ and $x(t) = 0.5$ both with $x(0) = 0$. The flux values for these boundary positions are 0.8 Wb and 0.6 Wb respectively. Figures 17a, b show the frequency voltage curves for these two cases with sinusoidal biasing inputs for both linear and nonlinear memristor models. The flux in positive half cycle for a sinusoidal input is obtained as

$$\varphi\left(\frac{T}{2}\right) = V\frac{T}{\pi} = 0.8 \tag{22}$$

Thus, the relation between frequency and voltage for full length boundary movement can be expressed as

$$V = \frac{0.8\pi}{T} \text{ or } V = 2.512f \tag{23}$$

The relation (23) between V and f is linear. Thus, the curve $V = mf$ is expected to be linear with slope $m = 2.512$. The nonlinear model is less compatible with the theoretical model but is showing more nearby values with practically measured data, as we will see later. The initial boundary value for these cases in nonlinear models is $x(0) = 0.01$. Similarly, the second case of $x(t) = 0.5$ is giving Frequency Voltage relationship $V = 1.884\,\mathrm{f}$, which is linear again as expected.

It is clear from Fig. 17 that the nonlinear model is showing less correlation with the theoretical model as compared to the linear model. While Fig. 17b shows that

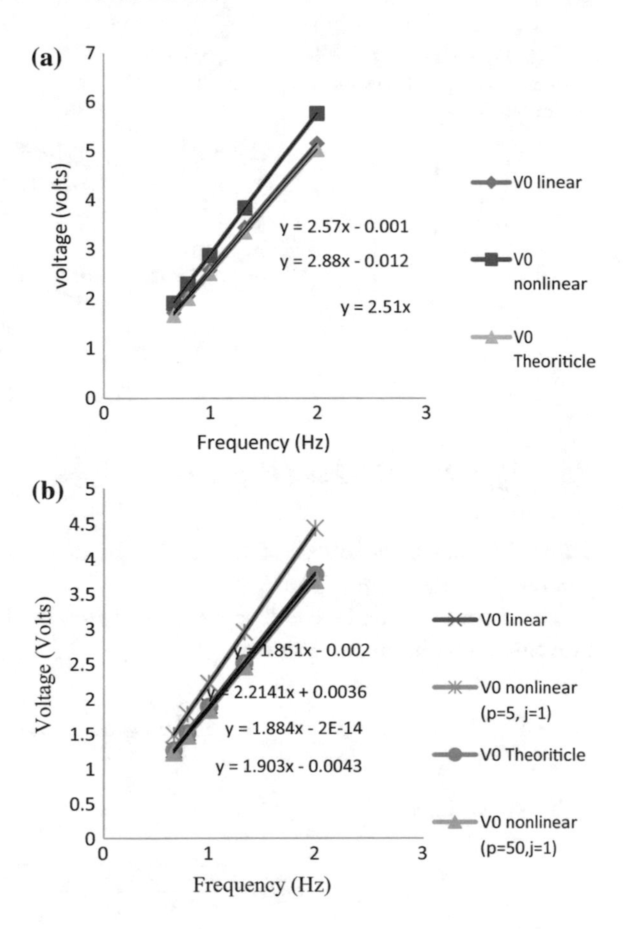

Fig. 17 Frequency voltage relationship for linear and nonlinear models for state variable value **a** $x(t) = 1$ and **b** $x(t) = 0.5$ with $\beta = 160$, $\mu = 10^{-14}$ m^2/v.sec

with higher values of *p*, the correlation nonlinear model response and theoretical model can be increased. The equation of linear curves are (shown in graphs above) more clearly showing the correctness of the model.

7 Effect of Parameters on F-V Curves

The reason behind the linear relationship between frequency and voltage is the dependence of flux only on the initial and final boundary position. The flux in fact depends upon the parameter values also i.e. the total device length (D), the off to on resistance ratio (β) and the mobility of charged dopant ions (μ). To observe the effect of these parameters on the flux or more clearly on frequency-voltage relationship, here one parameter is changing, keeping others constant.

Fig. 18 Dependency of frequency-voltage curves on device length D

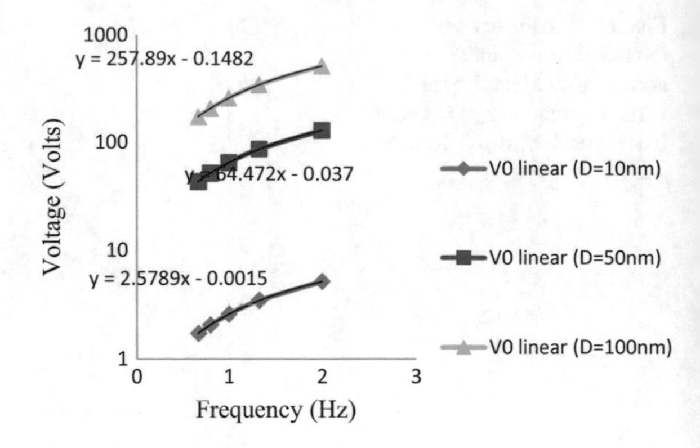

7.1 *Effect of Device Length (D)*

Figure 18 shows the effect of device length on frequency-voltage curves (with $\beta = 160$, $\mu = 10^{-14}$ m^2/v.sec).

For theoretical understanding of dependence on D for full length boundary movement condition, we can say

$$x(t) - \frac{[x(t)]^2}{2} = \frac{\mu}{D^2\beta}\varphi(t) \tag{24}$$

Thus, we have

$$\varphi(t) = \frac{0.5D^2(160)}{10^{-14}} = V\left(\frac{T}{\pi}\right) \tag{25}$$

or

$$V = (251.2)D^2 f \times 10^{14} \tag{26}$$

For $D = 10$ nm, 50 nm, 100 nm, the frequency-voltage relationship in Eq. (26) becomes $V = 2.51$ f, $V = 62.75$ f, $V = 251$ f, respectively.

The graphs of curves in Fig. 18 show good correlation with these theoretical equations defining the relation between voltage and frequency.

7.2 *Effect of R_{Off} to R_{On} Ratio (β)*

Figure 19 shows the effect of R_{Off} to R_{On} Ratio on frequency-voltage curves (D = 10 nm, $\mu = 10^{-14}$ m^2/v.sec). For theoretical understanding of dependence on β for full length boundary movement condition, we can say

Fig. 19 Dependency of frequency-voltage curves on R_{Off} to R_{On} Ratio (β)

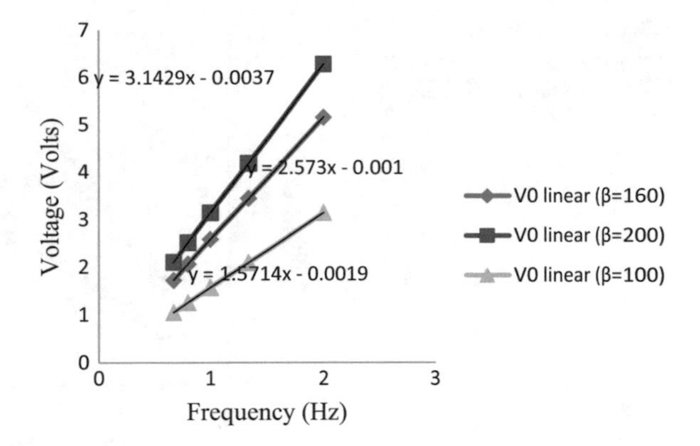

$$x(t) - \frac{[x(t)]^2}{2} = \frac{\mu}{D^2\beta}\varphi(t) \tag{27}$$

Thus, we have

$$\varphi(t) = \frac{0.5D^2(10^{-16})}{10^{-14}} = V\left(\frac{T}{\pi}\right) \tag{28}$$

or

$$V = (0.0157)\beta f \tag{29}$$

For $\beta = 160$, 100, 200, the frequency-voltage relationship in Eq. (29) becomes $V = 2.51\,\mathrm{f}$, $V = 1.57\,\mathrm{f}$, $V = 3.14\,\mathrm{f}$, respectively.

The graphs of curves in Fig. 19 show good correlation with these theoretical equations defining the relation between voltage and frequency.

7.3 *Effect of Charged Dopant Mobility (μ)*

Figure 20 shows the effect of Charged Dopant Mobility (μ) on frequency-voltage curves ($D = 10$ nm, $\beta = 160$). For theoretical understanding of dependence on μ for full length boundary movement condition, we can say

$$x(t) - \frac{[x(t)]^2}{2} = \frac{\mu}{D^2\beta}\varphi(t) \tag{30}$$

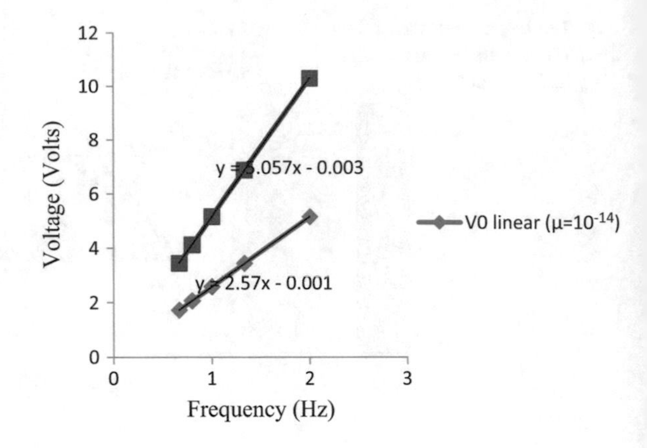

Fig. 20 Dependency of frequency-voltage curves on Charged Dopant Mobility (μ)

Thus, we have

$$\varphi(t) = \frac{0.5 \times 160 \times 10^{-16}}{\mu} = V\left(\frac{T}{\pi}\right) \tag{31}$$

or

$$V = \left(\frac{251.2 \times 10^{-16}}{\mu}\right) f \tag{32}$$

For $\mu = 10^{-14}$ m^2/v.sec, 5×10^{-15} m^2/v.sec, the frequency-voltage relationship in Eq. (32) becomes $V = 2.51$ f, $V = 5.02$ f, respectively. The graphs of curves in Fig. 20 show good correlation with these theoretical equations defining the relation between voltage and frequency.

7.4 *Effect of Driving Voltage Wave*

This is clear from state variable response for all waveform types of Driving Voltage that full length boundary movement can be achieved with a certain flux value. This Flux value can be calculated for given set of parameter ($\beta = 160, D = 10$ nm, $\mu = 10^{-14}$ m^2/v.sec) values. The overview of results for all above types of waveforms is given in Fig. 21 including maximum attainable state variable value and the Flux applied (Area under the Driving Voltage Curve) for the same.

The chart shows that for full length boundary movement, with given set of parameter values, the required flux value is $\varphi(t) = 0.8$ wb, but the maximum boundary position $x_{\max}$ cannot be equal to 1 for all types of waves due to some nonlinearities present in the device for those specific waves.

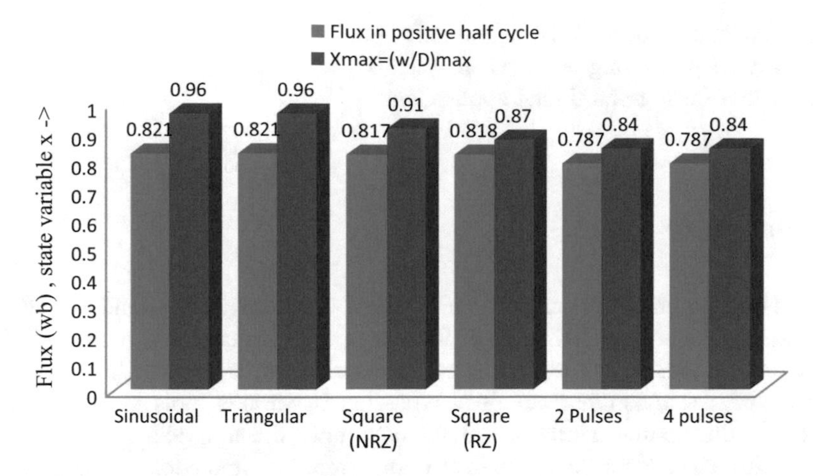

Fig. 21 Effect of driving voltage wave shape on maximum attainable boundary position

The required flux for achievable maximum boundary position can be defined by the following equation

$$x(t) - \frac{[x(t)]^2}{2} = \frac{\mu}{D^2\beta}\varphi(t) \tag{33}$$

For 2 pulse type biasing voltage, $x_{\max} = 0.84$ and the required flux for the same is $\varphi(t) = 0.78$. Also, the simulated flux value in the chart is in good coordination with the theoretical value.

8 Applications of Memristor

Recently there has been an increased interest in research on memristors due to the demonstration of memristor manufacturing as well as their potential applications. Research is in full swing to use memristors in computer memory, analog circuits, sensors, and digital logic. Memristor models need to be made available for the design engineers to use the memristor as a circuit element during design exploration. The following areas are where researchers getting interest and so getting familiar with memristor-

(i) Application of memristor in programmable logic designing
(ii) Memristor crossbar array formation
(iii) Memristor based nonvolatile memory designing and analysis
(iv) Artificial intelligence
(v) Thinking machine

(vi) Realization of artificial neural networks
(vii) Signal processing and control systems
(viii) Other analog and digital applications

9 Summary

A brief background of memristor and its development as a device and logic design are discussed first in this chapter. A linear model of memristor is proposed. Next, mathematical modelling and simulation results are presented for the linear model of memristor. The memristor has been modelled in various tools in VLSI Design including the nonlinear effects but till now only linear model is available for MATLAB. The conclusions have been drawn from the simulation results. The final boundary position depends on the flux passed through the device and initial boundary position for constant parameter (β, μ, D) values. So the boundary will definitely come to its initial position if the net applied flux is zero. The final state variable value does not depend upon the driving voltage wave shape if the flux corresponding to all the wave shapes is same. The terminal state problem can be successfully overcome by a lesser value of scaling parameter, both for hard current and soft current applications. Memristor is new device having both linear and non-linear behaviour. Linear model for memristor is discussed in detail in this chapter and non-linear model may considered as future work for memristor development.

References

Bhushan, S., Khandelwal, S., & Raj, B. (2013). Analyzing different mode FinFET based memory cell at different power supply for leakage reduction, *Proceedings of Seventh International Conference on Bio-Inspired Computing*.

Biolek, Z., Biolek, D., & Biolkova, V. (2009). SPICE model of memristor with non-linear dopant drift. *Radioengineering, 18*(2), 210–214.

Chua, L. O. (1971). Memristor-the missing circuit element. *IEEE Transactions on Circuit Theory, 18*(5).

Chua, L. O., & Kang, S. (1976). Memristive devices and systems. *Proceedings of the IEEE, 64*(2), 209–223.

da Costa, H. J. B., de Assis Brito Filho, F., & de Araujo do Nascimento, P. I. (2012). Memristor behavioural modeling and simulations using verilog-AMS, *Third Latin American Symposium on Circuits and Systems IEEE*, (pp. 1–4).

Dennard, R. (1968). Field-effect transistor memory, US 3387286, issued 4 June 1968 (filed 14 July 1967).

Fuechsle, M., Miwa, J. A., Mahapatra, S., Ryu, H., Lee, S., Warschkow, O., Hollenberg, L. C. L., Klimeck, G., & Simmons, M. Y. (2012). A single-atom transistor, *Nature Nanotechnology*, *7*, 242–246.

Gergel-Hackett, N., Hamadani, B., Dunlap, B., Suehle, J., Richter, C., Hacker, C., et al. (2009). A flexible solution-processed memristor. *IEEE Electron Device Letters, 30*(7), 706–708.

Jain et al. K. (1982). Ultrafast deep-UV lithography with excimer lasers, *IEEE Electron Device Letter, EDL-3*(53).

Joglekar, Y. N., & Wolf, S. J. (2009). The elusive memristor: Properties of basic electrical circuits. *European Journal of Physics, 30*(4), 661–675.

Johnson, D. (2010). Junctionless transistor fabricated from nanowires. *IEEE Spectrum*. Retrieved 04-20-2010.

La Fontaine, B. (2010). Lasers and Moore's Law, SPIE Professional, p. 20, Oct. 2010.

Liu, G., Fang, L., Li, N., Sui, B., & Duan, Z. (2010). New behavioral modeling method for crossbar-based memristor, *Asia Pacific Conference on Postgraduate Research in Microelectronics and Electronics*, (pp. 356–359).

Larrieu, G., & Han, X. L. (2013). Vertical nanowire array-based field effect transistors for ultimate scaling. *Nanoscale, 5*(6), 2437–2441.

Mohanty, S. P. (2013). Memristor: from basics to deployment. *IEEE Potentials, 32*(3), 34–39.

Mohsin, F. (2010). A Multivalued Storage System Using Memristor, *Proceedings of 13th International Conference on Computer and Information Technology*, (pp. 343–346).

Noyce, R. (1961). Semiconductor device-and-lead structure, US 2981877, issued 25 April 1961 (filed 30 July 1959).

Pattanaik, M., Raj, B., Sharma, S., & Kumar, A. (2012). Diode based trimode multi-threshold CMOS technique for ground bounce noise reduction in static CMOS adders. *Advanced Materials Research, 548,* 885–889.

Prodromakis, T., & Papavassiliou, C. (2011). A Versatile Memristor Model With nonlinear Dopant Kinetics. *IEEE Transactions on Electron Devices, 58*(9), 3099–3105.

Raj, B. (2014). Quantum mechanical potential modeling of FinFET. *Towards Quantum FinFET*, (Vol. 17, pp 81–97). Springer. (ISBN 978-3-319-02021-1).

Raj, B., Saxena, A. K., & Dasgupta, S. (2008). A compact drain current and threshold voltage quantum mechanical analytical modeling for FinFETs. *Journal of Nanoelectronics and Optoelectronics (JNO) USA, 3*(2), 163–170.

Raj, B., Saxena, A. K., & Dasgupta, S. (2009). Analytical modeling for the estimation of leakage current and subthreshold swing factor of nanoscale double gate finfet device. *Microelectronics International, UK, 26,* 53–63.

Raj, B., Saxena, A. K., & Dasgupta, S. (2011a). Nanoscale FinFET Based SRAM Cell Design: Analysis of Performance metric, Process variation, Underlapped FinFET and Temperature effect. *IEEE Circuits and System Magazine, 11*(2), 38–50.

Raj, B., Mitra, J., Bihani, D. K., Rangharajan, V, Saxena, A. K., & Dasgupta, S. (2011). Process variation tolerant FinFET based robust low power sram cell design at 32 nm technology. *Journal of Low Power Electronics (JOLPE), Academy Publisher, FINLAND, 7*(2), 163–171.

Raj, B., Saxena, A. K., & Dasgupta, S. (2011c). High performance double gate FinFET SRAM cell design for low power application. *International Journal of VLSI and Signal Processing Applications., 1*(1), 12–20.

Raj, B., Saxena, A. K., & Dasgupta, S. (2013). Quantum mechanical analytical modeling of nanoscale DG FinFET: evaluation of potential, threshold voltage and source/drain resistance. *Elsevier's Journal of Material Science in Semiconductor Processing, Elsevier, 16*(4), 1131–1137.

Raja, T., & Mourad, S. (2010). Digital logic implementation in memristor-based crossbars—a tutorial, *Proceedings. Fifth IEEE International Symposium on Electronic Design, Test & Applications*, (pp. 303–309).

Sharma, V. K., Pattanaik, M., & Raj, B. (2014). PVT variations aware low leakage INDEP approach for nanoscale CMOS circuits. *Microelectronics Reliability, 54*(1), 90–99.

Sharma, V. K., Pattanaik, M., & Raj, B. (2015). INDEP approach for leakage reduction in nanoscale CMOS circuits. *International Journal of Electronics, 102*(2), 200–215.

Shin, S., Kim, K., & Kang, S. M. (2010). Compact models for memristors based on charge–flux constitutive relationships. *IEEE Transactions on Computer-Aided Design of Integrated Circuits and Systems*, *29*(4).

Strukov, D. B., Snider, G. S., Stewart, D. R., & Williams, R. S. (2008). The missing memristor found. *Nature, 453,* 80–83.

Varghese, D., & Gandhi, G. (2009). Memristor based high linear range differential pair, *International Conference on Communications, Circuits and Systems*, (pp. 935–938).

Vishvakarma, S. K., Agrawal, V., Raj, B., Dasgupta, S., & Saxena, A. K. (2007). Two dimensional analytical potential modeling of Nanoscale Symmetric Double Gate (SDG) MOSFET with Ultra Thin Body (UTB). *Journal of Computational and Theoretical Nanoscience, 4*(6), 1144–1148.

Volos, Ch. K., & Kyprianidis, I. M., Stouboulos1, I. N., Tlelo-Cuautle2, E., Vaidyanathan, S. (2015). Memristor: a new concept in synchronization of coupled neuromorphic circuits, *Journal of Engineering Science and Technology Review*, *8*(2), 157–173.

Wanlass, F. (1967). Low stand-by power complementary field effect circuitry", US 3356858, issued 5 December 1967 (filed 18 June 1963).

Dynamics of Delayed Memristive Systems in Combination Chaotic Circuits

O.A. Adelakun, S.T. Ogunjo and I.A. Fuwape

Abstract The use of memristor in the realization of chaotic circuits has gained popularity in recent times. This can be attributed to its simplicity over the traditional Chua's diode. The memristor as a nanometer-scale passive circuit element which can be described as a resistor with memory and possesses nonlinear characteristics. In this chapter, the numerical and experimental dynamics of non-autonomous time delay memristive oscillator which consists of negative conductance and smooth-cubic memristor are reported. Diffusive and negative feed back coupling of combination-combination arrays of the electronic circuits are also presented. The viability of both numerical and electronic simulation are also presented.

1 Introduction

The discovery of a chaotic attractor by Lorenz (1963) brought attention to the study of chaos and chaotic systems. Several chaotic attractors such as Duffing oscillators, Rossler Chaotic attractors, (Pehlivan and Uyaroglu 2010) etc. have proposed for the study of chao s by researchers. Over the years, the study chaos has been extended to the study of time series data. Many natural and physical systems such as population of beetles (2013), BZ chemical reaction, economic data (Fuwape and Ogunjo 2013, 2015), atmospheric data (Fuwape et al. 2016) have been found to be chaotic. The pioneering work of Pecora and Carroll (1990) on synchronization showed that two similar or different systems following different trajectories can be made to track one another.

O.A. Adelakun · S.T. Ogunjo (✉) · I.A. Fuwape
Federal University of Technology, Akure, Ondo State, Nigeria
e-mail: stogunjo@futa.edu.ng

O.A. Adelakun
e-mail: aoadelakun@futa.edu.ng

I.A. Fuwape
e-mail: iafuwape@futa.edu.ng

© Springer International Publishing AG 2017

S. Vaidyanathan and C. Volos (eds.), *Advances in Memristors, Memristive Devices and Systems*, Studies in Computational Intelligence 701,
DOI 10.1007/978-3-319-51724-7_19

Synchronization is a process wherein two (or many) chaotic systems (either equivalent or nonequivalent) adjust a given property of their motion to a common behavior due to a coupling or to a forcing (periodical or noisy) (Boccaletti et al. 2002). Chaos synchronization has practical applications in secure communication (Ojo and Ogunjo 2012; Adelakun et al. 2014b), neuronal dynamics (Wang et al. 2011), chemical reactions, etc. Different synchronization schemes have been proposed and implemented for synchronization between chaotic systems, these include: active control, adaptive control, active backstepping, feedback control. Performance of different synchronization schemes have been investigated for integer order (Ojo et al. 2013) and fractional order (Ogunjo et al. 2017). The need for a multiuser communication scheme has led to the development of communication systems with multiple drives. In combination synchronization, synchronization is achieved between three similar or dissimilar chaotic systems (Ojo et al. 2016) and combination-combination synchronization between multiple drives and multiple slaves (Ojo et al. 2015a, b). Increased or reduced order synchronization is the synchronization of two or more systems with different order. Different order synchronization has been achieved between two systems (Ogunjo 2013) and multiple systems (Ojo et al. 2014a, b).

The prospect of practical application of chaos theory in secure communication has led to circuit implementation of chaotic circuits. Practical chaos based communication scheme has been achieved using fibre optics (Argyris et al. 2005) and semiconductor lasers (Mengue and Essimbi 2012). Synchronization of systems for secure communication has been implemented for discrete systems (Nagaraj and Vaidya 2009), integer order systems (Wang et al. 2012) and fractional order systems (El-Sayed et al. 2016). Rigorous testing and breaking of chaos based secure communication scheme has led to continuous development of more robust chaos based communication schemes (Jinfeng and Jingbo 2008; Li et al. 2012). Elhadj and Sprott (2008) reported that the attractors by Chua et al. (1986) has more complex dynamics than the Lorenz type attractors. This could be attributed to the presence of a memristor.

The memory resistor or memristor is a nonlinear device. It was proposed by Chua (1971) as the fourth circuit element after resistor, capacitor and inductor. The first practical memristor based on titanium dioxide thin films was developed by Strukov et al. (2008). A flux controlled memristor is characterized by a memductance ($W(\phi)$) that describes the flux dependent rate of change of charge (El-Sayed et al. 2013) as

$$W(\phi) = \frac{dq(\phi)}{d\phi} \tag{1}$$

using the relationship $i = \frac{dq}{dt}$ and $v = \frac{d\phi}{dt}$, the current through a flux controlled memristor can be written as

$$i(t) = W(\phi)v(t) \tag{2}$$

Various expressions proposed for the flux-dependent rate of change of charge include: $q(\phi) = a_1\phi + a_2\phi^3$ (El-Sayed et al. 2013; Adelakun et al. 2014a), $q(\phi) = -a\phi + 0.5b\phi^2 sgn(\phi)$ (Bo-Cheng et al. 2011), $q(\phi) = \phi^3 + a\phi^2 + b\phi + c$ (Messias et al. 2010) and

$$q(\phi) = \begin{cases} b\phi + b - a, & \text{if } \phi < -1; \\ a\phi, & \text{if } -1 < \phi \leq 1; \\ b\phi + a - b, & \text{if } x > 1. \end{cases} \quad (3)$$

where $a, b, c\, \mathbb{R}$ (Zuo and Cao 2015).

Memristors have found applications in neuronal spike event generation (Shin et al. 2012) and artificial neuron modelling (Aihara 1991). Memristor based Chua circuit for the generation of multiple attractors have been implemented (Xu et al. 2016). The dynamics of different memristor based nonlinear oscillators have been derived and explore (Itoh and Chua 2008). Theoretical analysis of memristive systems have also gain attention over the years (Adelakun 2013, 2014).

2 Related Works

The pioneering work of Pecora and Carroll (1990) in synchronizing two chaotic systems has led to new innovations such as increased order synchronization (Ogunjo 2013), compound synchronization (Ojo et al. 2016), compound-compound synchronization (Ojo et al. 2015a), fractional order synchronization (Ogunjo et al. 2017) and others. Chua (1971) predicted the memristor as a circuit element. Significant advances has been made in the development of the memristor and practical applications in secure communications (Xu et al. 2016; Adelakun 2013, 2014; Aihara 1991). Recent advances in the field of chaos, intelligent and control systems have been discussed extensively (Vaidyanathan and Volos 2016a, b; Azar and Vaidyanathan 2016).

3 System Description

Papadopoulou et al. (2008) proposed a non-autonomous chaotic system with double bell attractor. The systems is described as

$$\begin{aligned} \frac{dV_{C1}}{dt} &= \frac{1}{C1}(i_{L1} - i) \\ \frac{dV_{C2}}{dt} &= -\frac{1}{C2}(G_n V_{c2} + i_{L2} - i_{L1}) \\ \frac{di_{L1}}{dt} &= \frac{1}{L1}(V_{c2} - V_{c1} - i_{L1} R_1) \\ \frac{di_{L2}}{dt} &= \frac{1}{L1}(V_{c2} + i_{L2} R_2 + V_s(t)) \end{aligned} \quad (4)$$

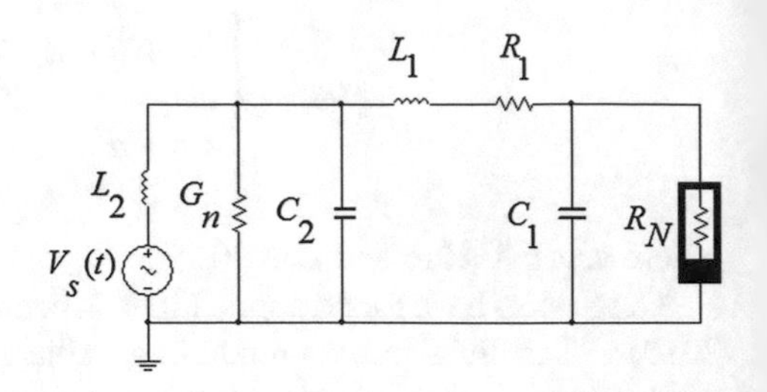

Fig. 1 Fourth order chaotic system using memristor (Papadopoulou et al. 2008)

Papadopoulou et al. (2008) used a nonlinear function i of the form described in Eq. 3. Synchronization of system (4) was reported by Mamat et al. (2012). This work extends these earlier work done by introducing an extra dimension, computing the dynamics of the system and combination-combination synchronization of the system (Fig. 1).

A time delayed memristive system described by Eq. 5 is used in this work.

$$\begin{aligned} C_1\frac{dV_{C1}}{dt} &= i_{L1} - i_M \\ C_2\frac{dV_{C2}}{dt} &= -g_n V_{c2} + i_{L2} - i_{L1} \\ L_1\frac{di_{L1}}{dt} &= V_{c2} - V_{c1} - i_{L1}r_1 \\ L_2\frac{di_{L2}}{dt} &= -V_{c2} - i_{L2}r_2 + V_s(t) \\ \frac{d\phi}{dt} &= V_{c1} \end{aligned} \tag{5}$$

where the current through the memristor $i_m = W(\phi)v_{c1}(t-\tau)$. The memductance of the form $W(\phi) = \frac{dq(\phi)}{d\phi} = a + 3b\phi^2$ is used in this research work. The IV characteristics of the memristive element is shown in Fig. 5. Negative conductance $g = -0.475$. Input sinusoidal signal $V_s = v_p \sin 2\pi ft$ with amplitude $v_p = 1$ V, frequency $f = 1$ kHz, internal resistance $R_{10} = r_2 = 1\Omega$. The circuit realization of the system described in Eq. 5 is shown in Fig. 2.

The smooth-cubic function Memristor $W(\phi) = a + 3b\phi^2$ and Negative conductance (G) were implemented using the circuits in Figs. 3 and 4 respectively. The $i - v$ characteristics of the smooth cubic function memristor and negative conductance are shown in Fig. 5.

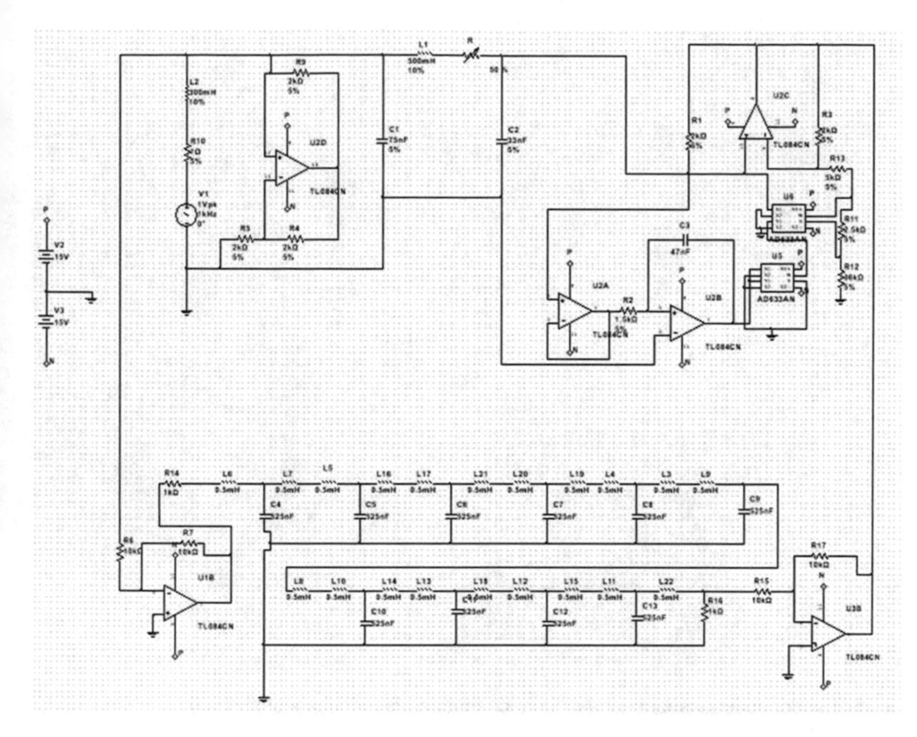

Fig. 2 Schematic circuit of fifth order non-autonomous Time Delayed Memristor (NTDM) chaotic system. $L3 = L22 = 3.5\,\text{mH}$, $L1 = 500\,\text{mH}$, $L2 = 300\,\text{mH}$, $C1 = 33\,\text{nF}$, $C2 = 75\,\text{nF}$, $C3 = 47\,\text{nF}$, $C4 - C13 = 525\,\text{nF}$, $R1 = R3 = R4 = R5 = R7 = R8 = R9 = 2\ \text{k}\Omega$, $R2 = 1.5\ \text{k}\Omega$, $R6 = R7 = R15 = 10\ \text{k}\Omega$, $R10 = 1\Omega$, $R11 = 2.5\ \text{k}\Omega$, $R12 = 36\ \text{k}\Omega$, $R13 = 5\ \text{k}\Omega$, $R14 = R16 = 1\ \text{k}\Omega$, Multiplier AD633AN, Operational Amplifier TL084CN, $RA = 1.1\,\text{k}\Omega$, $RB = 1\,\text{k}\Omega$, $RC = 1.14\,\text{k}\Omega$, $RD = 500\ \Omega$, AC Voltage source = 1 Vpk (1 kHz), Power supply = V1 = +15 V and V2 = −15 V

3.1 Equilibrium and Stability of the System

The equilibrium points of the system given in Eq. 5 can be found by equating the left hand side of the equation to zero. One possible equilibrium point for the system is, where ϕ is uncertain but constant (Bo-Cheng et al. 2011)

$$E = \{(v_{C1}, v_{C2}, i_{L1}, i_{L2}, \phi)|(v_{C1} = v_{C2} = i_{L1} = i_{L2} = 0, \phi = \phi_0\} \tag{6}$$

The Jacobian matrix of system (5) was obtained as

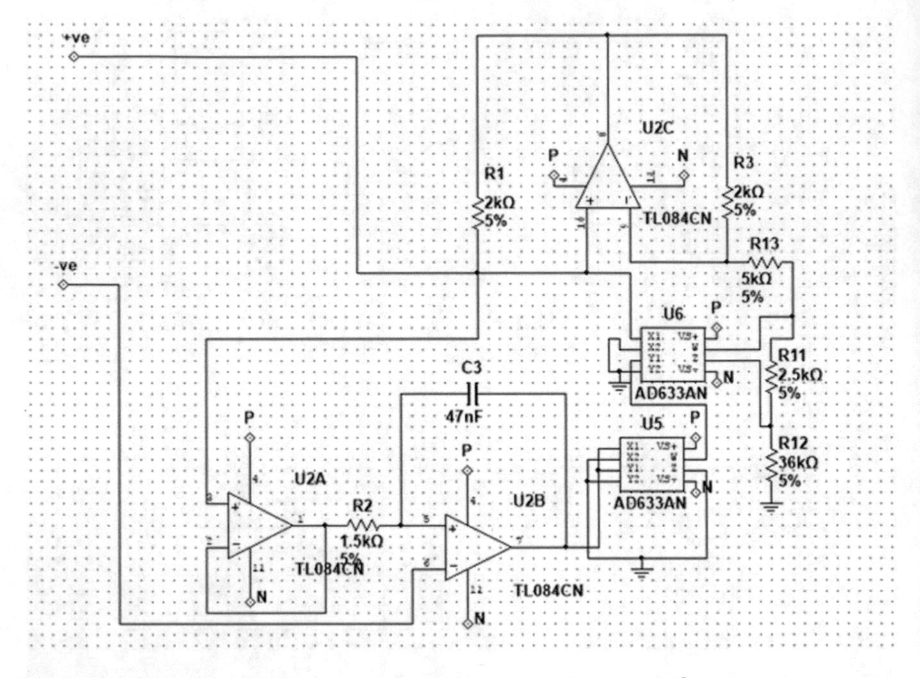

Fig. 3 Circuit realization of the smooth cubic function $W(\phi) = a + 3b\phi^2$

Fig. 4 Circuit realization of the negative conductance $G = -\frac{1}{R4}$

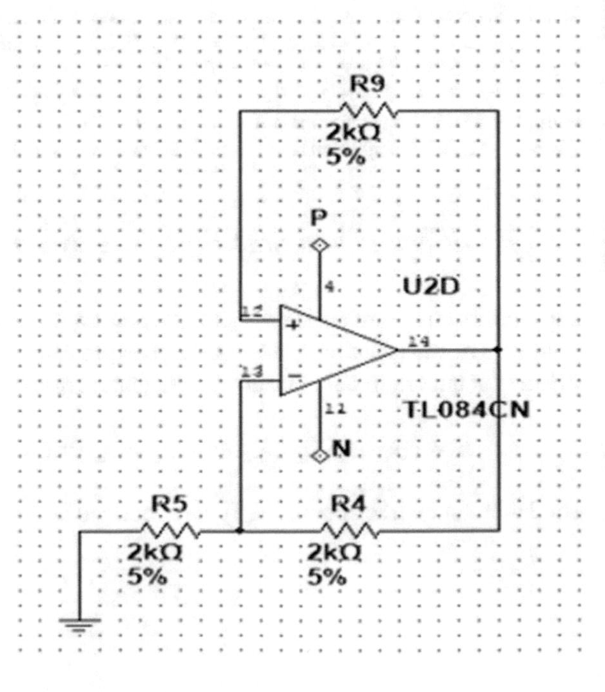

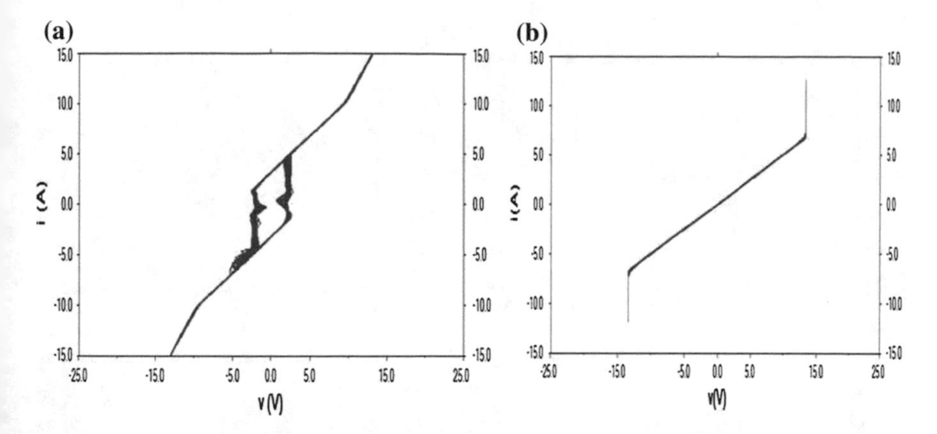

Fig. 5 $i-v$ characteristics of **a** smooth cubic nonlinear memristive element **b** negative conductance

$$J=\begin{bmatrix} -\frac{a}{C1}-\frac{3b\phi^2}{C1} & 0 & \frac{1}{C1} & 0 & -\frac{6bV_{C1}\phi}{C1} \\ 0 & -\frac{g_n}{C2} & -\frac{1}{C2} & \frac{1}{C2} & 0 \\ -\frac{1}{L1} & \frac{1}{L1} & -\frac{r1}{L1} & 0 & 0 \\ 0 & -\frac{1}{L2} & 0 & -\frac{r^2}{L2} & 0 \\ 1 & 0 & 0 & 0 & 0 \end{bmatrix} \tag{7}$$

At equilibrium point E, the Jacobian matrix $J(E)$ becomes

$$J(E)=\begin{bmatrix} -\frac{a}{C1}-\frac{3b\phi_0^2}{C1} & 0 & \frac{1}{C1} & 0 & 0 \\ 0 & -\frac{g_n}{C2} & -\frac{1}{C2} & \frac{1}{C2} & 0 \\ -\frac{1}{L1} & \frac{1}{L1} & -\frac{r1}{L1} & 0 & 0 \\ 0 & -\frac{1}{L2} & 0 & -\frac{r^2}{L2} & 0 \\ 1 & 0 & 0 & 0 & 0 \end{bmatrix} \tag{8}$$

The characteristic equation of the system can be written as

$$\lambda(\lambda^4+a_1\lambda^3+a_2\lambda^2+a_3\lambda+a_4)=0 \tag{9}$$

where

$$\begin{aligned} a_1 &= \frac{r^2}{L2}+\frac{r_1}{L1}+\frac{g_n}{C1}+\frac{1}{C1}W(\phi_0) \\ a_2 &= \frac{1}{C1}W(\phi_0)\left(\frac{r^2}{L2}+\frac{r_1}{L1}-\frac{g_n}{C1}\right)+\frac{g_n}{C1}\left(\frac{r_1}{L1}+\frac{1}{C1}-\frac{r^2}{L2}\right)+\frac{1}{C2}\left(\frac{1}{L1}-\frac{1}{L2}\right)+\frac{r1}{L1}\cdot\frac{r^2}{L2} \\ a_3 &= \frac{1}{C1}W(\phi_0)\Gamma+\frac{1}{C2L1}\left(\frac{r^2}{L2}-\frac{1}{L2C1}\right)+\frac{r1}{L1}\left(\frac{g_n}{C2}-\frac{1}{L2C2}\right)-\frac{1}{C1L1}\left(\frac{-g_n}{C2}-\frac{r^2}{L2}\right) \end{aligned}$$

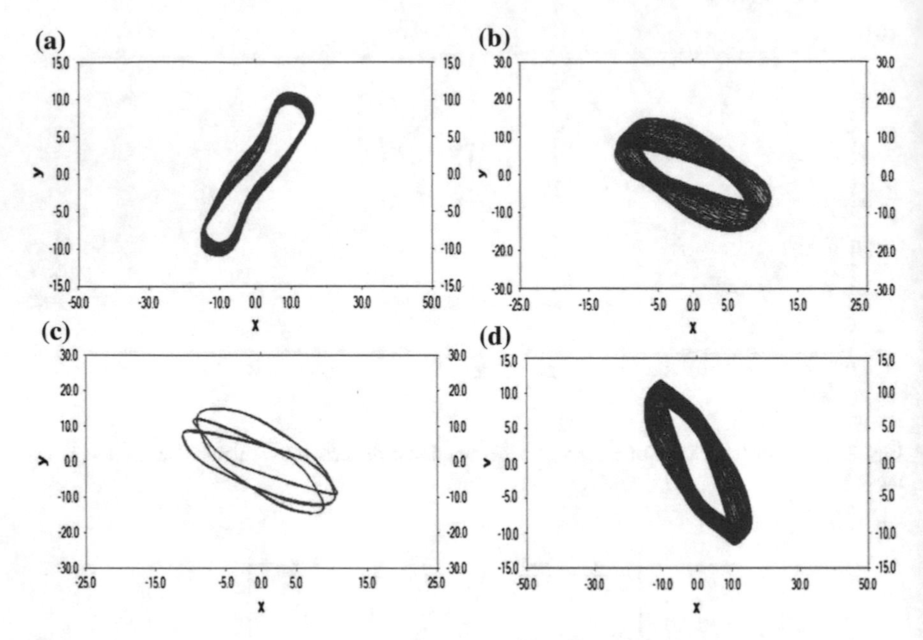

Fig. 6 Phase portraits $(x(t), y(t))$ of time-delayed unsaturated attractors of memristive systems when **a** 1.14 kΩ **b** 1.1 kΩ **c** 1 kΩ and **d** 500 Ω

$$a_4 = \frac{1}{C1}W(\phi_0)\left(\frac{-g_n r1}{C2L1} - \frac{r^2}{C2L1L2} + \frac{r1}{L1C2L2}\right) + \frac{-g_n}{C1C2}\left(\frac{r^2}{L1L2} + \frac{1}{L1L2C2}\right)$$

$$\Gamma = \left(\frac{1}{C2L2} - \frac{g_n r^2}{C2L2} - \frac{g_n r_1}{C2L1} - \frac{r1 r^2}{L1L2} - \frac{1}{C2L1}\right)$$

The equilibrium point E of the system is an unstable equilibrium when at least one root of Eq. 9 has real part greater than zero. It can be shown that the coefficients of the quartic polynomial in Eq. 9 are all non-zero. It can be shown that one or more roots of this polynomial have positive real parts using system parameters in Fig. 2 and the Routh-Hurwitz criteria. For a quartic polynomial the Routh-Hurwitz criteria are given as:

$$\begin{aligned} a_1 &> 0 \\ a_3 &> 0 \\ a_4 &> 0 \\ a_1 a_2 a_3 &> a_3^2 + a_1^2 a_4 \end{aligned} \tag{10}$$

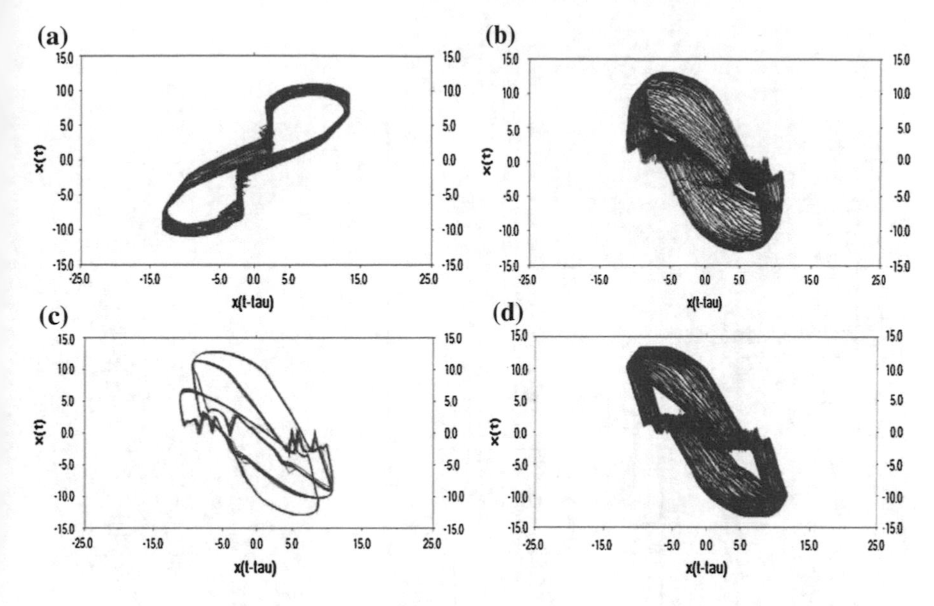

Fig. 7 Phase portraits $(x(t)x(t-\tau)$ of chaotic attractors for the delayed memristive systems when **a** 1.14 kΩ **b** 1.1 kΩ **c** 1 kΩ and **d** 500 Ω

Analysis of Eq. 10 suggests that the system has no stable for all positive real numbers, hence, the system can be said to be sensitive to initial conditions. The attractors of the fifth-order time delayed systems under different system parameters are shown in Figs. 6 and 7.

4 Coupling Schemes

Two different coupling schemes, diffusive and negative feedback, are used in this work. The circuit implementation of the diffusive and negative coupling scheme are shown in Figs. 8 and 9. In the combination-combination scheme considered in this work, two drive systems (A and B) were coupled to two slave systems (C and D). Circuit diagram for one each of the drive and response system are shown in Figs. 10 and 11 respectively.

5 Results and Discussion

Electronic simulation of the proposed synchronization was carried out by combining two drive systems (Fig. 10) and two response system (Fig. 11) using two different coupling schemes: diffusive (Fig. 8) and negative feedback coupling (Fig. 9).

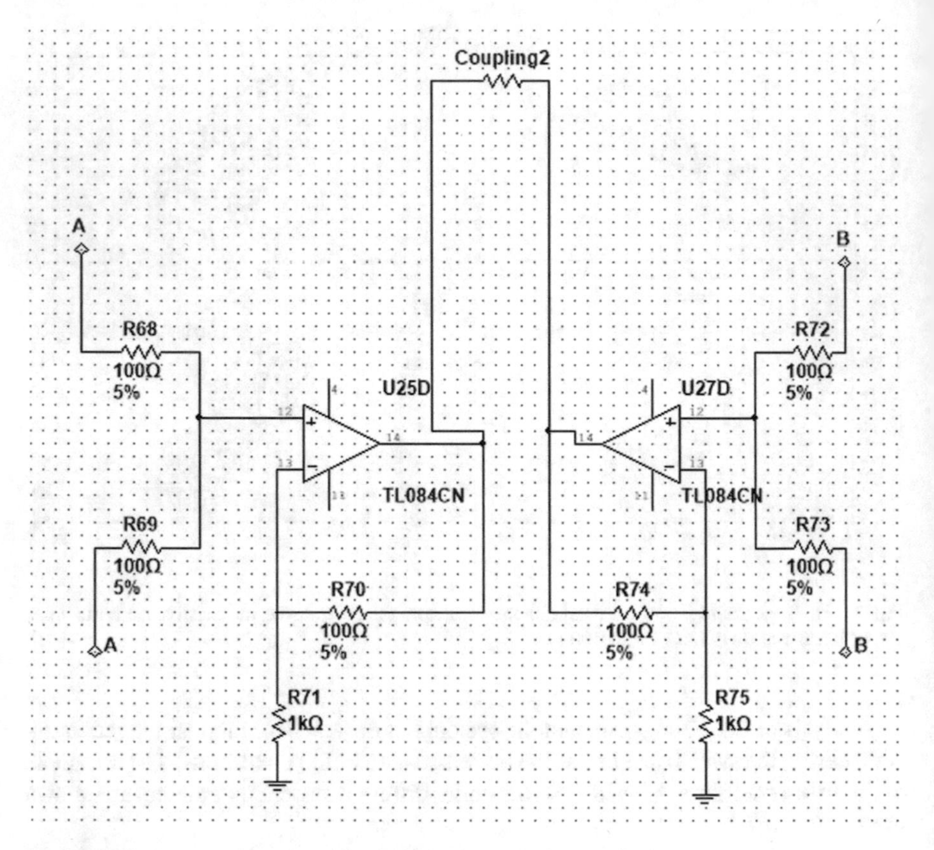

Fig. 8 Diffusive coupling scheme for combination-combination oscillators

The phase space of the two drive systems and two response systems without coupling is shown in Fig. 12. After the diffusive coupling of the drive and response systems, complete synchronization was observed in the phase space and time series as seen in Figs. 13 and 14 respectively. A similar result (Figs. 15 and 16) was obtained when negative feedback coupling was used to couple the two drive systems and two response system.

6 Conclusion

In this paper, a fourth order non-autonomous chaotic memristive system was modified into a fifth order time delayed chaotic memristive system. The equilibrium and stability of the new chaotic system shows that it has no stable region at equilibrium. This shows that the stability of the new system is sensitive to initial conditions, implying that it has complex dynamics. Combination-combination coupling involv-

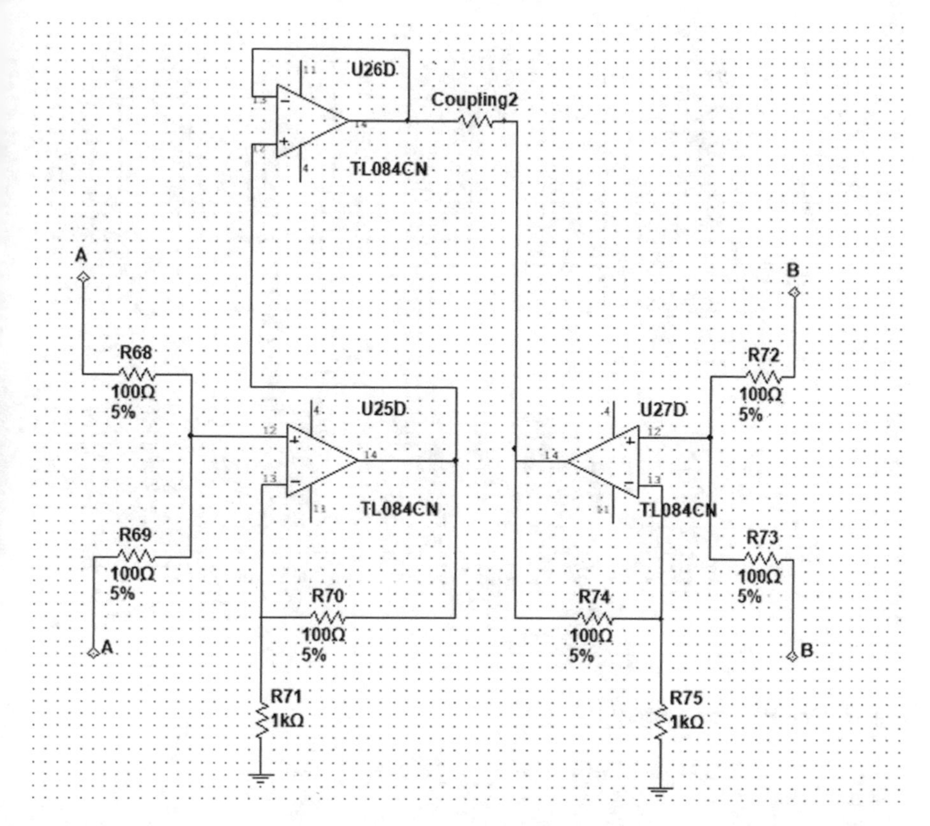

Fig. 9 Negative feedback coupling scheme for combination-combination oscillators

ing two drive systems and two response system of the same system was carried out using diffusive and negative feedback coupling. Electronic simulation of the coupled system was carried out and results presented.

Further research based on the success of this current work can be carried out. Practical implementation and deployment of this system in real-life application is suggested to test performance under field conditions. It is expedient to examine the possibility of signal switching between drives during transmission and between response systems at reception for improved security. The system needs to be subjected to actual known hacking methods such as brute force to determine reliability. Attempts should be made to test the possibility of different coupling schemes other than the ones used here. Furthermore, the use of different memductance in the drive to increase the complexity and security of the communication scheme. Since most communications systems use digital signals, it will be noteworthy to design the synchronization scheme for digital signal transmission.

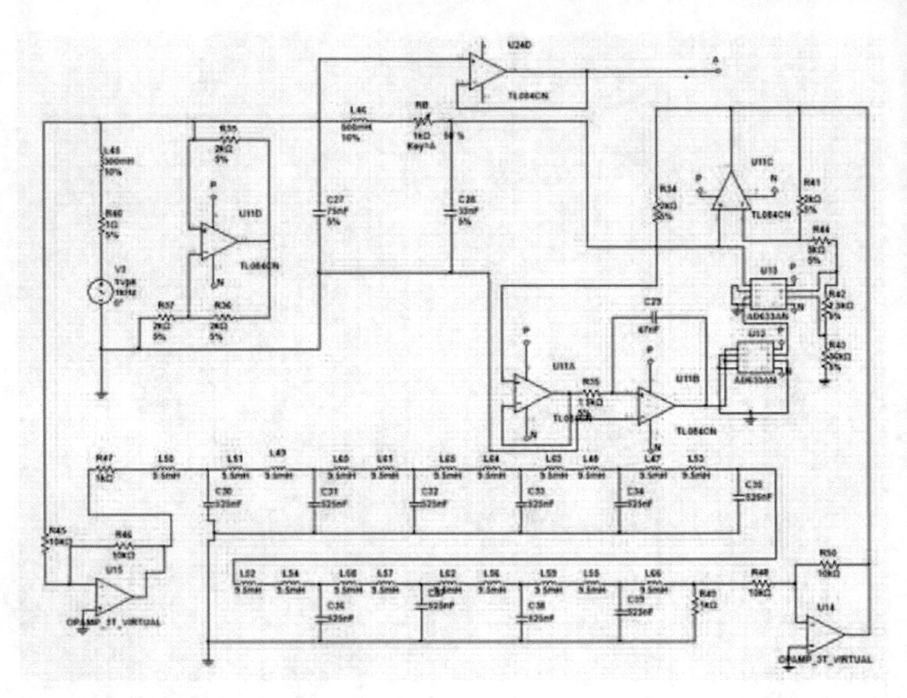

Fig. 10 One of the drive system

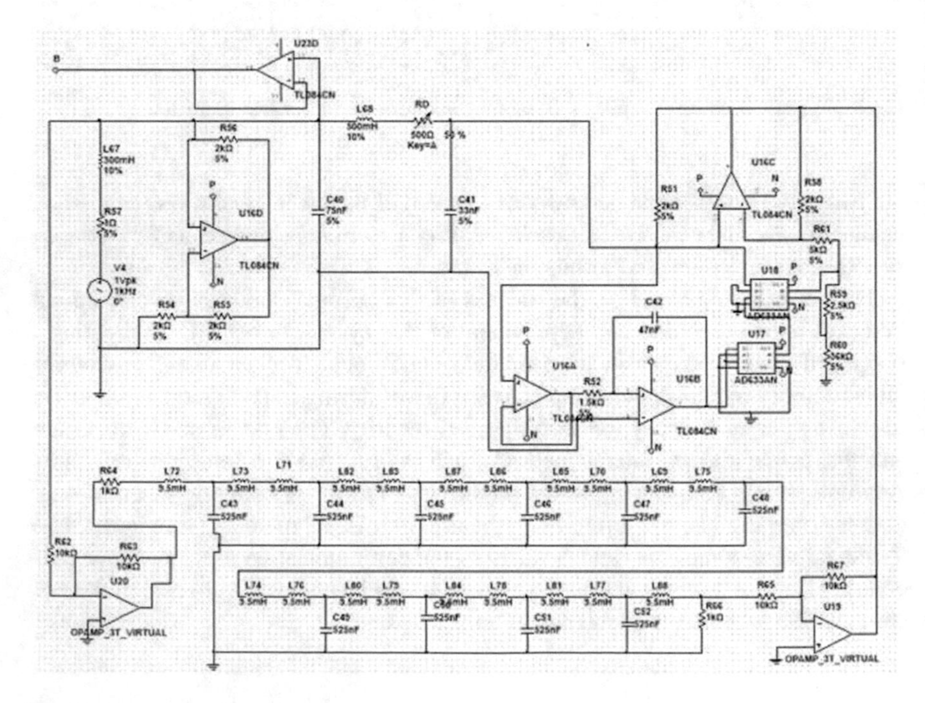

Fig. 11 One of the response system

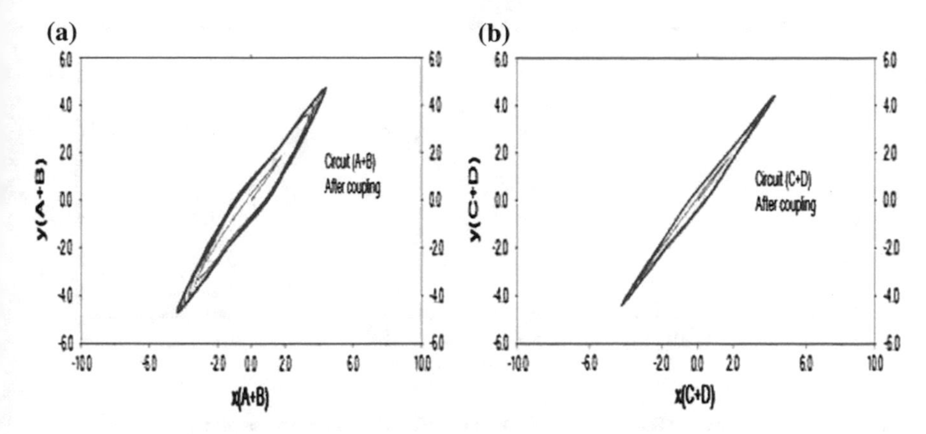

Fig. 12 **a** Combined attractor of the two drive systems when coupling R=1k **b** Combined attractor of the two response systems when coupling R = 1 k

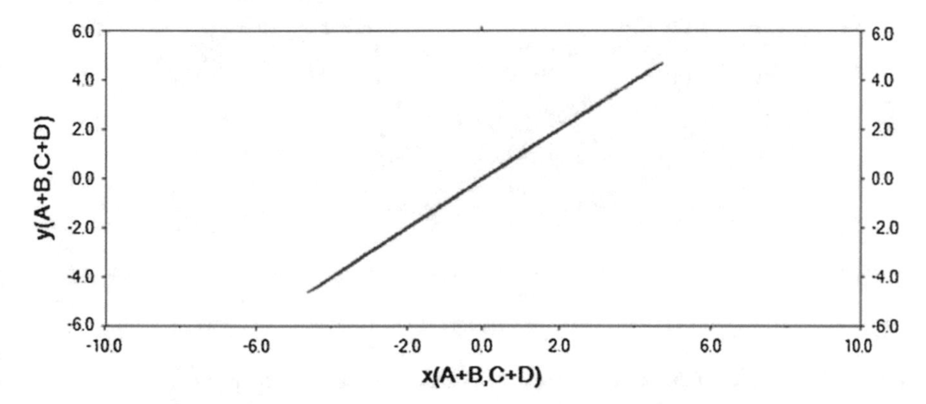

Fig. 13 Phase space of the drive and response systems after diffusive coupling

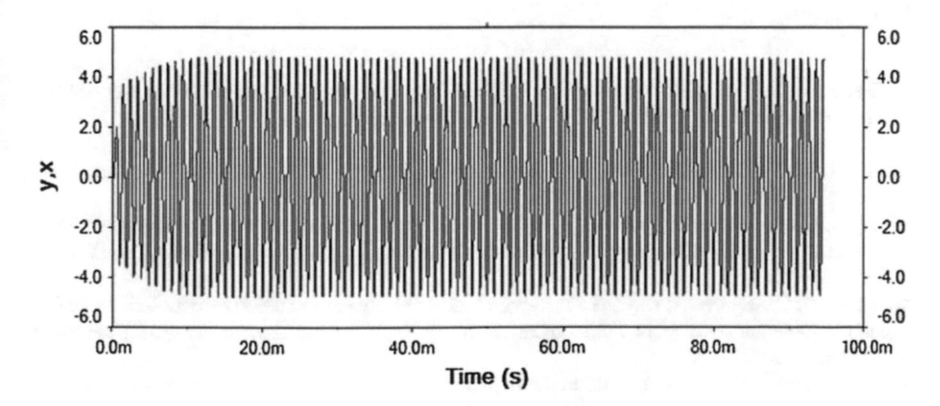

Fig. 14 Time series of the drive and response systems after diffusive coupling

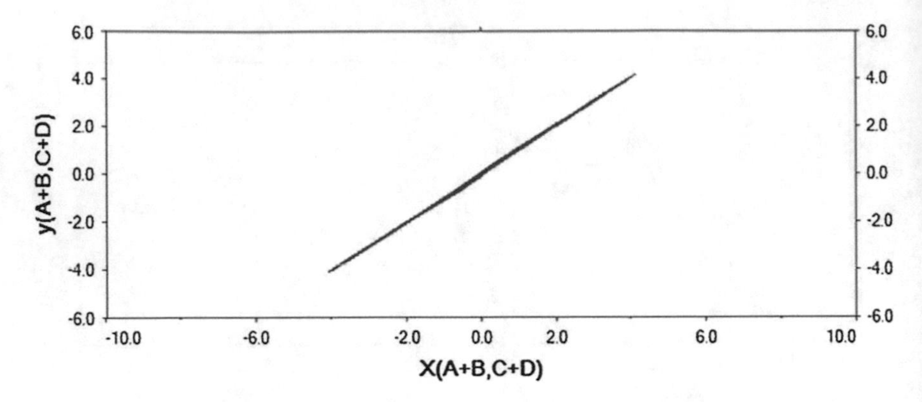

Fig. 15 Phase space of the drive and response systems after diffusive coupling

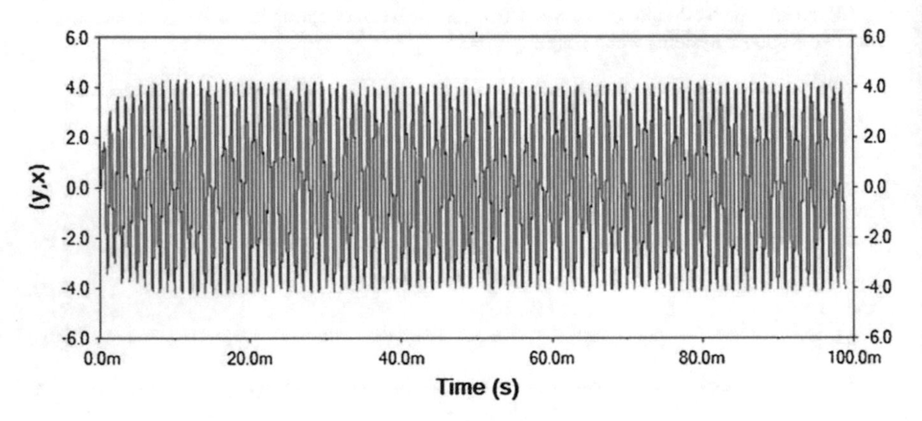

Fig. 16 Time series of the drive and response systems after diffusive coupling

References

Adelakun, A. O. (2013). Solutions of semistate duffing van der pol and bonhoeffer van der pol electronic circuits with memristor. *Journal of Nigerian Association of Mathematical Physics*, *25*(2).

Adelakun, A. O. (2014). Purely imaginary eigenvalues (pie) in extended circuits with mem-devices. *Journal of Electrical and Electronics Engineering*, *9*(3), 1–8.

Adelakun, A. O., Adeiza, O. F., & Oketayo, O. (2014a). Unidirectional synchronization of two identical jerk oscillators with memristor. *Journal of Nigerian Association of Mathematical Physics*, *28*(2), 35–42.

Adelakun, A. O., Egunjobi, I. A., & Oketayo, O. (2014b). Synchronization of new jerk oscillator and it's application to secure communication. *Journal of Nigerian Association of Mathematical Physics*, *28*(2), 27–34.

Aihara, K. (1991). Chaotic dynamics in nerve membranes and its modelling with an artificial neuron. In *IEEE international sympoisum on circuits and systems, 1991* (Vol. 3, pp. 1457–1460). doi:10.1109/ISCAS.1991.176649.

Argyris, A., Syvridis, D., Larger, L., Annovazzi-Lodi, V., Colet, P., Fischer, I., et al. (2005). Chaos-based communications at high bit rates using commercial fibre-optic links. *Nature*, *438*(7066), 343–346. doi:10.1038/nature04275.

Azar, A. T., & Vaidyanathan, S. (2016). *Advances in Chaos Theory and Intelligent Control*. Berlin, Germany: Springer.

Bo-Cheng, B., Jian-Ping, X., Guo-Hua, Z., Zheng-Hua, M., & Ling, Z. (2011). Chaotic memristive circuit: Equivalent circuit realization and dynamical analysis. *Chinese Physics B*, *20*(12), 120502.

Boccaletti, S., Kurths, J., Osipov, G., Valladares, D., & Zhou, C. S. (2002). The synchronization of chaotic systems. *Physics Reports*, *366*, 1–101.

Chua, L. O. (1971). Memristor the missing circuit element. *IEEE Transactions on Circuit Theory*, *18*, 507–519.

Chua, L. O., Komuro, M., & Matsumoto, T. (1986). The double scroll family, part I and II. *IEEE Transaction in Circuit and System*, CAS-*33*, 1073–1118.

El-Sayed, A., Elsaid, A., Nour, H., & Elsonbaty, A. (2013). Dynamical behavior, chaos control and synchronization of a memristor-based ADVP circuit. *Communications in Nonlinear Science and Numerical Simulation*, *18*(1), 148–170. doi:10.1016/j.cnsns.2012.06.011.

El-Sayed, A., Nour, H., Elsaid, A., Matouk, A., & Elsonbaty, A. (2016). Dynamical behaviors, circuit realization, chaos control, and synchronization of a new fractional order hyperchaotic system. *Applied Mathematical Modelling*, *40*(5), 3516–3534. doi:10.1016/j.apm.2015.10.010.

Elhadj, Z., & Sprott, J. C. (2008). On the robustness of chaos in dynamical systems: Theories and applications. *Frontiers of Physics in China*, *3*, 195. doi:10.1007/s11467-008-0017-z.

Fuwape, I. A., & Ogunjo, S. T. (2015). Fractal and entropy analysis of nigerian all share index (ASI) and gross domestic product (GDP). In *2nd international conference and exhibition (OWSD-FUTA)* (pp. 330–333).

Fuwape, I. A., Ogunjo, S. T., Oluyamo, S. S., & Rabiu, A. B. (2016). Spatial variation of deterministic chaos in mean daily temperature and rainfall over Nigeria. *Theoretical and Applied Climatology*. doi:10.1007/s00704-016-1867-x.

Fuwape, I. I. A., & Ogunjo, S. T. (2013). Investigating Chaos in the Nigerian Asset and Resource Management (ARM) Discovery Fund. *CBN Journal of Applied Statistics*, *4*(2), 129–140.

Itoh, M., & Chua, L. O. (2008). Memristor oscillators. *Interntional Journal of Bifurcation and Chaos*, *18*(11), 3183–3206.

Jinfeng, H., & Jingbo, G. (2008). Breaking a chaotic secure communication scheme. *Chaos*, *18*(013), 121. doi:10.1063/1.2885388.

Li, S., Álvarez, G., Chen, G., & Mou, X. (2012). Breaking a chaos-noise-based secure communication scheme. *Chaos*, *15*(013), 703. doi:10.1063/1.1856711.

Lorenz, E. N. (1963). Deterministic Nonperiodic Flow. *Journal of the Atmospheric Sciences*, *20*(2), 130–141. doi:10.1175/1520-0469(1963)020<0130:DNF>2.0.CO;2.

Mamat, M., Salleh, Z., Sanjaya, M. W. S., Noor, N. M. M., & Ahmad, M. F. (2012). Numerical simulation of unidirectional chaotic synchronization of non-autonomous circuit and its application for secure communication. *Advanced Studies in Theoretical Physics*, *6*(10), 497–509.

Mengue, A. D., & Essimbi, B. Z. (2012). Secure communication using chaotic synchronization in mutually coupled semiconductor lasers. *Nonlinear Dynamics*, *70*, 1241–1253. doi:10.1007/s11071-012-0528-6.

Messias, M., Nespoli, C., & Botta, V. A. (2010). Hopf bifurcation from lines of equilibria without parameters in memristor oscillators. *Interntional Journal of Bifurcation and Chaos*, *20*(2), 437–450.

Nagaraj, N., & Vaidya, P. G. (2009). Multiplexing of discrete chaotic signals in presence of noise. *Chaos*, *19*(033), 102. doi:10.1063/1.3157183.

Ogunjo, S. T. (2013). Increased and Reduced Order Synchronization of 2D and 3D Dynamical Systems. *International Journal of Nonlinear Science*, *16*(2), 105–112.

Ogunjo, S. T., Fuwape, I. A., & Olufemi, O. I. (2013). Chaotic Dynamics in a Population of Tribolium. *FUTA Journal of Research in Sciences*, *9*(2), 186–193.

Ogunjo, S. T., Ojo, K. S., & Fuwape, I. A. (2017). *Fractional order control and synchronization of chaotic systems: Studies in computational intelligence*. Germany: Springer. chap Comparison of Three Different Synchronization Scheme for Fractional Chaotic Systems.

Ojo, K. S., & Ogunjo, S. T. (2012). Synchronization of 4D Rabinovich Hyperchaotic System for Secure Communication. *Journal of Nigerian Association of Mathematical Physics*, *21*, 35–40.

Ojo, K. S., Njah, A., & Ogunjo, S. T. (2013). Comparison of backstepping and modified active control in projective synchronization of chaos in an extended Bonhoffer van der Pol oscillator. *Pramana*, *80*(5), 825–835, http://link.springer.com/article/10.1007/s12043-013-0526-3.

Ojo, K., Njah, A., Ogunjo, S., & Olusola, O. (2014a). Reduced order hybrid function projective combination synchronization of three Josephson junctions. *Archives of Control*, http://www.degruyter.com/view/j/acsc.2014.24.issue-1/acsc-2014-0007/acsc-2014-0007.xml.

Ojo, K. S., Njah, A., Ogunjo, S. T., & Olusola, O. I. (2014b). Reduced order function projective combination synchronization of three Josephson junctions using backstepping technique. *Nonlinear Dynamics and System Theory*, *14*(2), 119.

Ojo, K., Njah, A., & Olusola, O. (2015a). Compound-combination synchronization of chaos in identical and different orders chaotic systems. *Archives of Control Sciences*, *25*(4), 463–490.

Ojo, K., Njah, A., & Olusola, O. (2015b). Generalized function projective combination-combination synchronization of chaos in third order chaotic systems. *Chinese Journal of Physics*, *53*(3), 11–16.

Ojo, K., Njah, A., & Olusola, O. (2016). Generalized compound synchronization of chaos in different orders chaotic josephson junctions. *International Journal of Dynamics and Control*, *4*(1), 31–39.

Papadopoulou, M. S., Kyprianidis, I. M., & Stouboulos, I. N. (2008). Complex chaotic dynamics of the double-bell attractor. *WSEAS Transactions on Circuits and Systems*, *7*(1), 13–21.

Pecora, L. M., & Carroll, T. L. (1990). Synchronization in chaotic systems. *Physical Review Letters*, *64*, 821.

Pehlivan, I., & Uyaroglu, Y. (2010). A new chaotic attractor from general Lorenz system family and its electronic experimental implementation. *Turkish Journal of Electrical Engineering and Computer Sciences*, *18*(2), 171–184. doi:10.3906/elk-0906-67.

Shin, S., Sacchetto, D., Leblebici, Y., & Kang, S. M. S. (2012). Neuronal spike event generation by memristors. In *2012 13th international workshop on cellular nanoscale networks and their applications* (pp. 1–4). doi:10.1109/CNNA.2012.6331427.

Strukov, D., Snider, G., Stewart, D., & Williams, R. (2008). The missing memristor found. *Nature*, *453*, 80–83.

Vaidyanathan, S., Volos, C. (2016a). *Advances and Applications in Chaotic Systems*. Berlin, Germany: Springer.

Vaidyanathan S, Volos C (2016b) *Advances and Applications in Nonlinear Control Systems*. Berlin, Germany: Springer.

Wang, H., Wang, Q., & Lu, Q. (2011). Bursting oscillations, bifurcation and synchronization neuronal systems. *Chaos, Solitons and Fractals*, *44*, 667–675.

Wang, Z., Cang, S., Ochola, E. O., & Sun, Y. (2012). A hyperchaotic system without equilibrium. *Nonlinear Dynamics*, *69*(1–2), 531–537. doi:10.1007/s11071-011-0284-z.

Xu, Q., Lin, Y., Bao, B., & Chen, M. (2016). Multiple attractors in a non-ideal active voltage-controlled memristor based chua's circuit. *Chaos, Solitons and Fractals*, *83*, 186–200.

Zuo, C., & Cao, H. (2015). One of signatures of a memristor. *Communications in Nonlinear Science and Numerical Simulation*, *30*, 128–138.

A Novel Flux-Controlled Memristive Emulator for Analog Applications

Abdullah G. Alharbi, Mohammed E. Fouda and Masud H. Chowdhury

Abstract Emerging memristor technology is drawing widespread attention during the recent time due to its potential diverse applications in nanoelectronic memories, logic and neuromorphic computer architectures. Due to the absence of a practical memristive device, most of the research works in this area are still based on memristor emulator circuits that can be of current-controlled or voltage-controlled type. In this chapter, we introduce two emulator circuits for flux-controlled memductor and memristor. These emulator circuits have been built based on second generation current conveyer (CCII+), one multiplier and a square circuit to mimic the hysteresis behavior of the memristor. The proposed memristor emulator circuits can not only emulate memristive and plasticity function but also can be configured for floating configurations characteristic. Furthermore, we present the mathematical modeling, SPICE simulation and experimental results of the proposed emulator circuits. The series and parallel connectivity of these emulator circuits have been also studied, In addition to frequency analysis of their behavior.

Keywords Memrisitor · Non-linear Circuit · Memristor emulator · Hysteresis · Floating emulator · CCII+

A.G. Alharbi (✉) · M.H. Chowdhury
Computer Science and Electrical Engineering, University of Missouri-Kansas City, Kansas City, MO 64110, USA
e-mail: a.g.alharbi@ieee.org

M.H. Chowdhury
e-mail: masud@ieee.org

A.G. Alharbi
Electrical Engineering Department, Faculty of Engineering, AlJouf University, Sakaka 42421, Saudi Arabia

M.E. Fouda
Engineering Mathematics and Physics Department, Faculty of Engineering, Cairo University, Giza 12613, Egypt
e-mail: m_elneanaei@ieee.org

© Springer International Publishing AG 2017

S. Vaidyanathan and C. Volos (eds.), *Advances in Memristors, Memristive Devices and Systems*, Studies in Computational Intelligence 701,
DOI 10.1007/978-3-319-51724-7_20

1 Introduction

Memristor is a two terminal non-linear passive element, which was first introduced theoretically by Leon Chua in 1971 (Chua 1971; Chua and Kang 1976). He showed that memistor is the only non-linear passive element, which can relate the missing link between flux and electrical charges. In the I-V plane, it shows a unique pinched hysteresis loop that shrinks at higher frequency. Later in 2008, the two terminal nanoscale memristor device has been successfully fabricated by HewlettPackard (HP) lab (Strukov et al. 2008). Since then, memristor has been getting widespread attention from the scientific community due to its revolutionary potential in memory and various other systems. The most important physical property of the memristive device is that it does not discharge even after the applied voltage is removed and remains in its pre-charged state. This property can be utilized for memory application. Memristor can be of two types: charge-dependent or current-controlled and flux-dependent or voltage-controlled. However, in order to call any device a memristor, some substantial fingerprints are required that distinguish it from other devices (Adhikari et al. 2013). These fingerprints are discussed in details in Sect. 2. Moreover, in the last decade, many research projects and publications mentioned the dynamic nature and the potential applications of memristor. Some of the applications include implementation of high-speed memory arrays like resistive random access memory (RRAM), analog and digital circuits, sinusoidal and relaxation oscillators, neuromorphic circuits, adaptive filters (Zidan et al. 2014; Ascoli et al. 2014; Radwan and Fouda 2015; Vourkas and Sirakoulis 2016) and Chaotic oscillator (Vaidyanathan and Volos 2016a, b).

Since there is no physical memristive device available in the experimental labs or commercial design houses, most of the research are still at the theoretical stage. To validate the applications of memristor, we need precise behavior and SPICE macro models. In fact, numerous micro models are being developed using the equations of memristor proposed by the HP lab (Shin et al. 2010; Biolek et al. 2009; Garcia-Redondo 2016; Batas and Fiedler 2011; Berdan et al. 2014; Abdalla and Pickett 2011). In addition, a comparison between some of these model can be found in Ascoli et al. 2013. However, these models have many limitations for which it cannot mimic the physically developed memristor. Most of these models are only applicable to computer aided simulation of ideal memristor. In the absence of any practical memristive device, research community focuses on developing emulator circuits to mimic the dynamic behavior of memristor to explore the design issues and potential applications. Therefore, many emulator circuits have been proposed and designed based on different design methodologies using off-the-shelf active and passive devices that are commercially available. Some of them are implemented using analog components like Op-amps, second generation current conveyer (CCII), transistors, analog multiplier, floating capacitor, JFET, zener diodes, BJTs, diodes, and Differential Difference Current Conveyors, microcontroller unit, analog to digital converter and digital to analog converter. For instance, the memristor emulator circuit introduced in Pershin and Ventra 2010 uses a microcontroller unit and analog-to-digital, digital to analog converters. This emulator is topologically complex which limits their application in connecting with active and passive devices. In addition, a memristor emulator circuit based on Operational Transconductance Amplifier has been proposed in Kumngern and Moungnoul 2015. However, the experimental results of this emulator circuit do not satisfy the condition of memristor. A memristor emulator circuit based on an exponential amplifier and a CCII is presented in Alharbi et al. 2015b. The memristor emulator presented in Sánchez-López et al. 2014 uses five second generation current conveyors (CCII+), analog multiplier, five resistors and one capacitor. In addition, the emulator circuit proposed in Yeşil et al. 2014 requires four resistors, one Differential Difference Current Conveyors (DDCC) blocks, one grounded capacitor and one analog multiplier. Another memristor emulator introduced in Biolek et al. 2011 uses two current-feedback operational amplifiers (CFOAs) and one voltage-feedback operational amplifier, a large number of passive elements and a light-dependent resistor (LDR). Moreover, the emulator circuit presented in Abuelmaatti and Khalifa 2014 uses two current-feedback operational amplifiers (CFOAs), one diode, four resistors, two grounded capacitors. Voltage and current controlled mermsitor emulator is presented in Elwakil et al. 2013. Also, a CMOS based memristor emulator has been introduced in Hussein and Fouda 2013; Yener and Kuntman 2014. In addition, the memristor emulator uses two second generation current conveyer (CCIIs), two diode connected transistors and one resistor is provided in Alharbi et al. 2015d. This emulator has been improved in Alharbi et al. 2015c. A floating memristor emulator based relaxation oscillator is presented in Yu et al. 2014. Another floating emulator circuit is introduced in Shin et al. 2013. The emulator circuit uses an operational transconductance amplifiers (OTAs) and second generation current conveyors (CCIIs) is introduced in Sözen and Çam 2016. Simple floating and grounded voltage-controlled emuator are presented (Fouda and Radwan 2014; Alharbi et al. 2016). Furthermore, in Kim et al. 2012 an emulator for the memristor has been proposed, which is comprised of an adder, ten transistors, five OP-AMPs and eight resistors. Also, Electromechanical Emulator of Memristive Systems is introduced in Asapu and Pershin 2015. In addition, a cubic flux-controlled memristor is introduced in Liu et al. 2015 based on the cubic nonlinearity in Zhong 1994. However, most of these emulators have some drawbacks. Some are very complex and require rigid conditions. Some emulators do not exhibit or satisfy the three characteristic fingerprints of a memristor as discussed in Adhikari et al. 2013. The emulator circuit can be customized for different memristor models by selecting appropriate circuit elements is introduced in Alharbi et al. 2015a.

The rest of the chapter is organized as follows. In Sect. 2, a review of the fundamental properties of memristor is presented. In Sect. 3, we present our model and the emulator circuit development approach. The experimental results are shown in Sect. 4. Section 5 presents the frequency analysis of the flux-controlled memductor. In section Sect. 6, flux-controlled memristor circuit is introduced. Section 7 demonstrates the results and the analysis. Finally, Sect. 8 concludes the chapter with a highlight of future work.

2 Memristor Properties

For a device to be considered as a memristor it must have three significant fingerprints (Adhikari et al. 2013). Therefore, any memristor emulator circuit must also comply with these three fingerprints or defining characteristics. In this section, we have briefly summarized these fingerprints to establish the intellectual merits of the proposed emulator circuit.

- Memristor Fingerprint 1: Pinched Hysteresis Loop
  The first significant signature of the memristor is its unique pinched hysteresis loop which distinguishes it from any device that is not memristive in the (I-V plane).

(a) In I-V plane, the Lissajous figure of all memristors, having positive memristance and operated by sinusoidal signal of any amplitude and frequency, have to go through the origin.
(b) The value of $v(t)$ and $i(t)$ in the Lissajous figure should be same only when it will pass through origin, however, for rest of the times, V-I should have different values.

- Memristor Fingerprint 2: Hysteresis loop area decreases as frequency increases.
  The second vital signature of the memristor is the inversely proportional relationship between the frequency of periodic operating signal and memristors hysteresis lobe area. It states that with the increment of frequency, the lobe area will decrease.
- Memristor Fingerprint 3: No loop at infinite frequency.
  As we keep increasing the frequency, at some point, the lobe area will be reduced so much that there will no longer remain any loop, which means the memristors will behave as a linear device like resistor. At a very high frequency, memristor loses its unique non-linearity and the value of V and I remain same for all times in the I-V plane.

3 Proposed Flux-Contolled Memductor Emulator

The proposed emulator circuit for flux controlled memrductor has been designed with voltage difference circuit, voltage integrator and analog multiplier as shown in Fig. 1. In addition, the voltage difference circuit and the integrator are built based on the second generation current conveyer (CCII+).

3.1 Mathematical Analysis of the Proposed Emulator

The characteristics of an ideal CCII+ can be represented as in (1).

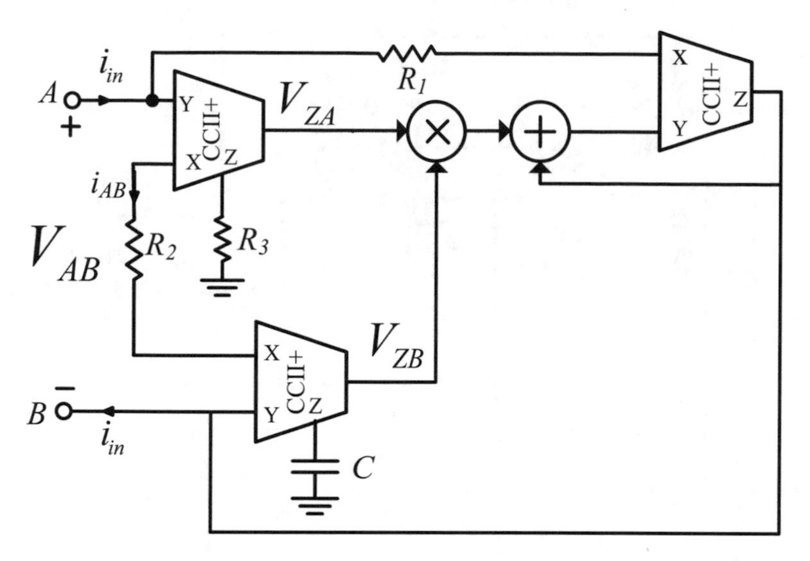

Fig. 1 The Proposed floating emulator circuit for flux controlled memductor

$$V_Y\ (t) = \ V_X\,(t)\ \text{and} \quad i_X\,(t) = i_Z\,(t) \tag{1}$$

The input current to the circuit can be written as in (2)

$$i_{AB}(t) = \frac{v_A - v_B}{R_2} \tag{2}$$

Based on (1), the output voltage of first CCII+, V_{ZA}, is given by (3)

$$V_{ZA} = V_{AB}\frac{R_3}{R_2} \tag{3}$$

and the second CCII+ works as integrator where the current i_{AB} is integrated through the capacitor. Hence, V_{ZB} is given by (4)

$$V_{ZB} = \frac{-\alpha}{R_2 C}\int_0^t V_{AB}(\tau)d\tau + V_Z \tag{4}$$

The voltages V_{ZA} and V_{ZB} are multiplied, α is the multiplier constant, and summed to V_Z of the third CCII which represents V_B. Consequently, the voltage of Y terminal of third CCII is $V_Y = V_{ZA} + V_{ZB} + V_B$. The input current, I_{in}, is given by (5)

$$i_{in}(t) = \frac{V_A - (V_{ZA}V_{ZB} + V_B)}{R_1} \tag{5}$$

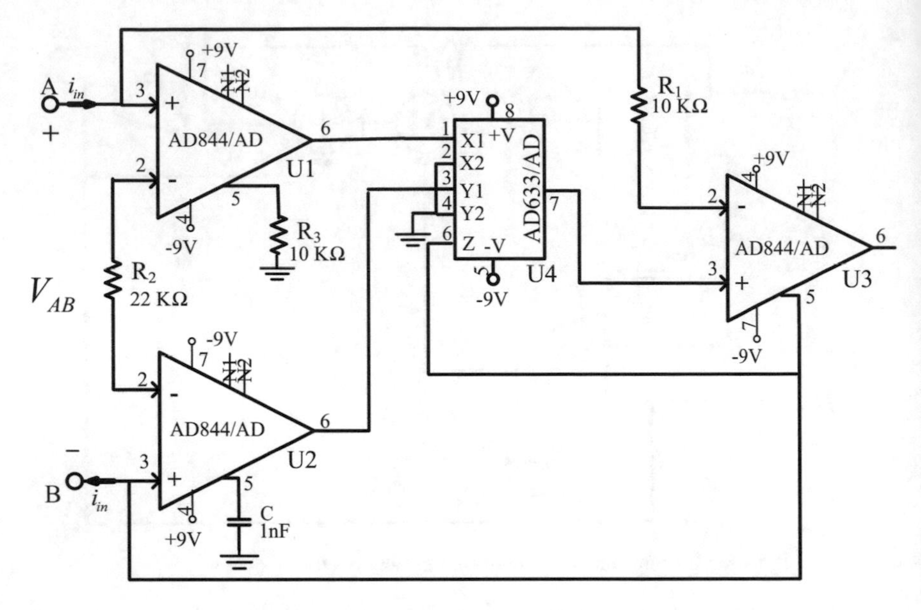

Fig. 2 Emulator circuit implementation of the flux controlled memductor model

By rearranging this equation, we obtain (6)

$$i_{in}(t) = \frac{V_{\mathrm{AB}} - V_{ZA}V_{ZB}}{R_1} \tag{6}$$

This current is mirrored to the input terminal V_B, so the input current to node A is the same as the output current of node B. Hence, this emulator represents a floating memristor emulator. So the current-voltage relation is given as in (7)

$$i_{in}(t) = V_{AB}(\frac{1}{R_1} + \frac{\alpha R_3}{R_2^2 R_1 C}\int_0^t V_{AB}(\tau)d\tau) \tag{7}$$

Thus, the input transconductance of the emulator, $G_m = i_{in}(t)/V(t)_{in} = i_{in}/V_{AB}$ representing memductance (memory transconductance) can be given by (8)

$$G_m = \frac{1}{R_1} + \frac{\alpha R_3}{R_2^2 R_1 C}\varphi(t) \tag{8}$$

It is clear that the memductance G_M is a function of the flux $\varphi(t)$. So, this model is referred as flux-controlled memductance

4 Circuit Realization and Experimental Validation

In order to ensure the validity and efficiency of the proposed memristor emulator circuit, the circuit shown in Fig. 1 has been realized and implemented in the lab from off-the-shelf components using AD844AN (constructed by the commercial AD844 current feedback operational amplifiers CFOA) as second-generation current conveyor (CCII+) and AD633 as a multiplier. Here, the used values of the passive elements are $R_1 = 10\,\mathrm{k}\Omega, R_2 = 22\,\mathrm{k}\Omega, R_3 = 10\,\mathrm{k}\Omega$ and $C = 1nF$ as shown in Fig. 2. We used DC supply voltages $= \pm 9V$. In order to plot the I-V curves, the current was sensed using an instrumentation amplifier sensing the differential voltage across the resistance R_1. We have used the Digilent Electronics Explorer board (EE board) and from PC-based WaveFormsTM software the experimental data has been exported and redrawn using MATLAB without alteration, to draw the hysteresis loop.

Figure 3 shows nonlinearity and hysteresis loop in the I-V plane for the floating voltage controlled memristor emulator. it is clearly evident that the emulator circuit exhibits the unique the fingerprints of a real memristor. In addition, at low frequencies, the circuit shows nonlinearity in the (I-V) plane. However, at high frequency this nonlinearity gradually decreases. Moreover, our findings indicate that the proposed emulator circuit exhibits the fingerprints of a memristor as introduced in Adhikari et al. 2013. In addition, it is seen that the memristor emulator acts as a non-linear device and pinched hysteresis loop is found in the I-V plane for a particular range of frequency. However, it is clearly evident that as we keep increasing the frequency, the lobe area of the hysteresis loop tends to shrink. At a certain point, the loop becomes a straight line and the emulator acts like a linear resistor as shown in Fig. 3a. Moreover, Fig. 3b shows the behavior of the proposed emulator under triangular input for 5 kHz and 8 kHz in the I-V plane.

We have also analyzed the memductance of the proposed circuit of Fig. 1. It is observed from Fig. 4a that the memductance varies with time for applied sinusoidal

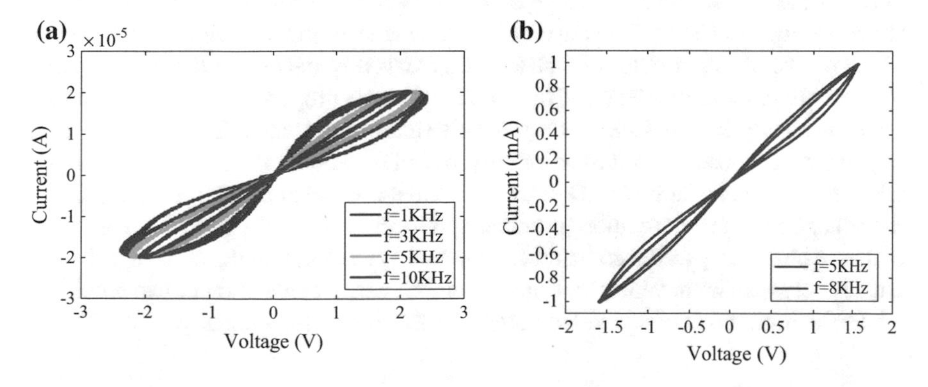

Fig. 3 Experimental results of the pinched hysteresis loop of the emulator circuit with various frequencies **a** sinusoidal input and **b** triangular input

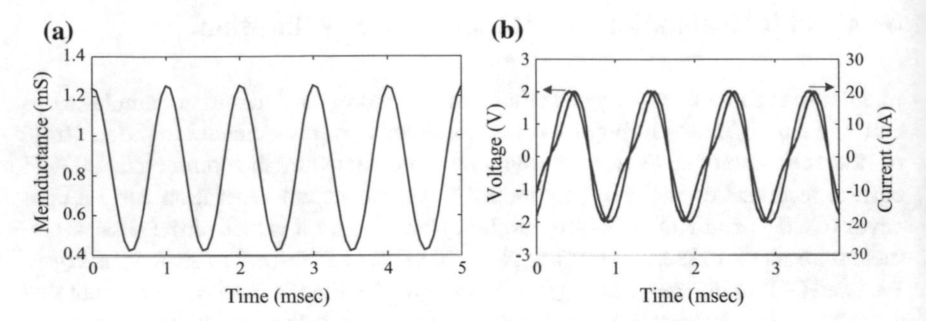

Fig. 4 SPICE transient simulation of proposed emulator signals at 1 kHz **a** Transient waveforms of memductance and **b** waveform of input voltage $v(t)$ (*blue*) and the input current $i(t)$ (*red*)

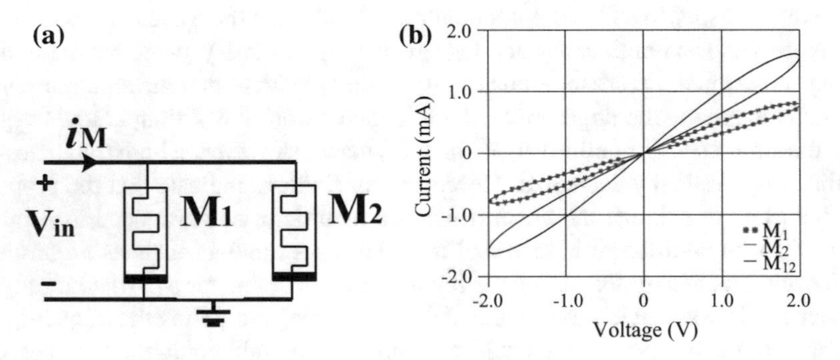

Fig. 5 Pinched hysteresis loops of two parallel connected emulators at 3 kHz

signal with amplitude 2 V and frequency 1 kHz. Furthermore, we can observe that the memductance changes from 0.42 mS to 1.23 mS.

As it is well known that the memristor is resistive so there is no phase shift between the current and the voltage, which is clear in the proposed emulator as shown in Fig. 4b. Clearly, the current is zero whenever the voltage is zero, which is the signature of a memristor. If a phase shift exists this means that there is a reactive element attached to the device. This implies that the proposed memristor emulator circuit is clearly resistive without any reactive element attached. In addition, in order to prove the proposed circuit functionality, two of the emulators are connected in parallel/series to make sure that the current/voltage is divided equally. Figure 5 shows the voltage and current relation (pinched hysteresis) in the I-V plane of the proposed emulator circuit in parallel connection. It is clearly evident that the current is divided equally between them. Whereas, Fig. 6 shows the series connection of two emulators where the input voltage is divided across the two memristors equally.

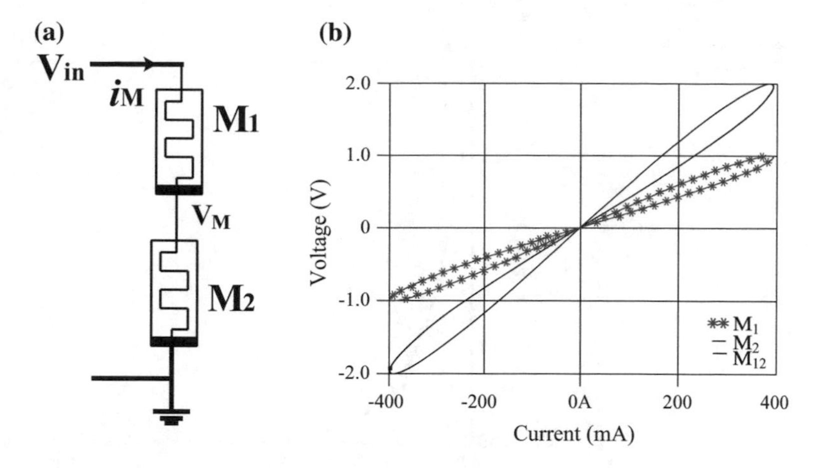

Fig. 6 Pinched hysteresis loops of two series connected emulators at 3 kHz

5 Frequency Analysis of the Flux-Controlled Memductor Emulator

In order to ensure the accuracy of the proposed emulator, we have studied the frequency analysis of the proposed emulator as in Sánchez-López et al. 2014. By applying a sinusoidal signal $V_{in} = Asin(\omega t)$, the memductance is given by (9)

$$G_m = \frac{1}{R_1}\left(1 + \frac{\alpha R_3 A(1 - cos(\omega t))}{R_2^2 C\omega}\right) \tag{9}$$

According to this equation, the minimum and maximum achievable memductances are given as follows (10)

$$G_{min} = 1/R_1 \; and \qquad G_{max} = \frac{1}{R_1}\left(1 + \frac{2\alpha R_3 A}{R_2^2 C\omega}\right) \tag{10}$$

As shown, the more the frequency, ω, increases, the more memristance decreases (Fingerprint 2). When ω tends to ∞, R_{max} tends to R_{min} which is a constant value meaning there is no hysteric behavior (Fingerprint 3). However, when ω tends to 0, R_{max}tends to ∞ which is not practical since the R_{max} saturates to certain value due to supply voltages which is corresponding to R_{on} and R_{off} in the fabricated devices.

It is obvious from (9) that the memductance equation is based on two terms, the first terms is constant resistance which is time invariant and the second one is time-varying resistor. The time varying term changes with function of the frequency and time constants of the integrator. The ratio between magnitude of both terms, β, can be defined as in (11):

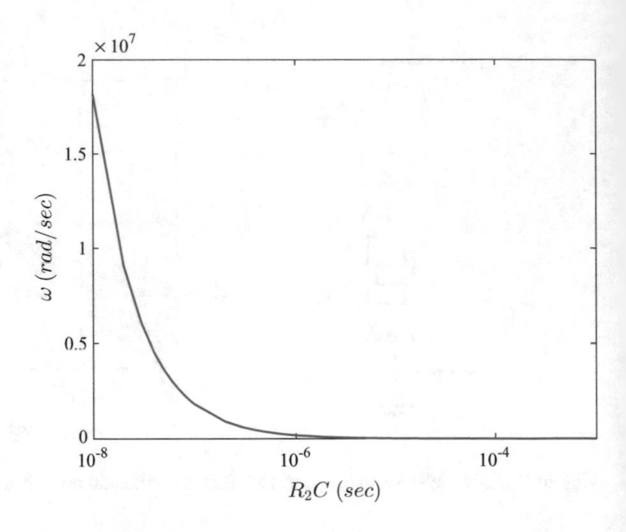

Fig. 7 Frequency behavior for $\beta = 0.5$ and $A = 1$

$$\beta = \frac{2\alpha R_3 A}{R_2^2 C\omega} = \beta_o \frac{T}{\tau} \tag{11}$$

where $\beta_o = \alpha R_3 A/\pi R_2$, $\tau = R_2 C$ and $T = 2\pi/\omega$.

It is clear with increasing the frequency, the ratio β decreases. By studying β, we can observe that β tends to zero if the frequency tends to ∞ where the memristor behavior is dominated by the linear term which is R_1. Also, hysteresis loop disappears when the time constant of the integrator τ is much greater than T.

Figure 7 shows the effect of changing the time constants τ with the frequency while maintaining the same ratio $\beta = 0.5$. The more τ decreases, the more operating frequency is needed for the same β_o.

6 Proposed Flux-Controlled Memristor Emulator Circuit

The proposed emulator circuit is the modification and improved version of the original emulator circuit in Abuelmaatti and Khalifa 2015. The emulator circuit shown in Fig. 8 consists of a practical integrator, differentiator and square function. The practical integrator and differentiator are built by using two second-generation current conveyors (CCII+s). Here, the square function is used to achieve the required non-linearity for memristive behavior.

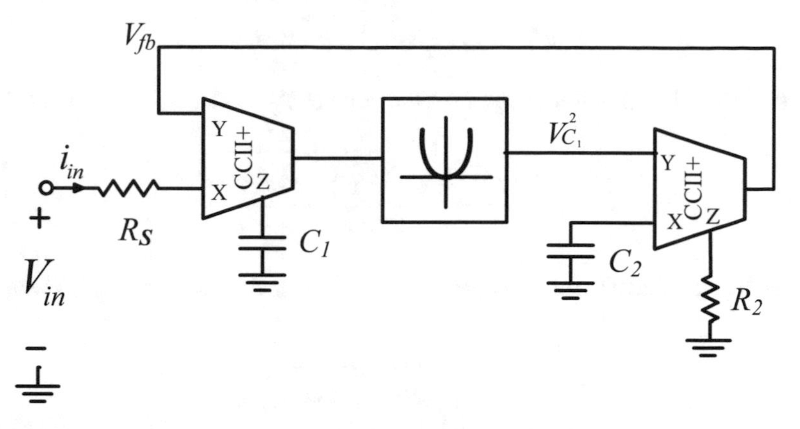

Fig. 8 Flux-controlled Memristor emulator circuit diagram with square function

6.1 Mathematical Analysis of the Proposed Emulator

The input current, i_{in}, is created by subtracting the feedback voltage, V_{fb} from the input voltage, V_{in} which can be written as in (12)

$$i_{in} = \frac{V_{in} - V_{fb}}{R_s} \tag{12}$$

Then, this current is imposed in the capacitor C_1, so the voltage across the capacitor is given by (13)

$$V_C = \frac{-1}{R_s C_1} \int_0^t (V_{in} - V_{fb}) d\tau \tag{13}$$

where V_{fb} represents the feedback voltage (output of second CCII+). This voltage is squared using squarer circuit, or multiplied by itself using multiplier as done in our circuit. Then, the output voltage of the multiplier is differentiated using the second CCII given the feedback voltage (14).

$$V_{fb} = \alpha R_2 C_2 \frac{dV_C^2}{dt} \tag{14}$$

where α is the multiplier gain. By substituting into (13), V_C is given as in (15)

$$V_C = \frac{-1}{R_S C_1} \int_0^t (V_{in} - \alpha R_2 C_2 \frac{dV_C^2}{dt}) d\tau = \frac{1}{R_S C_1} \left(\alpha R_2 C_2 V_C^2 - \varphi_{in} \right) \tag{15}$$

By rearranging the equation,

$$\alpha R_2 C_2 V_C^2 - R_S C_1 V_C - \varphi_{in} = 0 \tag{16}$$

It is second order equation of V_C, then the voltage V_C can be written as in (17)

$$V_C = \frac{R_S C_1 \pm \sqrt{R_S^2 C_1^2 + 4\alpha R_S C_2 \varphi_{in}}}{2\alpha R_2 C_2} \tag{17}$$

The feedback voltage is function of dV_C^2/dt where the derivative of V_C is given as in (18)

$$\frac{dV_C}{dt} = \pm \frac{V_{in}}{\sqrt{R_S^2 C_1^2 + 4\alpha R_2 C_2 \varphi_{in}}} \tag{18}$$

by applying the chain rule, the derivative of V_C^2 is given as follows (19):

$$\frac{dV_C^2}{dt} = 2V_C \frac{dV_C}{dt} = \pm 2V_C \frac{V_{in}}{\sqrt{R_S^2 C_1^2 + 4\alpha R_2 C_2 \varphi_{in}}}, \tag{19}$$

and the feedback voltage is given by (20)

$$V_{fb} = \alpha R_2 C_2 \frac{dV_C^2}{dt} = \left(1 \pm \frac{R_S C_1}{\sqrt{R_S^2 C_1^2 + 4\alpha R_2 C_2 \varphi_{in}}} \right) V_{in} \tag{20}$$

By subistuting into (12), and simplifying the expression, the input current can written as in (21)

$$i_{in} = \frac{C_1}{\sqrt{R_S^2 C_1^2 + 4\alpha R_2 C_2 \varphi_{in}}} V_{in} \tag{21}$$

and the input memristance is given by (22)

$$R_m = R_S \sqrt{1 + \frac{4\alpha R_2 C_2 \varphi_{in}}{R_S^2 C_1^2}} \tag{22}$$

As we can see that this equation is the same closed form solution of HP model. By comparing this equation and HP model solution, the initial memristance is R_s and the memristor speed term, k', is

$$k' = \frac{\alpha R_2 C_2}{C_1^2} \tag{23}$$

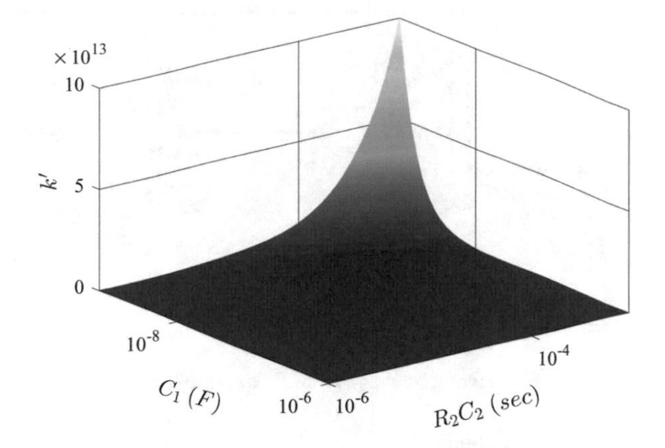

Fig. 9 Effect of changing circuit parameters on the memristor speed for $\alpha = 0.1$

As clear that the Memristor speed decreases quadratically with increasing C_1 and increases with the increasing the differentiator time constant R_2C_2. Figure 9 shows a 3D plot with changing the circuit components values.

6.2 *Implementation of the Emulator Circuit*

The proposed emulator circuit is simple and designed from the off-the-shelf components. This emulator circuit has been realized and implemented using AD844AN as second-generation current conveyors (CCII+), and the square function is implemented using a commercial AD633 (voltage multiplier) and some passive elements: $R_s = 1.5\,\mathrm{k}\Omega, R_2 = 2\,\mathrm{k}\Omega,\ C_1 = \mu F$ and $C_2 = 1\mu F$ as shown in Figs. 10a and 11. We

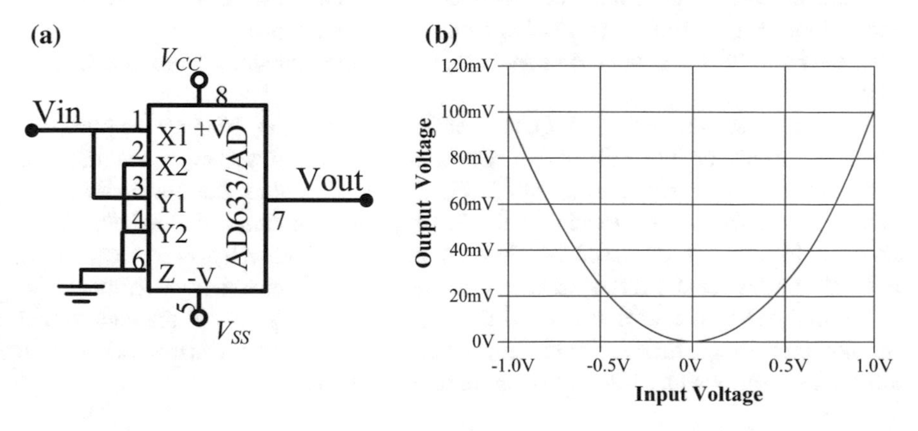

Fig. 10 Proposed implementation of the square function: **a** circuit diagram. **b** SPICE simulation of the Lissajous curve of the proposed square function

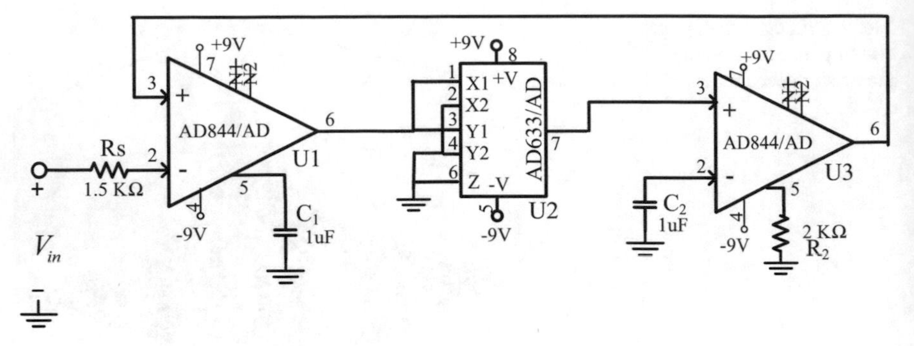

Fig. 11 The schematic diagram of flux controlled emulator circuit of memristor

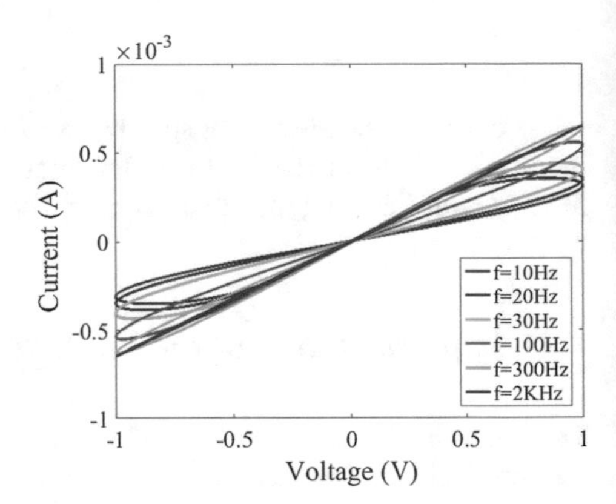

Fig. 12 Experimental results of the pinched hysteresis loop of the proposed memristor emulator circuit

have used $\pm 9V$ as supply voltages. In addition, we have used the Digilent Electronics Explorer Board and WaveFormsTM software to perform the experimentation. The data has been exported directly to MATLAB without alteration to draw the hysteresis loop. Figure 10b shows the Lissajous curves of the proposed square function obtained from SPICE simulation conducted on the implemented square function circuit.

Now, if we use the proposed square function circuit of Fig. 10a into the proposed emulator circuit model of Fig. 8, we achieve a emulator circuit implementation for flux controlled Memristor as shown in Fig. 11. Furthermore, the hysteresis loops obtained from the experimental data for the proposed emulator circuit of Fig. 11 are shown in Fig. 12. It is observed that the emulator has a pinched hysteresis loop in the I-V plane as expected. Moreover, at low frequencies the circuit shows nonlinear hysteresis in the (I-V) plane. However, with the increase of frequency of the input signal this nonlinearity gradually shrinks. Beyond 2 kHz, the emulator starts to behave like a linear resistor, which satisfies Chuas condition in Chua 2014.

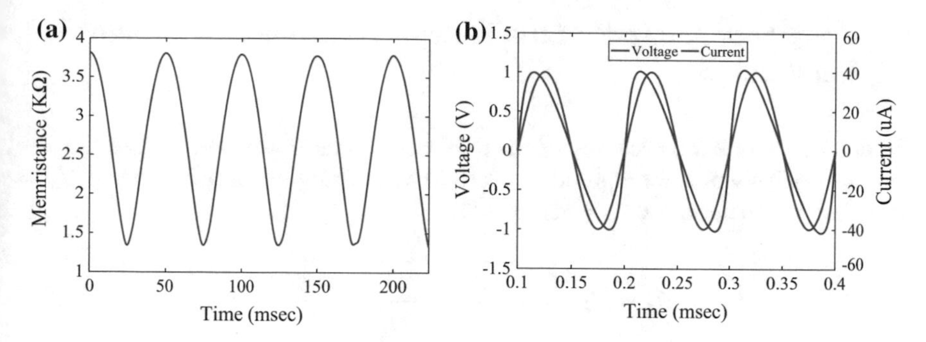

Fig. 13 **a** Transient memristance at 30 Hz **b** input voltage *v*(*t*) (*blue*) and the input current *i*(*t*) (*red*) for the proposed memristor at 30 Hz

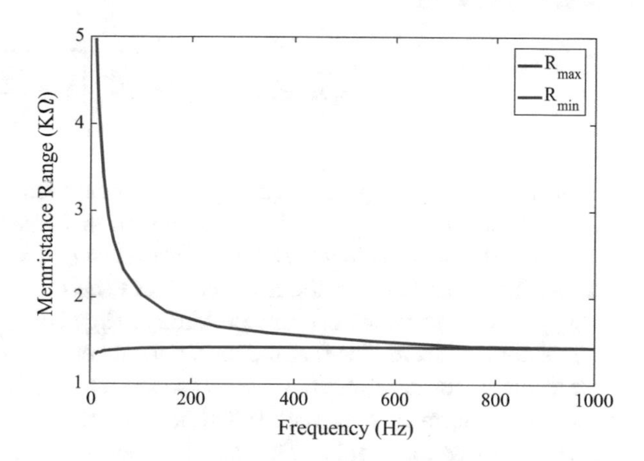

Fig. 14 Maximum and Minimum achievable memristance versus the frequency

It can be observed from Fig. 13a that the memristance changes with time as we apply the sinusoidal signal having 30 Hz of frequency and 1 V of amplitude. In addition, it is also significant to notice that the nature of changing the memristance is also sinusoidal. Furthermore, In addition, Fig. 13b demonstrates the waveform of the voltage and current of the proposed memristor emulator circuit. Moreover, there is no phase shift between the current and voltage. It is also seen that no current exists when the voltage is zero which validates a significant property of the memristor. Hence, we can conclude that our proposed emulator circuit is purely resistive element. Figure 14 shows the maximum and the minimum achievable memristance, which changes with the frequency. Here, Rmin is almost constant and it represents Rs. R_{max} decreases gradually with the applied frequency of the sinusoidal input signal and at one point it coincides with R_{min}. This indicates that beyond certain frequency the emulator circuit starts to behave like a linear resistor.

7 Frequency Analysis of the Voltage-Controlled Memristor Emulator

In order to ensure the accuracy of the proposed emulator, we have studied the frequency analysis of the proposed emulator. By applying a sinusoidal signal $V_{in} = Asin(\omega t)$, the memristance is given by (24)

$$R_m = R_S\sqrt{1 + \frac{4\alpha R_2 C_2 A(1 - \cos(\omega t))}{\omega R_S^2 C_1^2}} \tag{24}$$

According to this equation, the minimum and maximum acheivable memristance are given as follows (25)

$$R_{min} = R_S \ and \quad R_{max} = R_S\sqrt{1 + \frac{8\alpha R_2 C_2 A}{\omega R_S^2 C_1^2}} \tag{25}$$

As clear with increasing the frequency ω, the memristance decreases (Fingerprint 2). When ω tends to ∞, R_{max} tends to R_{min} which is a constant value meaning there is no hysteric behavior (Fingerprint 3). However, when ω tends to 0, R_{max} tends to ∞ which is not practical since the R_{max} saturates to certain value as shown in Fig. 14 due to supply voltages which is corresponding to R_{on} and R_{off} in the fabricated devices.

It is obvious from (24) that the memristance equation is based on two terms, the first terms is constant resistance and the second one is time-varying resistor. The time varying term changes with function of the frequency and time constants of the differentiator and integrator. The ratio between magnitude of both terms, β, can be defined as follows (26):

$$\beta = \frac{8\alpha R_2 C_2}{R_s^2 C_1^2}\frac{A}{\omega} = \beta_o\frac{\tau_2 T}{\tau_1^2} \tag{26}$$

where $\beta_o = 8\alpha A/2\pi$, $\tau 1 = R_s C_1$, $\tau_2 = R_2 C_2$, and $T = 2\pi/\omega$.

It is clear with increasing the frequency, the ratio β decreases. By studying β, we can observe that β tends to zero if the frequency tends to ∞ where the memristor behavior is dominated by the linear term which is Rs. Also, hysteresis loop disappears when the time constant of the integrator τ_1 is greater than $\sqrt{T}$. However, by increasing the differentiator time constant τ, the hysteresis becomes larger.

Figure 15 shows the effect of changing the time constants τ_1 and τ_2 with the frequency while maintaining the same ratio $\beta = 0.5$. The more τ_1 decreases, the more operating frequency is needed for the same τ_2 since they have quadratic relation. However, the more τ_2 decreases, a lower operating frequency is needed for the sam τ_1.

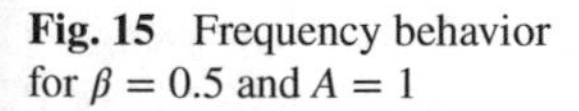
Fig. 15 Frequency behavior for $\beta = 0.5$ and $A = 1$

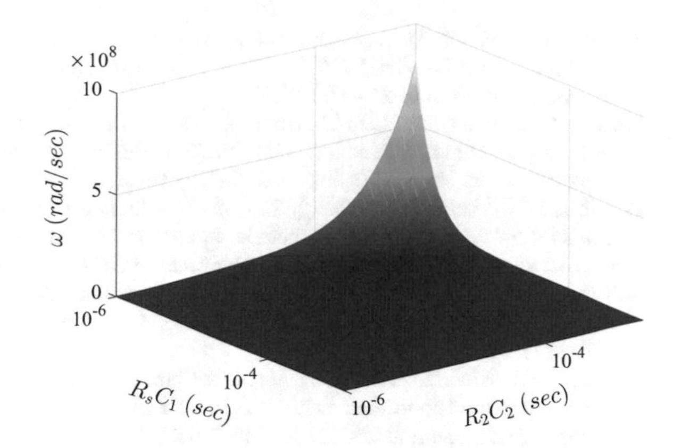

8 Conclusion and Future Work

We have presented a practical memristor emulator circuit development technique to mimic the nonlinear behavior of the memristor. We have demonstrated two different emulator circuits for the flux-controlled memristor. Our numerical analysis and simulation using SPICE and the experimental testing match very well, which indicate that the proposed circuits can accurately imitate the behavior of a memristor and satisfy all the three fingerprints of a memristor. The proposed circuit model of Fig. 1 is a floating memristor emulator, which is suitable for use in many digital and analog applications as a 2-terminal device. Moreover, in the absence of a real solid-state device for the memristor, these emulator circuits will be very useful to investigate the properties and potential applications of memristors. Hence, our emulator circuits have the potential to be used in many practical applications in the analog and digital world. The proposed circuits are practical and simple to design compared to many other emulator circuits proposed by different groups. Therefore, the proposed emulator circuit development technique would have significant impact on the development and educational aspects of this new direction of research.

References

Abdalla, H., & Pickett, M. D. (2011). Spice modeling of memristors. In *2011 IEEE International Symposium of Circuits and Systems (ISCAS)*.

Abuelmaatti, M. T., & Khalifa, Z. J. (2014). A new memristor emulator and its application in digital modulation. *Analog Integrated Circuits and Signal Processing*, *80*(3), 577–584.

Abuelmaatti, M. T., & Khalifa, Z. J. (2015). A continuous-level memristor emulator and its application in a multivibrator circuit. *AEU-International Journal of Electronics and Communications*, *69*(4), 771–775.

Adhikari, S. P., Sah, M. P., Kim, H., & Chua, L. O. (2013). Three fingerprints of memristor. *IEEE Transactions on Circuits and Systems I: Regular Papers*, *60*(11), 3008–3021.

Alharbi, A. G., Fouda, M. E., & Chowdhury, M. H. (2015a). Memristor emulator based on practical current controlled model. In *2015 IEEE 58th International Midwest Symposium on Circuits and Systems (MWSCAS)* (pp. 1–4). IEEE.

Alharbi, A. G., Fouda, M. E., & Chowdhury, M. H. (2015b). A novel memristor emulator based only on an exponential amplifier and ccii+. In *2015 IEEE International Conference on Electronics, Circuits, and Systems (ICECS)* (pp. 376–379). IEEE.

Alharbi, A. G., Fouda, M. E., Khalifa, Z. J., & Chowdhury, M. H. (2016). Simple generic memristor emulator for voltage-controlled models. In *2016 IEEE 59th International Midwest Symposium on, Circuits and Systems (MWSCAS),Abu Dhabi, UAE* (pp. 29–32). IEEE.

Alharbi, A. G., Khalifa, Z. J., Fouda, M. E., & Chowdhury, M. H. (2015c). Memristor emulator based on single ccii. In *2015 27th International Conference on Microelectronics (ICM)* (pp. 174–177). IEEE.

Alharbi, A. G., Khalifa, Z. J., Fouda, M. E., & Chowdhury, M. H. (2015d). A new simple emulator circuit for current controlled memristor. In *2015 IEEE International Conference on Electronics, Circuits, and Systems (ICECS)* (pp. 288–291). IEEE.

Asapu, S., & Pershin, Y. V. (2015). Electromechanical emulator of memristive systems and devices. *IEEE Transactions on Electron Devices*, *62*(11), 3678–3684.

Ascoli, A., Corinto, F., Gilli, M., & Tetzlaff, R. (2014). Memristor for neuromorphic applications: Models and circuit implementations. In *Memristors and Memristive Systems* (pp. 379–403). Springer.

Ascoli, A., Corinto, F., Senger, V., & Tetzlaff, R. (2013). Memristor model comparison. *IEEE Circuits and Systems Magazine*, *13*(2), 89–105.

Batas, D., & Fiedler, H. (2011). A memristor spice implementation and a new approach for magnetic flux-controlled memristor modeling. *IEEE Transactions on Nanotechnology*, *10*(2), 250–255.

Berdan, R., Lim, C., Khiat, A., Papavassiliou, C., & Prodromakis, T. (2014). A memristor spice model accounting for volatile characteristics of practical reram. *IEEE Electron Device Letters*, *35*(1), 135–137.

Biolek, D., Bajer, J., Biolkova, V., & Kolka, Z. (2011). Mutators for transforming nonlinear resistor into memristor. In *2011 20th European Conference on Circuit Theory and Design (ECCTD)* (pp. 488–491). IEEE.

Biolek, D., Biolkova, V., & Biolek, Z. (2009). Spice model of memristor with nonlinear dopant drift. *Radioengineering*.

Chua, L. (1971). Memristor-the missing circuit element. *IEEE Transactions on circuit theory*, *18*(5), 507–519.

Chua, L. (2014). If its pinched itsa memristor. *Semiconductor Science and Technology*, *29*(10), 104001.

Chua, L. O., & Kang, S. M. (1976). Memristive devices and systems. *Proceedings of the IEEE*, *64*(2), 209–223.

Elwakil, A. S., Fouda, M. E., & Radwan, A. G. (2013). A simple model of double-loop hysteresis behavior in memristive elements. *IEEE Transactions on Circuits and Systems*, *60*(8), 487–491.

Fouda, M., & Radwan, A. (2014). Simple floating voltage-controlled memductor emulator for analog applications. *Radioengineering*.

Garcia-Redondo, F., Gowers, R., Crespo-Yepes, A., Lopez-Vallejo, M., & Jiang, L. (2016). Spice compact modeling of bipolar/unipolar memristor switching governed by electrical thresholds. *IEEE Transactions on Circuits and Systems I Regular Papers* (pp. 1–10).

Hussein, A. I., & Fouda, M. E. (2013). A simple mos realization of current controlled memristor emulator. In *2013 25th International Conference on Microelectronics (ICM)* (pp. 1–4). IEEE.

Kim, H., Sah, M. P., Yang, C., Cho, S., & Chua, L. O. (2012). Memristor emulator for memristor circuit applications. *IEEE Transactions on Circuits and Systems I: Regular Papers*, *59*(10), 2422–2431.

Kumngern, M., & Moungnoul, P. (2015). A memristor emulator circuit based on operational transconductance amplifiers. In *2015 12th International Conference on Electrical Engineer-*

ing/Electronics, Computer, Telecommunications and Information Technology (ECTI-CON) (pp. 1–5). IEEE.

Liu, W., Wang, F.-Q., & Ma, X.-K. (2015). A unified cubic flux-controlled memristor: Theoretical analysis, simulation and circuit experiment. *International Journal of Numerical Modelling: Electronic Networks, Devices and Fields*, *28*(3), 335–345.

Pershin, Y. V., & Di Ventra, M. (2010). Practical approach to programmable analog circuits with memristors. *IEEE Transactions on Circuits and Systems I: Regular Papers*, *57*(8), 1857–1864.

Radwan, A. G., & Fouda, M. E. (2015). Memristor mathematical models and emulators. In *On the Mathematical Modeling of Memristor, Memcapacitor, and Meminductor* (pp. 51–84). Springer.

Sánchez-López, C., Mendoza-Lopez, J., Carrasco-Aguilar, M., & Muñiz-Montero, C. (2014). A floating analog memristor emulator circuit. *IEEE Transactions on Circuits and Systems II: Express Briefs*, *61*(5), 309–313.

Shin, S., Kim, K., & Kang, S.-M. (2010). Compact models for memristors based on charge-flux constitutive relationships. *IEEE Transactions on Computer-Aided Design of Integrated Circuits and Systems*, *29*(4), 590–598.

Shin, S., Zheng, L., Weickhardt, G., Cho, S., & Kang, S.-M. S. (2013). Compact circuit model and hardware emulation for floating memristor devices. *IEEE Circuits and Systems Magazine*, *13*(2), 42–55.

Sözen, H., & Çam, U. (2016). Electronically tunable memristor emulator circuit. *Analog Integrated Circuits and Signal Processing* (pp. 1–9).

Strukov, D., Snider, G., Stewart, D., & Williams, R. (2008). The missing memristor found. *Nature*, *453*(7191), 80–83.

Vaidyanathan, S. & Volos, C. (2016a). *Advances and Applications in Chaotic Systems*, (Vol. 636). Springer.

Vaidyanathan, S., & Volos, C. (2016b). *Advances and Applications in Nonlinear Control Systems*, (Vol. 635). Springer.

Vourkas, I., & Sirakoulis, G. C. (2016). *Memristor-Based Nanoelectronic Computing Circuits and Architectures*. Springer.

Yener, S. C., & Kuntman, H. H. (2014). Fully cmos memristor based chaotic circuit. *Radioengineering*.

Yeşil, A., Babacan, Y., & Kaçar, F. (2014). A new ddcc based memristor emulator circuit and its applications. *Microelectronics Journal*, *45*(3), 282–287.

Yu, D., Iu, H. H.-C., Fitch, A. L., & Liang, Y. (2014). A floating memristor emulator based relaxation oscillator. *IEEE Transactions on Circuits and Systems I: Regular Papers*, *61*(10), 2888–2896.

Zhong, G.-Q. (1994). Implementation of chua's circuit with a cubic nonlinearity. *IEEE Transactions on Circuits and Systems-Part I-Fundamental Theory and Applications*, *41*(12), 934–940.

Zidan, M. A., Omran, H., Smith, C., Syed, A., Radwan, A. G., & Salama, K. N. (2014). A family of memristor-based reactance-less oscillators. *International Journal of Circuit Theory and Applications*, *42*(11), 1103–1122.

Printed in the United States
By Bookmasters